INTRODUCTORY
ALGEBRA

AN INTERACTIVE APPROACH

INTRODUCTORY

ALGEBRA

AN INTERACTIVE APPROACH

Third Edition

Linda Pulsinelli   Patricia Hooper

Department of Mathematics, Western Kentucky University

MACMILLAN PUBLISHING COMPANY
NEW YORK

COLLIER MACMILLAN CANADA
TORONTO

MAXWELL MACMILLAN INTERNATIONAL

NEW YORK   OXFORD   SINGAPORE   SYDNEY

Copyright © 1991, Macmillan Publishing Company, a division of Macmillan, Inc.

Printed in the United States of America.

Earlier editions copyright © 1983, 1987 by Macmillan Publishing Company.

Macmillan Publishing Company
866 Third Avenue, New York, New York 10022

Collier Macmillan Canada, Inc.

Library of Congress Cataloging in Publication Data

Pulsinelli, Linda Ritter.
    Introductory algebra : an interactive approach / Linda Pulsinelli,
Patricia Hooper. — 3rd ed.
        p.   cm.
    ISBN 0-02-396984-9
    1. Algebra.   I. Hooper, Patricia (Patricia I.)   II. Title.
QA152.2.P85   1991   90-35847
512.9—dc20   CIP

Printing: 3 4 5 6 7 8   Year: 3 4 5 6 7 8 9

ISBN 0-02-396984-9

Preface

As its title indicates, *Introductory Algebra: An Interactive Approach* is intended to teach students the fundamental concepts and skills of algebra by involving them continually in the learning process, allowing them to see their own progress and therefore motivating them to continue.

The vocabulary level of a book of this type is crucial to its success and we have attempted to word our explanations in simple, straightforward language. The writing style is informal without being condescending, and important ideas are summarized in table form rather than paragraph form whenever appropriate.

Our experience with students at this level leads us to believe that they must practice each new skill as soon as it has been presented. For this reason, the text includes several unique features that are designed to provide maximum reinforcement. Each chapter in the book follows the same basic structure.

Motivational Applied Problem
At the beginning of each chapter we have presented an applied problem that can be solved after the student has mastered the skills in that chapter. Its solution appears within the chapter.

Explanations
We have tried to avoid the "cookbook" approach to algebra by including a straightforward and readable explanation of each new concept. Realizing that students at the introductory level become easily bogged down in reading lengthy explanations, we have attempted to make our explanations as brief as possible without sacrificing rigor.

Highlighting
Definitions, properties, theorems, and formulas are highlighted in boxes throughout the book for easy student reference. In most cases a rephrasing of a generalization in words accompanies the symbolic statement, and it is also highlighted in a box.

Examples
Immediately following the presentation of a new idea, several completely worked-out examples appear together with several partially worked-out examples to be finished by the student. These examples are completed correctly at the end of each section and the student is advised to check his or her work immediately.

Trial Runs
Sprinkled throughout each section are several Trial Runs, short lists of problems to check on the student's grasp of a new skill. The answers appear at the end of the section.

Exercise Sets
Each section concludes with an extensive Exercise Set in which each odd-numbered problem corresponds closely to the following even-numbered problem.

Stretching the Topics
At the end of each Exercise Set there are several problems designed to challenge students by extending to the next level of difficulty the skills learned in that section.

Checkups
Following each Exercise Set, a list of about 10 problems checks on the student's mastery of the most important concepts in the section. Each Checkup problem is keyed to comparable examples in the section for restudy if necessary.

Problem Solving
One section of almost every chapter involves switching from words to algebra. By including such a section in each chapter we are attempting to treat problem solving as a natural outgrowth of acquiring algebraic skills.

Chapter Summaries
Each chapter concludes with a summary in which the important ideas are again highlighted, in tables when possible. New concepts are presented in symbolic form and verbal form, accompanied by a typical example.

Speaking the Language of Algebra and Writing About Mathematics
Following the summary, we have included a group of sentences to be completed *with words* and questions to be answered *in sentences* by the student. Algebra students (especially those in self-paced programs) often lack the opportunity to "speak mathematics." We hope that these short sections will help them develop a better mathematics vocabulary and become more comfortable expressing themselves in mathematical terms.

Review Exercises
A list of exercises reviewing all the chapter's important concepts serves to give the student an overview of the content. Each problem is keyed to the appropriate section and examples.

Practice Test
A Practice Test is included to help the student prepare for a test over the material in the chapter. Once again, each problem is keyed to the appropriate chapter section and examples.

Sharpening Your Skills
Finally, we have included a short list of exercises that will provide a cumulative review of concepts and skills from earlier chapters. Retention seems to be a very real problem with students at this level and we hope that these exercises will serve to minimize that problem. Each cumulative review exercise is keyed to the appropriate chapter and section.

Throughout the book we have adhered to a rather standard order of topics, making an attempt to connect new concepts to old ones whenever appropriate. This modified spiraling technique is designed to help students maintain an overview of the content. Success in future courses seems to us to hinge on students' seeing that algebra is a logical progression of ideas rather than a set of unrelated skills to be memorized and forgotten.

Chapter 1 contains a review of the arithmetic of whole numbers, fractions, decimals, and percents. Students who have recently completed an arithmetic course or whose arithmetic skills are not weak may be allowed to skip this chapter. All the properties of real numbers are included in Chapter 2, so the omission of Chapter 1 will not handicap the student who is proficient in computation with fractional and decimal numbers. Problems involving such numbers appear throughout the remaining chapters.

A section on negative exponents is included in Chapter 5 for those instructors who consider this an appropriate point at which to introduce this topic to introductory algebra students. For those who do not, the section can be omitted or inserted at a later time without disturbing the continuity of the remainder of the book.

In preparing the third edition, we wished to make the text even more readable for students. We have eliminated wasted space, added more annotation to many examples, and increased the number of applied problems.

The answers to the odd-numbered exercises in the Exercise Sets appear in the back of the book together with answers for *all* items in Stretching the Topics, Checkups, Speaking the Language of Algebra, Review Exercises, Practice Tests, and Sharpening Your Skills.

More assistance for students and instructors can be found among the supplementary materials that accompany this book.

Instructor's Manual with Test Bank
The Instructor's Manual contains the answers for all exercises in the Exercise Sets and Stretching the Topics. In addition there are eight Chapter Tests (four open-ended and four multiple choice) for each chapter and four Final Examinations (two open-ended and two multiple choice). Answers to these tests and examinations also appear in the Instructor's Manual. The chapter tests are also available on computer disks.

Student's Solutions Manual
The Student's Solutions Manual, written by Rebecca Stamper, contains step-by-step solutions for the even-numbered exercises in the Exercise Sets and for *all* items in the Review Exercises, Practice Tests, Sharpening Your Skills, and exercises involving word problems. Using the same style as appears in the text, these solutions emphasize the procedure as well as the answer.

Video Tapes
A series of video tapes (each 20 to 30 minutes in length) provides explanations for many topics in the course.

Computerized Test Generator
A set of computer-generated tests is available for producing tests of any length for each chapter. Cumulative tests and final exams may also be constructed.

Interactive Software Program
This computer-assisted program is available for use with Apple and IBM systems and contains a series of lessons including problems at differing levels of difficulty.

Acknowledgments
The writing of this book would not have been possible without the assistance of many people. We express our appreciation to our typist Maxine Worthington, to Rebecca Stamper of Western Kentucky University and Kathy Rodgers of the University of Southern Indiana for carefully working all our problems, to our families for tolerating our obsessive work schedules, to our editors Gary Ostedt and Bob Pirtle for their enthusiastic support, and to Ed Burke and the production staff at Hudson River Studio for their expert guidance and efficient supervision.

We also thank our reviewers for their careful scrutiny and helpful comments.

John R. Martin, Tarrant County Junior College, Northeast Campus
Veronica L. Schreiber, DeVry Institute of Technology
Richard A. Langlie, North Hennepin Community College
Dan Streeter, Portland State University
Diane Geerken, SUNY, Cobleskill
Marshall Cates, California State University, Los Angeles
Judith L. Willoughby, Minneapolis Community College
James A. Wood, Virginia Commonwealth University
Jack W. Rotman, Lansing Community College.

The cover painting, *Final Point*, was executed in acrylic, gauze, and dyes by Frances Wells, a Kentucky artist and teacher. Her prints, oils, and watercolors appear in numerous galleries and in corporate and private collections throughout the United States.

Contents

INTRODUCTORY
ALGEBRA

AN INTERACTIVE APPROACH

1

Working with the Numbers of Arithmetic

Each month, Estelle saves $\frac{1}{10}$ of her paycheck in her savings bank and $\frac{1}{9}$ of her paycheck in her credit union.

What fractional part of her paycheck does Estelle save each month?

When you have completed this chapter, you will be able to solve this problem using the numbers of arithmetic. Although we assume that your basic arithmetic facts are in good shape, we must be sure that you recall the techniques required for working with whole numbers, fractions, and decimals.

In this chapter we

1. Perform operations with whole numbers.
2. Observe some properties of zero and 1.
3. Work with fractions.
4. Work with decimal numbers and percents.
5. Switch from word expressions to number expressions.

1.1 Working with Whole Numbers

As a child, you learned to count ''one, two, three,'' and so on. These counting numbers are also called **natural numbers**.

> Natural numbers: {1, 2, 3, 4, 5, . . . }

The dots after the 5 mean ''and so on''; in other words, this set continues indefinitely. If we include the number zero in this set, we obtain a new set of numbers called the **whole numbers**.

> Whole numbers: {0, 1, 2, 3, 4, 5, . . . }

The Number Line

To picture the whole numbers, we will use a **number line**. We draw a line and choose a zero point and a length to represent 1 unit. Then all points spaced 1 unit apart to the right of zero are labeled with the whole numbers in order. Many points between the whole numbers are named by fractions and decimal numbers.

The arrow shows that this line goes on indefinitely, so that numbers such as 24 and 793 also correspond to points.

To show a whole number on the number line, we put a solid dot at the point corresponding to that number. This is called **plotting a point** on the number line. The point is called the **graph** of the number and the number is called the **coordinate** of the point.

Example 1. Graph 4 on the number line.

Solution

You try Example 2.

Example 2. Graph 0 on the number line.

Solution

Check your work on page 15. ▶

The number line also gives us a handy way of comparing numbers. For example, if we look at the number line, we notice that all the numbers to the right of 3 are greater than 3. All numbers to the left of 3 are less than 3.

Mathematicians use symbols to compare numbers.

Symbol	Meaning	Example
>	is greater than	$4 > 1$
<	is less than	$6 < 20$

Such statements are called **inequalities** and we summarize them as follows, for any numbers A and B.

> Less than: $A < B$ if A lies to the left of B on the number line
> Greater than: $A > B$ if A lies to the right of B on the number line

Example 3. Place a $<$ or $>$ symbol between the numbers.

$$12 \underline{\qquad} 3$$
$$0 \underline{\qquad} 1$$

Solution

$12 > 3$ because   12 lies to the right of 3

$0 < 1$ because   0 lies to the left of 1

You try Example 4.

Example 4. Place a $<$ or $>$ symbol between the numbers.

$$5 \underline{\qquad} 7$$
$$110 \underline{\qquad} 99$$

Solution

$$5 \underline{\qquad} 7$$
$$110 \underline{\qquad} 99$$

Check your work on page 15. ▶

▥➡ Trial Run

1. Graph 6 on the number line.
0

Compare the numbers using the symbols $<$ or $>$.

2. $7 \underline{\qquad} 5$ **3.** $0 \underline{\qquad} 2$

4. $186 \underline{\qquad} 187$ **5.** $2 \underline{\qquad} 0$

Answers are on page 17.

Performing Operations with Whole Numbers

The four basic operations of arithmetic are *addition, subtraction, multiplication,* and *division.* It will help you in the study of algebra if you learn to name the parts of each of the four types of arithmetic problems now.

Addition

$$\overset{\displaystyle \text{term}}{\underset{}{\downarrow}}$$
$$12 \ + \ 3 \ = 15 \leftarrow \text{sum}$$
$$\underset{\text{term}}{\uparrow}$$

Subtraction

$$\overset{\displaystyle \text{term}}{\underset{}{\downarrow}}$$
$$12 \ - \ 3 \ = 9 \leftarrow \text{difference}$$
$$\underset{\text{term}}{\uparrow}$$

Multiplication

$$\begin{array}{c} \text{factor} \\ \downarrow \\ 12 \quad \cdot \quad 3 \quad = 36 \leftarrow \text{product} \\ \uparrow \\ \text{factor} \end{array}$$

Division

$$\begin{array}{c} \text{divisor} \\ \downarrow \\ 12 \quad \div \quad 3 \quad = 4 \leftarrow \text{quotient} \\ \uparrow \\ \text{dividend} \end{array}$$

Example 5. Find the sum of 10 and 9.

Solution

$$10 + 9 = 19$$

The sum is 19.

Example 6. Find the product of 10 and 9.

Solution

$$10 \cdot 9 = 90$$

The product is 90.

If you ever used your fingers to find a sum, you should agree that *addition is just a form of counting*. When you learned your basic addition facts in elementary school, you mastered a shortcut for finding sums without counting.

To learn your basic subtraction facts, you discovered that *every subtraction statement corresponds to an addition statement*.

$$15 - 8 = 7 \qquad \text{because} \qquad 15 = 8 + 7$$
$$9 - 0 = 9 \qquad \text{because} \qquad 9 = 0 + 9$$

Relating Subtraction to Addition. For whole numbers A, B, and C

$$A - B = C \quad \text{means} \quad A = B + C$$

There are several ways to indicate that two numbers are to be multiplied. For example, to multiply 4 times 3 we may write

$$4 \times 3 \qquad 4 \cdot 3 \qquad 4(3) \qquad (4)3 \qquad (4)(3)$$

Each of these expressions says the same thing, but we will not use the first notation 4×3, because of the confusion that might occur when we begin to use the letter x to mean something else.

One way to find the product of two numbers is to realize that *multiplication is just repeated addition*. Your knowledge of addition helped you learn your multiplication facts.

$$4 \cdot 3 \qquad \text{means} \qquad 3 + 3 + 3 + 3 \qquad \text{so} \qquad 4 \cdot 3 = 12$$
$$3 \cdot 4 \qquad \text{means} \qquad 4 + 4 + 4 \qquad \text{so} \qquad 3 \cdot 4 = 12$$
$$2 \cdot 5 \qquad \text{means} \qquad 5 + 5 \qquad \text{so} \qquad 2 \cdot 5 = 10$$

Relating Multiplication to Addition. For whole numbers A and B

$$A \cdot B \quad \text{means} \quad \underbrace{B + B + \cdots + B}_{A \text{ terms}}$$

Example 7. Write the addition statement that corresponds to the subtraction statement $25 - 12 = 13$.

Solution

$$25 - 12 = 13$$
$$\text{means} \quad 25 = 12 + 13$$

You try Example 8.

Example 8. Write the product $5 \cdot 7$ as a sum.

Solution

$$5 \cdot 7 = \underline{\quad} + \underline{\quad} + \underline{\quad}$$
$$+ \underline{\quad} + \underline{\quad}$$

Check your answer on page 15. ▶

There are several ways to indicate the operation of division. To show 6 divided by 3, we may write

$$6 \div 3 \quad 3\overline{)6} \quad 6/3 \quad \frac{6}{3}$$

Each of these expressions is equal to 2.

To understand the operation of division, you should remember that *every division statement corresponds to a multiplication statement.*

$$\frac{6}{3} = 2 \quad \text{because} \quad 6 = 3 \cdot 2$$

$$\frac{40}{8} = 5 \quad \text{because} \quad 40 = 8 \cdot 5$$

$$\frac{19}{1} = 19 \quad \text{because} \quad 19 = 1 \cdot 19$$

Relating Division to Multiplication. For whole numbers A, B, and C

$$\frac{A}{B} = C \quad \text{means} \quad A = B \cdot C$$

where B does *not* equal 0 ($B \neq 0$).

Example 9. Write the multiplication statement that corresponds to the division statement $\frac{30}{5} = 6$.

Solution

$$\frac{30}{5} = 6$$
$$\text{means} \quad 30 = 5 \cdot 6$$

You complete Example 10.

Example 10. Write the multiplication statement that corresponds to the division statement $\frac{23}{23} = 1$.

Solution

$$\frac{23}{23} = 1$$
$$\text{means} \quad \underline{\quad} = \underline{\quad} \cdot \underline{\quad}$$

Check your answer on page 15. ▶

⫸ Trial Run

_____ **1.** Write the addition statement that corresponds to the subtraction statement $11 - 0 = 11$.

_____ **2.** Write the product $4 \cdot 2$ as a sum.

_____ **3.** Write the multiplication statement that corresponds to the division statement $\dfrac{18}{6} = 3$.

_____ **4.** Write the multiplication statement that corresponds to the division statement $\dfrac{40}{8} = 5$.

Answers are on page 17.

Working with Zero and One

When we operate with the whole numbers zero and 1, we observe some interesting facts. Consider the following sums:

$$13 + 0 = 13$$
$$0 + 7 = 7$$

Notice that

> If zero is added to any whole number, the answer is that whole number.

This law is called the **addition property of zero** (or the **identity property for addition**).

> **Addition Property of Zero.** For any whole number A
> $$A + 0 = A$$
> $$0 + A = A$$

We call zero the **identity for addition** because when zero is added to any number, the answer is identical to the original number.

Similarly, the differences

$$17 - 0 = 17$$
$$1 - 0 = 1$$

should help you see that

> If zero is subtracted from any whole number, the answer is that whole number.

We call this law the **subtraction property of zero**.

Subtraction Property of Zero. For any whole number A

$$A - 0 = A$$

It is also interesting to note here that if any whole number is subtracted from itself, the answer is zero.

$$7 - 7 = 0 \quad \text{because} \quad 7 = 7 + 0$$
$$96 - 96 = 0 \quad \text{because} \quad 96 = 96 + 0$$

For any whole number A

$$A - A = 0$$

What happens when we multiply any number times zero? Recalling that multiplication is repeated addition, look at this product.

$$5 \cdot 0 = 0 + 0 + 0 + 0 + 0$$
$$= 0$$

If any whole number is multiplied times zero, the answer is 0.

This law is called the **multiplication property of zero**.

Multiplication Property of Zero. For any whole number A

$$A \cdot 0 = 0$$
$$0 \cdot A = 0$$

Use the properties of zero to complete Example 11.

Example 11. Complete each statement.

$$13 + 0 = \underline{\hspace{1cm}} \qquad 167 - 0 = \underline{\hspace{1cm}}$$
$$73 + \underline{\hspace{1cm}} = 73 \qquad 29 \cdot 0 = \underline{\hspace{1cm}}$$
$$19 - \underline{\hspace{1cm}} = 19 \qquad 0 \cdot 15 = \underline{\hspace{1cm}}$$
$$19 - \underline{\hspace{1cm}} = 0 \qquad 1 \cdot \underline{\hspace{1cm}} = 0$$

Check your work on page 16. ▶

Let's investigate what happens when zero appears as the divisor or dividend in a division problem.

What happens if we try to divide zero by any *other* whole number? You should agree that

$$\frac{0}{6} = 0 \quad \text{because} \quad 0 = 6 \cdot 0$$

$$\frac{0}{1} = 0 \quad \text{because} \quad 0 = 1 \cdot 0$$

$$\frac{0}{795} = 0 \quad \text{because} \quad 0 = 795 \cdot 0$$

> If zero is divided by any *nonzero* number, the answer is zero.

Now let's try to divide a whole number *by* zero. Consider the quotient $\frac{6}{0}$. Remember that every division statement corresponds to a multiplication statement. Can we say

$$\frac{6}{0} = 0 \quad \text{because} \quad 6 = 0 \cdot 0?$$

or

$$\frac{6}{0} = 6 \quad \text{because} \quad 6 = 0 \cdot 6?$$

Of course, *neither* of these answers is correct. In fact, there is *no answer* for the problem $\frac{6}{0}$.

On the other hand, for the quotient $\frac{0}{0}$ the answer could be *any* number. For instance,

$$\frac{0}{0} = 3 \quad \text{because} \quad 0 = 0 \cdot 3$$

$$\frac{0}{0} = 0 \quad \text{because} \quad 0 = 0 \cdot 0$$

In this case there are *too many answers*.

Because division *by* zero always gives us no answer or too many answers, we agree that *division by zero is impossible*, or

> Division by zero is undefined.

There is no way to define division by zero to give us *one and only one* answer.

To avoid confusion between division *by* zero and division *of* zero, we summarize the two situations here for any whole number A.

Division of Zero	Division by Zero
$\dfrac{0}{A} = 0$ provided that $A \neq 0$	$\dfrac{A}{0}$ is undefined

Use the facts about zero in division to complete Example 12.

Example 12. Complete each statement.

$$\frac{0}{36} = \underline{\qquad} \qquad \frac{?}{17} = 0$$

$$\frac{36}{0} = \underline{\qquad} \qquad \frac{0}{?} \text{ is undefined}$$

$$\frac{0}{0} = \underline{\qquad} \qquad \frac{17}{?} \text{ is undefined}$$

Check your work on page 16. ▶

Now let us consider what happens when we multiply any whole number by the whole number 1. Look at this product.

$$5 \cdot 1 = 1 + 1 + 1 + 1 + 1$$

$$5 \cdot 1 = 5$$

If any whole number is multiplied by 1, the answer is that whole number.

This law is called the **multiplication property of 1** (or the **identity property for multiplication**).

Multiplication Property of 1. For any whole number A

$$A \cdot 1 = A$$

$$1 \cdot A = A$$

We call 1 the **identity for multiplication** because when we multiply any number by 1, the answer is identical to the original number.

Similarly, the quotients

$$\frac{26}{1} = 26$$

$$\frac{103}{1} = 103$$

should help you see that

> If any whole number is divided by 1, the answer is that whole number.

We call this law the **division property of 1**.

Division Property of 1. For any whole number A

$$\frac{A}{1} = A$$

It is also interesting to note here that if any nonzero whole number is divided by itself, the answer is 1.

$$\frac{13}{13} = 1 \qquad \text{because} \qquad 13 = 13 \cdot 1$$

$$\frac{149}{149} = 1 \qquad \text{because} \qquad 149 = 149 \cdot 1$$

For any whole number A

$$\frac{A}{A} = 1 \qquad \text{provided that } A \neq 0$$

Use the properties of 1 to complete Example 13.

Example 13. Complete each statement.

$$78 \cdot 1 = \underline{\hspace{1cm}} \qquad\qquad \frac{?}{1} = 8$$

$$\underline{\hspace{1cm}} \cdot 1 = 16 \qquad\qquad \frac{33}{33} = \underline{\hspace{1cm}}$$

$$\underline{\hspace{1cm}} \cdot 10 = 10 \qquad\qquad \frac{159}{?} = 1$$

$$\frac{98}{1} = \underline{\hspace{1cm}}$$

Check your answers on page 16. ▶

⫸ Trial Run

Complete each statement.

_____ **1.** $12 + 0 = $ ____   _____ **2.** $46 - $ ____ $= 46$

_____ **3.** $56 - \text{\_\_\_\_} = 0$ _____ **4.** $29 \cdot 1 = \text{\_\_\_\_}$

_____ **5.** $\dfrac{100}{1} = \text{\_\_\_\_}$ _____ **6.** $\dfrac{?}{17} = 1$

Answers are on page 17.

Working with Exponents

When a whole number is repeated as a factor in a product, there is a handy notation that can be used.

Product	Notation	Read as
$4 \cdot 4$	4^2	4 squared
$2 \cdot 2 \cdot 2$	2^3	2 cubed
$3 \cdot 3 \cdot 3 \cdot 3 \cdot 3$	3^5	3 to the fifth power

In these expressions the small raised whole number is called the **exponent** and it tells us how many times the other number, called the **base**, is to be used as a factor in a product. The product itself is called a **power**.

Example 14. Find 2^5.

Solution

$$2^5 = 2 \cdot 2 \cdot 2 \cdot 2 \cdot 2$$
$$= 32$$

You try Example 15.

Example 15. Find 4^3.

Solution

$4^3 = \text{\_\_\_\_} \cdot \text{\_\_\_\_} \cdot \text{\_\_\_\_}$

$= \text{\_\_\_\_}$

Check your work on page 16. ▶

Rewriting Whole Numbers As Products

The whole number 12 can be written as a product of whole-number factors in several ways. For instance,

$$12 = 12 \cdot 1 \qquad 12 = 4 \cdot 3$$
$$12 = 6 \cdot 2 \qquad 12 = 2 \cdot 2 \cdot 3$$

The whole number 13, however, can be written only as a product of the whole numbers 1 and 13.

$$13 = 1 \cdot 13$$

A whole number greater than 1 (such as 13) that can *only* be written as a product of whole numbers using the factors 1 and itself is called a **prime number**. A whole number (such as 12) that is *not* a prime number is called a **composite number**. The whole numbers 0 and 1 are the only whole numbers that are not classified as prime or composite. Every whole number greater than 1 must be either prime or composite.

For future work we would like to practice rewriting some whole numbers as products of *prime number factors*. In writing the different factored forms of 12, only the last version was a product of prime numbers.

$$12 = 2 \cdot 2 \cdot 3$$
$$= 2^2 \cdot 3$$

Every whole number that is not prime can be written in just one way as a product of prime number factors. (The order in which those factors are written is not important.)

The process of rewriting a whole number as a product of prime number factors is called **factoring a whole number into a product of primes** or finding its prime factorization.

Example 16. Factor 75 into a product of primes.

Solution

$$75 = 3 \cdot 25$$
$$= 3 \cdot 5 \cdot 5$$
$$= 3 \cdot 5^2$$

You try Example 17.

Example 17. Factor 36 into a product of primes.

Solution

$$36 = 4 \cdot 9$$
$$=$$
$$=$$

Check your work on page 16. ▶

You may not see immediately the prime factors of a larger composite number. Here are a few hints to help you get started.

Factor Tests

If the *last digit* of a whole number is even (0, 2, 4, 6, or 8), then **2** must be a factor of that whole number.

If the *last digit* of a whole number is 0 or 5, then **5** must be a factor of that whole number.

If the *sum* of all the *digits* of a whole number is divisible by 3, then **3** must be a factor of that whole number.

Example 18. Write 234 as a product of prime factors.

Solution

Observation	Conclusion	Factorization
The number 234 ends in 4.	A factor of 234 is 2.	$234 = 2 \cdot 117$
The sum of the digits of 117 $(1 + 1 + 7 = 9)$ is divisible by 3.	A factor of 117 is 3.	$234 = 2 \cdot 3 \cdot 39$
The sum of the digits of 39 $(3 + 9 = 12)$ is divisible by 3.	A factor of 39 is 3.	$234 = 2 \cdot 3 \cdot 3 \cdot 13$
All factors are prime numbers.	The factorization is complete.	$234 = 2 \cdot 3^2 \cdot 13$

⫘➡ **Trial Run** ▬▬▬▬▬▬▬▬▬▬▬▬▬▬▬▬▬▬▬▬▬▬▬▬▬▬▬▬▬▬▬▬▬▬▬▬▬▬

Write each number as a product of prime factors.

_____ **1.** 42   _____ **2.** 60

_____ **3.** 315   _____ **4.** 198

Answers are on page 17.

Using Symbols of Grouping

When *parentheses* appear in an expression, they indicate the order in which operations are to be performed. Parentheses tell us to "do this first".

Example 19. Find $(3 + 2) \cdot 6$.

Solution

$$(3 + 2) \cdot 6$$
$$= \quad 5 \quad \cdot 6$$
$$= 30$$

Example 20. Find $3 + (2 \cdot 6)$.

Solution

$$3 + (2 \cdot 6)$$
$$= 3 + \quad 12$$
$$= 15$$

Notice in Examples 19 and 20 that the location of the parentheses made the problems completely different.

Parentheses ( ), brackets [ ], and braces { } are symbols of grouping that give you directions about the order in which operations are to be done. In problems containing parentheses, braces, and brackets, you must remember to *work inside the innermost symbols of grouping first.*

Example 21. Find

$$[2 + (8 \cdot 9) + 1] + 5$$

Solution

$$[2 + (8 \cdot 9) + 1] + 5$$
$$= [2 + 72 + 1] + 5$$
$$= 75 + 5$$
$$= 80$$

You complete Example 22.

Example 22. Find

$$(3 + 2)(8 + 1)$$

Solution

$$(3 + 2)(8 + 1)$$
$$= 5 \cdot \text{\_\_\_\_}$$
$$= \text{\_\_\_\_}$$

Check your work on page 16. ▶

The fraction bar acts like parentheses in division problems, telling us that we should simplify the dividend (top) and simplify the divisor (bottom) before performing the division.

Example 23. Find $\dfrac{7 + 8}{2 + 3}$.

Solution

$$\frac{7 + 8}{2 + 3} = \frac{15}{5}$$
$$= 3$$

Now you complete Example 24.

Example 24. Find $\dfrac{5 + 19}{2 \cdot 3}$.

Solution

$$\frac{5 + 19}{2 \cdot 3} = \frac{}{6}$$
$$= \underline{}$$

Check your work on page 16. ▶

But suppose that parentheses are not used in a problem that involves several operations. To avoid confusion, mathematicians have agreed to follow these rules.

Order of Operations. If there are no parentheses to indicate order of operations

1. Deal with exponents first.
2. Perform multiplications and/or divisions from left to right.
3. Perform additions and/or subtractions from left to right.

Example 25. Find $3^2 + \dfrac{42}{7}$.

Solution

$$3^2 + \frac{42}{7} = 9 + \frac{42}{7}$$
$$= 9 + 6$$
$$= 15$$

You try Example 26.

Example 26. Find $5 \cdot 2^3 + 6 \cdot 7$.

Solution

$$5 \cdot 2^3 + 6 \cdot 7$$
$$= 5 \cdot \underline{} + 6 \cdot 7$$
$$= \underline{} + \underline{}$$
$$= \underline{}$$

Check your work on page 16. ▶

Example 27. Find $6\left(9 + \dfrac{9}{3}\right) - 2$.

Solution

$$6\left(9 + \frac{9}{3}\right) - 2$$
$$= 6(9 + 3) - 2$$
$$= 6(12) - 2$$
$$= 72 - 2$$
$$= 70$$

You complete Example 28.

Example 28. Find $\dfrac{20 - 3 \cdot 4}{2(5 - 3)}$.

Solution

$$\frac{20 - 3 \cdot 4}{2(5 - 3)}$$
$$= \frac{20 - 12}{2(\underline{})}$$
$$= \frac{}{4}$$
$$= \underline{}$$

Check your work on page 16. ▶

⫸ Trial Run

Perform the indicated operations.

 _____ 1. $\dfrac{9 + 7}{4}$

_____ 2. $4[5 + (3 \cdot 2)]$

_____ 3. $(7 + 2)(3 + 1)$

_____ 4. $3^2(7 + 0)$

_____ 5. $5(2 + 7 \cdot 0)$

_____ 6. $\dfrac{4 \cdot 9 - 6}{2(8 - 5)}$

_____ 7. $5\left(6 + \dfrac{8}{2}\right) - 3$

_____ 8. $7 + \dfrac{9 + 23}{4} + 2$

Answers are on page 17.

▶ Examples You Completed

Example 2. Graph 0 on the number line.

Solution

```
0   1   2   3   4
```

Example 8. Write the product $5 \cdot 7$ as a sum.

Solution

$$5 \cdot 7 = 7 + 7 + 7 + 7 + 7$$

Example 4. Place a $<$ or $>$ symbol between the numbers.

$$5 \_\_\_\_ 7$$
$$110 \_\_\_\_ 99$$

Solution

$$5 < 7$$
$$110 > 99$$

Example 10. Write the multiplication statement that corresponds to the division statement $\frac{23}{23} = 1$.

Solution

$$\frac{23}{23} = 1$$

means $23 = 23 \cdot 1$

Example 11. Complete each statement.

$$13 + 0 = 13 \qquad 167 - 0 = 167$$
$$73 + 0 = 73 \qquad 29 \cdot 0 = 0$$
$$19 - 0 = 19 \qquad 0 \cdot 15 = 0$$
$$19 - 19 = 0 \qquad 1 \cdot 0 = 0$$

Example 12. Complete each statement.

$$\frac{0}{36} = 0 \qquad\qquad \frac{0}{17} = 0$$

$$\frac{36}{0} \text{ is undefined} \qquad \frac{0}{0} \text{ is undefined}$$

$$\frac{0}{0} \text{ is undefined} \qquad \frac{17}{0} \text{ is undefined}$$

Example 13. Complete each statement.

$$78 \cdot 1 = 78 \qquad \frac{8}{1} = 8$$

$$16 \cdot 1 = 16 \qquad \frac{33}{33} = 1$$

$$1 \cdot 10 = 10 \qquad \frac{159}{159} = 1$$

$$\frac{98}{1} = 98$$

Example 15. Find 4^3.

Solution

$$4^3 = 4 \cdot 4 \cdot 4$$
$$= 64$$

Example 17. Factor 36 into a product of primes.

Solution

$$36 = 4 \cdot 9$$
$$= 2 \cdot 2 \cdot 3 \cdot 3$$
$$= 2^2 \cdot 3^2$$

Example 22. Find $(3 + 2)(8 + 1)$.

Solution

$$(3 + 2)(8 + 1)$$
$$= 5 \cdot 9$$
$$= 45$$

Example 24. Find $\dfrac{5 + 19}{2 \cdot 3}$.

Solution

$$\frac{5 + 19}{2 \cdot 3} = \frac{24}{6}$$
$$= 4$$

Example 26. Find $5 \cdot 2^3 + 6 \cdot 7$.

Solution

$$5 \cdot 2^3 + 6 \cdot 7$$
$$= 5 \cdot 8 + 6 \cdot 7$$
$$= 40 + 42$$
$$= 82$$

Example 28. Find $\dfrac{20 - 3 \cdot 4}{2(5 - 3)}$.

Solution

$$\frac{20 - 3 \cdot 4}{2(5 - 3)} = \frac{20 - 12}{2(2)}$$
$$= \frac{8}{4}$$
$$= 2$$

Answers to Trial Runs

page 3   **1.**

$\longmapsto\!\!\!-\!\!\!+\!\!\!-\!\!\!+\!\!\!-\!\!\!+\!\!\!-\!\!\!+\!\!\!-\!\!\!\bullet\!\!\!-\!\!\!+\!\!\!-\!\!\!+\!\!\!\longrightarrow$

       0   6   **2.** $7 > 5$ **3.** $0 < 2$

           4. $186 < 187$ **5.** $2 > 0$

page 6   **1.** $11 = 0 + 11$ **2.** $2 + 2 + 2 + 2$ **3.** $18 = 6 \cdot 3$ **4.** $40 = 8 \cdot 5$

page 10   **1.** $12 + 0 = 12$ **2.** $46 - 0 = 46$ **3.** $56 - 56 = 0$ **4.** $29 \cdot 1 = 29$ **5.** $\dfrac{100}{1} = 100$

           6. $\dfrac{17}{17} = 1$

page 13   **1.** $2 \cdot 3 \cdot 7$ **2.** $2^2 \cdot 3 \cdot 5$ **3.** $3^2 \cdot 5 \cdot 7$ **4.** $2 \cdot 3^2 \cdot 11$

page 15   **1.** 4   **2.** 44   **3.** 36   **4.** 63   **5.** 10   **6.** 5   **7.** 47   **8.** 17

EXERCISE SET 1.1

1. Graph the points corresponding to these numbers on a number line.
   (a) 3   (b) 0   (c) 11   (d) 1

0

2. Graph the points corresponding to these numbers on a number line.
   (a) 5   (b) 1   (c) 12   (d) 0

0

3. Tell what coordinate corresponds to each letter on the number line.

4. Tell what coordinate corresponds to each letter on the number line.

Compare the numbers using the symbols < or >.

5. 12 _____ 7   6. 23 _____ 15

7. 0 _____ 9   8. 1 _____ 10

9. 22 _____ 55   10. 43 _____ 17

_____ 11. Write the addition statement that corresponds to the subtraction statement
             $32 - 15 = 17$.

_____ 12. Write the addition statement that corresponds to the subtraction statement
             $28 - 19 = 9$.

_____ 13. Write the multiplication statement that corresponds to the division statement
             $\dfrac{35}{7} = 5$.

_____ 14. Write the multiplication statement that corresponds to the division statement
             $\dfrac{24}{6} = 4$.

Perform the indicated operations.

_____ 15. $3 + 2 + 1$ _____ 16. $10 + 6 + 7$

_____ 17. $15 - 3$ _____ 18. $14 - 11$

_____ 19. $2 \cdot 3 \cdot 8$ _____ 20. $5 \cdot 6 \cdot 4$

_____ 21. $8(5 \cdot 4)$

_____ 22. $9(7 \cdot 2)$

_____ 23. $\dfrac{8 + 6}{7}$

_____ 24. $\dfrac{9 + 3}{4}$

_____ 25. $6 \cdot 7 + 5$

_____ 26. $4 + 9 \cdot 8$

_____ 27. $17 - (14 - 2)$

_____ 28. $27 - (9 - 3)$

_____ 29. $3 \cdot 5 + 7^2$

_____ 30. $6 \cdot 12 + 5^2$

_____ 31. $(8 + 2^2) - (6 + 1)$

_____ 32. $(13 + 1^3) - (2 + 3)$

_____ 33. $5(9 - 3)$

_____ 34. $7(8 - 1)$

_____ 35. $(9 + 7)(3 + 1)$

_____ 36. $(7 + 2)(3 + 3)$

_____ 37. $6^2 + \dfrac{5}{5}$

_____ 38. $9^2 + \dfrac{7}{7}$

_____ 39. $\dfrac{11 + 23}{13 + 4}$

_____ 40. $\dfrac{19 + 11}{5 + 10}$

_____ 41. $6 + 2(15 - 4)$

_____ 42. $7 + 3(14 - 1)$

_____ 43. $\dfrac{37 - 22}{8 - 8}$

_____ 44. $\dfrac{27 + 13}{11 - 11}$

_____ 45. $2(8 + 3 \cdot 2)$

_____ 46. $3(10 + 5 \cdot 3)$

_____ 47. $4 + [3^2 + 4(1 + 5)]$

_____ 48. $7 + [6^2 + 2(1 + 3)]$

_____ 49. $2^3\left(7 + \dfrac{10}{2}\right)$

_____ 50. $3^2\left(11 + \dfrac{16}{4}\right)$

_____ 51. $\dfrac{7(18 - 2 \cdot 3^2)}{4}$

_____ 52. $\dfrac{9(12 - 3 \cdot 2^2)}{8}$

_____ 53. $\dfrac{2^5 + 6 \cdot 5}{7 - \dfrac{15}{3}}$

_____ 54. $\dfrac{3^4 - 9 \cdot 3}{\dfrac{42}{6} + 2}$

Rewrite each number as a product of prime factors.

_____ 55. 50

_____ 56. 51

_____ 57. 105

_____ 58. 154

_____ 59. 882

_____ 60. 520

☆ Stretching the Topics

Perform the indicated operations.

_____ 1. $\dfrac{5\left[2^2(3) - \dfrac{40}{5}\right]}{3^3 - 7}$

_____ 2. $15 - \{5[16 - (5 - 2)^2] - 4[(9 - 7)^3 - 3]\}$

_____ 3. When 520 and 7371 are written as products of prime factors, what is the only prime factor that they have in common?

Check your answers in the back of your book.

If you can simplify the expressions in **Checkup 1.1**, you are ready to go on to Section 1.2.

✓ **CHECKUP 1.1**

Perform the indicated operations.

_____ **1.** $(2^5 - 12) \cdot 4$

_____ **2.** $4^3 + (3 \cdot 12)$

_____ **3.** $(8 - 3)(15 - 7)$

_____ **4.** $\dfrac{19 + 3}{17 + 5}$

_____ **5.** $\dfrac{8 + 28}{3 \cdot 2}$

_____ **6.** $[(4 \cdot 6) - 3^2] + \dfrac{48}{6}$

_____ **7.** $4 \cdot 3^2 + 8 \cdot 7$

_____ **8.** $8\left(15 - \dfrac{12}{2}\right) + 13$

_____ **9.** $\dfrac{13(8 - 4 \cdot 2)}{5 \cdot 6 - 4 \cdot 5}$

_____ **10.** $\dfrac{5 + 8 \cdot 4}{28 - 4 \cdot 7}$

Check your answers in the back of your book.

If You Missed Problems:	You Should Review Examples:
1, 2	14–17
3	19
4	13, 20
5	24
6	21, 25
7, 8	26, 27
9, 10	12, 28

1.2 Working with Fractions

So far we have learned the rules for adding, subtracting, multiplying, numbers, but there are many useful numbers that are *not* whole number $\frac{1}{2}$, $\frac{5}{3}$, and $\frac{17}{25}$ are called **fractions**. If the numerator of a fraction is smaller th it is called a **proper fraction**. If not, it is called an **improper fraction**. Every improper fraction can be written as a whole number or as a **mixed number**. A mixed number contains a whole number part and a fractional part.

Example 1. Complete each statement.

$\frac{1}{2}$ is a(n) _____ fraction.

$\frac{5}{3}$ is a(n) _____ fraction.

Check your answers on page 33. ▶

Example 2. Write $\frac{5}{3}$ as a mixed number.

Solution

$$\frac{5}{3} = \frac{3}{3} + \frac{2}{3}$$

$$= 1 + \frac{2}{3}$$

$$= 1\frac{2}{3}$$

Reducing Fractions

Sometimes we can find a new fraction that is equivalent to a certain fraction by a process called **reducing the fraction to lowest terms**. To reduce a fraction, we first look for any number(s) that will divide exactly into both the numerator and denominator (with no remainder). Such numbers are called **common factors** for the numerator and denominator. We then divide the numerator and denominator by all common factors.

For instance, to reduce $\frac{21}{33}$ we first look for common factors. One way to do this is to rewrite the numerator and denominator as products of prime factors.

$$\frac{21}{33} = \frac{3 \cdot 7}{3 \cdot 11}$$ Rewrite numerator and denominator.

$$= \frac{\overset{1}{\cancel{3}} \cdot 7}{\underset{1}{\cancel{3}} \cdot 11}$$ Divide numerator and denominator by the common factor 3. Remember, $\frac{3}{3} = 1$.

$$= \frac{7}{11}$$ Write the equivalent fraction.

> **Reducing Fractions**
>
> **1.** Write the numerator and denominator as products of prime factors.
> **2.** Divide the numerator and denominator by all common factors.
> **3.** Multiply remaining factors in the numerator; multiply remaining factors in the denominator.
>
> The final fraction is equivalent to the original fraction.

Example 3. Reduce $\dfrac{24}{84}$.

Solution

$$\dfrac{24}{84} = \dfrac{2 \cdot 2 \cdot 2 \cdot 3}{2 \cdot 2 \cdot 3 \cdot 7}$$ Write prime factorizations.

$$= \dfrac{\overset{1}{\cancel{2}} \cdot \overset{1}{\cancel{2}} \cdot 2 \cdot \overset{1}{\cancel{3}}}{\underset{1}{\cancel{2}} \cdot \underset{1}{\cancel{2}} \cdot \underset{1}{\cancel{3}} \cdot 7}$$ Divide out all common factors.

$$= \dfrac{2}{7}$$ Reduced form.

Now try Example 4.

Example 4. Reduce $\dfrac{110}{25}$.

Solution

$$\dfrac{110}{25} =$$

Check your work on page 33. ▶

⟶ Trial Run

Describe each fraction as proper or improper.

_____ 1. $\dfrac{16}{29}$ _____ 2. $\dfrac{93}{92}$

Reduce each fraction.

_____ 3. $\dfrac{28}{44}$ _____ 4. $\dfrac{70}{42}$

_____ 5. $\dfrac{51}{93}$ _____ 6. $\dfrac{63}{180}$

Answers are on page 34.

Multiplying and Dividing Fractions

From previous work in arithmetic we remember that to multiply fractions, we multiply the numerators and multiply the denominators. Answers should always be reduced if possible.

> **Multiplying Fractions**
>
> $$\dfrac{A}{B} \cdot \dfrac{C}{D} = \dfrac{A \cdot C}{B \cdot D} \qquad B \neq 0, D \neq 0$$

You try multiplying the fractions in Example 5.

Example 5. Find $\dfrac{1}{3} \cdot \dfrac{2}{5}$.

Solution

$$\frac{1}{3} \cdot \frac{2}{5} = \frac{1 \cdot 2}{3 \cdot 5}$$

$$= \underline{\qquad}$$

Example 6. Find $\dfrac{11}{3} \cdot \left(\dfrac{2}{3}\right)^2$.

Solution

$$\frac{11}{3} \cdot \left(\frac{2}{3}\right)^2 = \frac{11}{3} \left(\frac{2}{3} \cdot \frac{2}{3}\right)$$

$$= \frac{11}{3} \cdot \frac{4}{9}$$

$$= \frac{11 \cdot 4}{3 \cdot 9}$$

$$= \frac{44}{27}$$

Check your work on page 33. ▶

Notice that we leave our answers in fractional form. For work in algebra, fractional form is always preferred to mixed number form.

Now let's use two approaches to find the product $\frac{3}{8} \cdot \frac{4}{9}$.

APPROACH 1

$$\frac{3}{8} \cdot \frac{4}{9} = \frac{12}{72}$$ Multiply numerators; multiply denominators.

$$= \frac{2 \cdot 2 \cdot 3}{2 \cdot 2 \cdot 2 \cdot 3 \cdot 3}$$ Write prime factorizations.

$$= \frac{\overset{1}{\cancel{2}} \cdot \overset{1}{\cancel{2}} \cdot \overset{1}{\cancel{3}}}{\underset{1}{\cancel{2}} \cdot \underset{1}{\cancel{2}} \cdot 2 \cdot \underset{1}{\cancel{3}} \cdot 3}$$ Divide out common factors.

$$= \frac{1}{6}$$ Multiply remaining factors.

APPROACH 2

$$\frac{3}{8} \cdot \frac{4}{9} = \frac{3}{2 \cdot 2 \cdot 2} \cdot \frac{2 \cdot 2}{3 \cdot 3}$$ Write prime factorizations.

$$= \frac{\overset{1}{\cancel{3}}}{\underset{1}{\cancel{2}} \cdot \underset{1}{\cancel{2}} \cdot 2} \cdot \frac{\overset{1}{\cancel{2}} \cdot \overset{1}{\cancel{2}}}{\underset{1}{\cancel{3}} \cdot 3}$$ Divide out common factors.

$$= \frac{1}{6}$$ Multiply remaining factors.

Both of these approaches give the same answer. Since the second approach is more direct, it will be our preferred method for multiplying fractions. From now on, we will always *reduce first, then multiply.*

Example 7. Multiply $\dfrac{15}{36} \cdot \dfrac{44}{75}$.

Solution

$$\frac{15}{36} \cdot \frac{44}{75} = \frac{\overset{1}{\cancel{3}} \cdot \overset{1}{\cancel{5}}}{\underset{1}{\cancel{4}} \cdot 9} \cdot \frac{\overset{1}{\cancel{4}} \cdot 11}{\underset{1}{\cancel{3}} \cdot \underset{1}{\cancel{5}} \cdot 5}$$

$$= \frac{1 \cdot 11}{9 \cdot 5}$$

$$= \frac{11}{45}$$

You complete Example 8.

Example 8. Multiply $\dfrac{3}{2} \cdot \dfrac{5}{9} \cdot \dfrac{4}{15}$.

Solution

$$\frac{3}{2} \cdot \frac{5}{9} \cdot \frac{4}{15} = \frac{3}{2} \cdot \frac{5}{3 \cdot 3} \cdot \frac{2 \cdot 2}{3 \cdot 5}$$

Check your work on page 33. ▶

In a division problem, the number being divided is called the *dividend*; the number doing the dividing is called the *divisor*; the answer is called the *quotient*. Recall the steps for dividing fractions.

Dividing Fractions

1. Keep the dividend as it is.
2. Invert the divisor (turn it upside down).
3. Multiply, reducing if possible.

$$\frac{A}{B} \div \frac{C}{D} = \frac{A}{B} \cdot \frac{D}{C} \qquad B, C, D \neq 0$$

The inverted form of a fraction is called its **reciprocal**.

Example 9. Find $\dfrac{9}{16} \div \dfrac{3}{8}$.

Solution

$$\frac{9}{16} \div \frac{3}{8} = \frac{9}{16} \cdot \frac{8}{3}$$

$$= \frac{3 \cdot \overset{1}{\cancel{3}}}{2 \cdot \underset{1}{\cancel{8}}} \cdot \frac{\overset{1}{\cancel{8}}}{\underset{1}{\cancel{3}}}$$

$$= \frac{3}{2}$$

Now you complete Example 10.

Example 10. Find $\dfrac{2}{3} \div 6$.

Solution

$$\frac{2}{3} \div 6 = \frac{2}{3} \div \frac{6}{1}$$

$$= \frac{2}{3} \cdot \frac{1}{6}$$

Check your work on page 33. ▶

Symbols of grouping continue to tell us what to do first.

Example 11. Find $\dfrac{1}{7} \div \left(\dfrac{2}{3} \cdot \dfrac{9}{14} \right)$.

Solution

$$\frac{1}{7} \div \left(\frac{2}{3} \cdot \frac{9}{14} \right) = \frac{1}{7} \div \left(\frac{\overset{1}{\cancel{2}}}{\cancel{3}} \cdot \frac{\overset{1}{\cancel{3}} \cdot 3}{\underset{1}{\cancel{2}} \cdot 7} \right) \qquad \text{Reduce within parentheses.}$$

$$= \frac{1}{7} \div \frac{3}{7} \qquad \text{Find the product.}$$

$$= \frac{1}{\cancel{7}} \cdot \frac{\overset{1}{\cancel{7}}}{3} \qquad \text{Invert the divisor; change divison to multiplication.}$$

$$= \frac{1}{3} \qquad \text{Reduce and multiply.}$$

▥➡ Trial Run

Perform the indicated operations. Write answers in reduced form.

_____ 1. $\dfrac{20}{21} \cdot \dfrac{14}{15}$

_____ 2. $\dfrac{4}{5} \cdot 15$

_____ 3. $\dfrac{3}{8} \cdot \left(\dfrac{4}{3}\right)^2$

_____ 4. $\dfrac{2}{3} \cdot \dfrac{5}{14} \cdot \dfrac{33}{25}$

_____ 5. $\dfrac{8}{17} \div \dfrac{4}{5}$

_____ 6. $\dfrac{4}{5} \div 3$

_____ 7. $\dfrac{1}{15} \div \dfrac{5}{9}$

_____ 8. $\left(\dfrac{2}{3} \cdot \dfrac{15}{22}\right) \div \dfrac{5}{11}$

Answers are on page 34.

Building Fractions

We have spent some time reducing fractions to simplest form by *dividing* numerator and denominator by common factors. Now we must learn how to "build up" a fraction so that it will have a specified denominator. We do this by *multiplying* numerator and denominator by the same factor.

Suppose that we wish to write the fraction $\frac{1}{2}$ as a new fraction with a denominator of 10.

$\dfrac{1}{2} = \dfrac{?}{10}$ We must find a numerator for the fraction on the right. By what factor must 2 be multiplied to give 10?

$\dfrac{1}{2} = \dfrac{?}{2 \cdot 5}$ The factor is 5.

$\dfrac{1}{2} = \dfrac{1 \cdot 5}{2 \cdot 5}$ Since we multiplied the original denominator by 5, we must also multiply the original numerator by 5, because $\dfrac{5}{5} = 1$.

$\dfrac{1}{2} = \dfrac{5}{10}$ We have found our new fraction.

Example 12. Find the new fraction.

$$\frac{2}{3} = \frac{?}{21}$$

Solution

$$\frac{2}{3} = \frac{?}{21}$$

$$\frac{2}{3} = \frac{?}{3 \cdot 7}$$

$$\frac{2}{3} = \frac{2 \cdot 7}{3 \cdot 7}$$

$$\frac{2}{3} = \frac{14}{21}$$

Now you complete Example 13.

Example 13. Find the new fraction.

$$\frac{5}{9} = \frac{?}{180}$$

Solution

$$\frac{5}{9} = \frac{?}{180}$$

$$\frac{5}{9} = \frac{?}{9 \cdot 20}$$

$$\frac{5}{9} = \frac{5 \cdot}{9 \cdot 20}$$

$$\frac{5}{9} = \frac{}{180}$$

Check your work on page 33. ▶

⫸ Trial Run

Find the new fraction.

_____ **1.** $\dfrac{4}{9} = \dfrac{?}{36}$ _____ **2.** $\dfrac{3}{8} = \dfrac{?}{24}$

_____ **3.** $\dfrac{5}{12} = \dfrac{?}{84}$ _____ **4.** $\dfrac{2}{7} = \dfrac{?}{91}$

Answers are on page 34.

Adding and Subtracting Fractions with the Same Denominator

To add (or subtract) fractions with the same denominator, we must add (or subtract) the numerators and keep the same denominator.

Adding Fractions
$\dfrac{A}{C} + \dfrac{B}{C} = \dfrac{A + B}{C}$

Subtracting Fractions
$\dfrac{A}{C} - \dfrac{B}{C} = \dfrac{A - B}{C}$

Example 14. Find $\frac{1}{9} + \frac{4}{9}$.

Solution

$$\frac{1}{9} + \frac{4}{9} = \frac{1 + 4}{9}$$

$$= \frac{5}{9}$$

You try Example 15.

Example 15. Find $\frac{5}{3} - \frac{2}{3}$.

Solution

     $$\frac{5}{3} - \frac{2}{3} =$$

Did you remember to reduce? Check your work on page 34. ▶

Example 16. Find $\frac{3}{29} + \frac{9}{29} - \frac{12}{29}$.

Solution

$$\frac{3}{29} + \frac{9}{29} - \frac{12}{29}$$

$$= \frac{3 + 9 - 12}{29}$$

$$= \frac{12 - 12}{29}$$

$$= \frac{0}{29}$$

$$= 0$$

You complete Example 17.

Example 17. Find $\frac{8}{45} + \frac{14}{45} - \frac{7}{45}$.

Solution

     $$\frac{8}{45} + \frac{14}{45} - \frac{7}{45}$$

$$= \frac{8 + 14 - 7}{45}$$

Did you remember to reduce? Check your work on page 34. ▶

▌▌▌▶ Trial Run

Find the sums or differences.

_____ 1. $\frac{3}{8} + \frac{2}{8}$

_____ 2. $\frac{7}{15} - \frac{4}{15}$

_____ 3. $\frac{16}{21} - \frac{5}{21} - \frac{11}{21}$

_____ 4. $\frac{21}{20} + \frac{3}{20} - \frac{7}{20}$

Answers are on page 34.

Adding and Subtracting Fractions with Different Denominators

If we wish to add or subtract fractions with different denominators, we must make all the denominators *match*. First we decide on a good choice for a new denominator and then *build up* each of the original fractions to a new fraction having the new denominator.

Consider the sum

$$\frac{2}{3} + \frac{1}{2} + \frac{5}{6}$$

What would be a good choice for a new denominator? We may choose any number that is a multiple of 3 and 2 and 6. The numbers, 6, 12, 18, 24, 30, and so on, are all good choices, but the *best* choice is the *smallest* number in that list. We will use 6.

$\frac{2}{3} + \frac{1}{2} + \frac{5}{6}$	From each original fraction we must build a new fraction with a denominator of 6.
$= \frac{?}{6} + \frac{?}{6} + \frac{?}{6}$	
$= \frac{?}{3 \cdot 2} + \frac{?}{2 \cdot 3} + \frac{?}{6 \cdot 1}$	Decide by what factor each original denominator was multiplied.
$= \frac{2 \cdot 2}{3 \cdot 2} + \frac{1 \cdot 3}{2 \cdot 3} + \frac{5 \cdot 1}{6 \cdot 1}$	Now multiply the original numerator by the same factor.
$= \frac{4}{6} + \frac{3}{6} + \frac{5}{6}$	Simplify each numerator and denominator by multiplying the factors.
$= \frac{4 + 3 + 5}{6}$	Combine the numerators over the same denominator.
$= \frac{12}{6}$	Simplify.
$= 2$	Reduce.

In this example, the number 6 is called the **lowest common denominator**.

Lowest Common Denominator (LCD). The lowest common denominator for a set of fractions is the smallest number that is a multiple of each of the denominators of the original fractions.

To find the LCD for a set of fractions, we must consider the prime factors of the original denominators. *The LCD must contain every prime factor that appears in any of the original denominators.*

Consider the following sum.

$$\frac{6}{7} + \frac{4}{5} - \frac{3}{10}$$

$= \frac{6}{7} + \frac{4}{5} - \frac{3}{2 \cdot 5}$	Factor the original denominators. Select the common denominator. LCD: $7 \cdot 5 \cdot 2$.

$$= \frac{?}{7 \cdot 5 \cdot 2} + \frac{?}{7 \cdot 5 \cdot 2} - \frac{?}{7 \cdot 5 \cdot 2}$$

Notice by what building factor each denominator was multiplied to get the LCD.

$$= \frac{6 \cdot 5 \cdot 2}{7 \cdot 5 \cdot 2} + \frac{4 \cdot 7 \cdot 2}{7 \cdot 5 \cdot 2} - \frac{3 \cdot 7}{7 \cdot 5 \cdot 2}$$

Multiply each numerator by that building factor.

$$= \frac{60}{70} + \frac{56}{70} - \frac{21}{70}$$

Simplify each numerator and denominator by multiplying the factors.

$$= \frac{60 + 56 - 21}{70}$$

Combine numerators over the common denominator.

$$= \frac{95}{70}$$

Simplify the numerator.

$$= \frac{5 \cdot 19}{5 \cdot 14}$$

Factor numerator and denominator.

$$= \frac{19}{14}$$

Reduce the fraction.

Let's summarize the steps we must use to find the LCD.

Finding the LCD

1. Factor the denominators of the original fractions.
2. Decide what different prime factors appear in the denominators.
3. Notice the most times each prime factor appears in any single denominator.
4. The LCD is the product of all the different prime factors, with each prime factor used the *most* times that it appears in any *single* denominator.

Example 18. Find $\frac{5}{9} + \frac{1}{12}$.

Solution

$$\frac{5}{9} + \frac{1}{12}$$

$$= \frac{5}{3 \cdot 3} + \frac{1}{3 \cdot 2 \cdot 2}$$

To form the LCD, we must use any factor the *most* number of times that it appears in any *single* denominator; LCD: $3 \cdot 3 \cdot 2 \cdot 2$.

$$= \frac{5 \cdot 2 \cdot 2}{3 \cdot 3 \cdot 2 \cdot 2} + \frac{1 \cdot 3}{3 \cdot 3 \cdot 2 \cdot 2}$$

Build each fraction to a new fraction having the LCD.

$$= \frac{20}{36} + \frac{3}{36}$$

Simplify numerators and denominators.

$$= \frac{20 + 3}{36}$$

Combine numerators over one denominator.

$$= \frac{23}{36}$$

Simplify the numerator.

Now you try to complete Example 19.

Example 19. Find $\dfrac{7}{25} + \dfrac{3}{2} - \dfrac{1}{8}$.

Solution

$$\dfrac{7}{25} + \dfrac{3}{2} - \dfrac{1}{8}$$

$$= \dfrac{7}{5 \cdot 5} + \dfrac{3}{2} - \dfrac{1}{2 \cdot 2 \cdot 2} \qquad \text{LCD: } 5 \cdot 5 \cdot 2 \cdot 2 \cdot 2 = 200$$

$$= \dfrac{7 \cdot 2 \cdot 2 \cdot 2}{5 \cdot 5 \cdot 2 \cdot 2 \cdot 2} + \dfrac{3 \cdot 5 \cdot 5 \cdot 2 \cdot 2}{5 \cdot 5 \cdot 2 \cdot 2 \cdot 2} - \dfrac{1 \cdot 5 \cdot 5}{5 \cdot 5 \cdot 2 \cdot 2 \cdot 2}$$

$$= \dfrac{}{200} + \dfrac{}{200} - \dfrac{}{200}$$

Check your work on page 34. ▶

⟫ **Trial Run**

Find the sums or differences.

_____ **1.** $\dfrac{3}{10} + \dfrac{5}{14}$ _____ **2.** $\dfrac{3}{4} + \dfrac{2}{5} - \dfrac{5}{6}$

_____ **3.** $\dfrac{5}{2} - \dfrac{1}{4} - \dfrac{11}{18}$ _____ **4.** $\dfrac{9}{16} + \dfrac{7}{36} - \dfrac{2}{3}$

_____ **5.** $\dfrac{11}{45} - \dfrac{7}{30}$ _____ **6.** $\dfrac{7}{6} - \dfrac{1}{3} - \dfrac{5}{7}$

Answers are on page 34.

▶ Examples You Completed _____

Example 1. Complete each statement.

$\dfrac{1}{2}$ is a proper fraction.

$\dfrac{5}{3}$ is an improper fraction.

Example 4. Reduce $\dfrac{110}{25}$.

Solution

$$\dfrac{110}{25} = \dfrac{2 \cdot 5 \cdot 11}{5 \cdot 5}$$

$$= \dfrac{2 \cdot \overset{1}{\cancel{5}} \cdot 11}{5 \cdot \underset{1}{\cancel{5}}}$$

$$= \dfrac{22}{5}$$

Example 5. Find $\dfrac{1}{3} \cdot \dfrac{2}{5}$.

Solution

$$\dfrac{1}{3} \cdot \dfrac{2}{5} = \dfrac{1 \cdot 2}{3 \cdot 5}$$

$$= \dfrac{2}{15}$$

Example 8. Multiply $\dfrac{3}{2} \cdot \dfrac{5}{9} \cdot \dfrac{4}{15}$.

Solution

$$\dfrac{3}{2} \cdot \dfrac{5}{9} \cdot \dfrac{4}{15} = \dfrac{3}{2} \cdot \dfrac{5}{3 \cdot 3} \cdot \dfrac{2 \cdot 2}{3 \cdot 5}$$

$$= \dfrac{\overset{1}{\cancel{3}}}{\underset{1}{\cancel{2}}} \cdot \dfrac{\overset{1}{\cancel{5}}}{\underset{1}{\cancel{3}} \cdot 3} \cdot \dfrac{\overset{1}{\cancel{2}} \cdot 2}{3 \cdot \underset{1}{\cancel{5}}}$$

$$= \dfrac{2}{9}$$

Example 10. Find $\dfrac{2}{3} \div 6$.

Solution

$$\dfrac{2}{3} \div 6 = \dfrac{2}{3} \div \dfrac{6}{1}$$

$$= \dfrac{2}{3} \cdot \dfrac{1}{6}$$

$$= \dfrac{\overset{1}{\cancel{2}}}{3} \cdot \dfrac{1}{\underset{1}{\cancel{2}} \cdot 3}$$

$$= \dfrac{1}{9}$$

Example 13. Find the new fraction.

$$\dfrac{5}{9} = \dfrac{?}{180}$$

Solution

$$\dfrac{5}{9} = \dfrac{?}{180}$$

$$\dfrac{5}{9} = \dfrac{?}{9 \cdot 20}$$

$$\dfrac{5}{9} = \dfrac{5 \cdot 20}{9 \cdot 20}$$

$$\dfrac{5}{9} = \dfrac{100}{180}$$

Example 15. Find $\dfrac{5}{3} - \dfrac{2}{3}$.

Solution

$$\frac{5}{3} - \frac{2}{3} = \frac{5-2}{3}$$

$$= \frac{3}{3}$$

$$= 1$$

Example 17. Find $\dfrac{8}{45} + \dfrac{14}{45} - \dfrac{7}{45}$.

Solution

$$\frac{8}{45} + \frac{14}{45} - \frac{7}{45} = \frac{8 + 14 - 7}{45}$$

$$= \frac{22 - 7}{45}$$

$$= \frac{15}{45}$$

$$= \frac{1}{3}$$

Example 19. Find $\dfrac{7}{25} + \dfrac{3}{2} - \dfrac{1}{8}$.

Solution

$$\frac{7}{25} + \frac{3}{2} - \frac{1}{8}$$

$$= \frac{7}{5 \cdot 5} + \frac{3}{2} - \frac{1}{2 \cdot 2 \cdot 2} \qquad \text{LCD: } 5 \cdot 5 \cdot 2 \cdot 2 \cdot 2 = 200$$

$$= \frac{7 \cdot 2 \cdot 2 \cdot 2}{5 \cdot 5 \cdot 2 \cdot 2 \cdot 2} + \frac{3 \cdot 5 \cdot 5 \cdot 2 \cdot 2}{5 \cdot 5 \cdot 2 \cdot 2 \cdot 2} - \frac{1 \cdot 5 \cdot 5}{5 \cdot 5 \cdot 2 \cdot 2 \cdot 2}$$

$$= \frac{56}{200} + \frac{300}{200} - \frac{25}{200}$$

$$= \frac{56 + 300 - 25}{200}$$

$$= \frac{331}{200}$$

Answers to Trial Runs

page 24 1. Proper 2. Improper 3. $\dfrac{7}{11}$ 4. $\dfrac{5}{3}$ 5. $\dfrac{17}{31}$ 6. $\dfrac{7}{20}$

page 27 1. $\dfrac{8}{9}$ 2. 12 3. $\dfrac{2}{3}$ 4. $\dfrac{11}{35}$ 5. $\dfrac{10}{17}$ 6. $\dfrac{4}{15}$ 7. $\dfrac{3}{25}$ 8. 1

page 28 1. $\dfrac{16}{36}$ 2. $\dfrac{9}{24}$ 3. $\dfrac{35}{84}$ 4. $\dfrac{26}{91}$

page 29 1. $\dfrac{5}{8}$ 2. $\dfrac{1}{5}$ 3. 0 4. $\dfrac{17}{20}$

page 32 1. $\dfrac{23}{35}$ 2. $\dfrac{19}{60}$ 3. $\dfrac{59}{36}$ 4. $\dfrac{13}{144}$ 5. $\dfrac{1}{90}$ 6. $\dfrac{5}{42}$

EXERCISE SET 1.2

Reduce the fractions.

_____ 1. $\dfrac{3}{15}$

_____ 2. $\dfrac{6}{18}$

_____ 3. $\dfrac{21}{15}$

_____ 4. $\dfrac{30}{25}$

_____ 5. $\dfrac{9}{16}$

_____ 6. $\dfrac{25}{36}$

_____ 7. $\dfrac{25}{35}$

_____ 8. $\dfrac{49}{63}$

_____ 9. $\dfrac{18}{12}$

_____ 10. $\dfrac{24}{21}$

_____ 11. $\dfrac{7}{49}$

_____ 12. $\dfrac{9}{81}$

Perform the indicated operations. Write answers in reduced form.

_____ 13. $\dfrac{1}{3} \cdot \dfrac{2}{7}$

_____ 14. $\dfrac{2}{5} \cdot \dfrac{1}{3}$

_____ 15. $\dfrac{1}{2} \cdot 6$

_____ 16. $\dfrac{1}{5} \cdot 10$

_____ 17. $\left(\dfrac{2}{3}\right)^2 \cdot \dfrac{9}{10}$

_____ 18. $\left(\dfrac{3}{5}\right)^2 \cdot \dfrac{25}{33}$

_____ 19. $\dfrac{2}{3} \cdot \dfrac{3}{5} \cdot \dfrac{1}{6}$

_____ 20. $\dfrac{4}{7} \cdot \dfrac{7}{8} \cdot \dfrac{1}{3}$

_____ 21. $\dfrac{1}{4} \div \left(\dfrac{1}{6}\right)^2$

_____ 22. $\dfrac{1}{9} \div \left(\dfrac{1}{12}\right)^2$

_____ 23. $\dfrac{2}{3} \div \dfrac{5}{7}$

_____ 24. $\dfrac{2}{5} \div \dfrac{7}{9}$

_____ 25. $\dfrac{4}{5} \div \dfrac{12}{35}$

_____ 26. $\dfrac{3}{8} \div \dfrac{15}{16}$

_____ 27. $\dfrac{3}{5} \div 9$

_____ 28. $\dfrac{7}{8} \div 21$

_____ 29. $\dfrac{20}{7} \div \dfrac{16}{77}$

_____ 30. $\dfrac{18}{5} \div \dfrac{27}{55}$

_____ 31. $\dfrac{8}{25} \div \left(\dfrac{24}{35} \cdot \dfrac{14}{27}\right)$

_____ 32. $\dfrac{10}{66} \cdot \left(\dfrac{8}{9} \div \dfrac{25}{33}\right)$

Find the new fractions.

_____ 33. $\dfrac{6}{7} = \dfrac{?}{49}$ _____ 34. $\dfrac{3}{5} = \dfrac{?}{25}$

_____ 35. $\dfrac{8}{3} = \dfrac{?}{36}$ _____ 36. $\dfrac{9}{4} = \dfrac{?}{48}$

_____ 37. $\dfrac{4}{11} = \dfrac{?}{99}$ _____ 38. $\dfrac{2}{13} = \dfrac{?}{65}$

_____ 39. $\dfrac{17}{15} = \dfrac{?}{45}$ _____ 40. $\dfrac{13}{12} = \dfrac{?}{60}$

_____ 41. $\dfrac{7}{8} = \dfrac{?}{72}$ _____ 42. $\dfrac{7}{10} = \dfrac{?}{120}$

_____ 43. $9 = \dfrac{?}{5}$ _____ 44. $7 = \dfrac{?}{4}$

_____ 45. $\dfrac{31}{84} = \dfrac{?}{252}$ _____ 46. $\dfrac{29}{132} = \dfrac{?}{396}$

Find the sum or difference and write the answers in reduced form.

_____ 47. $\dfrac{3}{5} + \dfrac{1}{5}$ _____ 48. $\dfrac{4}{5} - \dfrac{1}{5}$

_____ 49. $\dfrac{2}{27} + \dfrac{7}{27}$ _____ 50. $\dfrac{3}{8} + \dfrac{1}{8}$

_____ 51. $\dfrac{7}{13} - \dfrac{2}{13}$ _____ 52. $\dfrac{11}{15} - \dfrac{4}{15}$

_____ 53. $\dfrac{3}{11} + \dfrac{2}{11} - \dfrac{4}{11}$ _____ 54. $\dfrac{9}{13} + \dfrac{2}{13} - \dfrac{10}{13}$

_____ 55. $\dfrac{9}{11} - \dfrac{2}{11} + \dfrac{4}{11}$ _____ 56. $\dfrac{10}{17} - \dfrac{5}{17} + \dfrac{12}{17}$

_____ 57. $\dfrac{4}{5} - \dfrac{3}{5} + \dfrac{2}{5} - \dfrac{3}{5}$ _____ 58. $\dfrac{4}{7} - \dfrac{2}{7} + \dfrac{3}{7} - \dfrac{5}{7}$

_____ 59. $\dfrac{1}{3} + \dfrac{1}{5}$ _____ 60. $\dfrac{1}{7} + \dfrac{1}{9}$

_____ 61. $\dfrac{2}{3} - \dfrac{4}{9}$ _____ 62. $\dfrac{3}{5} - \dfrac{11}{25}$

_____ 63. $3 - \dfrac{4}{5}$ _____ 64. $2 - \dfrac{3}{7}$

_____ **65.** $\dfrac{7}{12} - \dfrac{11}{28} + \dfrac{17}{21}$ _____ **66.** $\dfrac{3}{10} + \dfrac{12}{35} + \dfrac{5}{14}$

_____ **67.** $\dfrac{29}{45} - \dfrac{17}{36} + \dfrac{7}{20}$ _____ **68.** $\dfrac{11}{18} - \dfrac{37}{63} + \dfrac{3}{14}$

_____ **69.** $\dfrac{8}{15} + \dfrac{24}{35} + \dfrac{1}{21}$ _____ **70.** $\dfrac{20}{21} - \dfrac{13}{28} + \dfrac{7}{12}$

☆ Stretching the Topics _____

Perform the indicated operations. Write answers in reduced form.

_____ **1.** $\left(\dfrac{3}{5} \cdot \dfrac{7}{9}\right) + \left(\dfrac{9}{10} \cdot \dfrac{5}{18}\right)$

_____ **2.** $\dfrac{4}{5}\left[\left(\dfrac{3}{4} + \dfrac{2}{5}\right) - \left(\dfrac{3}{8} + \dfrac{5}{12}\right)\right]$

_____ **3.** $\dfrac{1}{2}\left\{\left[\dfrac{5}{6} - \left(\dfrac{2}{3}\right)^2\right] \div \left(\dfrac{5}{6} \cdot \dfrac{7}{3}\right)\right\}$

Check your answers in the back of your book.

If you can complete **Checkup 1.2**, you are ready to go on to Section 1.3.

✓ CHECKUP 1.2

_____ 1. Reduce $\dfrac{56}{63}$.

_____ 2. Find the new fraction. $\dfrac{5}{7} = \dfrac{?}{56}$

Perform the indicated operations. Write answers in reduced form.

_____ 3. $\dfrac{5}{9} \cdot \dfrac{6}{35}$

_____ 4. $\dfrac{2}{3} \div \dfrac{10}{9}$

_____ 5. $12 \div \dfrac{4}{5}$

_____ 6. $\dfrac{9}{25} \div \left(\dfrac{12}{35} \cdot \dfrac{28}{39} \right)$

_____ 7. $\dfrac{2}{3} - \dfrac{4}{21}$

_____ 8. $5 - \dfrac{2}{3}$

_____ 9. $\dfrac{1}{7} + \dfrac{3}{8}$

_____ 10. $\dfrac{8}{15} + \dfrac{24}{35} - \dfrac{1}{21}$

Check your answers in the back of your book.

If You Missed Problems:	You Should Review Examples:
1	3
2	12
3	5
4, 5	9, 10
6	11
7–9	14, 15
10	17

1.3 Working with Decimals and Percents

When you receive a check for $73.28 or when you read that unemployment has reached 10.2 percent, you are being asked to deal with **decimal numbers**. Every decimal number has two parts, separated by the decimal point. The part to the left of the decimal point is the **whole-number part**, and the part to the right of the decimal point is called the **fractional part**.

Relating Decimal Numbers and Fractions

Each digit in a decimal number has a certain **place value**. The value of each place is *ten times* the value of the place immediately to its *right*. Or, stated in another way, each place has *one-tenth* the value of the place immediately to its *left*.

$$\underline{4}\ \underline{6}\ \underline{9}\ \underline{8}.\underline{2}\ \underline{3}\ \underline{1}\ \underline{5}$$

thousands hundreds tens ones . tenths hundredths thousandths ten thousandths

To read the fractional part of a decimal number, we note the position in which the last digit appears. The place value of that position tells us whether we are dealing with tenths or hundredths and so on. Every decimal number can then be rewritten as a fraction. If a decimal number has no fractional part, it is a whole number.

Decimal Number	Meaning	Fraction
0.3	3 tenths	$\dfrac{3}{10}$
0.19	19 hundredths	$\dfrac{19}{100}$
0.257	257 thousandths	$\dfrac{257}{1000}$
73.29	73 and 29 hundredths	$73\dfrac{29}{100}$ or $\dfrac{7329}{100}$
16	16	$\dfrac{16}{1}$

Example 1. Write 0.013 as a fraction.

Solution

$$0.013 = \frac{13}{1000}$$

You try Example 2.

Example 2. Write $\dfrac{37}{10,000}$ as a decimal number.

Solution. $\dfrac{37}{10,000}$ means 37 ten thousandths.

$$\frac{37}{10,000} = 0.\underline{\hspace{2cm}}$$

Check your work on page 46. ▶

In the fractional part of a decimal number, zeros may be attached *after* the last nonzero digit without changing the value of the number. We may write

$$0.5 = 0.50 = 0.500 \quad \text{because} \quad \frac{5}{10} = \frac{50}{100} = \frac{500}{1000}$$

Example 3. Rewrite 1.2 as a decimal number that ends in the thousandths place.

Solution. We attach the necessary zeros.

$$1.2 = 1.200$$

You try Example 4.

Example 4. Rewrite $4 as a decimal number that ends in the hundredths place.

Solution. We attach the necessary zeros.

$$\$4 = \$4.\underline{}$$

Check your work on page 46. ▶

To change any fraction to a decimal number, we must divide the numerator by the denominator, placing a decimal point and attaching zeros where necessary in the dividend. For example, let's change $\frac{1}{4}$ to a decimal number.

$$4\overline{)1.00}$$ Place the decimal point after the 1 and attach zeros.

$$\begin{array}{r} .25 \\ 4\overline{)1.00} \\ \underline{8} \\ 20 \\ \underline{20} \\ 0 \end{array}$$ Divide in the usual way.

Therefore, $\dfrac{1}{4} = 0.25$.

Example 5. Change $\frac{1}{3}$ to a decimal number.

Solution

$$\begin{array}{r} .333 \\ 3\overline{)1.000} \\ \underline{9} \\ 10 \\ \underline{9} \\ 10 \\ \underline{9} \\ 1 \end{array}$$

We will continue to obtain the digit 3 in the quotient. This **repeating decimal** number can be written $0.3\overline{3}$, where the bar shows the repeating digit. Thus,

$$\frac{1}{3} = 0.3\overline{3}$$

Now try Example 6.

Example 6. Change $\frac{5}{8}$ to a decimal number.

Solution

$$8\overline{)5.000}$$

Therefore, $\dfrac{5}{8} = \underline{}$.

Check your work on page 46 ▶

Sometimes it is necessary to **round** decimal numbers to some particular place. We agree to use the following procedure.

Rounding Decimal Numbers

1. Look at the digit in the place immediately to the right of the place to which you wish to round the number.
2. If the digit in the place to the right is less than 5, leave the digit in the rounding place as it is. Drop all digits to the right of the rounding place.
3. If the digit in the place to the right is 5 or more, add 1 to the digit in the rounding place. Drop all digits to the right of the rounding place.
4. State that your original decimal number is approximately equal to ($\doteq$) your rounded decimal number.

Example 7. Round 17.261 to the tenths place.

Solution

$$17.261 \doteq 17.3$$

Now you try Example 8.

Example 8. Round 17.261 to the hundredths place.

Solution

$$17.261 \doteq \underline{\hspace{2cm}}$$

Check your work on page 46. ▶

Example 9. Round $0.6\overline{6}$ to the thousandths place.

Solution. Remember that this repeating decimal number is

$$0.66666. . .$$

so in rounded form we have

$$0.6\overline{6} \doteq 0.667$$

Now try Example 10.

Example 10. Round $2.7\overline{373}$ to the hundredths place.

Solution

$$2.7\overline{373} \doteq \underline{\hspace{1.5cm}}$$

Check your work on page 46. ▶

⫸ Trial Run

_____ **1.** Write 0.377 as a fraction.

_____ **2.** Write $\dfrac{49}{100}$ as a decimal number.

_____ **3.** Write $\dfrac{4}{5}$ as a decimal number.

_____ **4.** Round 1.3219 to the hundredths place.

_____ **5.** Change $\dfrac{1}{6}$ to a decimal number and round to the thousandths place.

Answers are on page 46.

Operating with Decimal Numbers

To **add** or **subtract** decimal numbers, we follow this procedure.

> **Adding or Subtracting Decimal Numbers**
>
> **1.** Write the problem vertically, with decimal points lined up beneath each other.
> **2.** Attach necessary zeros to make all fractional parts contain the same number of digits.
> **3.** Add or subtract the columns of numbers in the usual way, keeping the decimal point in its original position.

Example 11. Find 3.62 + 2.8 + 0.317.

Solution

$$
\begin{array}{r}
3.620 \\
2.800 \\
+0.317 \\
\hline
6.737
\end{array}
$$

Now try Example 12.

Example 12. Find $76 − $15.34.

Solution

$$
\begin{array}{r}
\$76.00 \\
-\ 15.34 \\
\hline
\$\rule{1cm}{0.4pt}
\end{array}
$$

Check your work on page 46. ▶

In **multiplying** decimal numbers, the location of the decimal point in the answer depends on the location of the decimal points in the factors being multiplied.

> **Multiplying Decimal Numbers**
>
> **1.** Multiply the digits (ignoring the decimal points).
> **2.** Add the number of places to the right of the decimal point in all the numbers being multiplied. This sum tells how many places there will be to the right of the decimal point in the product.
> **3.** Locate the decimal point in the answer by "counting off" the places from the right.

Example 13. Find (0.12)(1.8).

Solution

$$
\begin{array}{r}
0.12 \quad \text{(2 places)}\\
\times\ \underline{1.8} \quad \text{(1 place)}\\
96 \\
\underline{12} \\
.216 \quad \text{(3 places)}
\end{array}
$$

(0.12)(1.8) = 0.216

Example 14. Find (0.182)(0.05).

Solution

$$
\begin{array}{r}
0.182 \quad \text{(3 places)}\\
\times\ \underline{0.05} \quad \text{(2 places)}\\
.00910 \quad \text{(5 places)}
\end{array}
$$

(0.182)(0.05) = 0.00910

In Example 14, notice that we attached zeros in front of our product in order to reach the correct location of the decimal point.

To **divide** decimal numbers, *we always want the divisor to be a whole number.*

Dividing Decimal Numbers

1. Change the divisor to a whole number by moving the decimal point to the right as many places as necessary.
2. Move the decimal point in the dividend the *same* number of places to the right, attaching zeros if necessary.
3. Locate the decimal point in the quotient *directly above* the new decimal point location in the dividend.
4. Ignore the decimal points and divide as usual.

Example 15. Find $1.3939 \div 0.53$.

Solution

$$
\begin{array}{r}
2.63 \\
0.53\overline{)1.39\,39} \\
\underline{1\,06} \\
33\,3 \\
\underline{31\,8} \\
1\,59 \\
\underline{1\,59} \\
0
\end{array}
$$

Therefore, $1.3939 \div 0.53 = 2.63$.

Example 16. Find $0.464 \div 0.000116$.

Solution

$$
\begin{array}{r}
4000. \\
0.000116\overline{)0.464000} \\
\underline{464} \\
00 \\
\underline{00} \\
00 \\
\underline{00} \\
00 \\
\underline{00} \\
0
\end{array}
$$

Therefore, $0.464 \div 0.000116 = 4000$.

In Example 16, notice that we added zeros in the dividend so that we could move the decimal point the required number of places to the right.

⟴ Trial Run

Perform the indicated operation.

_____ 1. $7.23 + 0.591$ _____ 2. $\$6.82 + \$3 + \$9.29$

_____ 3. $0.317 - 0.25$ _____ 4. $\$19 - \$18.83$

_____ 5. $(10.21)(3.051)$ _____ 6. $(0.0035)(0.28)$

_____ 7. $1.1607 \div 7.3$ _____ 8. $1.065 \div 0.00213$

Answers are on page 46.

Working with Percents

Decimals are used extensively in work with percents. The word **percent** means "per hundred." Understanding this meaning makes it easy to change a percent to a fraction or a decimal number.

Percent	Meaning	Fraction	Decimal Number
31%	31 per hundred	$\dfrac{31}{100}$	0.31
9%	9 per hundred	$\dfrac{9}{100}$	0.09
113%	113 per hundred	$\dfrac{113}{100}$	1.13

Looking closely at the first and last columns in this chart, you should agree to the following method for changing a percent to a decimal number.

Changing a Percent to a Decimal Number. To change a percent to a decimal number, drop the % symbol and move the decimal point *two* places to the *left*.

You try Example 17.

Example 17. Change 69% to a decimal number.

Solution

$$69\% = 0.\text{\_\_\_\_\_}$$

Check your work on page 46. ▶

Example 18. Change 10.2% to a decimal number.

Solution

$$10.2\% = 0.102$$

Of course, we can reverse this procedure if we wish to change a decimal number to a percent.

Changing a Decimal Number to a Percent. To change a decimal number to a percent, move the decimal point *two* places to the *right* and attach a % symbol.

Example 19. Change 0.73 to a percent.

Solution

$$0.73 = 73\%$$

Now try Example 20.

Example 20. Change 1.2 to a percent.

Solution

$$1.2 = \text{\_\_\_\_\_}\%$$

Check your work on page 46. ▶

Suppose that your boss promises you a raise that is 8 percent of your current salary of $300 per week. How can you calculate your dollar raise? You must find 8 percent of $300. Recall that the word *of* usually tells us to *multiply*. But in order to multiply by a percent, we must change it to a decimal number (or to a fraction).

$$8\% \text{ of } 300 = 8\% \cdot 300$$
$$= (0.08)(300)$$
$$= 24$$

Your raise should be $24 per week.

> **Finding a Percentage of a Number.** To find a percentage of some number, change the percent to a decimal number and multiply.

You try Example 21.

Example 21. Find 110% of 56.

Solution

$$110\% \text{ of } 56 = 110\% \cdot 56$$
$$= (1.10)(56)$$
$$= \underline{\hspace{1cm}}$$

Therefore, 110% of 56 is _____ .

Check your work on page 46. ▶

Example 22. Find 6.5% of $200.

Solution

$$6.5\% \text{ of } 200 = 6.5\% \cdot 200$$
$$= (0.065)(200)$$
$$= 13.0$$

Therefore, 6.5% of $200 is $13.

⫸ Trial Run

Change each percent to a decimal number.

_____ **1.** 53%   _____ **2.** 150%

Change each decimal number to a percent.

_____ **3.** 0.18   _____ **4.** 0.071

Find each percentage.

_____ **5.** 12% of 80   _____ **6.** 7.8% of $120

Answers are on page 46.

▶ Examples You Completed

Example 2. Write $\dfrac{37}{10,000}$ as a decimal number.

Solution. $\dfrac{37}{10,000}$ means 37 ten thousandths.

$$\frac{37}{10,000} = 0.0037$$

Example 6. Change $\dfrac{5}{8}$ to a decimal number.

Solution

$$\begin{array}{r} .625 \\ 8\overline{)5.000} \\ \underline{4\,8} \\ 20 \\ \underline{16} \\ 40 \\ \underline{40} \\ 0 \end{array}$$

Therefore, $\dfrac{5}{8} = 0.625$.

Example 12. Find $76 - \$15.34$.

Solution

$$\begin{array}{r} \$76.00 \\ -\ \underline{15.34} \\ \$60.66 \end{array}$$

Example 20. Change 1.2 to a percent.

Solution

$$1.2 = 120\%$$

Example 4. Rewrite $4 as a decimal number that ends in the hundredths place.

Solution. We attach the necessary zeros.

$$\$4 = \$4.00$$

Example 8. Round 17.261 to the hundredths place.

Solution

$$17.261 \doteq 17.26$$

Example 10. Round $2.7\overline{373}$ to the hundredths place.

Solution

$$2.7\overline{373} \doteq 2.74$$

Example 17. Change 69% to a decimal number.

Solution

$$69\% = 0.69$$

Example 21. Find 110% of 56.

Solution

$$\begin{aligned} 110\% \text{ of } 56 &= 110\% \cdot 56 \\ &= (1.10)(56) \\ &= 61.6 \end{aligned}$$

Therefore, 110% of 56 is 61.6.

Answers to Trial Runs

page 41   1. $\dfrac{377}{1000}$ 2. 0.49   3. 0.8   4. 1.32   5. 0.167

page 43   1. 7.821   2. $19.11   3. 0.067   4. $0.17   5. 31.15071   6. 0.00098   7. 0.159
8. 500

page 45   1. 0.53   2. 1.5   3. 18%   4. 7.1%   5. 9.6   6. $9.36

EXERCISE SET 1.3

Write each decimal number as a fraction. Reduce to lowest terms.

_____ 1. 0.8

_____ 2. 0.6

_____ 3. 0.55

_____ 4. 0.75

_____ 5. 0.175

_____ 6. 0.112

_____ 7. 0.012

_____ 8. 0.028

_____ 9. Rewrite 2.3 as a decimal number that ends in the thousandths place.

_____ 10. Rewrite 4.7 as a decimal number that ends in the thousandths place.

_____ 11. Rewrite $15 as a decimal number that ends in the hundredths place.

_____ 12. Rewrite $12 as a decimal number that ends in the hundredths place.

_____ 13. Rewrite 13.2 as a decimal number that ends in the thousandths place.

_____ 14. Rewrite 10.6 as a decimal number that ends in the hundredths place.

Change each fraction to a decimal number. Round to the hundredths place.

_____ 15. $\dfrac{3}{4}$

_____ 16. $\dfrac{1}{4}$

_____ 17. $\dfrac{5}{8}$

_____ 18. $\dfrac{7}{16}$

_____ 19. $\dfrac{5}{6}$

_____ 20. $\dfrac{4}{9}$

_____ 21. $\dfrac{8}{15}$

_____ 22. $\dfrac{12}{19}$

_____ 23. $\dfrac{9}{25}$

_____ 24. $\dfrac{18}{25}$

Perform the indicated operations.

_____ 25. 0.83 + 0.7

_____ 26. 49.6 + 3.002

_____ 27. $9.75 + $13 + $3.05

_____ 28. $20 − $15.78

_____ 29. 9.005 − 3.72

_____ 30. 13.003 − 4.139

_____ 31. 3.089 + 0.845 − 1.32

_____ 32. $50 − $7.85 − $19.46

_____ 33. (0.027)(0.78)

_____ 34. (56.9)(3.24)

_____ **35.** (3.27)(0.008)   _____ **36.** (7.85)(0.007)

_____ **37.** 0.273 ÷ 0.07   _____ **38.** 2.58 ÷ 0.0043

_____ **39.** 1.6849 ÷ 8.3   _____ **40.** 305.2 ÷ 8.72

Change each percent to a decimal number.

_____ **41.** 19%   _____ **42.** 64%

_____ **43.** 37.1%   _____ **44.** 28.2%

_____ **45.** 150%   _____ **46.** 220%

_____ **47.** 0.5%   _____ **48.** 0.2%

Change each decimal number to a percent.

_____ **49.** 0.09   _____ **50.** 0.12

_____ **51.** 7.1   _____ **52.** 5.6

_____ **53.** 0.032   _____ **54.** 0.054

_____ **55.** 1.75   _____ **56.** 2.50

Find each percentage.

_____ **57.** 13% of 470   _____ **58.** 12% of 560

_____ **59.** 9% of 72   _____ **60.** 6% of 84

_____ **61.** 125% of 3240   _____ **62.** 175% of 4760

_____ **63.** 8.3% of 6000   _____ **64.** 9.4% of 5000

_____ **65.** 300% of 74   _____ **66.** 500% of 68

_____ **67.** 10% of 9378   _____ **68.** 20% of 8476

_____ **69.** 25% of 1250   _____ **70.** 50% of 2740

☆ Stretching the Topics

_____ **1.** Find $\dfrac{0.5[11.36 + (7.2 - 3.065)]}{0.002}$.

_____ **2.** Find $12\frac{1}{2}\%$ of \$5000.

_____ **3.** Find 0.25% of \$600.

Check your answers in the back of your book.

If you can do the problems in **Checkup 1.3**, you are ready to go on to Section 1.4.

Name _____ Date _____

✔ CHECKUP 1.3

_____ **1.** Write 0.024 as a fraction and reduce to lowest terms.

_____ **2.** Rewrite 43.5 as a decimal number that ends in the thousandths place.

_____ **3.** Change $\frac{7}{8}$ to a decimal number.

_____ **4.** Change 3.25% to a decimal number.

_____ **5.** Change 0.125 to a percent.

Perform the indicated operations.

_____ **6.** 2.032 + 7.85 − 4.02   _____ **7.** $25 − $17.98

_____ **8.** (0.32)(0.7)   _____ **9.** 0.02616 ÷ 3.27

_____ **10.** Find 8.5% of 325.

Check your answers in the back of your book.

If You Missed Problems:	You Should Review Examples:
1	1
2	3
3	5, 6
4	17, 18
5	19, 20
6, 7	11, 12
8, 9	13–16
10	21, 22

1.4 Switching from Word Expressions to Number Expressions

It is important for algebra students to be able to change word statements into number statements. Such an ability will allow us to solve many ''real-life'' problems that could otherwise be solved only by trial and error.

Let's see if we can practice switching some word expressions into number expressions. There are several key phrases that will help you switch from words to numbers.

Words	Numbers
the sum of 7 and 9	$7 + 9$
6 more than 25	$25 + 6$
1.6 increased by 25.4	$1.6 + 25.4$
24 decreased by 8	$24 - 8$
5 fewer than 12	$12 - 5$
0.3 less than 1.7	$1.7 - 0.3$
the product of 6 and 11	$6 \cdot 11$
twice as much as 7.3	$2(7.3)$
14 doubled	$2 \cdot 14$
5 times as large as 10	$5 \cdot 10$
24 divided by 0.8	$\dfrac{24}{0.8}$
half of $\dfrac{6}{7}$	$\dfrac{1}{2} \cdot \dfrac{6}{7}$

See if you can complete the following table.

Words	Numbers
13 increased by 7	_____
_____	$\dfrac{32}{2.8}$
the sum of 12 and 3	_____
the product of 12 and 3	_____
5 fewer than 18	_____
twice as much as 10.6	_____
_____	$42 \div 15$
19 decreased by 15	_____
two-thirds of 24	_____
3 times as large as 25	_____

Check your work on page 56. ▶

Example 1. Nancy's hourly wage is $3. Next week she will receive a raise of $1.10 per hour. Write an expression for her new hourly wage.

Solution

$3 *increased by* $1.10

$3 + $1.10

Complete Example 2.

Example 2. At noon the temperature was 18°. By midnight, the temperature had decreased by 12°. Write an expression for the temperature at midnight.

Solution

18° *decreased by* 12°

18° _____ 12°

Check your work on page 56. ▶

Example 3. Tom weighs 20 pounds more than twice his son's weight. If Tom's son weighs 73 pounds, write an expression for Tom's weight.

Solution

son's weight	73
twice son's weight	2(73)
20 *more than* twice son's weight	2(73) + 20

The expression 2(73) + 20 represents Tom's weight (in pounds).

Now let's solve the problem stated at the beginning of this chapter.

Example 4. Each month Estelle saves $\frac{1}{10}$ of her paycheck in her savings bank and $\frac{1}{9}$ of her paycheck in her credit union. What fractional part of her paycheck does Estelle save each month?

Solution

We Think: The total of Estelle's savings is the *sum* of the two fractions.

We Write

$$\frac{1}{10} + \frac{1}{9} = \frac{1}{2 \cdot 5} + \frac{1}{3 \cdot 3} \qquad \text{LCD: } 2 \cdot 5 \cdot 3 \cdot 3 = 90$$

$$= \frac{1 \cdot 3 \cdot 3}{2 \cdot 5 \cdot 3 \cdot 3} + \frac{1 \cdot 2 \cdot 5}{2 \cdot 5 \cdot 3 \cdot 3}$$

$$= \frac{9}{90} + \frac{10}{90}$$

$$= \frac{19}{90}$$

Estelle saves $\frac{19}{90}$ of her paycheck each month.

Example 5. The sales tax in a certain state is 6.5 percent. How much tax would a customer pay on a record selling for $8? How much would the customer pay altogether?

Solution

We Think	We Write
The tax is found by *multiplying* the percentage rate times the price of the record.	6.5% of $8 = 0.065($8) = $0.52
The total paid is the *sum* of the price of the record and the tax.	$8 + $0.52 = $8.52

Altogether the customer will pay $8.52.

Now you complete Example 6.

Example 6. A real estate developer owns a large plot of land with four sides measuring 2.3 miles, 1.4 miles, 4.25 miles, and 3.5 miles. If the developer drives around the boundary of her land, how many miles will she travel?

Solution. The distance traveled is found by _____ the four distances.

_____ + _____ + _____ + _____

= _____ miles

The developer will travel _____ miles.

Check your work on page 56. ▶

Working with Averages

Even when you are no longer a student, you will see the word **average** in the newspaper or on television. You may remember that an average is found by dividing a *total* by the *number of equal parts* that the total is to contain.

You complete Example 7.

Example 7. The Colemans spent $186 on groceries and $240 on rent during 4 weeks. What was the Colemans' *average* weekly expense for groceries and rent?

Solution

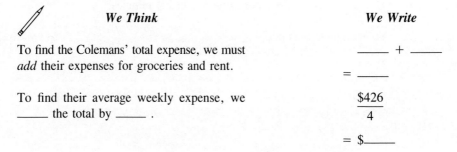

We Think	We Write
To find the Colemans' total expense, we must *add* their expenses for groceries and rent.	_____ + _____ = _____
To find their average weekly expense, we _____ the total by _____ .	$\dfrac{$426}{4}$ = $_____

The Colemans' average expense for groceries and rent was $ _____ each week.

Check your work on page 56. ▶

Working with Rates

One type of problem that you often will be asked to solve is that involving some sort of **rate**, such as

miles *per* hour

dollars *per* hour

miles *per* gallon

revolutions *per* minute

Each of these rates is described as "so many of one thing *per* one of something else" and tells us how many units of the first thing we can expect for *one* unit of the other thing. For instance:

Rate	Meaning
23 miles per gallon	We can travel 23 miles on *one* gallon of gas.
$3 per hour	We are paid $3 for every *one* hour worked.
55 miles per hour	We will travel 55 miles in *one* hour of driving.

If we know a rate and if we know how many of the second thing we have, we can find how many of the first thing we have by *multiplying*. For example,

At 55 miles per hour for 3 hours, we will travel $55 \cdot 3 = 165$ miles

At $3 per hour for 40 hours, we will earn $3 \cdot 40 = 120$ dollars

Example 8. How many revolutions will a record make during a 2.4-minute song if it is spinning at a rate of 45 revolutions per minute (rpm)?

Solution

We Think	*We Write*
We must *multiply* the rate times the number of minutes.	45(2.4)
	= 108 revolutions

The record will make 108 revolutions in 2.4 minutes.

You complete Example 9.

Example 9. Juan drove at a rate of 55 miles per hour (mph) for 3 hours and at a rate of 50 mph for 2 hours. How far did Juan drive?

Solution

We Think	*We Write*
To find the first distance, we *multiply* the rate (55 mph) by the time (3 hours).	55 · _____ = _____ miles
To find the second distance, we *multiply* the rate (50 mph) by the time (2 hours).	50 · _____ = _____ miles
Juan's total distance is the *sum* of the two distances.	_____ + _____ = _____ miles

Juan drove a total of _____ miles.

Check your work on page 56. ▶

Now try Example 10.

Example 10. On a business trip Harvey's car mileage gauge changed from 21623 to 22007. If Harvey's company pays him 25 cents per mile, what will the company pay Harvey for car expenses on this trip?

Solution

We Think	*We Write*
First we must find the number of miles driven by finding the *difference* in the mileage readings.	22007 − _____ = _____ miles
To find the company's payment, we multiply the rate (_____ per _____) times the number of _____ .	25(_____) = _____ cents = $_____

Harvey's company will pay him $ _____ .

Check your work on page 56. ▶

▶ **Examples You Completed** ───────────────────────

Words	Numbers
13 increased by 7	$13 + 7$
32 divided by 2.8	$\dfrac{32}{2.8}$
the sum of 12 and 3	$12 + 3$
the product of 12 and 3	$12 \cdot 3$
5 fewer than 18	$18 - 5$
twice as much as 10.6	$2(10.6)$
42 divided by 15	$42 \div 15$
19 decreased by 15	$19 - 15$
two-thirds of 24	$\dfrac{2}{3} \cdot 24$
3 times as large as 25	$3 \cdot 25$

Example 2. At noon the temperature was 18°. By midnight the temperature had decreased by 12°. Write an expression for the temperature at midnight.

   Solution

$$18° \text{ decreased by } 12°$$
$$18° - 12°$$

Example 6. A real estate developer owns a large plot of land with four sides measuring 2.3 miles, 1.4 miles, 4.25 miles, and 3.5 miles. If the developer drives around the boundary of her land, how many miles will she travel?

   Solution. The distance traveled is found by adding the four distances.

$$2.3 + 1.4 + 4.25 + 3.5$$
$$= 11.45 \text{ miles}$$

The developer will travel 11.45 miles.

Example 7. The Colemans spent $186 on groceries and $240 on rent during 4 weeks. What was the Colemans' *average* weekly expense for groceries and rent?

   Solution

We Think	*We Write*
To find the Colemans' total expense, we must *add* their expenses for groceries and rent.	$186 + 240$ $= 426$
To find their average weekly expense, we divide the total by 4.	$\dfrac{\$426}{4}$ $= \$106.50$

The Colemans' average weekly expense for groceries and rent was $106.50.

Example 9. Juan drove at a rate of 55 miles per hour (mph) for 3 hours and at a rate of 50 mph for 2 hours. How far did Juan drive?

Solution

We Think	*We Write*
To find the first distance, we *multiply* the rate (55 mph) by the time (3 hours).	$55 \cdot 3$ $= 165$ miles
To find the second distance, we *multiply* the rate (50 mph) by the time (2 hours).	$50 \cdot 2$ $= 100$ miles
Juan's total distance is the *sum* of the two distances.	$165 + 100$ $= 265$ miles

Juan drove a total of 265 miles.

Example 10. On a business trip Harvey's car mileage gauge changed from 21623 to 22007. If Harvey's company pays him 25 cents per mile, what will the company pay Harvey for car expenses on this trip?

Solution

We Think	*We Write*
First we must find the number of miles driven by finding the *difference* in the mileage readings.	$22007 - 21623$ $= 384$ miles
To find the company's payment, we multiply the rate (25¢ per mile) times the number of miles.	$25(384)$ $= 9600$ cents $= \$96$

Harvey's company will pay him $96.

EXERCISE SET 1.4

Change each word expression to a number expression and simplify.

_____ 1. Leroy's electric bill this month was $15.93 more than his bill last month. If his bill last month was $33.24, what is his bill this month?

_____ 2. If one calculator costs $18.95, how much will eight calculators cost?

_____ 3. If Carl's car averages 19 miles per gallon, how far can he drive when he has 15 gallons of gasoline?

_____ 4. Sherry, Sandra, and Jo have a job washing the windows of an office building. If the building has 48 windows and they divide the work evenly, how many windows will each wash?

_____ 5. Each month, Sam's wife earns $125 more than twice Sam's salary. If Sam's monthly salary is $685, what is his wife's monthly salary?

_____ 6. Debbie scored 10 points higher on her second history test than on her first test. If she scored 76 on the first test, what is her average on the two tests?

_____ 7. During 6 months, the Sturgis Cap Factory had a profit of $42,000. What was the average profit for 1 month?

_____ 8. If Mark weighed $235\frac{1}{2}$ pounds before his diet and now weighs $190\frac{3}{4}$ pounds, how much weight did he lose?

_____ 9. If $\frac{2}{3}$ of 645 homes surveyed had a microwave oven, how many homes in that area had a microwave?

_____ 10. Chris is putting a string of Christmas lights on his house. If the cord is 20 feet long and has a bulb every $\frac{1}{4}$ of a foot, how many bulbs will he need?

_____ 11. Rachel knows that it takes $1\frac{3}{4}$ yards of material for each costume she is sewing for the dance recital. If there are 12 dancers in the recital, how many yards of material will she need for the costumes?

_____ 12. Kelvin has a rectangular garden that measures $24\frac{3}{4}$ feet by $20\frac{1}{3}$ feet. Find the area of the garden.

_____ 13. Martin spends $\frac{1}{10}$ of his salary each month for gasoline. His car payment is $\frac{1}{5}$ of his salary. What fractional part of his monthly salary is spent on his car?

_____ 14. At Citizens Bank, savings certificates will earn $8\frac{3}{4}$ percent interest. Union Bank's interest rate is $10\frac{4}{5}$ percent. What is the difference in the interest rates?

_____ 15. Simon jogged $2\frac{1}{2}$ miles on Monday, $3\frac{1}{5}$ miles on Tuesday, and $1\frac{5}{6}$ miles on Wednesday. How many miles did Simon jog during the 3 days?

_____ **16.** Roy needs 8 cups of flour for his cake recipes. He finds $3\frac{2}{3}$ cups in the canister and he borrows $2\frac{1}{4}$ cups from his neighbors. How much more flour does he need?

_____ **17.** Ms. Humphrey is buying fringe to sew around the edge of a rectangular tablecloth. If the tablecloth is 2.3 meters long and 1.2 meters wide, how much fringe should she buy?

_____ **18.** On a shopping trip to Louisville, Mary Ruth wrote checks for $85.98, $33.79, $126.93, and $16.59. If her checking balance when she began was $934.17, how much did she have in her account after shopping?

_____ **19.** In 4 days of driving, Lana drove 200.9, 158.7, 369.4, and 276 miles. Find the average number of miles per day Lana drove on her trip.

_____ **20.** Wilson can throw the discus 58.7 meters, but Ted can throw it 60.26 meters. How much farther can Ted throw the discus than Wilson?

_____ **21.** The tips at Judy's Castle for one evening totaled $63.75. If the tips are split evenly among the three waitresses, how much did each waitress get in tips that evening?

_____ **22.** If Michelle's car averages 21.25 miles per gallon of gasoline, how far can she travel on a 16-gallon tank?

_____ **23.** A football field is 109.68 meters long and 48.72 meters wide. Find the area of the football field. (Round answer to the hundredths place.)

_____ **24.** Mr. Hunter sold 1542.8 bushels of soybeans for $9025.38. How much per bushel was he paid for his soybeans?

_____ **25.** At an after-Christmas sale, a department store advertises 15 percent off all merchandise. What will be the discount on a sweater that regularly sells for $35.95? (Round answer to the nearest cent.)

_____ **26.** A daily newspaper has a circulation of 528,200. If they wish to increase their circulation during the next year by 12 percent, how many more copies must they sell?

_____ **27.** The average yearly precipitation for a certain city is 28.6 inches, but last year it received 125 percent of its average precipitation. How many inches of precipitation did it receive last year?

_____ **28.** When 3000 athletes were tested for drug usage, 9.4 percent were found to have used steroids. How many athletes had used steroids?

_____ **29.** A theater seats 2400 persons. If it was filled to 75 percent capacity, how many seats were filled?

_____ **30.** Seventy-two percent of the employees at a coal mine voted to accept the union contract. If there are 450 miners at that mine, how many voted to accept the contract?

☆ Stretching the Topics _____

_____ **1.** The Lollipop Tree advertises 15 percent off on all blouses, 20 percent off on all jeans, and 25 percent off on all dresses. If Kathy bought two blouses that regularly sell for $25 each, a pair of jeans that regularly sells for $33.95, and a dress that regularly sells for $48, find the total amount she owes for the clothing.

_____ **2.** Bruce's car averages 21.5 miles per gallon on the interstate highway and 16.5 miles per gallon on two-lane roads. The trip from his home to college requires that he drive 129 miles on the interstate highway and 66 miles on two-lane roads. If gasoline costs $1.339 per gallon, how much will the trip cost? (Round answer to the nearest cent.)

Check your answers in the back of your book.

If you can complete **Checkup 1.4** correctly, you are ready to do the **Review Exercises** for Chapter 1.

✓ CHECKUP 1.4

Change each word expression to a number expression and simplify.

_____ **1.** Each spring, Natalie plants $\frac{1}{4}$ of her garden in green beans and $\frac{2}{5}$ in lima beans. What fractional part of her garden is used to raise beans?

_____ **2.** In a state that has a 5.5 percent sales tax, how much sales tax would be paid on an automobile that costs $12,248.82? (Round answer to the nearest cent.)

_____ **3.** The three sides of a triangular plot of land measure 13.6 feet, 8.92 feet, and 7.3 feet. Find (to the nearest tenth of a foot) how many feet of fence would be needed to enclose the plot.

_____ **4.** If Tom's monthly bills were $325 for rent, $73.86 for utilities, and $195 for groceries, find his average weekly living expenses. (Assume a 4-week month.)

_____ **5.** On a business trip Lydia averaged 62 miles per hour for 4 hours on the interstate highway and 53 miles per hour for 3 hours on two-lane roads. If Lydia is paid 25 cents per mile, what will the company pay for her expenses?

Check your answers in the back of your book.

If You Missed Problems:	You Should Review Examples:
1	4
2	5
3	6
4	11
5	8, 9

Summary

In this chapter we reviewed the arithmetic of **whole numbers**, **fractions**, and **decimal numbers**. We observed how the four basic operations of **addition**, **subtraction**, **multiplication**, and **division** are related to each other, and noted some special properties of zero and 1.

Property	Symbols	Examples
Addition property of zero	$A + 0 = A$	$3 + 0 = 3$
		$0 + 29 = 29$
Subtraction property of zero	$A - 0 = A$	$5 - 0 = 5$
Multiplication property of zero	$A \cdot 0 = 0$	$13 \cdot 0 = 0$
	$0 \cdot A = 0$	$0 \cdot 9 = 0$
Division of zero	$\dfrac{0}{A} = 0 \quad (A \neq 0)$	$\dfrac{0}{25} = 0$
Division by zero	$\dfrac{A}{0}$ is undefined	$\dfrac{19}{0}$ is undefined
		$\dfrac{0}{0}$ is undefined
Multiplication property of 1	$A \cdot 1 = A$	$6 \cdot 1 = 6$
	$1 \cdot A = A$	$1 \cdot 77 = 77$
Division property of 1	$\dfrac{A}{1} = A$	$\dfrac{769}{1} = 769$

We also noted that $A - A = 0$ and $\dfrac{A}{A} = 1$ (provided that $A \neq 0$).

The use of **exponents** was discussed as a way to indicate that a base is to be repeated as a factor in a product. For example,

$$5^3 = 5 \cdot 5 \cdot 5 = 125$$

We discovered that **symbols of grouping** (parentheses, braces, brackets, and fraction bars) help us simplify expressions by telling us which operations to perform first. In the absence of parentheses, we agreed to perform operations in the following order.

1. Deal with exponents.
2. Perform multiplications and/or divisions from left to right.
3. Perform additions and/or subtractions from left to right.

In our work with **fractions** we mastered several important techniques.

In Order to	We Must	Examples
Reduce a fraction	Divide numerator and denominator by all common factors.	$$\frac{28}{20} = \frac{\overset{1}{\cancel{2}} \cdot \overset{1}{\cancel{2}} \cdot 7}{\underset{1}{\cancel{2}} \cdot \underset{1}{\cancel{2}} \cdot 5} = \frac{7}{5}$$
Multiply fractions	Multiply numerators and multiply denominators.	$$\frac{5}{9} \cdot \frac{2}{3} = \frac{10}{27}$$
Divide fractions	Invert the divisor and multiply.	$$\frac{3}{11} \div \frac{1}{3} = \frac{3}{11} \cdot \frac{3}{1} = \frac{9}{11}$$
Build fractions	Multiply numerator and denominator by the same factor.	$$\frac{3}{4} = \frac{3 \cdot 7}{4 \cdot 7} = \frac{21}{28}$$
Find lowest common denomintor (LCD)	Find product of all different prime factors of original denominators, using each prime factor the most times that it appears in any single denominator.	For $\frac{1}{6}, \frac{2}{3}$, and $\frac{3}{4}$ LCD is $2 \cdot 2 \cdot 3 = 6$
Add (or subtract) fractions	Be sure that fractions have same denominator. Add (or subtract) numerators and keep same denominator.	$$\frac{5}{8} - \frac{2}{8} = \frac{5-2}{8} = \frac{3}{8}$$ $$\frac{1}{6} + \frac{2}{3} - \frac{3}{4} = \frac{2}{12} + \frac{8}{12} - \frac{9}{12}$$ $$= \frac{2+8-9}{12} = \frac{1}{12}$$

We reviewed the operations of addition, subtraction, multiplication, and division of **decimal numbers** and discovered how **percents** are related to fractions and decimal numbers by the fact that percent means "per hundred." For example,

$$23\% = \frac{23}{100} = 0.23$$

Finally, we learned to use whole numbers, fractions, and decimal numbers to switch from word expressions to number expressions.

❑ Speaking the Language of Algebra ─────────────

Complete each statement with the appropriate word or phrase.

1. The set of numbers {0, 1, 2, 3, . . .} is called the set of _____ _____ .

2. When we add two numbers, the result is called the _____ . When we subtract two numbers, the result is called the _____ .

3. In multiplication the numbers being multiplied are called the _____ and the result is called the _____ .

4. In division the number being divided is called the _____ , the number doing the dividing is called the _____ , and the result is called the _____ .

5. If the numerator of a fraction is smaller than the denominator, it is called a _____ fraction. If not, it is called an _____ fraction.

6. To reduce a fraction to lowest terms, we divide the numerator and denominator by all _____ _____ .

7. To divide fractions, we leave the dividend as it is, _____ the divisor, and _____ .

8. Fractions can be added only if they have the _____ _____ .

9. In a decimal number, the part to the left of the decimal point is called the _____ _____ part and the part to the right of the decimal point is called the _____ part.

10. The word "percent" means " _____ _____ ." To change a percent to a decimal number, we drop the % symbol and move the decimal point _____ places to the _____ .

Check your answers in the back of your book.

△ Writing About Mathematics ─────────────

Write your response to each question in complete sentences.

1. Describe the order in which you would perform the indicated operations in simplifying each of the following expressions:
   (a) $2 + 3(20 - 14 \div 2)$
   (b) $(2 + 3)[(20 - 14) \div 2]$

2. Why must we pay close attention to division problems in which the numerator and/or the denominator is zero?

3. Explain the process you would use to round the decimal number 7.3257 to the hundredths place.

4. Explain the procedure for finding the lowest common denominator for two or more fractions.

REVIEW EXERCISES for Chapter 1

_____ **1.** Use the symbol $<$ or $>$ to complete the statement: 3 _____ 19.

_____ **2.** Write the addition statement that corresponds to the subtraction statement
$43 - 32 = 11$.

_____ **3.** Write the multiplication statement that corresponds to the division statement
$\dfrac{91}{13} = 7$.

Perform the indicated operations. Write answers in reduced form.

_____ **4.** $3(5 + 2 \cdot 7)$ _____ **5.** $(7 + 2)(8 + 0)$

_____ **6.** $5 \cdot 2^3 + 4 \cdot 9$ _____ **7.** $[9 + (8 \cdot 7) + 1] + 6$

_____ **8.** $7\left(\dfrac{15}{3} - \dfrac{5 \cdot 0}{2}\right)$ _____ **9.** $\dfrac{80 - 8 \cdot 6}{9 - 5}$

_____ **10.** $\dfrac{9 \cdot 6 - 3^2}{56 - 7 \cdot 8}$ _____ **11.** $\dfrac{3 \cdot 8 - 6 \cdot 4}{19 - 12}$

_____ **12.** $\dfrac{15}{4} \cdot \left(\dfrac{2}{5}\right)^2$ _____ **13.** $\dfrac{5}{6} \cdot \dfrac{7}{25} \cdot \dfrac{15}{28}$

_____ **14.** $\dfrac{7}{3} \div 63$ _____ **15.** $\dfrac{1}{9} \div \left(\dfrac{4}{5} \cdot \dfrac{10}{27}\right)$

_____ **16.** $\dfrac{1}{15} + \dfrac{9}{25}$ _____ **17.** $\dfrac{10}{21} + \dfrac{11}{18} - \dfrac{9}{14}$

_____ **18.** Change $\dfrac{5}{16}$ to a decimal number.

_____ **19.** Round 13.072 to the tenths place.

_____ **20.** $13 + 4.0765 + 0.09$ _____ **21.** $276.35 - 98.86$

_____ **22.** $(2.75)(0.032)$ _____ **23.** $0.2236 \div 0.043$

_____ **24.** Change 0.09 to a percent.

_____ **25.** Change 12.7% to a decimal number.

_____ **26.** Find 13.2% of $5000.

Change each word expression to a number expression and simplify.

_____ **27.** Julie bought five tapes at $7 each and three records at $5 each. Find the total
amount she spent at the Music Mart.

_____ **28.** Greg receives a 15 percent commission on each house he sells. How much commission would he receive on a house selling for $48,500?

_____ **29.** In his will Uncle Lucian left $\frac{1}{4}$ of his estate to the United Way and $\frac{2}{5}$ of his estate to the Salvation Army. What fractional part of his estate did Uncle Lucian leave to charity?

_____ **30.** Over a period of 4 weeks, Jane's weight changed from 183 pounds to 157 pounds. What was her average weight loss each week?

Check your answers in the back of your book.

If You Missed Exercises:	You Should Review Examples:	
1	SECTION 1.1	3, 4
2		7
3		9, 10
4–7		19–22
8–11		23–28
12–15	SECTION 1.2	5–11
16, 17		14–17
18	SECTION 1.3	5, 6
19		7, 8
20, 21		11, 12
22, 23		13–16
24		19, 20
25		17, 18
26		21, 22
27	SECTION 1.4	8
28		5
29		4
30		10

If you have completed the **Review Exercises** and corrected your errors, you are ready to take the **Practice Test** for Chapter 1.

PRACTICE TEST for Chapter 1

			SECTION	EXAMPLES
_____	**1.**	Write the addition statement that corresponds to the subtraction statement $24 - 13 = 11$.	1.1	7
_____	**2.**	Write the multiplication statement that corresponds to the division statement $\dfrac{55}{11} = 5$.	1.1	9, 10

Perform the indicated operations. Write answers in reduced form.

_____	**3.**	$8(6 + 2 \cdot 9)$	1.1	19, 20
_____	**4.**	$8 \cdot 3^2 + 5 \cdot 6$	1.1	26
_____	**5.**	$[8 + (6 \cdot 3) + 2] + 9$	1.1	21
_____	**6.**	$\dfrac{8 \cdot 7 - 2^3}{72 - 9 \cdot 8}$	1.1	12, 26, 28
_____	**7.**	$\dfrac{9 \cdot 4 - 6 \cdot 6}{27 - 19}$	1.1	12, 28
_____	**8.**	$\left(\dfrac{45}{28}\right)\left(\dfrac{2}{3}\right)^3$	1.2	6
_____	**9.**	$\dfrac{8}{9} \div 16$	1.2	10
_____	**10.**	$\dfrac{4}{3} \div \left(\dfrac{5}{18} \cdot \dfrac{8}{25}\right)$	1.2	11
_____	**11.**	$\dfrac{1}{12} + \dfrac{7}{9}$	1.2	18
_____	**12.**	$\dfrac{7}{10} + \dfrac{9}{12} - \dfrac{8}{15}$	1.2	19
_____	**13.**	Change $\dfrac{3}{8}$ to a decimal number.	1.3	6
_____	**14.**	Round 7.1354 to the hundredths place.	1.3	8
_____	**15.**	$8 + 11.35 + 0.135$	1.3	11
_____	**16.**	$(3.29)(0.25)$	1.3	13, 14
_____	**17.**	$6.3252 \div 1.26$	1.3	15, 16
_____	**18.**	Find 8.5% of $270.	1.3	21, 22

	SECTION	EXAMPLES

Change each word expression to a number expression and simplify.

_____ **19.** Pansy bought 2 pounds of ground beef at $1.79 per pound and 1 pound of ground pork at $1.89 per pound. Find the total amount she spent for the meat.   1.4   8, 9

_____ **20.** During the first 4 days of March, Doug jogged 3.7, 4.2, 3.8, and 4.5 miles. Find the average number of miles Doug jogged each day.   1.4   6, 7

_____ **21.** During its back-to-school sale, the Jeans Depot advertised 15 percent off all jeans. Find the discount on a pair of jeans that regularly sold for $29.95. (Round to the nearest cent.)   1.4   5

_____ **22.** Find the area of a rectangular patio that measures $16\frac{2}{3}$ feet by $20\frac{1}{4}$ feet.   1.4   4

Check your answers in the back of your book.

2

Working with
Real Numbers

In one series of plays, the State University football team completed two
7-yard passes and ran for 5 yards. Then the quarterback was sacked
twice, for a 3-yard loss each time, and the team received a 15-yard
penalty. What was the team's total gain or loss?

When you have completed this chapter, you will be able to solve this problem using
arithmetic of **real numbers**.
   In this chapter we

1. Introduce integers.
2. Add, subtract, multiply, and divide integers.
3. Introduce real numbers.
4. Observe some properties of real numbers.
5. Switch from word expressions to real number expressions.

2.1 Understanding Integers

On a winter day, have you ever watched the thermometer drop to a temperature of 3° below zero? Have you ever seen your favorite football team lose 6 yards on a play? Have you ever borrowed $10?

In each of these situations the set of whole numbers does not provide us with a way to make the necessary measurement. We know that temperatures of 3° above zero and 3° below zero are very different. A gain of 6 yards is very different from a loss of 6 yards, and a credit of $10 is very different from a debt of $10. We can use a **negative sign** to help us, and

3° *below* zero	is written	$-3°$
a *loss* of 6 yards	is written	-6 yards
a *debt* of 10 dollars	is written	-10 dollars

Locating Integers on the Number Line

On a number line we illustrate these negative numbers by extending our line to the left, marking off each unit in a leftward direction from zero.

This new set of numbers, which includes all the whole numbers *and* their negative counterparts, is called the set of **integers**.

> Integers: $\{\ldots, -6, -5, -4, -3, -2, -1, 0, 1, 2, 3, 4, \ldots\}$

Notice that this list continues indefinitely in either direction, as shown by the arrows at both ends of the number line. Units to the *right* of zero correspond to the natural numbers (also called **positive integers**) and units to the *left* of zero correspond to the **negative integers**.

> Positive integers: $\{1, 2, 3, 4, 5, \ldots\}$
> Negative integers: $\{\ldots, -4, -3, -2, -1\}$

The set of integers contains all the positive integers, all the negative integers, and the integer zero. Each integer corresponds to one and only one point on the number line.

The number line again gives us a handy way to compare integers. If one integer lies to the *left* of another integer, we say that the first integer is **less than** ($<$) the second integer. If one integer lies to the *right* of another integer, we say that the first integer is **greater than** ($>$) the second integer.

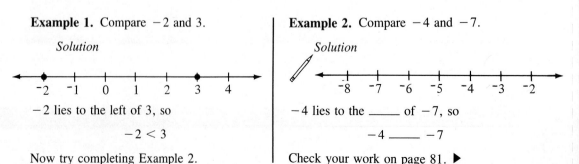

Example 1. Compare -2 and 3.

Solution

-2 lies to the left of 3, so

$$-2 < 3$$

Now try completing Example 1.

Example 2. Compare -4 and -7.

Solution

-4 lies to the _____ of -7, so

$$-4 \underline{\quad} -7$$

Check your work on page 81. ▶

Example 3. Locate -3 and locate 3 on the number line.

Solution

If we were to take a ruler and measure the distance between -3 and 0 on our number line and then measure the distance between 0 and 3 on our number line, what would we observe? Clearly, those distances are the same. Each of the points, -3 and 3, measures 3 units from zero. Their locations are on *opposite* sides of zero, but their distances represent the same number of units. For this reason, we say

$$-3 \text{ is the } opposite \text{ of } 3$$

$$3 \text{ is the } opposite \text{ of } -3$$

When two numbers measure the same distance from zero but the two numbers are on opposite sides of zero, we say that one number is the **opposite** of the other.

Example 4. What is the opposite of 16?

Solution. -16 is the opposite of 16.

You try Example 5.

Example 5. What is the opposite of -5?

Solution. _____ is the opposite of -5.

Check your answer on page 81. ▶

Sometimes we refer to a positive integer, such as 5, as $+5$, but the positive sign is not required. You should understand that if there is no sign in front of a number, it is considered positive. We also agree here that the opposite of 0 is 0.

We can use a *negative sign* in front of a number to denote the opposite of that number.

"The opposite of 3 is -3" can be written $-(3) = -3$

"The opposite of -3 is 3" can be written $-(-3) = 3$

Writing Opposites

Opposite of a: $-(a)   = -a$
Opposite of $-a$: $-(-a) = a$
Opposite of 0: $-(0)   = 0$

Example 6. Find $-(23)$.

Solution

$$-(23) = -23$$

Example 7. Find $-(-11)$.

Solution

$$-(-11) = 11$$

When we measure the distance between 0 and the point on the number line corresponding to an integer, we are finding the **absolute value** of that integer. The absolute value of a number, a, is symbolized by $|a|$.

> **Absolute Value.** If a is a number, then $|a|$ is the distance between 0 and a on the number line.

Because absolute value represents distance, we note that

> The absolute value of a number is never negative.

Example 8. Find $|-6|$.

Solution
$$|-6| = 6$$

You try Example 9.

Example 9. Find $|23|$.

Solution

$$|23| = \underline{\qquad}$$

Check your work on page 81. ▶

Then try Example 10.

Example 10. Find $|0|$.

Solution
$$|0| = \underline{\qquad}$$

Check your work on page 81. ▶

Example 11. Find $-|-10|$.

Solution. Here we must find the opposite of the absolute value of -10.
$$-|-10| = -(10)$$
$$= -10$$

Adding Integers

How do we go about adding integers? First, let us recall how to add integers that are positive or zero. Remember, these are the whole numbers 0, 1, 2, 3, 4, . . ., which we have already discussed. Let's use the number line.

Example 12. Use the number line to find $5 + 2$.

Solution. We start at 5 and move 2 units to the right.

$$5 + 2 = 7$$

Notice that whenever we add two *positive* integers, the answer will be a *positive* integer.

> To add two *positive* integers, add the units (absolute values) and give the answer a *positive* sign.

What happens when we add two *negative* integers? The number line will be a help here. In locating a negative number, we must move *left* rather than right.

Example 13. Use the number line to add -3 and -2.

Solution. We start at -3 and move 2 units to the *left*.

Since we ended up at -5, the sum of -3 and -2 must be -5. We write this as

$$-3 + (-2) = -5$$

where parentheses are used around -2 for clarity. Notice that in this example of adding two *negative* integers, the number of units was the *sum* of the units in the original numbers, but the *sign* of the answer was *negative*. We conclude:

> To add two *negative* integers, add the units (absolute values) and give the answer a *negative* sign.

Example 14. Find $-13 + (-76)$.

Solution. The answer will have a *negative* sign.

$$-13 + (-76) = -89$$

You complete Example 15.

Example 15. Find $[-31 + (-8)] + (-12)$.

Solution. The answer will have a _____ sign.

$$[-31 + (-8)] + (-12)$$

$$= \underline{\hphantom{xx}} + (-12)$$

$$= \underline{\hphantom{xx}}$$

Check your work on page 81. ▶

What happens if we try to add two numbers when one of the numbers is *positive* and one of the numbers is *negative*? The number line will give us a clue if we remember to move *right* for *positive* numbers and move *left* for *negative* numbers.

Example 16. Use the number line to find $-6 + 4$.

Solution. We start at -6 and move 4 units to the *right*.

We end up at -2, so

$$-6 + 4 = -2$$

Example 17. Use the number line to find $-3 + 8$.

Solution. We start at -3 and move 8 units to the *right*.

We end up at 5, so

$$-3 + 8 = 5$$

Do you see how we could add two numbers with opposite signs without using the number line? In each example, the *units* in our answer represented the *difference* between our original units. The *sign* of the answer was the sign of the original number having *more* units. Remember, the number of units for an integer is that integer's distance from zero on the number line (absolute value).

> To add a *positive* integer and a *negative* integer, find the *difference* between their units (absolute values) and give the answer the sign of the original integer having more units (larger absolute value).

Example 18. Find $-11 + 26$.

Solution. Our answer will have a positive sign. The difference between 11 units and 26 units is 15 units.

$$-11 + 26 = 15$$

You try Example 19.

Example 19. Find $20 + (-30)$.

Solution. Our answer will have a _____ sign. The difference between 30 units and 20 units is _____ units.

$$20 + (-30) = \text{\_\_\_\_\_}$$

Check your work on page 81. ▶

As was true for whole numbers, fractions, and decimal numbers, *parentheses* give us directions for the order in which to operate with integers.

Example 20. Find $-10 + (-17 + 2)$.

Solution. Parentheses still say "do this first."

$$-10 + (-17 + 2)$$
$$= -10 + (-15)$$
$$= -25$$

You try completing Example 21.

Example 21. Find $[5 + (-16)] + 3$.

Solution

$$[5 + (-16)] + 3$$
$$= \text{\_\_\_\_\_} + 3$$
$$= \text{\_\_\_\_\_}$$

Check your work on page 81. ▶

Example 22. Find

$$9\left[\frac{18}{2} + (-7 + 2)\right]$$

Solution

$$9\left[\frac{18}{2} + (-7 + 2)\right]$$

$$= 9[9 + (-5)]$$

$$= 9 \cdot 4$$

$$= 36$$

You try Example 23.

Example 23. Find $\dfrac{7(-6 + 6)}{6 + (-3)}$.

Solution

$$\frac{7(-6 + 6)}{6 + (-3)}$$

$$=$$

Check your work on page 81. ▶

⟫ Trial Run

Perform the indicated operations.

_____ **1.** $8 + (-6)$

_____ **2.** $-11 + 5$

_____ **3.** $-3 + (-3)$

_____ **4.** $[-5 + (-6)] + 10$

_____ **5.** $(-12 + 2) + (-9)$

_____ **6.** $\dfrac{3[9 + (-4)]}{-2 + 7}$

Answers are on page 81.

Using Some Properties of Addition

There are several laws or properties of addition that we may observe from our work with integers. Note these sums.

$$-15 + 8 = -7$$
$$8 + (-15) = -7$$

> It does not matter in which order we add two integers. The sum will be the same.

This law is called the **commutative property for addition**, and we state it in symbols as

> **Commutative Property for Addition.** For any integers A and B
> $$A + B = B + A$$

Now consider these sums involving *three* integers.

$$(-6 + 8) + (-3) = 2 + (-3) = -1$$
$$-6 + [8 + (-3)] = -6 + 5 \ = -1$$

In these two sums, the order in which the integers appeared was the same, but the position of the symbols of grouping was different. In both cases the sum was the same.

> When adding three integers, it does not matter if we add the first two and then add the third, *or* add the second two and then add the first. The sum will be the same.

This law is called the **associative property for addition**.

> **Associative Property for Addition.** For any integers A, B, and C
> $$(A + B) + C = A + (B + C)$$

Complete Examples 24 and 25.

Example 24. Use the commutative property to rewrite the sum $-5 + (-9)$.

Solution

$$-5 + (-9)$$
$$= \underline{} + (\underline{})$$

Check your answer on page 81. ▶

Example 25. Use the associative property to rewrite the sum $[4 + (-2)] + (-17)$.

Solution

$$[4 + (-2)] + (-17)$$
$$= \underline{} + [\underline{} + (\underline{})]$$

Check your answer on page 81. ▶

To add several integers when there are no brackets, the commutative and associative properties of addition allow us to arrange them in any order we choose. It is often easier to *add all the numbers with matching signs first*.

Example 26. Find the sum $-22 + (-11) + 6 + (-30) + 50$.

Solution

$-22 + (-11) + 6 + (-30) + 50$

$= [-22 + (-11) + (-30)] + [6 + 50]$ Rearrange terms (commutative and associative properties).

$= -63 + 56$ Add negative integers; add positive integers.

$= -7$ Find the sum.

You try Example 27.

Example 27. Find the sum $16 + (-23) + 4 + (-8) + (-10)$.

Solution

$16 + (-23) + 4 + (-8) + (-10)$

$= [16 + \underline{\quad}] + [-23 + (\underline{\quad}) + (\underline{\quad})]$ Rearrange terms.

$= \underline{\quad} + (\underline{\quad})$ Add positive integers; add negative integers.

$= \underline{\quad}$ Find the sum.

Check your work on page 81. ▶

What happens when we add *zero* to an integer? Look at these sums.

$$13 + 0 = 13$$
$$-26 + 0 = -26$$
$$0 + (-93) = -93$$

As was true with whole numbers, when zero is added to any integer, the sum is that integer.

> **Addition Property of Zero.** For any integer A
> $$A + 0 = A$$

Recall that zero is called the **identity for addition**.

Consider the result when an integer is added to its *opposite*.

$$9 + (-9) = 0$$
$$-62 + 62 = 0$$

These examples should help you see that

The sum of an integer and its opposite is 0.

The *opposite* of a number is also called its **additive inverse** and we can state this new law as follows.

Property of Opposites *(or Additive Inverses).* For any integer A

$$A + (-A) = 0$$

Example 28. Find the sum $-15 + 12 + (-11) + 14$.

Solution

$$-15 + 12 + (-11) + 14$$
$$= [-15 + (-11)] + [12 + 14] \qquad \text{Rearrange terms.}$$
$$= -26 + 26 \qquad \text{Add negative integers; add positive integers.}$$
$$= 0 \qquad \text{Property of opposites.}$$

⫸ Trial Run

Complete each statement.

1. By the commutative property for addition, $-8 + 7 =$ _____ .

2. By the associative property for addition, $-5 + [3 + (-9)] =$ _____ .

Find the sums.

_____ 3. $8 + (-6) + (-2)$ _____ 4. $(17 + 0) + [0 + (-3)]$

_____ 5. $-22 + (-7) + 3 + (-20) + 40$

_____ 6. $16 + (-23) + 4 + (-8) + (-10)$

Answers are on page 81.

▶ **Examples You Completed** _____

Example 2. Compare -4 and -7.

Solution

-4 lies to the right of -7, so

$$-4 > -7$$

Example 5. What is the opposite of -5?

Solution. 5 is the opposite of -5.

Example 9. Find $|23|$.

Solution

$$|23| = 23$$

Example 10. Find $|0|$.

Solution

$$|0| = 0$$

Example 15. Find $[-31 + (-8)] + (-12)$.

Solution. The answer will have a negative sign.

$$[-31 + (-8)] + (-12)$$
$$= -39 + (-12)$$
$$= -51$$

Example 19. Find $20 + (-30)$.

Solution. Our answer will have a negative sign. The difference between 30 units and 20 units is 10 units.

$$20 + (-30) = -10$$

Example 21. Find $[5 + (-16)] + 3$.

Solution

$$[5 + (-16)] + 3$$
$$= -11 + 3$$
$$= -8$$

Example 23. Find $\dfrac{7(-6 + 6)}{6 + (-3)}$.

Solution

$$\frac{7(-6 + 6)}{6 + (-3)} = \frac{7(0)}{3}$$
$$= \frac{0}{3}$$
$$= 0$$

Example 24. Use the commutative property to rewrite the sum $-5 + (-9)$.

Solution

$$-5 + (-9)$$
$$= -9 + (-5)$$

Example 25. Use the associative property to rewrite the sum $[4 + (-2)] + (-17)$.

Solution

$$[4 + (-2)] + (-17)$$
$$= 4 + [-2 + (-17)]$$

Example 27. Find the sum $16 + (-23) + 4 + (-8) + (-10)$.

Solution

$$16 + (-23) + 4 + (-8) + (-10)$$
$$= [16 + 4] + [-23 + (-8) + (-10)]$$
$$= 20 + (-41)$$
$$= -21$$

Answers to Trial Runs _____

page 77   **1.** 2   **2.** -6 **3.** -6 **4.** -1 **5.** -19 **6.** 3

page 80   **1.** $7 + (-8)$ **2.** $(-5 + 3) + (-9)$ **3.** 0   **4.** 14   **5.** -6 **6.** -21

EXERCISE SET 2.1

1. Graph the points corresponding to these integers on the number line.
   (a) -7 **(b)** 4   **(c)** 0   **(d)** -3

2. Graph the points corresponding to these integers on the number line.
   (a) -5 **(b)** 3   **(c)** 1   **(d)** -8

3. Identify by number each letter on the number line.

4. Identify by number each letter on the number line.

5. Compare each pair of numbers using < or >.

   _____ **(a)** 5, 12   _____ **(b)** $-4, -10$

   _____ **(c)** 8, -11 _____ **(d)** $-3, 7$

6. Compare each pair of numbers using < or >.

   _____ **(a)** 3, 9   _____ **(b)** $-12, -2$

   _____ **(c)** 10, -15 _____ **(d)** $-2, 9$

7. Use the commutative property for addition to rewrite each sum.

   _____ **(a)** $-7 + 2$ _____ **(b)** $6 + (-9)$

   _____ **(c)** $-20 + (-6)$ _____ **(d)** $8 + 10$

8. Use the commutative property for addition to rewrite each sum.

   _____ **(a)** $19 + 0$ _____ **(b)** $-3 + (-11)$

   _____ **(c)** $-12 + 5$ _____ **(d)** $9 + (-3)$

9. Use the associative property for addition to rewrite each sum.

   _____ **(a)** $(6 + 3) + (-2)$ _____ **(b)** $-5 + [4 + (-3)]$

   _____ **(c)** $[(-4) + 9] + (-2)$ _____ **(d)** $7 + [(-2) + 11]$

10. Use the associative property for addition to rewrite each sum.

_____ **(a)** $7 + (2 + 5)$ _____ **(b)** $[(-5) + 8] + 9$

_____ **(c)** $13 + [(-2) + (-8)]$ _____ **(d)** $[12 + (-5)] + (-3)$

Simplify by removing absolute value bars.

_____ **11.** $|3|$ _____ **12.** $|10|$

_____ **13.** $|-5|$ _____ **14.** $|-7|$

_____ **15.** $-|9|$ _____ **16.** $-|15|$

_____ **17.** $-|-8|$ _____ **18.** $-|-17|$

_____ **19.** $|0|$ _____ **20.** $-|0|$

Perform the indicated operations.

_____ **21.** $4 + 18$ _____ **22.** $12 + 5$

_____ **23.** $-5 + (-7)$ _____ **24.** $-9 + (-12)$

_____ **25.** $15 + (-3)$ _____ **26.** $19 + (-4)$

_____ **27.** $3 + (-9)$ _____ **28.** $2 + (-15)$

_____ **29.** $-11 + 4$ _____ **30.** $-12 + 7$

_____ **31.** $-2 + 15$ _____ **32.** $-3 + 22$

_____ **33.** $14 + [8 + (-2)]$ _____ **34.** $20 + [10 + (-3)]$

_____ **35.** $[9 + (-7)] + 6$ _____ **36.** $[15 + (-8)] + 7$

_____ **37.** $[-15 + (-3)] + 1$ _____ **38.** $[-41 + (-9)] + 1$

_____ **39.** $-12 + (-7 + 5)$ _____ **40.** $-13 + (-5 + 12)$

_____ **41.** $21 + (-12) + 9$ _____ **42.** $32 + (-24) + 7$

_____ **43.** $-15 + 7 + (-13)$ _____ **44.** $-26 + 10 + (-4)$

_____ **45.** $13 + (-8) + (-9) + 4$ _____ **46.** $18 + (-5) + (-17) + 4$

_____ **47.** $-19 + (-9) + 12 + 7 + (-2)$ _____ **48.** $-25 + (-11) + 9 + 8 + (-15)$

_____ **49.** $2[8 + (-3)]$ _____ **50.** $3[13 + (-10)]$

_____ 51. $4.3 + (-17)$ _____ 52. $5.6 + (-20)$

_____ 53. $(-12) + (5 \cdot 0)$ _____ 54. $(-13) + (8 \cdot 0)$

_____ 55. $2[(-8) + 5 \cdot 4]$ _____ 56. $3[(-9) + 3 \cdot 6]$

_____ 57. $\dfrac{24 + (-3)}{-8 + 15}$ _____ 58. $\dfrac{33 + (-7)}{-2 + 15}$

_____ 59. $\dfrac{-3 + 30}{9 + 0}$ _____ 60. $\dfrac{-4 + 20}{8 + 0}$

_____ 61. $\dfrac{-7 + 16}{-4 + 4}$ _____ 62. $\dfrac{-9 + 13}{-5 + 5}$

_____ 63. $-8 + (-9) + \dfrac{20}{2}$ _____ 64. $-10 + (-3) + \dfrac{25}{5}$

_____ 65. $\dfrac{5(-7 + 7)}{9 + (-4)}$ _____ 66. $\dfrac{6(-8 + 8)}{18 + (-9)}$

_____ 67. $-18 + \left(6 + \dfrac{16}{4}\right)$ _____ 68. $-23 + \left(9 + \dfrac{15}{3}\right)$

_____ 69. $3\left[\dfrac{21}{3} + (-9 + 2)\right]$ _____ 70. $4\left[\dfrac{15}{3} + (-6 + 1)\right]$

☆ Stretching the Topics _____

Perform the indicated operations.

_____ 1. $5\{(-11) + 6[7 + (-3)]\}$

_____ 2. $-4 + 2\left\{\dfrac{18}{9} + 3[7 + (-1)]\right\}$

_____ 3. $10\left\{2 \cdot 4 + \dfrac{2[13 + (-5)]}{4}\right\}$

Check your answers in the back of your book.

If you can complete **Checkup 2.1**, you are ready to go on to Section 2.2.

✓ CHECKUP 2.1

_____ **1.** Compare 3 and -15 using $<$ or $>$.

_____ **2.** Use the commutative property to rewrite the sum $17 + (-5)$.

_____ **3.** Use the associative property to rewrite the sum $-9 + (-3 + 7)$.

_____ **4.** Simplify $-|-12|$ by removing absolute value bars.

Perform the indicated operations.

_____ **5.** $-9 + 4$ _____ **6.** $10 + (-2)$

_____ **7.** $[-3 + (-5)] + 7$ _____ **8.** $-8 + (-9) + 5 + (-3)$

_____ **9.** $6\left[\dfrac{18}{2} + (-12 + 3)\right]$ _____ **10.** $\dfrac{2(-3 + 5)}{6 + (-2)}$

Check your answers in the back of your book.

If You Missed Problems:	You Should Review Examples:
1	1, 2
2, 3	24, 25
4	11
5, 6	18, 19
7	15, 21
8	27
9	22
10	23

2.2 Subtracting Integers

When we discussed the subtraction of whole numbers, we agreed that

$$9 - 3 = 6$$

In the section on the addition of integers, we agreed that

$$9 + (-3) = 6$$

so it seems that $9 - 3$ and $9 + (-3)$ are two different ways of saying the same thing.

Subtraction		*Addition*
$9 - 3$	means the same as	$9 + (-3)$
9 *minus* 3	means the same as	9 *plus* the *opposite* of 3

In fact, this is the way in which *subtraction* is defined using *addition*. Since we know how to *add* integers, subtraction should proceed smoothly.

Example 1. Write $10 - 1$ using addition, and simplify.

Solution

$$10 - 1 = 10 + (-1)$$
$$= 9$$

Now you complete Example 2.

Example 2. Write $6 - 6$ using addition, and simplify.

Solution

$$6 - 6 = 6 + (-6)$$
$$= \underline{\qquad}$$

Check your work on page 89. ▶

Every subtraction problem can be written as an addition problem.

To subtract two integers, take the *opposite* of the integer being subtracted, and *add*.

If A and B represent any integers, we write

Definition of Subtraction
$A - B = A + (-B)$

Example 3. Find $-7 - (-5)$.

Solution

$-7 - (-5)$	Subtraction problem.
$= -7 + 5$	Take the opposite of -5, and *add*.
$= -2$	

You try Example 4.

Example 4. Find $-11 - 19$.

Solution

$-11 - 19$	Subtraction problem.
$= -11 + (\underline{\quad})$	Take the opposite of 19, and *add*.
$= \underline{\quad}$	

Check your work on page 89. ▶

When expressions contain symbols of grouping, we continue to perform the operations within parentheses first and to work from the innermost symbols of grouping outward. If you are impatient and try to hurry through these problems, you will find yourself making mistakes. It is important to *get in the habit of dealing carefully with symbols of grouping*.

Example 5. Find $(7 - 10) - (8 - 15)$.

Solution

$$(7 - 10) - (8 - 15)$$

$= [7 + (-10)] - [8 + (-15)]$ Rewrite inner subtractions as additions.

$= -3 - [-7]$ Simplify within brackets.

$= -3 + 7$ Rewrite subtraction as addition.

$= 4$ Find the sum.

You complete Example 6.

Example 6. Find $8 - (10 - 23)$.

Solution

$$8 - (10 - 23)$$

$= 8 - [10 + (\text{\_\_\_})]$ Rewrite inner subtraction as addition.

$= 8 - [\text{\_\_\_}]$ Simplify within brackets.

$= 8 + \text{\_\_\_}$ Rewrite subtraction as addition.

$= \text{\_\_\_}$

Check your work on page 89. ▶

Example 7. Find $(8 - 13) - (10 - 16) - 4$.

Solution

$$(8 - 13) - (10 - 16) - 4$$

$= [8 + (-13)] - [10 + (-16)] - 4$ Rewrite inner subtractions as additions.

$= -5 - [-6] - 4$ Simplify within brackets.

$= -5 + 6 + (-4)$ Rewrite subtractions as additions.

$= -5 + (-4) + 6$ Rearrange terms.

$= -9 + 6$ Add negative numbers.

$= -3$ Find the sum.

IIII➡ Trial Run

Find the differences.

_____ 1. $7 - 12$ _____ 2. $-15 - 8$

_____ 3. $-9 - (-17)$ _____ 4. $9 - (3 + 12)$

_____ **5.** $(3 - 7) - (9 - 12)$ _____ **6.** $10 - (9 - 21)$

Answers are below.

A few examples might help us decide whether the operation of subtraction is *commutative* or *associative*.

Is subtraction commutative?

$$-3 - (-7) \overset{?}{=} -7 - (-3)$$
$$-3 + 7 \overset{?}{=} -7 + 3$$
$$4 \neq -4$$

There is *no* commutative property for *subtraction*.

Is subtraction associative?

$$10 - (8 - 3) \overset{?}{=} (10 - 8) - 3$$
$$10 - 5 \overset{?}{=} 2 - 3$$
$$5 \neq -1$$

There is *no* associative property for *subtraction*.

▶ **Examples You Completed** _____

Example 2. Write $6 - 6$ using addition, and simplify.

Solution

$$6 - 6 = 6 + (-6)$$
$$= 0$$

Example 6. Find $8 - (10 - 23)$.

Solution

$$8 - (10 - 23)$$
$$= 8 - [10 + (-23)]$$
$$= 8 - [-13]$$
$$= 8 + 13$$
$$= 21$$

Example 4. Find $-11 - 19$.

Solution

$$-11 - 19$$
$$= -11 + (-19)$$
$$= -30$$

Answers to Trial Run _____

page 89   **1.** -5 **2.** -23 **3.** 8   **4.** -6 **5.** -1 **6.** 22

Name _____ **Date** _____

EXERCISE SET 2.2

Write each expression using addition and simplify.

_____ **1.** $12 - 9$ _____ **2.** $23 - 8$

_____ **3.** $11 - 27$ _____ **4.** $9 - 18$

_____ **5.** $-15 - 7$ _____ **6.** $-19 - 8$

_____ **7.** $12 - (-3)$ _____ **8.** $17 - (-5)$

_____ **9.** $-9 - (-7)$ _____ **10.** $-11 - (-8)$

Perform the indicated operations.

_____ **11.** $15 + (8 - 13)$ _____ **12.** $17 + (9 - 15)$

_____ **13.** $21 - (3 + 7)$ _____ **14.** $11 - (5 - 12)$

_____ **15.** $(8 - 7) - 12$ _____ **16.** $(13 - 10) - 20$

_____ **17.** $(7 - 14) - 2$ _____ **18.** $(9 - 16) - 5$

_____ **19.** $(5 - 4) - (-2)$ _____ **20.** $(10 - 3) - (-4)$

_____ **21.** $-7 - (10 - 5)$ _____ **22.** $-13 - (12 - 6)$

_____ **23.** $[(-5) - 2] - 3$ _____ **24.** $[(-8) - 4] - 9$

_____ **25.** $[-5 - (-2)] - 3$ _____ **26.** $[-9 - (-4)] - 5$

_____ **27.** $[-10 - 5] - (-15)$ _____ **28.** $[-12 - 6] - (-18)$

_____ **29.** $-13 - [(-3) - 11]$ _____ **30.** $-12 - [(-5) - 8]$

_____ **31.** $3[(-2) - (-7)]$ _____ **32.** $4[(-3) - (-8)]$

_____ **33.** $3 \cdot 5 - 17$ _____ **34.** $2 \cdot 6 - 21$

_____ **35.** $-6 - (8 \cdot 0)$ _____ **36.** $-9 - (6 \cdot 0)$

_____ **37.** $2(7 \cdot 4 - 8)$ _____ **38.** $5(6 \cdot 5 - 10)$

_____ **39.** $(12 - 15) - (7 - 13)$ _____ **40.** $(8 - 13) - (5 - 9)$

_____ **41.** $(-3 - 9) - (7 - 3)$ _____ **42.** $(-4 - 11) - (10 - 7)$

_____ **43.** $(9 - 15) - (17 - 21) - 5$ _____ **44.** $(4 - 12) - (21 - 15) - 7$

_____ **45.** $[7 - (-3)] - [-12 - (-9)]$ _____ **46.** $[8 - (-5)] - [-16 - (-7)]$

90

_____ 47. $\dfrac{25 - 3}{4 - (-7)}$

_____ 48. $\dfrac{33 - 9}{6 - (-6)}$

_____ 49. $\dfrac{3 - (-15)}{9 - 0}$

_____ 50. $\dfrac{4 - (-17)}{7 - 0}$

_____ 51. $\dfrac{-3 - (-12)}{5 - 5}$

_____ 52. $\dfrac{17 - (-2)}{7 - 7}$

_____ 53. $-8 - 9 - \dfrac{16}{4}$

_____ 54. $-10 - 7 - \dfrac{20}{10}$

_____ 55. $\dfrac{4(8 - 8)}{-4 - (-9)}$

_____ 56. $\dfrac{3(11 - 11)}{-5 - (-10)}$

_____ 57. $-13 - 8 - \dfrac{10}{2}$

_____ 58. $-17 - 9 - \dfrac{15}{3}$

☆ Stretching the Topics _____

Perform the indicated operations.

_____ 1. $-9 - \{10 + 3[1 - (-2)]\}$

_____ 2. $2\{3[10 - (-7)] - 4(-2 + 5)\}$

_____ 3. $9\left[4.1 - \dfrac{14 - (-12)}{6 - (-7)}\right]$

Check your answers in the back of your book.

If you can simplify the expressions in **Checkup 2.2**, you are ready to go on to Section 2.3.

CHECKUP 2.2

Perform the indicated operations.

_____ **1.** 12 − 5   _____ **2.** 8 − (−13)   _____ **3.** −17 − 5

_____ **4.** −19 − (−9)   _____ **5.** 10 − 28   _____ **6.** −7 − (10 − 23)

_____ **7.** (4 − 11) − (11 − 19)   _____ **8.** (−3 − 7) − (9 − 11)

_____ **9.** (9 − 12) − (21 − 30) − 7   _____ **10.** [9 − (−5)] − [−13 − (−4)]

Check your answers in the back of your book.

If You Missed Problems:	You Should Review Examples:
1–5	1–4
6	6
7, 8	5
9, 10	7

2.3 Multiplying and Dividing Integers

Multiplying Integers

When multiplying whole numbers, we observed that multiplication was a way of performing repeated addition. Whenever we multiply two *positive* integers, we are repeatedly adding positive numbers, so the answer must be *positive*.

$$4 \cdot 3 = 3 + 3 + 3 + 3 = 12$$
$$5 \cdot 1 = 1 + 1 + 1 + 1 + 1 = 5$$

> To find the product of *two positive* integers, multiply the units (absolute values) and give the answer a *positive* sign.

Look at the product of a *positive* integer and a *negative* integer, again using repeated addition.

$$5(-3) = -3 + (-3) + (-3) + (-3) + (-3) = -15$$
$$4(-1) = -1 + (-1) + (-1) + (-1) = -4$$

In each case, we were repeatedly adding negative numbers, so the answer was *negative*.

> To find the product of a *positive* integer and a *negative* integer, multiply the units (absolute values) and give the answer a *negative* sign.

Example 1. Find $3(-6)$.

Solution

$$3(-6) = -18$$

You do Example 2.

Example 2. Find $-4 \cdot 11$.

Solution

$$-4 \cdot 11 = \underline{\qquad}$$

Check your work on page 102. ▶

What happens when we multiply two *negative* integers? Look at the pattern that occurs when we consider some multiples of -3.

$$4(-3) = -12$$
$$3(-3) = -9$$
$$2(-3) = -6$$
$$1(-3) = -3$$
$$0(-3) = 0$$

As the multipliers of -3 on the left are *decreasing by 1*, notice that the products on the right are *increasing by 3*. To continue this pattern, we must let the multipliers continue to *decrease by 1*. At the same time, the products must continue to *increase by 3*.

$$-1(-3) = 3$$
$$-2(-3) = 6$$
$$-3(-3) = 9$$
$$-4(-3) = 12$$

Observe that the product of *two negative* numbers turns out to be a *positive* number.

To find the product of *two negative* integers, multiply the units (absolute values) and give the answer a *positive* sign.

Example 3. Find $(-1)(-1)$.

Solution

$$(-1)(-1) = 1$$

You try Example 4.

Example 4. Find $(-8)(-6)$.

Solution

$$(-8)(-6) = \underline{\hspace{1cm}}$$

Check your work on page 102. ▶

Symbols of grouping still give us directions regarding the order of operations.

Example 5. Find $[-3(-2)](-7)$.

Solution

$$[-3(-2)](-7)$$
$$= [6](-7)$$
$$= -42$$

You complete Example 6.

Example 6. Find $-5(3) + (-2)(-1)$.

Solution

$$-5(3) + (-2)(-1)$$
$$= -15 + \underline{\hspace{1cm}}$$
$$= \underline{\hspace{1cm}}$$

Check your work on page 102. ▶

Example 7. Find $-3[-5 - (10 - 27)]$.

Solution

$-3[-5 - (10 - 27)]$	
$= -3\{-5 - [10 + (-27)]\}$	Rewrite inner subtraction as addition.
$= -3\{-5 - [-17]\}$	Add within brackets.
$= -3\{-5 + 17\}$	Rewrite subtraction as addition.
$= -3 \cdot 12$	Add within braces.
$= -36$	Multiply.

||||➡ **Trial Run**

Perform the indicated operations.

_____ **1.** $-7(-8)$

_____ **2.** $-6(12)$

_____ **3.** $4(-3) + 5$

_____ **4.** $-5(8 - 17)$

_____ **5.** $-2[-7 - (13 - 16)]$ _____ **6.** $[-7(-3)][-1(2)]$

Answers are on page 103.

Using Some Properties of Multiplication

There are several properties of multiplication that are similar to the properties of addition. From the products

$$-6(5) = -30$$
$$5(-6) = -30$$

you might agree that

> It does not matter in which order we multiply two integers. The product is the same.

This law is the **commutative property for multiplication**.

> **Commutative Property for Multiplication.** For integers A and B
> $$A \cdot B = B \cdot A$$

Now consider these products involving three factors.

$$[-2 \cdot 3](-4) = -6(-4) = 24$$
$$-2[3(-4)] = -2(-12) = 24$$

In these products, the order in which the factors appeared was the same, but the position of the symbols of grouping was different. In both cases, the product was the same.

> When multiplying three integers, it does not matter if we multiply the product of the first two times the third, or multiply the first times the product of the second two. The product will be the same.

As you may have guessed, this law is called the **associative property for multiplication.**

> **Associative Property for Multiplication.** For any integers A, B, and C
> $$(A \cdot B) \cdot C = A \cdot (B \cdot C)$$

Complete Examples 8 and 9.

Example 8. Use the commutative property to rewrite the product $-2(-11)$.

Solution

$$-2(-11)$$
$$= \underline{\quad} (\underline{\quad})$$

Check your answer on page 102. ▶

Example 9. Use the associative property to rewrite the product $5[-1(-8)]$.

Solution

$$5[-1(-8)]$$
$$= [5(\underline{\quad})](\underline{\quad})$$

Check your answer on page 102. ▶

To multiply several integers when there are no brackets, the commutative and associative properties of multiplication allow us to arrange them in any order we choose.

Example 10. Find $3(-2)(-1)(-5)$.

Solution

$$3(-2)(-1)(-5)$$
$$= -6(5) \qquad \text{Multiply } 3(-2) \text{ and multiply } (-1)(-5).$$
$$= -30 \qquad \text{Multiply.}$$

There are two ways to work a problem such as $8(-2 + 5)$.

FIRST METHOD	SECOND METHOD
$8(-2 + 5)$	$8(-2 + 5)$
$= 8 \cdot 3$	$= 8(-2) + 8 \cdot 5$
$= 24$	$= -16 + 40$
	$= 24$

Notice that the results are the same. It seems that the first method is easier, but later we shall have to use the second method, so we note that

To multiply a number times the *sum* of numbers in parentheses, we may multiply the number outside the parentheses times each of the numbers inside the parentheses and add those products together.

This new law is called the **distributive property for multiplication over addition**.

Distributive Property for Multiplication over Addition and Subtraction.
For integers A, B, and C

$$A(B + C) = A \cdot B + A \cdot C$$
$$A(B - C) = A \cdot B - A \cdot C$$

Example 11. Use the distributive property to find the product $-5(7 - 1)$.

Solution

$$-5(7 - 1)$$

$$= -5(7) - (-5)(1)$$

$$= -35 - (-5)$$

$$= -35 + 5$$

$$= -30$$

Now try Example 12.

Example 12. Use the distributive property to find the product $-10(-17 + 19)$.

Solution

$$-10(-17 + 19)$$

Check your work on page 102. ▶

What happens when we multiply any integer by 1? Consider these products.

$$-3 \cdot 1 = -3$$

$$1 \cdot 18 = 18$$

As with whole numbers, when 1 is multiplied times any integer, the product is that integer.

Multiplication Property of 1. For any integer A

$$A \cdot 1 = A$$

Recall that 1 is called the **identity for multiplication**.

⫸ Trial Run

Complete each statement.

1. According to the commutative property for multiplication, $10(-7) = $ _____ .

2. According to the associative property for multiplication, $[6(-5)] \cdot 9 = $ _____ .

3. According to the distributive property for multiplication over addition,
$-3(2 + 8) = $ _____ .

Find each product.

_____ **4.** $-10(-3)(-7)$ _____ **5.** $6(5)(-1)(-1)$

Answers are on page 103.

Finding Powers of Integers

In Chapter 1 we discovered that **exponents** could be used to write products in which a factor is repeated.

Definition of a^n

$$a^n = \underbrace{a \cdot a \cdot a \cdot \ldots \cdot a}_{n \text{ factors}}$$

This definition also "works" when the base (a) is an *integer* and the exponent (n) is a natural number.

Example 13. Find $(-3)^3$.

Solution

$$(-3)^3 = (-3)(-3)(-3)$$
$$= 9(-3)$$
$$= -27$$

You try Example 14.

Example 14. Find $(-2)^4$.

Solution

$$(-2)^4 = (-2)(-2)(-2)(-2)$$
$$= 4(\underline{})$$
$$= \underline{}$$

Check your work on page 102. ▶

We note here that the sign of a power, a^n, depends on the sign of the base, a, and whether the exponent, n, is *odd* or *even*.

---

1. If a is *positive*, then a^n is always *positive*.
2. If a is *negative* and n is *even*, then a^n is *positive*.
3. If a is *negative* and n is *odd*, then a^n is *negative*.

For example,

$$(-1)^6 = 1 \quad \text{because} \quad 6 \text{ is even}$$

but

$$(-1)^{13} = -1 \quad \text{because} \quad 13 \text{ is odd}$$

Remember that in our rules for the order of operations, we deal with *exponents* first.

Example 15. Find -2^4.

Solution. Here we must find the *opposite* of 2^4.

$$-2^4 = -1 \cdot 2^4$$
$$= -1 \cdot 16$$
$$= -16$$

Contrast this example with Example 14 where parentheses indicated that the base was -2.

You try Example 16.

Example 16. Find $(-2)^3(-5)^2$.

Solution

$$(-2)^3(-5)^2$$
$$= (-2)(-2)(-2)(-5)(-5)$$
$$= (-8)(\underline{})$$
$$= \underline{}$$

Check your work on page 102. ▶

⫸ Trial Run

Simplify.

_____ **1.** $(-7)^2$ _____ **2.** $(-5)^3$ _____ **3.** -7^2

_____ **4.** $-(-4)^3$ _____ **5.** $(-2)^2(-6)^2$ _____ **6.** $-2(-1)^6$

Answers are on page 103.

Dividing Integers

Remember that every division statement corresponds to a multiplication statement. To work a division problem, we may look at the corresponding multiplication problem.

Example 17. Find $\dfrac{123}{3}$ and write the corresponding multiplication statement.

Solution

$$\frac{123}{3} = 41$$

because $3 \cdot 41 = 123$

You try Example 18.

Example 18. Find $\dfrac{99}{9}$ and write the corresponding multiplication statement.

Solution

$$\frac{99}{9} = \underline{\quad\quad}$$

because $9 \cdot \underline{\quad\quad} = 99$

Check your work on page 102. ▶

In each of these examples we found that the quotient of *two positive* integers was a *positive integer*.

> To divide a *positive* integer by a *positive* integer, divide the units (absolute values) and give the answer a *positive* sign.

How do we find the quotient of two *negative* integers? We can use multiplication of integers to help us see that

$$\frac{-36}{-9} = 4 \qquad \text{because} \qquad -9(4) = -36$$

and that

$$\frac{-18}{-6} = 3 \qquad \text{because} \qquad -6(3) = -18$$

In each case, we found that the quotient of *two negative* integers was a *positive* integer.

> To divide a *negative* integer by a *negative* integer, divide the units (absolute values) and give the answer a *positive* sign.

Example 19. Find $\dfrac{-75}{-3}$.

Solution

$$\frac{-75}{-3} = 25$$

because $-3(25) = -75$

Complete Example 20.

Example 20. Find $\dfrac{-6}{-6}$.

Solution

$$\frac{-6}{-6} = \underline{\quad\quad}$$

because $-6(\underline{\quad\quad}) = -6$

Check your work on page 102. ▶

What happens if the dividend and divisor have *different* signs? Again we must remember our multiplication facts for integers. You should agree that

$$\frac{-35}{7} = -5 \quad \text{because} \quad 7(-5) = -35$$

and

$$\frac{39}{-3} = -13 \quad \text{because} \quad -3(-13) = 39$$

In both cases, we found that the quotient of a *positive* integer and a *negative* integer was a *negative* integer.

To divide a *positive* integer by a *negative* integer (or a *negative* integer by a *positive* integer), divide the units (absolute values) and give the answer a *negative* sign.

Example 21. Find $\dfrac{-16}{2}$.

Solution

$$\frac{-16}{2} = -8$$

Complete Example 22.

Example 22. Find $\dfrac{17}{-17}$.

Solution

$$\frac{17}{-17} = \underline{\quad\quad}$$

Check your work on page 103. ▶

Some people prefer to learn the rules for multiplying and dividing integers in a slightly different form.

Multiplying or Dividing Two Integers

1. If the signs are the *same*, the answer will be *positive*.
2. If the signs are *different*, the answer will be *negative*.

The rules for working with *zero* in a division problem continue to apply here.

Example 23. Find $\dfrac{0}{-13}$.

Solution

$$\frac{0}{-13} = 0$$

because $-13(0) = 0$

Example 24. Find $\dfrac{-37}{0}$.

Solution

$$\frac{-37}{0} \text{ is undefined.}$$

By using symbols of grouping we may combine *all* the integer operations we have learned.

Example 25. Find $\dfrac{3(-1) - \dfrac{24}{-8}}{17 + (-2)(5)}$.

Solution

$$\frac{3(-1) - \dfrac{24}{-8}}{17 + (-2)(5)} = \frac{-3 - (-3)}{17 + (-10)}$$

$$= \frac{-3 + 3}{7}$$

$$= \frac{0}{7}$$

$$= 0$$

Now you complete Example 26.

Example 26. Find $\dfrac{5[18 + 2(-3)]}{-6}$.

Solution

$$\frac{5[18 + 2(-3)]}{-6} = \frac{5[18 + (-6)]}{-6}$$

$$= \frac{5[\underline{}]}{-6}$$

$$= \frac{\overline{}}{-6}$$

$$= \underline{}$$

Check your work on page 103. ▶

Trial Run

Perform the indicated operations.

_____ 1. $\dfrac{-16}{2}$

_____ 2. $\dfrac{-24}{-8}$

_____ 3. $\dfrac{-9}{0}$

_____ 4. $\dfrac{-4 + 10}{-3}$

_____ 5. $\dfrac{3(-6) + 8}{5}$

_____ 6. $\dfrac{-3[2(-4) - (-2)]}{-11 + 2}$

Answers are on page 103.

We mention here that

There is *no* commutative property for division.
There is *no* associative property for division.

The following quotients should convince you.

$$\frac{6}{3} \neq \frac{3}{6} \qquad -16 \div (4 \div 2) \neq (-16 \div 4) \div 2$$

$$-16 \div 2 \neq -4 \div 2$$

$$2 \neq \frac{1}{2} \qquad\qquad -8 \neq -2$$

▶ Examples You Completed _____

Example 2. Find $-4 \cdot 11$.

Solution

$$-4 \cdot 11 = -44$$

Example 4. Find $(-8)(-6)$.

Solution

$$(-8)(-6) = 48$$

Example 6. Find $-5(3) + (-2)(-1)$.

Solution

$$-5(3) + (-2)(-1)$$
$$= -15 + 2$$
$$= -13$$

Example 8. Use the commutative property to rewrite the product $-2(-11)$.

Solution

$$-2(-11)$$
$$= -11(-2)$$

Example 9. Use the associative property to rewrite the product $5[-1(-8)]$.

Solution

$$5[-1(-8)]$$
$$= [5(-1)](-8)$$

Example 12. Use the distributive property to find the product $-10(-17 + 19)$.

Solution

$$-10(-17 + 19)$$
$$= -10(-17) + (-10)(19)$$
$$= 170 + (-190)$$
$$= -20$$

Example 14. Find $(-2)^4$.

Solution

$$(-2)^4 = (-2)(-2)(-2)(-2)$$
$$= 4(4)$$
$$= 16$$

Example 16. Find $(-2)^3(-5)^2$.

Solution

$$(-2)^3(-5)^2$$
$$= (-2)(-2)(-2)(-5)(-5)$$
$$= (-8)(25)$$
$$= -200$$

Example 18. Find $\frac{99}{9}$ and write the corresponding multiplication statement.

Solution

$$\frac{99}{9} = 11$$

because $9 \cdot 11 = 99$

Example 20. Find $\frac{-6}{-6}$.

Solution

$$\frac{-6}{-6} = 1$$

because $-6(1) = -6$

Example 22. Find $\dfrac{17}{-17}$.

Solution

$$\frac{17}{-17} = -1$$

Example 26. Find $\dfrac{5[18 + 2(-3)]}{-6}$.

Solution

$$\frac{5[18 + 2(-3)]}{-6} = \frac{5[18 + (-6)]}{-6}$$

$$= \frac{5(12)}{-6}$$

$$= \frac{60}{-6}$$

$$= -10$$

Answers to Trial Runs

page 94   **1.** 56   **2.** -72 **3.** -7 **4.** 45   **5.** 8   **6.** -42

page 97   **1.** $-7(10)$ **2.** $6[-5(9)]$ **3.** $-3(2) + (-3)(8)$ **4.** -210 **5.** 30

page 99   **1.** 49   **2.** -125 **3.** -49 **4.** 64   **5.** 144   **6.** -2

page 101   **1.** -8 **2.** 3   **3.** Undefined   **4.** -2 **5.** -2 **6.** -2

EXERCISE SET 2.3

1. Use the commutative property for multiplication to rewrite each product.

   _____ (a) $-9(-8)$ _____ (b) $5(-12)$

   _____ (c) $0 \cdot 6$ _____ (d) $-3(7)$

2. Use the commutative property for multiplication to rewrite each product.

   _____ (a) $-5(0)$ _____ (b) $-7(10)$

   _____ (c) $12(-8)$ _____ (d) $-3(-4)$

3. Use the associative property for multiplication to rewrite each product.

   _____ (a) $(3 \cdot 4)2$ _____ (b) $6[5(-7)]$

   _____ (c) $-7(-3 \cdot 10)$ _____ (d) $[9(-2)](-8)$

4. Use the associative property for multiplication to rewrite each product.

   _____ (a) $(5 \cdot 7)9$ _____ (b) $12[7(-9)]$

   _____ (c) $-8(-2 \cdot 5)$ _____ (d) $[3(-11)](-5)$

5. Use the distributive property for multiplication over addition or subtraction to rewrite each product.

   _____ (a) $-3(7 + 2)$ _____ (b) $5(-8 - 3)$

   _____ (c) $4[7 - (-2)]$ _____ (d) $-10(-9 + 8)$

6. Use the distributive property for multiplication over addition or subtraction to rewrite each product.

   _____ (a) $-9(8 + 10)$ _____ (b) $2(-11 - 4)$

   _____ (c) $5[6 - (-3)]$ _____ (d) $-4(-1 + 7)$

Perform the indicated operations.

_____ 7. $5(-6)$ _____ 8. $3(-4)$

_____ 9. $-6(-7)$ _____ 10. $-4(-8)$

_____ 11. $-3(8)$ _____ 12. $-7(9)$

_____ 13. $-10(3)(-5)$ _____ 14. $-5(7)(-3)$

_____ 15. $4(-3)(0)(9)$ _____ 16. $6(-5)(0)(11)$

_____ 17. $-9(3 - 8)$ _____ 18. $-10(5 - 12)$

———— **19.** $6(-10) - 20$ ⠀⠀⠀⠀ ———— **20.** $8(-9) - 15$

———— **21.** $23 - 7(-2)$ ⠀⠀⠀⠀ ———— **22.** $29 - 3(-5)$

———— **23.** $(-8)^2$ ⠀⠀⠀⠀ ———— **24.** $(-2)^4$

———— **25.** -5^2 ⠀⠀⠀⠀ ———— **26.** -3^4

———— **27.** $(-3)^3$ ⠀⠀⠀⠀ ———— **28.** $(-4)^3$

———— **29.** -1^4 ⠀⠀⠀⠀ ———— **30.** -6^2

———— **31.** $(-2)^3(-6)^2$ ⠀⠀⠀⠀ ———— **32.** $(-5)^3(-1)^2$

———— **33.** $\dfrac{24}{-6}$ ⠀⠀⠀⠀ ———— **34.** $\dfrac{20}{-4}$

———— **35.** $\dfrac{-16}{8}$ ⠀⠀⠀⠀ ———— **36.** $\dfrac{-15}{5}$

———— **37.** $\dfrac{-35}{-7}$ ⠀⠀⠀⠀ ———— **38.** $\dfrac{-48}{-6}$

———— **39.** $\dfrac{11(-4)}{22}$ ⠀⠀⠀⠀ ———— **40.** $\dfrac{12(-3)}{4}$

———— **41.** $\dfrac{15}{-3} + \dfrac{-12}{6}$ ⠀⠀⠀⠀ ———— **42.** $\dfrac{21}{-7} + \dfrac{-28}{4}$

———— **43.** $\dfrac{-20}{5} - \dfrac{-30}{-6}$ ⠀⠀⠀⠀ ———— **44.** $\dfrac{-45}{9} - \dfrac{-55}{-11}$

———— **45.** $2\left(\dfrac{-18}{3}\right)$ ⠀⠀⠀⠀ ———— **46.** $5\left(\dfrac{-54}{9}\right)$

———— **47.** $\dfrac{-20 + 15}{7 - 7}$ ⠀⠀⠀⠀ ———— **48.** $\dfrac{-35 + 20}{5 - 5}$

———— **49.** $\dfrac{-40 + 15}{16 - (-9)}$ ⠀⠀⠀⠀ ———— **50.** $\dfrac{-17 + 10}{4 - (-3)}$

———— **51.** $\dfrac{3(-8) + (-2)(-4)}{4(-4)}$ ⠀⠀⠀⠀ ———— **52.** $\dfrac{5(-9) + (-3)(-2)}{3(-13)}$

———— **53.** $\dfrac{9(-1) + \dfrac{14}{-2}}{4}$ ⠀⠀⠀⠀ ———— **54.** $\dfrac{3(-9) + \dfrac{5}{-1}}{8}$

———— **55.** $\dfrac{7[18 + 5(-2)]}{-8}$ ⠀⠀⠀⠀ ———— **56.** $\dfrac{9[21 + 7(-2)]}{-21}$

_____ 57. $\dfrac{-4[3(-2) - (-7)]}{13 + 11}$

_____ 58. $\dfrac{-5[6(-3) - (-20)]}{17 + 12}$

_____ 59. $\dfrac{20(-4)}{10} + \dfrac{21 - 6}{-3}$

_____ 60. $\dfrac{36(-2)}{12} + \dfrac{27 - 9}{-6}$

_____ 61. $\dfrac{-9(7 - 1) - 8(-3)}{-5(-8 + 6)}$

_____ 62. $\dfrac{-8(10 - 3) - 9(-4)}{2(-9 + 4)}$

_____ 63. $-3\left[\dfrac{-14}{7} + 3(7 - 12)\right]$

_____ 64. $-5\left[\dfrac{-15}{3} + 2(9 - 12)\right]$

_____ 65. $10\left[3(6) + \dfrac{13 + (-5)}{4}\right]$

_____ 66. $5\left[-5(7) + \dfrac{14 + (-5)}{3}\right]$

_____ 67. $\dfrac{\dfrac{-15}{5} + \dfrac{0}{7}}{\dfrac{25}{-5} + 2}$

_____ 68. $\dfrac{\dfrac{-18}{6} + \dfrac{0}{9}}{\dfrac{-30}{5} + 3}$

_____ 69. $4\left[-9 + \dfrac{7 + (-3)}{6 - 8}\right] - 3^3$

_____ 70. $5\left[-10 + \dfrac{8 + (-3)}{15 - 20}\right] - 6^2$

☆ Stretching the Topics

Perform the indicated operations.

_____ 1. $-7\left[-2(5) - \dfrac{3^2 - (-2)^2}{6 - 11}\right]$

_____ 2. $(-2)^2\{4(-1)^3 - 3[8 - (-12)]\}$

_____ 3. $3^3[7 + (-2)^3] - [(-1)^5(-3) - 12] - (8 - 12)^2$

Check your answers in the back of your book.

If you can simplify the expressions in **Checkup 2.3**, you are ready to go on to Section 2.4.

CHECKUP 2.3

_____ **1.** Use the commutative property for multiplication to rewrite the product $(-3)(5)$.

_____ **2.** Use the associative property for multiplication to rewrite the product $-3[2(-7)]$.

_____ **3.** Use the distributive property for multiplication over addition to rewrite the product $-5(-2 + 8)$.

Perform the indicated operations.

_____ **4.** $6(-5)$ _____ **5.** $\dfrac{-35}{7}$ _____ **6.** $\dfrac{-50}{(-5)^2}$

_____ **7.** $[-4(-3)](-5)$ _____ **8.** $-2[-3 - (7 - 12)]$

_____ **9.** $\dfrac{5[13 + 3(-1)]}{-25}$ _____ **10.** $\dfrac{-3\left[2(-1) - \dfrac{24}{-6}\right]}{-10 + 19}$

Check your answers in the back of your book.

If You Missed Problems:	You Should Review Examples:
1	8
2	9
3	11, 12
4	1–4
5, 6	13, 19–22
7, 8	5–7
9, 10	25, 26

2.4 Understanding Real Numbers

So far we have discussed several important sets of numbers.

Natural numbers: $\{1, 2, 3, \ldots\}$

Whole numbers: $\{0, 1, 2, 3, \ldots\}$

Integers: $\{\ldots, -3, -2, -1, 0, 1, 2, 3, \ldots\}$

We have learned how to add, subtract, multiply, and divide such numbers. But in Chapter 1 we also discussed operations with fractional and decimal numbers that do *not* belong to these sets. In fact, we must define two more sets of numbers in order to include all fractions and decimal numbers in our study of algebra.

Rational Numbers. A rational number is any number that can be written as a *fraction* with an *integer* as the numerator and a *nonzero integer* as the denominator. Every repeating or terminating decimal number is a rational number.

Rational Number	$\frac{17}{37}$	$-\frac{5}{3}$	$3\frac{1}{2}$	0.9	$\sqrt{25}$	-0.73	4.31	2	-3	0
Fraction Form	$\frac{17}{37}$	$\frac{-5}{3}$	$\frac{7}{2}$	$\frac{9}{10}$	$\frac{5}{1}$	$\frac{-73}{100}$	$\frac{431}{100}$	$\frac{2}{1}$	$\frac{-3}{1}$	$\frac{0}{1}$

You should agree that every rational number corresponds to a point on the number line.

Notice also that *every integer is a rational number*.

Look at the following three rational numbers.

$$\frac{-6}{2} \qquad \frac{6}{-2} \qquad -\frac{6}{2}$$
$$= -3 \qquad = -3 \qquad = -3$$

Each of these fractions has the same value. The first form is useful when we wish to operate with fractions, but answers will usually be stated in the last form.

$$\frac{-a}{b} = \frac{a}{-b} = -\frac{a}{b}$$

all represent the same fraction

It may seem to you that all numbers must be rational. Indeed, most numbers that you work with every day *are* rational. However, the numbers π, $\sqrt{2}$, and $\sqrt{3}$, which you may have seen before, are examples of numbers that are *not* rational because they cannot be written as fractions of integers. Such numbers are called **irrational numbers**; they will be discussed more completely in Chapter 12.

Irrational Numbers. An irrational number is a number that corresponds to a point on the number line but is *not* a rational number. Every decimal number that does not repeat or terminate is an irrational number.

The set that includes *all* the numbers in the sets we have discussed is the set of **real numbers**.

Real Numbers. A real number is any number that corresponds to a point on the number line. Every real number is either rational or irrational.

We can illustrate the relationships among our sets of numbers in the following chart.

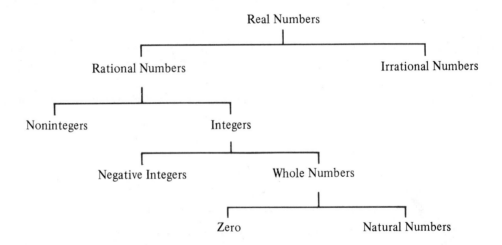

You will be relieved to know that *all the properties and definitions that you learned for operating with integers will continue to work for the real numbers*. The following examples review those properties and definitions.

Example 1. What is the opposite of $\frac{5}{6}$?

Solution. $-\frac{5}{6}$ is the opposite of $\frac{5}{6}$.

You try Example 2.

Example 2. What is the opposite of -2.3?

Solution. _____ is the opposite of -2.3.

Check your answer on page 115. ▶

Example 3. Find $|-7.35|$.

Solution

$$|-7.35| = 7.35$$

Now you try Example 4.

Example 4. Find $-\left|-\dfrac{7}{3}\right|$.

Solution

$$-\left|-\frac{7}{3}\right| = -\left(\frac{7}{3}\right)$$

$$= \underline{\qquad}$$

Check your work on page 115. ▶

Example 5. Find $-\dfrac{6}{11} + \dfrac{2}{11} + \left(-\dfrac{5}{11}\right)$.

Solution

$$-\frac{6}{11} + \frac{2}{11} + \left(-\frac{5}{11}\right)$$

$$= \frac{-6}{11} + \frac{-5}{11} + \frac{2}{11}$$

$$= \frac{-6 + (-5) + 2}{11}$$

$$= \frac{-11 + 2}{11}$$

$$= -\frac{9}{11}$$

Now you try Example 6.

Example 6. Find

$$-2.3 + (-5.1) + 12.8$$

Solution

$$-2.3 + (-5.1) + 12.8$$

$$= -7.4 + 12.8$$

$$= \underline{\qquad}$$

Check your work on page 115. ▶

Example 7. Find $\dfrac{5}{3} - \dfrac{11}{2}$.

Solution

$$\frac{5}{3} - \frac{11}{2}$$

$$= \frac{5}{3} + \frac{-11}{2} \qquad \text{LCD: } 3 \cdot 2$$

$$= \frac{5 \cdot 2}{3 \cdot 2} + \frac{-11 \cdot 3}{3 \cdot 2}$$

$$= \frac{10}{6} + \frac{-33}{6}$$

$$= \frac{10 + (-33)}{6}$$

$$= -\frac{23}{6}$$

You complete Example 8.

Example 8. Find $-\dfrac{1}{2} - \dfrac{1}{5}$.

Solution

$$-\frac{1}{2} - \frac{1}{5}$$

$$= \frac{-1}{2} + \frac{-1}{5} \qquad \text{LCD: } 2 \cdot 5$$

$$= \frac{-1 \cdot 5}{2 \cdot 5} + \frac{-1 \cdot 2}{2 \cdot 5}$$

$$=$$

Check your work on page 115. ▶

Example 9. Find $2.3 - (3.51 - 7.6)$.

Solution

$$2.3 - (3.51 - 7.6)$$
$$= 2.3 - [3.51 + (-7.6)]$$
$$= 2.3 - (-4.09)$$
$$= 2.3 + 4.09$$
$$= 6.39$$

You complete Example 10.

Example 11. Find $-\dfrac{2}{3} \cdot \dfrac{5}{6}$.

Solution

$$-\frac{2}{3} \cdot \frac{5}{6} = \frac{\overset{-1}{\cancel{-2}}}{3} \cdot \frac{5}{\underset{1}{\cancel{2}} \cdot 3}$$

$$= \frac{-1 \cdot 5}{3 \cdot 3}$$

$$= -\frac{5}{9}$$

Now you try Example 12.

Example 13. Find $-3(2.7 - 8.2)$.

Solution

$$-3(2.7 - 8.2)$$
$$= -3[2.7 + (-8.2)]$$
$$= -3(-5.5)$$
$$- 16.5$$

You complete Example 14.

Example 15. Find $\left(-\dfrac{2}{3}\right)^3$.

Solution

$$\left(-\frac{2}{3}\right)^3 = \left(-\frac{2}{3}\right)\left(-\frac{2}{3}\right)\left(-\frac{2}{3}\right)$$

$$= \frac{(-2)(-2)(-2)}{3 \cdot 3 \cdot 3}$$

$$= -\frac{8}{27}$$

You try Example 16.

Example 10. Find

$$(0.37 - 3.2) - (6.1 - 0.09)$$

Solution

$$(0.37 - 3.2) - (6.1 - 0.09)$$
$$= [0.37 + (-3.2)] - [6.1 + (-0.09)]$$

Check your work on page 115. ▶

Example 12. Find $-\dfrac{4}{9} \cdot \dfrac{3}{7}\left(-\dfrac{21}{2}\right)$.

Solution

$$-\frac{4}{9} \cdot \frac{3}{7}\left(-\frac{21}{2}\right) =$$

Check your work on page 116. ▶

Example 14. Find $0.5[-3.9 - (6 + 9.1)]$.

Solution

$$0.5[-3.9 - (6 + 9.1)]$$
$$= 0.5[-3.9 - (15.1)]$$
$$= 0.5[-3.9 + (-15.1)]$$
$$= 0.5[\underline{\quad\quad}]$$
$$= \underline{\quad\quad}$$

Check your work on page 116. ▶

Example 16. Find $(-1.1)^2$.

Solution

$$(-1.1)^2 = (-1.1)(-1.1)$$
$$= \underline{\quad\quad}$$

Check your work on page 116. ▶

Example 17. Find $-4.44 \div 3.7$.

Solution. The answer will be *negative*.

$$
\begin{array}{r}
1.2 \\
3.7_\wedge \overline{)4.4_\wedge 4} \\
\underline{3\ 7} \\
7\ 4 \\
\underline{7\ 4} \\
0
\end{array}
$$

$$-4.44 \div 3.7 = -1.2$$

Example 18. Find $-\dfrac{5}{7} \cdot \dfrac{3}{10} \div \left(-\dfrac{9}{20}\right)$.

Solution

$$-\frac{5}{7} \cdot \frac{3}{10} \div \left(-\frac{9}{20}\right)$$

$$= -\frac{5}{7} \cdot \frac{3}{10} \cdot \left(-\frac{20}{9}\right)$$

$$= \frac{\overset{-1}{\cancel{-5}}}{7} \cdot \frac{\overset{1}{\cancel{3}}}{\underset{1}{\cancel{2} \cdot \cancel{5}}} \cdot \frac{-2 \cdot \overset{1}{\cancel{2}} \cdot 5}{\underset{1}{\cancel{3} \cdot 3}}$$

$$= \frac{-1(-2)(5)}{7 \cdot 3}$$

$$= \frac{10}{21}$$

⫸ Trial Run

Perform the indicated operations.

———— **1.** $\dfrac{5}{8} - \dfrac{1}{2} - \dfrac{3}{4}$

———— **2.** $8.26 - (-3.5)$

———— **3.** $\dfrac{6}{7} - \left(\dfrac{1}{7} - \dfrac{3}{7}\right)$

———— **4.** $-1.7(-0.2)$

———— **5.** $-0.1[6.8 - (2 + 1.9)]$

———— **6.** $-\dfrac{8}{9} \cdot \dfrac{3}{4}$

———— **7.** $\dfrac{-2.5}{-0.5}$

———— **8.** $\dfrac{3}{4}\left(-\dfrac{5}{3}\right) \div 2$

Answers are on page 116.

In working examples throughout this chapter in which you added and multiplied real numbers, you have relied on two properties of which you were probably unaware.

> When two real numbers are added, their sum is a unique real number.
>
> When two real numbers are multiplied, their product is a unique real number.

The fact that sums and products of real numbers are *unique* real numbers means that there is *one and only one* answer when two certain numbers are added or multiplied. Think how confusing mathematics would be if this were not so!

These laws are called the **closure properties** for real numbers.

Closure Properties. For real numbers A and B

$A + B$ is real and unique

$A \cdot B$ is real and unique

We noted earlier that every real number has an *opposite* and that the sum of every number and its opposite is zero, the *additive identity*. A similar property exists for multiplication. Consider these products.

$$2 \cdot \frac{1}{2} = \frac{\overset{1}{\cancel{2}}}{1} \cdot \frac{1}{\cancel{2}} = 1 \qquad \frac{5}{3} \cdot \frac{3}{5} = \frac{\cancel{5}}{\cancel{3}} \cdot \frac{\cancel{3}}{\cancel{5}} = 1 \qquad -\frac{1}{4}\left(-\frac{4}{1}\right) = \frac{-1}{\cancel{4}} \cdot \frac{\overset{-1}{\cancel{-4}}}{1} = 1$$

In each product the second number is called the **reciprocal** of the first number, and their product is 1, the *identity for multiplication*. Every real number (except zero) has its own reciprocal (also called its **multiplicative inverse**).

Property of Reciprocals. For any real number A ($A \neq 0$)

$$A \cdot \frac{1}{A} = 1$$

Example 19. Find the reciprocal of $-\frac{8}{5}$.

Solution. The reciprocal of $-\frac{8}{5}$ is $-\frac{5}{8}$ because

$$-\frac{8}{5}\left(-\frac{5}{8}\right) = 1$$

You complete Example 20.

Example 20. Find the reciprocal of $\frac{1}{\pi}$.

Solution. The reciprocal of $\frac{1}{\pi}$ is _____ because

$$\frac{1}{\pi} \cdot \underline{\qquad} = 1$$

Check your work on page 116. ▶

The fact that every real number (except 0) has a reciprocal allows us to define the operation of division in terms of the operation of multiplication.

Definition of Division. For real numbers A and B ($B \neq 0$)

$$\frac{A}{B} = A \cdot \frac{1}{B}$$

Example 21. Rewrite the quotient $\dfrac{-63}{7}$ as a product.

Solution

$$\frac{-63}{7} = -63\left(\frac{1}{7}\right)$$

Example 22. Rewrite the product $\dfrac{1}{4}\pi$ as a quotient.

Solution

$$\frac{1}{4}\pi = \frac{\pi}{4}$$

▶ **Examples You Completed** ————————————————————

Example 2. What is the opposite of -2.3?

Solution. 2.3 is the opposite of -2.3.

Example 4. Find $-\left|-\dfrac{7}{3}\right|$.

Solution

$$-\left|-\frac{7}{3}\right| = -\left(\frac{7}{3}\right)$$

$$= -\frac{7}{3}$$

Example 6. Find $-2.3 + (-5.1) + 12.8$.

Solution

$$-2.3 + (-5.1) + 12.8$$
$$= -7.4 + 12.8$$
$$= 5.4$$

Example 10. Find

$$(0.37 - 3.2) - (6.1 - 0.09)$$

Solution

$$(0.37 - 3.2) - (6.1 - 0.09)$$
$$= [0.37 + (-3.2)] - [6.1 + (-0.09)]$$
$$= -2.83 - 6.01$$
$$= -2.83 + (-6.01)$$
$$= -8.84$$

Example 8. Find $-\dfrac{1}{2} - \dfrac{1}{5}$.

Solution

$$-\frac{1}{2} - \frac{1}{5}$$

$$= \frac{-1}{2} + \frac{-1}{5} \qquad \text{LCD: } 2 \cdot 5$$

$$= \frac{-1 \cdot 5}{2 \cdot 5} + \frac{-1 \cdot 2}{2 \cdot 5}$$

$$= \frac{-5}{10} + \frac{-2}{10}$$

$$= \frac{-5 + (-2)}{10}$$

$$= \frac{-7}{10}$$

Example 12. Find $-\dfrac{4}{9} \cdot \dfrac{3}{7}\left(-\dfrac{21}{2}\right)$.

Solution

$$-\frac{4}{9} \cdot \frac{3}{7}\left(-\frac{21}{2}\right)$$

$$= \frac{\overset{-1}{\cancel{-2}} \cdot 2}{\underset{1}{\cancel{3}} \cdot \underset{1}{\cancel{3}}} \cdot \frac{\overset{1}{\cancel{3}}}{\underset{1}{7}} \cdot \frac{\overset{-1}{\cancel{-3}} \cdot \overset{1}{\cancel{7}}}{\underset{1}{\cancel{2}}}$$

$$= 2$$

Example 14. Find $0.5[-3.9 - (6 + 9.1)]$.

Solution

$$0.5[-3.9 - (6 + 9.1)]$$

$$= 0.5[-3.9 - (15.1)]$$

$$= 0.5[-3.9 + (-15.1)]$$

$$= 0.5[-19]$$

$$= -9.5$$

Example 16. Find $(-1.1)^2$.

Solution

$$(-1.1)^2 = (-1.1)(-1.1)$$

$$= 1.21$$

Example 20. Find the reciprocal of $\dfrac{1}{\pi}$.

Solution. The reciprocal of $\dfrac{1}{\pi}$ is π because

$$\frac{1}{\pi} \cdot \pi = 1$$

Answers to Trial Run

page 113   **1.** $-\dfrac{5}{8}$ **2.** 11.76 **3.** $\dfrac{8}{7}$ **4.** 0.34 **5.** -0.29 **6.** $-\dfrac{2}{3}$ **7.** 5 **8.** $-\dfrac{5}{8}$

EXERCISE SET 2.4

Perform the indicated operations.

_____ **1.** $|-8.3|$

_____ **2.** $|-12.7|$

_____ **3.** $|3.2 - 8.9|$

_____ **4.** $|4.5 - 7.3|$

_____ **5.** $-\left|-\dfrac{2}{3}\right|$

_____ **6.** $-\left|-\dfrac{7}{8}\right|$

_____ **7.** $\left|\dfrac{1}{8}\right| - \left|-\dfrac{7}{8}\right|$

_____ **8.** $\left|\dfrac{3}{11}\right| - \left|-\dfrac{7}{11}\right|$

_____ **9.** $\dfrac{19}{21} - \left(\dfrac{18}{21} - \dfrac{7}{21}\right)$

_____ **10.** $\dfrac{16}{27} - \left(\dfrac{13}{27} - \dfrac{17}{27}\right)$

_____ **11.** $-\dfrac{5}{6} + \dfrac{2}{5} - \dfrac{1}{2}$

_____ **12.** $-\dfrac{10}{21} + \dfrac{11}{18} - \dfrac{9}{14}$

_____ **13.** $-13.5 + 7.5$

_____ **14.** $-18.6 + 21.6$

_____ **15.** $3.72 - (0.49 - 1.08)$

_____ **16.** $12.35 - (4.06 - 0.04)$

_____ **17.** $-\dfrac{4}{5} \cdot \dfrac{15}{28}$

_____ **18.** $-\dfrac{7}{8} \cdot \dfrac{12}{35}$

_____ **19.** $(-8.62)(1.03)$

_____ **20.** $(-2.74)(2.03)$

_____ **21.** $(-4.3)(-7.001)(1000)$

_____ **22.** $(-3.2)(-4.005)(1000)$

_____ **23.** $\left(-\dfrac{3}{4}\right)^3$

_____ **24.** $\left(-\dfrac{2}{5}\right)^3$

_____ **25.** $(-1.2)^2$

_____ **26.** $(-2.3)^2$

_____ **27.** $\left(\dfrac{1}{5}\right)^4$

_____ **28.** $\left(\dfrac{1}{3}\right)^5$

_____ **29.** $-0.035 \div 0.05$

_____ **30.** $-0.72 \div 0.06$

_____ **31.** $-1.606 \div (-7.3)$

_____ **32.** $-2.697 \div (-4.65)$

_____ **33.** $-\dfrac{12}{75}\left(-\dfrac{75}{84}\right)$

_____ **34.** $-\dfrac{27}{50}\left(-\dfrac{70}{81}\right)$

_____ **35.** $\dfrac{3}{5} \div \dfrac{6}{25}$

_____ **36.** $\dfrac{5}{7} \div \dfrac{20}{63}$

_____ **37.** $-0.3[-6.2 - (8 - 9.3)]$

_____ **38.** $-0.4[-5.3 - (7 - 8.6)]$

_____ **39.** $-\dfrac{3}{8} \cdot \dfrac{4}{9} \div \left(-\dfrac{5}{12}\right)$

_____ **40.** $-\dfrac{3}{7} \cdot \dfrac{14}{15} \div \left(-\dfrac{4}{25}\right)$

☆ Stretching the Topics

Perform the indicated operations.

_____ 1. $\left[\left(\dfrac{1}{2} - \dfrac{2}{3}\right) \div \left(\dfrac{5}{6} - \dfrac{5}{8}\right)\right]^2$

_____ 2. $(-0.1)^3[3.76(0.2)^2 + (-0.3)^3(-5)]$

_____ 3. $\left(\dfrac{3}{4}\right)^2 - \left(\dfrac{7}{12} \div \dfrac{2}{3}\right) + \left(\dfrac{2}{9} \cdot \dfrac{3}{4}\right)$

Check your answers in the back of your book.

If you can do the problems in **Checkup 2.4**, you are ready to go to Section 2.5.

✓ **CHECKUP 2.4**

Perform the indicated operations.

_____ 1. $\left|-\dfrac{4}{5}\right|$

_____ 2. $-\dfrac{2}{3} + \dfrac{1}{2} + \left(-\dfrac{1}{6}\right)$

_____ 3. $-0.3 - 0.25 + 2.26$

_____ 4. $\dfrac{3}{5} - \dfrac{7}{6}$

_____ 5. $-\dfrac{3}{5}\left(-\dfrac{7}{36}\right) \div \dfrac{14}{3}$

_____ 6. $-5(0.23 - 1.09)$

_____ 7. $\left(-\dfrac{3}{5}\right)^3$

_____ 8. $2.5(-0.2)^2$

_____ 9. $-0.1764 \div 0.063$

_____ 10. $0.2[-1.6 - (8.2 - 1.9)]$

Check your answers in the back of your book.

If You Missed Problems:	You Should Review Examples:
1	3, 4
2, 3	5, 6
4	7, 8
5	11, 12
6	13
7, 8	15, 16
9	17
10	14

2.5 Switching from Word Expressions to Real Number Expressions

Now that we understand how to operate with all the real numbers, we must practice switching from words to expressions involving positive and negative numbers.

Example 1. Yesterday, Amy's checking account showed a balance of $67.39. Today she wrote checks for $31.93 and $17.29, deposited $100, and wrote another check for $125. What is the condition of her account now?

Solution

We Think

Amy's balance and her deposit are *positive* numbers, but her checks are *negative* numbers. We must find the *sum* of these quantities.

We Write

$67.39 + (-31.93) + (-17.29) + (-125) + 100$ Write the sum.

$= 67.39 + 100 + (-31.93) + (-17.29) + (-125)$ Rearrange terms.

$= 167.39 + (-174.22)$ Add positive numbers; add negative numbers.

$= -6.83$ Find the sum.

Amy has *overdrawn* her account by $6.83.

You try Example 2, which is the problem stated at the beginning of this chapter.

Example 2. In one series of plays, the State University football team completed two 7-yard passes and ran for 5 yards. Then the quarterback was sacked twice, for a 3-yard loss each time, and the team received a 15-yard penalty. What was the team's total gain or loss?

Solution

We Think

Gains are represented by _____ numbers.

Losses are represented by _____ numbers.

To find the team's total gain or loss, we can _____ these numbers.

In this series of plays the team had a _____ of _____ yards.

Check your work on page 123. ▶

We Write

Gains from passes: 2(_____)
Gain from run: _____

Losses from sacks: 2(_____)
Loss from penalty: _____

$2(7) + \_\_\_ + 2(-3) + (\_\_\_)$
$=$
$=$

Example 3. On Monday morning, a share of Intral stock was selling for $27. The following changes occurred in the price of the stock during the week. On Monday the price rose by $\frac{1}{8}$; on Tuesday it rose by $\frac{1}{4}$; on Wednesday it fell by $\frac{5}{8}$; on Thursday it rose by $\frac{1}{8}$; and on Friday it fell by $\frac{1}{8}$. What was the price of a share of Intral stock on Friday afternoon?

Solution

We Think

A rise in price is represented by a *positive* number, but a drop in price is represented by a *negative* number. We must *add* the changes to the original price to arrive at the Friday price.

We Write

$$27 + \frac{1}{8} + \frac{1}{4} + \left(-\frac{5}{8}\right) + \frac{1}{8} + \left(-\frac{1}{8}\right) \qquad \text{LCD: } 8$$

$$= \frac{216}{8} + \frac{1}{8} + \frac{2}{8} + \frac{1}{8} + \frac{-5}{8} + \frac{-1}{8}$$

$$= \frac{216 + 1 + 2 + 1 + (-5) + (-1)}{8}$$

$$= \frac{220 + (-6)}{8}$$

$$= \frac{214}{8}$$

$$= 26\frac{6}{8}$$

$$= 26\frac{3}{4}$$

The price of a share of Intral stock was $26.75 on Friday.

Now you complete Example 4.

Example 4. Last week the daily temperatures at noon in Anytown were $-4°$, $-3°$, $0°$, $8°$, $11°$, $-2°$, and $-3°$. What was the *average* noon temperature last week?

Solution

We Think

First we must _____ the temperatures. To find the *average*, we must _____ that total by _____ .

We Write

$$\frac{-4° + (-3°) + 0° + 8° + 11° + (-2°) + (-3°)}{7}$$

$$= \frac{\rule{1cm}{0.4pt}}{7}$$

$$= \rule{1cm}{0.4pt}$$

The average noon temperature was _____ .

Check your work on page 123. ▶

▶ Examples You Completed _____

Example 2. In one series of plays, the State University football team completed two 7-yard passes and ran for 5 yards. Then the quarterback was sacked twice, for a 3-yard loss each time, and the team received a 15-yard penalty. What was the team's gain or loss?

Solution

We Think	*We Write*
Gains are represented by positive numbers.	Gains from passes: 2(7) Gain from run: 5
Losses are represented by negative numbers.	Losses from sacks: 2(-3) Loss from penalty: -15
To find the team's total gain or loss, we can add these numbers.	$2(7) + 5 + 2(-3) + (-15)$ $= 14 + 5 + (-6) + (-15)$ $= -2$

In this series of plays the team had a loss of 2 yards.

Example 4. Last week the daily temperatures at noon in Anytown were $-4°$, $-3°$, $0°$, $8°$, $11°$, $-2°$, and $-3°$. What was the *average* noon temperature last week?

Solution

We Think

First we must add the temperatures. To find the *average*, we must divide that total by 7.

We Write

$$\frac{-4° + (-3°) + 0° + 8° + 11° + (-2°) + (-3°)}{7}$$

$$= \frac{7°}{7}$$

$$= 1°$$

The average noon temperature was 1°.

EXERCISE SET 2.5

Change each word expression to a number expression and simplify.

_____ **1.** Darlene scored 87 on her first math test. If she scored 14 points lower on her next test, what score did she receive on her second test?

_____ **2.** At a summer sale, Bob bought a $39 stepladder, which had been reduced by $14. What was the sale price of the stepladder?

_____ **3.** The Momans used 312 fewer kilowatt-hours of electricity in March then they did in February. If they used 1605 kilowatt-hours in February, how many kilowatt-hours did they use in March?

_____ **4.** At the Key Market, Royce's total grocery bill was $63.78, but after the total value of his coupons was deducted, he paid only $55.28. Find how much he saved by using the coupons.

_____ **5.** A dinner for two at the Chicken King costs $5.86. If Jim and Maria split the expenses evenly, how much will each pay?

_____ **6.** If Carol's car averages 19.75 miles per gallon, how many gallons will she need to drive 316 miles?

_____ **7.** If Sarah drives at an average speed of 53 miles per hour, how far can she drive in 4.5 hours?

_____ **8.** Carol bought 15.6 gallons of gasoline at $1.69 per gallon. Find the cost of the gasoline.

_____ **9.** Ms. Fletcher uses $\frac{2}{3}$ cup of fertilizer for each shrub in her yard. If she has 25 shrubs, how many cups of fertilizer does she need?

_____ **10.** Pansy uses $2\frac{1}{2}$ cups of flour for each jam cake she bakes. If she is baking seven jam cakes for the Junior League Bake Sale, how many cups of flour does she need?

_____ **11.** Lester played the 25-cent slot machine seven times, winning the 75-cent jackpot three times. How much did Lester win or lose?

_____ **12.** If Harvey works for 8 hours at $9.85 an hour and has $12.73 deducted for taxes, what will be his take-home pay?

_____ **13.** If taxi fare is $1.75 for the first mile and $1.10 for each additional mile, what will be the fare for an 8-mile trip?

_____ **14.** A club sold 150 dance tickets for $4.25 each and had expenses of $275.85. What was the club's profit?

_____ **15.** Miranda borrowed $5 from her brother twice last month. This month she paid him $7 and borrowed $3. Where does Miranda's account with her brother now stand?

_____ **16.** The estate of wealthy Uncle Donald is divided equally among his 5 nieces. If the estate includes a house worth $150,000 and 3 automobiles worth $11,000 each and debts of $9000 and $29,000, how much will each niece inherit?

_____ 17. Earl is fencing his rectangular yard, which measures $150\frac{3}{4}$ feet by $250\frac{2}{3}$ feet, but there will be two gates where no fencing is needed. One gate is $3\frac{1}{3}$ feet wide and the other is $6\frac{1}{2}$ feet wide. How many feet of fencing will Earl need?

_____ 18. Vanessa uses $\frac{1}{5}$ of her salary for rent, $\frac{1}{8}$ for food, and $\frac{3}{10}$ for utilities. What fractional part of her salary is left for other expenses?

_____ 19. Kim works in the morning for 2 hours at $3.85 per hour and in the afternoon for 4 hours at $4.90 per hour. How much does she earn in 1 day?

_____ 20. Mrs. Curtis works 8 hours a day at $8.97 per hour and hires a baby-sitter for 9 hours a day at $2.75 per hour. After she pays the baby-sitter, how much does Mrs. Curtis earn each day?

_____ 21. For lunch Camelia had a Big Burger for $1.69, a salad for $1.25, and a
_____   carton of milk for $0.45. If the sales tax rate is 5.5 percent, find the tax on
_____   her lunch. Find the total cost of her lunch. How much change should she receive from a 10-dollar bill?

_____ 22. Henry and his roommate split rent and food expenses equally. This month they spent $225 on rent and $115.82 on food. If Henry started the month with $315.75, how much does he have left after paying his share of the expenses?

_____ 23. On an achievement test, a student receives 2 points for every correct answer, loses 2 points for every incorrect answer, and loses 1.5 points for every unanswered question. If Mark answered 27 questions correctly, answered 13 questions incorrectly, and left 10 questions unanswered, find his test score.

_____ 24. At the Plaza Theater an adult's ticket costs $4.50. A child's ticket costs half as much as an adult's ticket. A senior citizen's ticket costs $0.75 less than an adult's ticket. If 60 adults' tickets, 25 children's tickets, and 15 senior citizens' tickets were sold, find the amount received at the ticket window.

☆ Stretching the Topics _____

_____ 1. Profits or losses are split equally among the 5 owners of the Pizza Shack. During the past weeks, the shack took in $1873.95. Supplies cost $928.17, 8 workers were paid $150 each, and advertising cost $53.75. What was each owner's share of the profits or losses?

_____ 2. June Ricketts bought a lot and built a house on it. The total cost was $95,000. She sold the property a year later for $116,000. During the year she spent $787 for taxes and $585 for insurance. She paid 6 percent of the selling price to a real estate agent for selling the property. How much profit did she make on the transaction?

Check your answers in the back of your book.

If you can do **Checkup 2.5,** you are ready to do the **Review Exercises** for Chapter 2.

✓ **CHECKUP 2.5**

Change each word expression to a number expression and simplify.

_____ **1.** Atha has a balance of $785.28 in her checking account. If she writes checks at the grocery store for $119.61 and at the cleaners for $28.75, what is her balance?

_____ **2.** If a balloon rises from the ground at a rate of 3 feet per second for 20 seconds and then falls at a rate of 2 feet per second for 10 seconds, what is its final height?

_____ **3.** During one year a farm with three equal-sharing owners sold products totaling $625,823 and had expenses of $125,729. What is each owner's share of the profit? (Round to the nearest cent.)

_____ **4.** In a certain town the Ohio River reached flood stage at 48 feet. During the next 5 hours, readings indicated that the river rose $\frac{2}{9}$ foot, rose $\frac{1}{8}$ foot, remained steady, fell $\frac{1}{6}$ foot, and fell $\frac{2}{3}$ foot. What was the river stage after the last reading?

_____ **5.** At the local Bingo game, Betty lost $5 per week for 3 weeks, won $7 per week for 2 weeks, and lost $2.50 per week for 4 weeks. Find her average gain or loss per week.

Check your answers in the back of your book.

If You Missed Problems:	You Should Review Examples:
1	1
2	2
3	4
4	3
5	2, 4

Summary

In this chapter we learned that all the numbers that we have discussed belong to the set of **real numbers**. Every number that corresponds to a point on the number line is a real number.

Number Set	Description	Examples
Natural numbers	$\{1, 2, 3, \ldots\}$	7 is a natural number
Whole numbers	$\{0, 1, 2, 3, \ldots\}$	0 is a whole number 195 is a whole number
Integers	$\{\ldots, -3, -2, -1, 0, 1, 2, 3, \ldots\}$	11 is an integer -29 is an integer
Rational numbers	Numbers that can be written as fractions with integer numerators and nonzero integer denominators	$\frac{5}{9}$ is a rational number 2.7 is a rational number -16 is a rational number $\sqrt{9}$ is a rational number
Irrational numbers	Numbers on the number line that are *not* rational	$\sqrt{5}$ is an irrational number π is an irrational number
Real numbers	Numbers on the number line	6.3 is a real number -82 is a real number $\sqrt{2}$ is a real number

We developed several definitions and properties that apply to all real numbers. Suppose that we let A, B, and C represent any real numbers.

Symbol	Words	Meaning	Examples
$A < B$	A is less than B	A lies to the left of B on the number line	$0 < 3$ $-7 < -4$
$A > B$	A is greater than B	A lies to the right of B on the number line	$9 > 1$ $-3.7 > -4$
A^n	A to the nth power	$A^n = A \cdot A \cdot \ldots \cdot A$ n factors	$3^4 = 3 \cdot 3 \cdot 3 \cdot 3$ $\left(\frac{1}{2}\right)^3 = \frac{1}{2} \cdot \frac{1}{2} \cdot \frac{1}{2}$ $(-1.5)^2 = (-1.5)(-1.5)$
$\lvert A \rvert$	Absolute value of A	The distance between 0 and A on the number line	$\lvert -3 \rvert = 3$ $\left\lvert \frac{17}{5} \right\rvert = \frac{17}{5}$ $\lvert -6.84 \rvert = 6.84$

Property Name	Statement	Examples
Closure property for addition	$A + B$ is real and unique	$3 + 2.8$ is real $$-\frac{1}{6} + \frac{2}{5} \text{ is real}$$
Closure property for multiplication	$A \cdot B$ is real and unique	$-4.1(10)$ is real $$\frac{1}{5} \cdot \frac{2}{9} \text{ is real}$$
Commutative property for addition	$A + B = B + A$	$3 + (-6) = -6 + 3$ $$\frac{1}{8} + \frac{5}{4} = \frac{5}{4} + \frac{1}{8}$$
Commutative property for multiplication	$A \cdot B = B \cdot A$	$0.6(2.9) = 2.9(0.6)$ $$-\frac{5}{3} \cdot 6 = 6\left(-\frac{5}{3}\right)$$
Associative property for addition	$(A + B) + C = A + (B + C)$	$(6 + 3.1) + 0.9 = 6 + (3.1 + 0.9)$ $$\left(\frac{5}{7} + \frac{1}{3}\right) + \frac{1}{2} = \frac{5}{7} + \left(\frac{1}{3} + \frac{1}{2}\right)$$
Associative property for multiplication	$(A \cdot B)C = A(B \cdot C)$	$$(3 \cdot 2)\frac{1}{2} = 3\left(2 \cdot \frac{1}{2}\right)$$ $(-0.9 \cdot 5)2 = -0.9(5 \cdot 2)$
Distributive properties	$A(B + C) = A \cdot B + A \cdot C$ $A(B - C) = A \cdot B - A \cdot C$	$3(2 + 5) = 3 \cdot 2 + 3 \cdot 5$ $$\frac{1}{2}(8 - 0.4) = \frac{1}{2} \cdot 8 - \frac{1}{2}(0.4)$$
Opposites property	$A + (-A) = 0$	$-3.7 + 3.7 = 0$ $$\frac{5}{6} + \left(-\frac{5}{6}\right) = 0$$
Reciprocals property	$A \cdot \dfrac{1}{A} = 1 \quad (A \neq 0)$	$$\frac{2}{3} \cdot \frac{3}{2} = 1$$ $$-5\left(-\frac{1}{5}\right) = 1$$

We also continued to use the important properties of **zero** and **1**.

Property Name	Statement	Example
Addition property of zero	$A + 0 = A$	$2.3 + 0 = 2.3$
Subtraction property of zero	$A - 0 = A$	$\frac{17}{39} - 0 = \frac{17}{39}$
Multiplication property of 1	$A \cdot 1 = A$	$-629 \cdot 1 = -629$
Multiplication property of zero	$A \cdot 0 = 0$	$-\frac{2}{3} \cdot 0 = 0$
Division property of 1	$\frac{A}{1} = A$	$\frac{56}{1} = 56$
Division of zero	$\frac{0}{A} = 0 \quad (A \neq 0)$	$\frac{0}{6.91} = 0$
Division by zero	$\frac{A}{0}$ is undefined	$\frac{-0.23}{0}$ is undefined

☐ Speaking the Language of Algebra

Complete each statement with the appropriate word or phrase.

1. The set containing all the whole numbers and their opposites is called the set of _____ .

2. Every rational number can be expressed as a fraction with a numerator that is an _____ and a denominator that is a _____ _____ .

3. A real number that is *not* rational is called an _____ number.

4. The symbol $|a|$ represents the _____ _____ _____ _____ . On the number line, $|a|$ is found as the _____ between _____ and a.

5. The absolute value of a real number is never _____ .

6. On the number line, points to the right of zero correspond to _____ real numbers, and points to the left of zero correspond to _____ real numbers.

7. The sum of a real number and its opposite is always _____ .

8. The operations of _____ and _____ are commutative and associative, but the operations of _____ and _____ are not.

9. To find the difference $A - B$, we must find the sum of A and the _____ of B.

10. The sum of two positive numbers is a _____ number. The sum of two negative numbers is a _____ number.

11. The product of two positive numbers is a _____ number. The product of two negative numbers is a _____ number. The product of a positive number and a negative number is a _____ number.

12. The expression A^n tells us to use the base, _____ , as a factor in a product _____ times.

Check your answers in the back of your book.

△ Writing About Mathematics

Write your response to each question in complete sentences.

1. Explain how you would use the number line to find the sum $(-5 + 3) + (-6)$.

2. If a is a negative integer, explain why
   (a) a^n is positive when n is even;
   (b) a^n is negative when n is odd.

3. Write a word problem that would lead you to write the number expression $2(\$150) + 3(-\$75) + \$100$.

REVIEW EXERCISES for Chapter 2

1. Graph the point corresponding to each number on the number line.

    (a) -3 **(b)** 0 **(c)** $\dfrac{5}{2}$ **(d)** -0.6

_____ **2.** Use the commutative property for addition to rewrite the sum $-5 + (-6)$.

_____ **3.** Use the associative property for multiplication to rewrite the sum
         $[-3 + (-2)] + 5$.

_____ **4.** Use the commutative property for multiplication to rewrite the product
         $-4 \cdot 5$.

_____ **5.** Use the associative property for multiplication to rewrite the product
         $-7(-8 \cdot 9)$.

_____ **6.** Use the distributive property for multiplication over addition to rewrite
         $-2(-6 + 9)$.

7. Compare each of the following pairs of numbers using $<$ or $>$.

   _____ **(a)** $4, 14$ _____ **(b)** $-5, -11$

   _____ **(c)** $9, -12$ _____ **(d)** $-9, 3$

Perform the indicated operations.

_____ **8.** $|-13|$ _____ **9.** $-6 + (-12)$

_____ **10.** $8 + (-5)$ _____ **11.** $-6 + 3 + 8 + (-10)$

_____ **12.** $-3 + \left(-5 + \dfrac{24}{3}\right)$ _____ **13.** $32 - (-6)$

_____ **14.** $-9 - 7$ _____ **15.** $[7 + (-2)] - (-5 + 1)$

_____ **16.** $(-8)(-3)(5)$ _____ **17.** $-4(5) + (-3)(-7)$

_____ **18.** $7[4(8 - 5) - (-3)]$ _____ **19.** $(-3)^2(-1)^3$

_____ **20.** $\dfrac{-56}{8}$ _____ **21.** $\dfrac{18}{2} - \dfrac{-15}{3}$

_____ **22.** $\dfrac{-9 - (-3)}{8 - 10}$ _____ **23.** $\dfrac{3(-4) + (-6)(-2)}{-7 - 2}$

_____ **24.** $\dfrac{-3(8 - 2) - 2(-1)}{7 + (1 - 10)}$ _____ **25.** $\dfrac{\dfrac{-24}{3} + \dfrac{0}{2}}{\dfrac{10}{5} - \dfrac{8}{4}}$

_____ **26.** $-3.6 + (-4.02) + 9$ _____ **27.** $\dfrac{3}{4} - \dfrac{7}{6} - \dfrac{2}{3}$

_____ 28. $0.7[-4.6 - (4 - 8.3)]$ _____ 29. $14.77 \div (-0.35)$

_____ 30. $\left(-\dfrac{2}{3}\right)^3 \div \dfrac{16}{45}$

_____ 31. At the track Eddie bet $2 on each of 10 races. Twice he won $4 and once he won $15. How much did Eddie win or lose at the track?

_____ 32. Erica earns $253.93 per week, but deductions of $25.19, $15.32, and $6 are made before her check is written. She also earned a $25 bonus for an efficiency suggestion she made last week. What is the amount of her check?

_____ 33. One day recently the stock market reported that at the beginning of the day Xerox stock shares sold for $102\frac{1}{8}$. During the day the price rose by $\frac{1}{2}$, rose by $\frac{1}{4}$, then fell by $\frac{5}{8}$. What was the price of a share at the close of the day?

_____ 34. A crew washed $\frac{1}{9}$ of the windows in an office building during the first week, $\frac{3}{8}$ of the windows the second week, and $\frac{1}{6}$ of the windows the third week. What fractional part of the windows remain to be washed?

_____ 35. During the first 6 months of this year the rainfall was 1.2 inches above normal for 2 months, 2 inches above normal 1 month, and 0.8 inch below normal for 3 months. Find the average number of inches the precipitation was above or below normal for the first 6 months.

Check your answers in the back of your book.

If You Missed Exercises:	You Should Review Examples:	
1	SECTION 2.1	3
2, 3		24, 25
4, 5	SECTION 2.3	8, 9
6		11, 12
7	SECTION 2.1	1, 2
8		8
9, 10		14, 19
11		26
12		20, 22
13, 14	SECTION 2.2	3, 4
15		5
16, 17	SECTION 2.3	5, 6
18		7
19		16
20, 21		21
22–25		25, 26
26, 27	SECTION 2.4	6–8
28–30		14–18
31	SECTION 2.5	2
32		1
33, 34		3
35		4

If you have completed the **Review Exercises** and corrected your errors, you are ready to take the **Practice Test** for Chapter 2.

PRACTICE TEST for Chapter 2

		SECTION	EXAMPLES

1. Graph the point corresponding to each number on the number line. **2.1** **3**

   (a) $-\dfrac{2}{3}$ (b) 0.7 (c) -1.6 (d) $\dfrac{9}{4}$

_____ 2. Use the commutative property for multiplication to rewrite $-3(5)$. 2.3 4

_____ 3. Use the associative property for addition to rewrite $-4 + [9 + (-2)]$. 2.1 25

_____ 4. Use the distributive property for multiplication over addition to rewrite $-4[11 + (-2)]$. 2.3 6

5. Compare each pair of numbers using $<$ or $>$. 2.1 1, 2

   _____ (a) $-15, -1$ _____ (b) $0, -3$

   _____ (c) $6, 5.9$ _____ (d) $\dfrac{1}{3}, \dfrac{1}{2}$

Perform the indicated operations.

_____ 6. $-|-8|$ 2.1 11

_____ 7. $3 + (-11) + (-12) + 7$ 2.1 26

_____ 8. $-15 - (-22)$ 2.2 3

_____ 9. $[-10 + 6] - [5 + (-9)]$ 2.2 5

_____ 10. $(-7)(2)(-3)$ 2.3 5

_____ 11. $5[-2(7 - 3) - (-2)]$ 2.3 7

_____ 12. $(-2)^4(-1)^5$ 2.3 16

_____ 13. $\dfrac{-24}{6} - \dfrac{10}{-5}$ 2.3 21, 22

_____ 14. $\dfrac{-11 - 12}{-5 + 8}$ 2.3 23, 24

_____ 15. $\dfrac{-6(-3) + (18)(-1)}{-9 + 4}$ 2.3 23, 24

_____ 16. $\dfrac{-2(11 - 9) - 5(-2)}{6 + (5 - 11)}$ 2.3 23, 24

_____ 17. $5.7 + (-8.2) - 2.95$ 2.4 6

		SECTION	EXAMPLES
_____	**18.** $\dfrac{3}{8} + \left(-\dfrac{2}{3}\right) - \dfrac{7}{6}$	2.4	7, 8
_____	**19.** $0.3[-2.3 - (7 - 6.1)]$	2.4	14
_____	**20.** $-32.2 \div 0.23$	2.4	17
_____	**21.** $\left(-\dfrac{3}{4}\right)^2 \div \dfrac{40}{21}$	2.4	15, 18

_____ **22.** After attending 8 weekly meetings of Weight Worriers, Clyde's record showed that he had lost 2.5 pounds twice, gained 1.5 pounds twice, lost 4 pounds three times, and stayed the same once. What was Clyde's average weekly loss?   2.5   4

_____ **23.** Martha opened a quart of milk in the morning. She used $\frac{1}{2}$ cup on her cereal, toook $\frac{3}{4}$ cup to work in her thermos, made pudding with $1\frac{3}{4}$ cups, and drank $\frac{2}{3}$ cup with supper. How much milk was left after supper? (*Remember*: 1 quart = 4 cups.)   2.5   3

_____ **24.** For its 10-year reunion, the Class of 1980 rented a banquet hall for $150. Decorations cost $75.85. The caterer charged $5.75 per person. If each of the 87 persons attending paid $11, how much money will the class have left in its treasury to use for the next reunion?   2.5   1

Check your answers in the back of your book.

3

Working with Variable Expressions

The Internal Revenue Service allows a taxpayer to deduct $2000 for every dependent claimed. For an income of $12,000, write an expression describing the remaining taxable income after a taxpayer has made the deduction for *d* dependents.

Algebra can be thought of as the bridge between arithmetic and higher mathematics. A person with even a limited knowledge of algebra can solve many everyday problems that otherwise could be solved only by trial and error. In algebra we must work with **constants**, such as the numbers we have already studied, but we must also work with symbols that stand for numbers. Such symbols are called **variables** and are usually represented by letters.

In this chapter we learn how to

1. Switch from word expressions to algebraic expressions.
2. Evaluate algebraic expressions.
3. Add and subtract algebraic expressions.
4. Multiply constants times algebraic expressions.

3.1 Switching from Word Expressions to Algebraic Expressions

Combinations of constants and/or variables involving the operations of addition, subtraction, multiplication, and division are called **algebraic expressions**.

Words	Symbols
add 4 and the variable x	$4 + x$
subtract 9 from the variable y	$y - 9$
divide the variable a by 2	$\dfrac{a}{2}$
multiply 3 times the variable x	$3 \cdot x$ or $3x$
the sum of x and 5	$x + 5$
the product of 5 and q	$5 \cdot q$ or $5q$
a decreased by 1.5	$a - 1.5$
twice the variable x	$2 \cdot x$ or $2x$
7 less than n	$n - 7$
add 5 and y; then divide by 6	$\dfrac{5 + y}{6}$
7 times the sum of b and 3	$7(b + 3)$

Let's try switching from word expressions to algebraic expressions.

Example 1. The product of 7 and y.

Solution

$$7 \cdot y \quad \text{or} \quad 7y$$

Notice that the symbol for multiplication may be left out. We agree that a number written next to a variable will always mean *multiply*.

Example 2. 32 decreased by x.

Solution

$$32 - x \quad \text{or} \quad 32 + (-x)$$

Notice that when dealing with variables, we can use the same definition for subtraction that we use with integers.

You try Examples 3 and 4.

Example 3. From 17, subtract the product of 9 and y.

Solution

$$17 - \underline{\hspace{1cm}}$$

Check your work on page 141. ▶

Example 4. The sum of 13 and twice x.

Solution

$$13 + \underline{\hspace{1cm}}$$

Check your work on page 141. ▶

Example 5. Subtract x from 11, then divide by -4.

Solution

$$\dfrac{11 - x}{-4}$$

You try Example 6.

Example 6. Three times the sum of y and 5.

Solution

$$3(\underline{\hspace{1cm}} + \underline{\hspace{1cm}})$$

Check your work on page 141. ▶

⇒ Trial Run

Write an algebraic expression that means:

_____ **1.** Add 7 and x.   _____ **2.** Multiply -5 times y.

_____ **3.** Subtract 8 from a.   _____ **4.** Twice the sum of x and 13.

_____ **5.** Subtract x from 15; then multiply by -3.

_____ **6.** Add -12 to y; then divide by 10.

Answers are on page 142.

Evaluating Algebraic Expressions

In every algebraic expression containing variables, the numerical value of the expression depends on what number the variable stands for. For instance, to find the value of the expression $9 + x$, we must know what number x represents.

$$\text{if } x = 2 \quad \text{then} \quad 9 + x = 9 + 2 \quad = 11$$

$$\text{if } x = -4 \quad \text{then} \quad 9 + x = 9 + (-4) = 5$$

$$\text{if } x = -15 \quad \text{then} \quad 9 + x = 9 + (-15) = -6$$

This process is called **evaluating an expression**. To evaluate an algebraic expression for some value of the variable, we *substitute* (or ''plug in'') the value we are given for the variable and then perform the necessary arithmetic.

Example 7. Evaluate $x - 32y$ when $x = 4$ and $y = -1$.

   Solution. If $x = 4$ and $y = -1$, then

$$x - 32y = 4 - 32(-1)$$
$$= 4 - (-32)$$
$$= 4 + 32$$
$$= 36$$

Now you try Example 8.

Example 8. Evaluate $7y$ when $y = 3$; when $y = -1.1$; when $y = 0$.

   Solution

     If $y = 3$, then

$$7y = 7(\underline{\quad}) = \underline{\quad}$$

     If $y = -1.1$, then

$$7y = 7(\underline{\quad}) = \underline{\quad}$$

     If $y = 0$, then

$$7y = 7(\underline{\quad}) = \underline{\quad}$$

Check your work on page 142. ▶

Example 9. Evaluate $\dfrac{2b + a}{5}$ when $b = -3$ and $a = 6$.

Solution. If $b = -3$ and $a = 6$, then

$$\frac{2b + a}{5} = \frac{2(-3) + 6}{5}$$

$$= \frac{-6 + 6}{5}$$

$$= \frac{0}{5}$$

$$= 0$$

You try Example 10.

Example 10. Evaluate $4(11 - xy)$ when $x = 5$ and $y = -2$.

Solution. If $x = 5$ and $y = -2$, then

$$4(11 - xy) = 4[11 - 5(-2)]$$

$$= 4[11 + \underline{\quad}]$$

$$= 4[\underline{\quad}]$$

$$= \underline{\quad}$$

Check your work on page 142. ▶

⟱ **Trial Run**

Evaluate each expression.

_____ **1.** $6x$ when $x = -2$; when $x = 0$; when $x = 7$.

_____ **2.** $x - 23$ when $x = 50$; when $x = 10$; when $x = -5$.

_____ **3.** $\dfrac{60}{y}$ when $y = 6$; when $y = -12$; when $y = 0$.

_____ **4.** $\dfrac{a + b}{-6}$ when $a = 10$ and $b = 8$; when $a = -11$ and $b = 5$; when $a = -2$ and $b = 2$.

_____ **5.** $3(13 - m)$ when $m = 3$; when $m = -2$; when $m = 13$.

Answers are on page 142.

Using Algebraic Expressions

Variables and algebraic expressions give us a handy way of dealing with word expressions in which the value of one or more of the quantities is not known.

For example, suppose that movie tickets cost $3.50 each. We could use a variable expression to represent the cost of *any number* of tickets by letting x stand for the number of tickets sold. Then $3.50x$ would represent the cost of x tickets. For example,

$$\text{if } x = 2 \quad \text{then} \quad \$3.50x = \$3.50(2) \quad = \$7$$

$$\text{if } x = 10 \quad \text{then} \quad \$3.50x = \$3.50(10) \quad = \$35$$

$$\text{if } x = 173 \quad \text{then} \quad \$3.50x = \$3.50(173) = \$605.50$$

You try Example 11.

Example 11. If part-time students at a university are charged $25 per credit hour, write an algebraic expression describing the cost of h credit hours. Find the cost of 3 hours; of 8 hours.

Solution

We Think	**We Write**
The rate is $25 *per* credit hour, so the cost of h hours is found by *multiplying*.	Cost of h hours: 25 _____ dollars
To find the cost of 3 hours, we let $h = 3$.	$25h = 25(\underline{\quad})$
	$= \underline{\quad}$ dollars
To find the cost of 8 hours, we let $h = 8$.	$25h = 25(\underline{\quad})$
	$= \underline{\quad}$ dollars

Check your work on page 142. ▶

Example 12. If the population of a town is 25,230, write an algebraic expression for the number of people in the town after a change in population of x people has occurred. Then find the new population if 2900 people move in; if 1600 people move out.

Solution

We Think	**We Write**
The new population is found by *adding* the change to the current population.	New population: $25,230 + x$
An *increase* in population is represented by a *positive* number ($x = 2900$).	$25,230 + x = 25,230 + 2900$
	$= 28,130$
A *decrease* in population is represented by a *negative* number ($x = -1600$).	$25,230 + x = 25,230 + (-1600)$
	$= 23,630$

Now you try Example 13.

Example 13. A gasoline distributor supplies 10 local gas stations with equal allotments of gasoline each week. Write an algebraic expression to describe each station's share of the distributor's total number of gallons, G. Then find each station's share if the distributor has 60,000 gallons; 85,000 gallons; 0 gallons.

Solution

We Think	We Write
To find each station's equal share of the gasoline, we must _____ the total by _____ .	Each share: $\dfrac{G}{10}$
If the total is 60,000 gallons, we find each share by letting $G = 60,000$.	$\dfrac{G}{10} = \dfrac{(\quad)}{10}$ $= $ _____ gallons
If the total is 85,000 gallons, we find each share by letting $G = 85,000$.	$\dfrac{G}{10} = \dfrac{(\quad)}{10}$ $= $ _____ gallons
If the total is 0 gallons, we find each share by letting $G = 0$.	$\dfrac{G}{10} = \dfrac{(\quad)}{10}$ $= $ _____ gallons.

Check your work on page 142. ▶

Let's solve the problem stated at the beginning of this chapter.

Example 14. The Internal Revenue Service allows a taxpayer to deduct $2000 for each dependent claimed. For an income of $12,000, write an expression describing the remaining taxable income after a taxpayer has made the deduction for d dependents. Then find the remaining taxable income if a taxpayer claims one dependent; five dependents.

Solution

We Think	We Write
The tax deduction is $2000 *per* dependent, so we find the total deduction for d dependents by *multiplying*.	Total deduction: $2000d$
The taxable income is found by *subtracting* deductions from income.	Taxable income = income − deduction $= 12{,}000 - 2000d$
If the taxpayer claims one dependent, we let $d = 1$.	$12{,}000 - 1000d = 12{,}000 - 2000(1)$ $= 12{,}000 - 2000$ $= 10{,}000$ dollars
If the taxpayer claims five dependents, we let $d = 5$.	$12{,}000 - 1000d = 12{,}000 - 2000(5)$ $= 12{,}000 - 10{,}000$ $= 2000$ dollars

⇒ Trial Run

_____ **1.** If Arnold drives 55 miles per hour, write an algebraic expression for the distance that he can travel in h hours. Then find the distance he travels in 3 hours; in $4\frac{1}{2}$ hours.

_____ **2.** If there are 1252 students in a high school, write an algebraic expression for the number of students after a change in enrollment of x students has occurred. Then find the enrollment if 142 new students enroll; if 93 students drop out.

_____ **3.** The meat from a beef calf is to be evenly divided among five families. Write an algebraic expression for each family's share if the total weight of the calf is T pounds. Then find each family's share if the calf weighs 725 pounds; if the calf weighs 931 pounds.

_____ **4.** The cost of renting a garden tiller is $25, plus $8.50 for each day it is used. Write an algebraic expression for the cost of renting a tiller for d days. Then find the cost for 2 days; for 5 days.

Answers are on page 142.

▶ Examples You Completed

Example 3. From 17, subtract the product of 9 and y.

Solution

$$17 - 9y$$

Example 6. Three times the sum of y and 5.

Solution

$$3(y + 5)$$

Example 4. The sum of 13 and twice x.

Solution

$$13 + 2x$$

Example 8. Evaluate $7y$ when $y = 3$; when $y = -1.1$; when $y = 0$.

Solution

If $y = 3$, then $7y = 7(3) = 21$

If $y = -1.1$, then $7y = 7(-1.1) = -7.7$

If $y = 0$, then $7y = 7(0) = 0$

Example 10. Evaluate $4(11 - xy)$ when $x = 5$ and $y = -2$.

Solution. If $x = 5$ and $y = -2$, then

$$4(11 - xy) = 4[11 - 5(-2)]$$
$$= 4(11 + 10)$$
$$= 4(21)$$
$$= 84$$

Example 11. If part-time students at a university are charged $25 per credit hour, write an algebraic expression describing the cost of h credit hours. Find the cost of 3 hours; of 8 hours.

Solution

We Think	*We Write*
The rate is $25 *per* credit hour, so the cost of h hours is found by *multiplying*.	Cost of h hours: $25h$ dollars
To find the cost of 3 hours, we let $h = 3$.	$25h = 25(3)$
	$= 75$ dollars
To find the cost of 8 hours, we let $h = 8$.	$25h = 25(8)$
	$= 200$ dollars

Example 13. A gasoline distributor supplies 10 local gas stations with equal allotments of gasoline each week. Write an algebraic expression to describe each station's share of the distributor's total number of gallons, G. Then find each station's share if the distributor has 60,000 gallons; 85,000 gallons; 0 gallons.

Solution

We Think	*We Write*
To find each station's equal share of the gasoline, we must divide the total by 10.	Each share: $\dfrac{G}{10}$
If the total is 60,000 gallons, we find each share by letting $G = 60,000$.	$\dfrac{G}{10} = \dfrac{60,000}{10}$
	$= 6000$ gallons
If the total is 85,000 gallons, we find each share by letting $G = 85,000$.	$\dfrac{G}{10} = \dfrac{85,000}{10}$
	$= 8500$ gallons
If the total is 0 gallons, we find each share by letting $G = 0$.	$\dfrac{G}{10} = \dfrac{0}{10}$
	$= 0$ gallons

Answers to Trial Runs

page 137   **1.** $7 + x$ **2.** $-5y$ **3.** $a - 8$ **4.** $2(x + 13)$ **5.** $-3(15 - x)$ **6.** $\dfrac{-12 + y}{10}$

page 138   **1.** $-12; 0; 42$ **2.** $27; -13; -28$ **3.** $10; -5;$ undefined   **4.** $-3; 1; 0$ **5.** $30; 45; 0$

page 141   **1.** $55h$; 165 mi; 247.5 mi   **2.** $1252 + x$; 1394; 1159   **3.** $\dfrac{T}{5}$; 145 lb; 186.2 lb   **4.** $25 + 8.50d$; $42; $67.50

EXERCISE SET 3.1

Write an algebraic expression for each word statement or expression.

_____ 1. Add 10 to y.

_____ 2. Add -5 to x.

_____ 3. Multiply -3 times m.

_____ 4. Multiply 8 times n.

_____ 5. Subtract 6.5 from a.

_____ 6. Subtract 1.3 from b.

_____ 7. Divide t by -15.

_____ 8. Divide s by 12.

_____ 9. The sum of x and -9.

_____ 10. The sum of y and 7.

_____ 11. The product of q and 3.2.

_____ 12. The product of m and -1.3.

_____ 13. n decreased by 6.3.

_____ 14. d decreased by -9.5.

_____ 15. Twice x, divided by 7.

_____ 16. Five times y, divided by -11.

_____ 17. 9 more than the product of $\frac{1}{6}$ and k.

_____ 18. 10 more than the product of $\frac{2}{3}$ and m.

_____ 19. 5 less than the quotient of p divided by 7.

_____ 20. 13 less than the quotient of p divided by -3.

_____ 21. 4 times x is subtracted from 9.

_____ 22. 8 times y is subtracted from 23.

_____ 23. -4 times the sum of x and 1.5.

_____ 24. 9 times the sum of y and 2.5.

Evaluate each expression.

_____ 25. $y - 3$ when $y = 10$; when $y = -3$

_____ 26. $y - 8$ when $y = 15$; when $y = -8$

_____ 27. $\frac{m}{5}$ when $m = -15$; when $m = 0$

_____ 28. $\frac{m}{9}$ when $m = 36$; when $m = -18$

_____ 29. $-7a$ when $a = -6$; when $a = 7$

_____ 30. $-9a$ when $a = 10$; when $a = -3$

_____ 31. $4(x - y)$ when $x = -3$ and $y = 5$; when $x = 5$ and $y = -5$

_____ 32. $5(x - y)$ when $x = -1$ and $y = 12$; when $x = 3$ and $y = -3$

_____ 33. $2x + 4$ when $x = -4$; when $x = -2$

_____ **34.** $5x + 10$ when $x = -3$; when $x = 0$

_____ **35.** $\dfrac{a - b}{7}$ when $a = 13$ and $b = 6$; when $a = -14$ and $b = 7$

_____ **36.** $\dfrac{a - b}{5}$ when $a = 14$ and $b = 9$; when $a = -10$ and $b = 15$

Write an algebraic expression for each word expression and evaluate as indicated.

_____ **37.** If the Betas are selling lottery tickets on a car at $2.50 per ticket, write an algebraic expression for the cost of y tickets. Find the cost of buying 10 tickets; 25 tickets.

_____ **38.** The cost of a chartered bus to the ballgame is $570, to be divided evenly among the number of students who ride the bus. Write an algebraic expression for the cost to each passenger if x students ride the bus. Find the cost for each student if there are 60 passengers; 30 passengers.

_____ **39.** Write an algebraic expression for the number of liters remaining in a 75-liter tank if x liters have already been used. Find how many liters remain when 8.7 liters have been used; when 13.2 liters have been used.

_____ **40.** Irene has a part-time job paying $3.50 per hour. Write an algebraic expression for her earnings after she has worked x hours. Find her earnings when she works $14\frac{1}{4}$ hours; 22 hours.

_____ **41.** An oil tank that contains 200 barrels of oil begins to leak. Three barrels a day are leaking from the tank. Write an algebraic expression for the amount of oil in the tank after n days. How many barrels does the tank contain after $6\frac{1}{2}$ days; after 30 days?

_____ **42.** Jake earns $250 per week plus $7.25 per hour for each hour of overtime. Write an algebraic expression for Jake's pay for a week when he worked x hours of overtime. Find his pay when he worked 8 hours of overtime; 15 hours of overtime.

_____ **43.** Tony is saving money for a vacation during spring break. If he deposits $50 and then saves $12 a week for n weeks, write an algebraic expression for the amount he will have at the end of n weeks. Find his savings after 6 weeks; after 12 weeks.

_____ **44.** The winning candidate in an election received 50 more votes than twice the number of votes his opponent received. If his opponent received x votes, write an algebraic expression for the number of votes the winner received. Find the number of votes the winner received if the loser got 120 votes; if the loser got 500 votes.

_____ **45.** The balance owed on Maurice's charge card is $587.39. If he pays x dollars per month for 6 months, write an algebraic expression for the balance. Find his balance if he pays $25 a month; if he pays $50 a month.

_____ **46.** Maxine pays $300 a month rent. Mary pays x dollars per month more than Maxine pays. Write an algebraic expression for 3 months' rent for Mary. Find Mary's total rent for 3 months if she pays $25 per month more than Maxine; if she pays $37 more per month.

_____ **47.** The cost of the annual Christmas party will be shared by the 32 factory employees. Write an algebraic expression for the amount each employee will pay if the party costs x dollars. Find the amount each employee will pay when the party costs $96; when the party costs $176.

_____ **48.** A clothing store makes x dollars profit on each sweater sold. If the store wants to have a $600 profit from the sale of sweaters, write a variable expression for the number of sweaters the store must sell. Find the number of sweaters if the profit per sweater is $12; if the profit per sweater is $7.50.

_____ **49.** In a physical fitness class Carlos did three fewer than twice as many push-ups as his brother. If his brother did n push-ups, write an algebraic expression for the number of push-ups Carlos did. Find the number of push-ups Carlos did if his brother did 15 push-ups; if his brother did 25 push-ups.

_____ **50.** At Weight Worriers, Sue discovered she had lost 5 pounds more than half as many pounds as Sydney. If Sydney lost p pounds, write an algebraic expression for the number of pounds Sue lost. Find the number of pounds Sue had lost if Sydney had lost 4 pounds; if Sydney had lost 11 pounds.

☆ Stretching the Topics _____

_____ **1.** Write an algebraic expression for the quotient when 3 times x subtracted from 4 times y is divided by 15.

_____ **2.** Evaluate $\dfrac{7(2x - 5y)}{4z + y}$ when $x = -2$, $y = 1$, and $z = 2$.

_____ **3.** The total area of Kym's house is 2000 square feet. During remodeling, a sewing room with a total area of x square feet was removed. A family room was added which has a total area of 108 square feet more than $\frac{2}{3}$ the area of the sewing room. Write a variable expression for the area of Kym's house after remodeling. What will be the total area if the area of the sewing room was 72 square feet?

Check your answers in the back of your book.

If you can do **Checkup 3.1**, you are ready to go to Section 3.2.

✓ CHECKUP 3.1

Write an algebraic expression for each word expression.

_____ **1.** The sum of -3 and x.

_____ **2.** Eleven decreased by a.

_____ **3.** The product of m and -6.

_____ **4.** Five more than twice b.

Evaluate each expression.

_____ **5.** $x + 9$ when $x = -12$; when $x = -9$

_____ **6.** $\dfrac{24}{n}$ when $n = -6$; when $n = 0$

_____ **7.** $3(y - 5)$ when $y = -2$; when $y = 5$

_____ **8.** $\dfrac{m - n}{3}$ when $m = 13$ and $n = 10$; when $m = 3$ and $n = -6$

_____ **9.** The total budget for the athletic department will be divided equally among the three major sports. Write an algebraic expression for the share of each sport if the budget is x dollars. Find each sport's share if the budget is $150,000; if the budget is $360,000.

_____ **10.** A building contractor must pay a penalty of 6 percent of $2500, plus an additional $500 a day for each day over the deadline for completion of a building. Write an algebraic expression for the amount of the penalty after x days. Find the amount if the building is completed 3 days late; if it is completed 15 days late.

Check your answers in the back of your book.

If You Missed Problems:	You Should Review Examples:
1	4
2	2
3	1
4	4
5, 6	7, 8
7	10
8	9
9	13
10	14

3.2 Combining Like Terms

We turn our attention now to learning ways of making algebraic expressions as simple as possible. This process is called **simplifying** algebraic expressions.

Recognizing Like Terms

Let's look at some algebraic expressions and learn to call their parts by their correct mathematical names. In the expression **6x + 5x + 2x** we say that 6x is one **term**, 5x is another *term*, and 2x is another *term*.

> A **term** is a constant alone, or a product of a constant and one (or more) variables. Terms are separated from each other by the operation of addition.

Example 1. Identify the terms in the expression 4x + 7 − 3x.

Solution. We may write this expression as the sum

$$4x + 7 + (-3x)$$

Then we see that

4x is a term

7 is a term

−3x is a term

You try Example 2.

Example 2. Identify the terms in the expression 5x − 3y − 10.

Solution. We may write this expression as the sum

$$5x + (\underline{\hspace{1cm}}) + (\underline{\hspace{1cm}})$$

Then we see that

_____ is a term

_____ is a term

_____ is a term

Check your work on page 150. ▶

Terms that contain a number alone are called **constant terms**. Terms that contain a variable are called **variable terms**. The number part of a variable term is called the **numerical coefficient** for the variable in that term.

Expression	Constant Term	Variable Terms	Numerical Coefficients
x + 3	3	x (or 1x)	1
5x − 3y − 10	−10	5x −3y	5 −3
8x − 5x − x	None (or 0)	8x −5x −x (or −1x)	8 −5 −1

Example 3. Identify the numerical coefficient in each variable term of the expression $11x - 9x + 2$.

Solution. Note that 2 is the constant term.

For $11x$, 11 is the numerical coefficient of x.

For $-9x$, -9 is the numerical coefficient of x.

Now you try Example 4.

Example 4. Identify the numerical coefficient in each variable term of the expression $-8a - b + c$.

Solution

For $-8a$, _____ is the numerical coefficient of _____ .

For $-b$, -1 is the numerical coefficient of _____ .

For c, _____ is the numerical coefficient of c.

Check your work on page 150. ▶

Variable terms are called **like terms** if their variable parts are *exactly* the same. All constant terms are like terms. In the expression $6x + 5x + 2x$ the terms are all *like terms*, because the variable part of each term is x. But in the expression $-8a - b + c$ the terms are all *unlike terms*, because their variable parts are different.

Example 5. Identify the like terms in the expression

$$6x + 7y - 2 - 5x - 2y + 8$$

Solution

$6x$ and $-5x$ are like terms.

$7y$ and $-2y$ are like terms.

-2 and 8 are like terms.

You try Example 6.

Example 6. Identify the like terms in the expression

$$6A + 3a - 5A - a$$

Solution

$6A$ and _____ are like terms.

_____ and _____ are like terms.

Check your work on page 150. ▶

⫸ Trial Run

_____ 1. What are the terms in the expression $3x - 2y - 7$?

_____ 2. In the expression $7m - 9n + 11$, identify the constant term and the variable terms.

_____ 3. Identify the numerical coefficient in each variable term of the expression $-3a + b - 9$.

_____ 4. Identify the like terms in the expression $-10x + 5 + 6x - 9$.

Answers are on page 150.

Adding Terms

If you were asked, without any hints, to simplify the sum $6x + 5x + 2x$, what would your answer be? If you think the answer is $13x$, you are absolutely correct.

The key is to *look for like terms*. A sum of terms can be simplified *only if* it contains *like terms*. The process of adding like terms together is called **combining like terms**. The *distributive property* gives us the method for combining like terms.

$$6x + 5x + 2x$$
$$= (6 + 5 + 2)x \qquad \text{Use distributive property.}$$
$$= 13x \qquad \text{Simplify the sum.}$$

It is not necessary to show the step in which you use the distributive property if you realize that the process can be described as follows.

> **Combining Like Terms.** To combine like terms, we add the numerical coefficients and keep the same variable part.

Let's combine like terms in this expression.

$$5x + 12y + 3 + 8x - 6y - 1 \qquad \text{Notice which terms are like terms.}$$
$$= 5x + 8x + 12y - 6y + 3 - 1 \qquad \text{Rearrange, so that like terms are next to each other.}$$
$$= 13x + 6y + 2 \qquad \text{Combine the coefficients of the like terms.}$$

Try combining like terms in Example 7.

Example 7. Combine like terms in

$$17x + x - 20x$$

Solution

$$17x + x - 20x$$
$$= 17x + 1x + (\underline{\qquad}x)$$
$$= \underline{\qquad}x$$

Check your work on page 150. ▶

Example 8. Combine like terms in

$$x - (-2y) + 5 - 2x - 9y$$

Solution

$$x - (-2y) + 5 - 2x - 9y$$
$$= x + 2y + 5 + (-2x) + (-9y)$$
$$= 1x + (-2x) + 2y + (-9y) + 5$$
$$= -1x + (-7y) + 5$$
$$= -x - 7y + 5$$

Notice that it is customary to write simplified expressions with just *one sign* between the terms and that in our answers we usually write

$$1x \quad \text{as} \quad x$$
$$-1x \quad \text{as} \quad -x$$

We agree that 1 or -1 is "understood" to be the numerical coefficient in such cases.

⫸ **Trial Run**

Combine like terms.

_____ **1.** $3x - x + 12x$ _____ **2.** $-4x + 5 + 7x - 13$

_____ **3.** $-12x + 7y + 3x - 15y$ _____ **4.** $6A + 3a - 5A - a$

_____ **5.** $3a + 2b - 5a - 7b - 1$ _____ **6.** $3x - 4y + 9z - 2x + 4y + 6$

Answers are below.

▶ **Examples You Completed**

Example 2. Identify the terms in the expression
$$5x - 3y - 10$$
Solution. We may write this expression as the sum
$$5x + (-3y) + (-10)$$
Then we see that
$$5x \text{ is a term}$$
$$-3y \text{ is a term}$$
$$-10 \text{ is a term}$$

Example 4. Identify the numerical coefficient in each variable term of the expression
$-8a - b + c$.

Solution

For $-8a$, -8 is the numerical coefficient of a.
For $-b$, -1 is the numerical coefficient of b.
For c, 1 is the numerical coefficient of c.

Example 6. Identify the like terms in the expression
$$6A + 3a - 5A - a$$
Solution
$$6A \text{ and } -5A \text{ are like terms.}$$
$$3a \text{ and } -a \text{ are like terms.}$$

Example 7. Combine like terms in
$$17x + x - 20x$$
Solution
$$17x + x - 20x$$
$$= 17x + 1x + (-20x)$$
$$= -2x$$

Answers to Trial Runs

page 148   **1.** $3x, -2y, -7$ **2.** $7m$ and $-9n$ are variable terms; 11 is the constant term
3. -3 is the numerical coefficient of a; 1 is the numerical coefficient of b
4. $-10x$ and $6x$; 5 and -9

page 150   **1.** $14x$ **2.** $3x - 8$ **3.** $-9x - 8y$ **4.** $A + 2a$ **5.** $-2a - 5b - 1$ **6.** $x + 9z + 6$

EXERCISE SET 3.2

Identify the terms in the following expressions.

_____ 1. $3x - 5$ _____ 2. $4x - 7$

_____ 3. $\frac{1}{2}a - 7b + 9$ _____ 4. $\frac{1}{3}a - 2b - 2$

_____ 5. $1 - 5m + 6n$ _____ 6. $-1 - 9m + 4n$

_____ 7. $13x - 5y + 2z - 7$ _____ 8. $14x - 12y + 5z - 3$

Identify the constant terms and variable terms.

_____ 9. $4x - 19$ _____ 10. $7x + 23$

_____ 11. $2x + 3y$ _____ 12. $6x - 5y$

_____ 13. $-4a - 7b + 5.2$ _____ 14. $-5a + 11b - 8.4$

_____ 15. $1 + 7m - 5n + 14p - 4$ _____ 16. $2 + 5m - 9n + 3p - 5$

Identify the numerical coefficient of each variable term.

_____ 17. $3x + 7$ _____ 18. $5x - 13$

_____ 19. $\frac{1}{4}x - 12y$ _____ 20. $-4x + \frac{1}{5}y$

_____ 21. $3a + 4b - 2$ _____ 22. $8a - 5b + 9$

_____ 23. $0.3m - n - 2p$ _____ 24. $10m + n - 0.4p$

Identify like terms in each expression.

_____ 25. $8a - 3a$ _____ 26. $7a - 5a$

_____ 27. $3h - 2k + 5h + 5k$ _____ 28. $2h - 5k + 6h - 8k$

_____ 29. $\frac{2}{5}a - 6 + 3a - 9$ _____ 30. $\frac{2}{3}a - 8 + 2a - 15$

_____ 31. $3x - 0.24y + 1 + 4y - 1.2x - 7$ _____ 32. $x - 2.5y + 2 - 5y + 0.4x + 3$

Combine like terms.

_____ 33. $3a - 5a$ _____ 34. $5a - 7a$

_____ 35. $2x - 8x + 9x$ _____ 36. $3x - 11x + 10x$

_____ 37. $9 - 0.2y - 0.3y$

_____ 38. $7 - 0.4y - 0.5y$

_____ 39. $4m - 6 + 11m + 9$

_____ 40. $5m - 9 + 8m + 1$

_____ 41. $3h - k + 5 + 4h + 2k - 9$

_____ 42. $2h - 5k - 1 + 7h + 3k - 11$

_____ 43. $4x - 3y + 13z + 7x + 8y - 6z$

_____ 44. $2x + 5y - 6z + 3x - y - 3z$

_____ 45. $a + 10b - 9a + 7$

_____ 46. $2a - 11b - 6a + 3$

_____ 47. $\frac{2}{3}m - 5n + 4p + \frac{1}{3}m - 6n$

_____ 48. $\frac{3}{4}m - 3n + p + \frac{1}{4}m - 8n$

_____ 49. $5u - 3v + w + 2v - 5u - 7w$

_____ 50. $6u - 2v + 4w + 2v - 4u + 5w$

☆ Stretching the Topics _____

Combine like terms.

_____ 1. $0.7a - 3.21b - 0.23c - 1.5a + 5b + 0.45c$

_____ 2. $\frac{1}{2}x - 5y + \frac{4}{5}x - \frac{2}{3}y + 2z - 2x - \frac{1}{4}z$

_____ 3. $-25m + 16 - 73n + 24p - 58 + 32m + 93n - 29p + 42 + 13m$

Check your answers in the back of your book.

If you can complete **Checkup 3.2**, you are ready to go to Section 3.3.

CHECKUP 3.2

_____ **1.** Identify the terms in the expression $3 - 7x + 2y$.

_____ **2.** Identify the constant terms and the variable terms in the expression
$-4a + 11 + 7a - 13$.

_____ **3.** Identify the numerical coefficient of each variable term in the expression
$3x - y - 4z + 7$.

_____ **4.** Identify the like terms in the expression $5x - 2y + 4 - 7x + 2y - 1$.

Combine like terms.

_____ **5.** $3a - 7a$ _____ **6.** $x - 2x + 5x$

_____ **7.** $5 - 3y + 11y$ _____ **8.** $4m - 5 - 3m + 7$

_____ **9.** $2x + 7y - 3x - 6$ _____ **10.** $4a - 2b + c - 3b - 4a + c$

Check your answers in the back of your book.

If You Missed Problems:	You Should Review Examples:
1	1, 2
2, 3	3, 4
4	5, 6
5, 6	7
7–10	8

3.3 Working with Symbols of Grouping

Parentheses and brackets provide directions for operating with expressions, and they must be treated carefully.

Multiplying or Dividing a Term by a Constant

An expression such as $3(5x)$ means to multiply 3 times $5x$. The associative property for multiplication permits us to write

$$3(5x) = (3 \cdot 5)x \qquad \text{Use the associative property.}$$
$$= 15x \qquad \text{Simplify the product.}$$

> To multiply a constant times a variable term, multiply the constant times the numerical coefficient and keep the variable part.

Example 1. Multiply $3(-7x)$.

Solution

$$3(-7x) = [3 \, (-7)]x$$
$$= -21x$$

You try Example 2.

Example 2. Multiply $-2(-8a)$.

Solution

$$-2(-8a) =$$

Check your work on page 160. ▶

Example 3. Multiply $(-2)(3)(5x)$.

Solution

$$(-2)(3)(5x)$$
$$= (-2 \cdot 3)(5x)$$
$$= -6(5x)$$
$$= (-6 \cdot 5)x$$
$$= -30x$$

You try Example 4.

Example 4. Multiply $7(-4)(-x)$.

Solution

$$7(-4)(-x)$$
$$= 7(-4)(-1x)$$

Check your work on page 160. ▶

Example 5. Find $4x - 2(3x)$.

Solution

$$4x - 2(3x)$$
$$= 4x - (2 \cdot 3)x$$
$$= 4x - 6x$$
$$= 4x + (-6x)$$
$$= -2x$$

Now you try Example 6.

Example 6. Find $2(5a) + 8a$.

Solution

$$2(5a) + 8a$$
$$=$$

Check your work on page 160. ▶

To divide a variable term by a constant, we can rewrite the quotient.

$$\frac{45x}{5} = \frac{45}{5} \cdot x \quad = \quad 9x$$

$$\frac{-39x}{3} = \frac{-39}{3} \cdot x = -13x$$

> To divide a variable term by a constant, divide the numerical coefficient by the constant and keep the variable part.

Example 7. Find $\dfrac{16x}{-2}$.

Solution

$$\frac{16x}{-2} = \frac{16}{-2} \cdot x$$

$$= -8x$$

You try Example 8.

Example 8. Find $\dfrac{-28x}{-4}$.

Solution

$$\frac{-28x}{-4} =$$

Check your work on page 160. ▶

Example 9. Find $\dfrac{-x}{-1}$.

Solution

$$\frac{-x}{-1} = \frac{-1x}{-1}$$

$$= \frac{-1}{-1} \cdot x$$

$$= 1x$$

$$= x$$

You try Example 10.

Example 10. Find $\dfrac{7a}{7}$.

Solution

$$\frac{7a}{7} =$$

Check your work on page 160. ▶

⫸ Trial Run

Perform the indicated operations.

_____ **1.** $4(5x)$

_____ **2.** $-3(-8x)$

_____ **3.** $-11(3x)$

_____ **4.** $(-2)(5)(7x)$

_____ **5.** $3(-2x) + 10x$

_____ **6.** $\dfrac{-32x}{-4}$

_____ **7.** $\dfrac{x}{-1}$

_____ **8.** $\dfrac{-6a}{6}$

Answers are on page 160.

Multiplying a Constant Times a Sum

In working with variable expressions, the **distributive property for multiplication over addition (or subtraction)** will be one of our most valuable tools.

Distributive Property
$A(B + C) = A \cdot B + A \cdot C$
$A(B - C) = A \cdot B - A \cdot C$

Example 11. Find $4(x + 2)$.

Solution

$$4(x + 2)$$
$$= 4 \cdot x + 4 \cdot 2$$
$$= 4x + 8$$

Example 12. Find $\frac{1}{2}(2x - 6y)$.

Solution

$$\frac{1}{2}(2x - 6y)$$
$$= \frac{1}{2} \cdot 2x - \frac{1}{2} \cdot 6y$$
$$= \left(\frac{1}{2} \cdot 2\right)x - \left(\frac{1}{2} \cdot 6\right)y$$
$$= 1x - 3y$$
$$= x - 3y$$

Perhaps it is time to observe some patterns and take some shortcuts. Let's work some more examples and use *arrows* to indicate our thoughts.

Example 13. Find $-4(3x - 7y)$.

Solution

$$-4(3x - 7y) = -4(3x - 7y)$$
$$= -12x + 28y$$

You try arrows in Example 14.

Example 14. Find $5(6x - 3y)$.

Solution

$$5(6x - 3y) = 5(6x - 3y)$$
$$= \underline{}x - \underline{}y$$

Check your work on page 160. ▶

⇒ Trial Run

Find each product.

_____ 1. $3(x + y)$ 　　　　 _____ 2. $-2(x + 3y)$

_____ 3. $\frac{1}{4}(12x - 4y)$ 　　 _____ 4. $-2(3a - 2b)$

_____ 5. $3(-2m - n)$ 　　 _____ 6. $-5(-x + 6y)$

Answers are on page 160.

Using Symbols of Grouping

The algebraic expression $(4x + 5y)$ can be thought of as an example of the distributive property if we agree that 1 is understood to be in front of the parentheses. In other words,

$$(4x + 5y) = 1(4x + 5y)$$
$$= 1 \cdot 4x + 1 \cdot 5y$$
$$= 4x + 5y$$

We see, then, that to simplify $(4x + 5y)$, we merely "drop the parentheses."

Similarly, we should agree that we can use the indicated steps to simplify the following expression.

$$(4x + 5y) - (3x + 2y)$$

$= 1(4x + 5y) - 1(3x + 2y)$	Insert 1 before each quantity.
$= 4x + 5y - 3x - 2y$	Drop the first pair of parentheses. Multiply the second quantity by -1.
$= 4x - 3x + 5y - 2y$	Rearrange terms.
$= x + 3y$	Combine like terms.

Now try Example 15.

Example 15. Simplify

$$(8x + 7) - (2x + 3)$$

Solution

$$(8x + 7) - (2x + 3)$$
$$= 1(8x + 7) - 1(2x + 3)$$

Check your work on page 160. ▶

Example 16. Simplify

$$(9a - 6c) - (10a - 5c)$$

Solution

$$(9a - 6c) - (10a - 5c)$$
$$= 1(9a - 6c) - 1(10a - 5c)$$
$$= 9a - 6c - 10a + 5c$$
$$= 9a - 10a - 6c + 5c$$
$$= -a - c$$

Example 17. Simplify

$$-2(5a - 3b) - 7(a - 3b)$$

Solution

$$-2(5a - 3b) - 7(a - 3b)$$
$$= -10a + 6b - 7a + 21b$$
$$= -10a - 7a + 6b + 21b$$
$$= -17a + 27b$$

Now try Example 18.

Example 18. Simplify

$$3(x - 5y) + 4(-x + 3y)$$

Solution

$$3(x - 5y) + 4(-x + 3y)$$
$$=$$

Check your work on page 160. ▶

In simplifying expressions containing brackets and parentheses, we shall continue the practice of working with the innermost symbols of grouping first. We shall also use our shortcuts.

Example 19. Simplify $2[5x - 3(x - 1)]$.

Solution

$$2[5x - 3(x - 1)]$$

$$= 2[5x - 3x + 3] \qquad \text{Remove innermost parentheses (distributive property).}$$

$$= 2[2x + 3] \qquad \text{Find the sum within brackets.}$$

$$= 4x + 6 \qquad \text{Remove brackets (distributive property).}$$

Example 20. Simplify $-4[2(a - 3b) - (-3a - 2b)]$.

Solution

$$-4[2(a - 3b) - (-3a - 2b)]$$

$$= -4[2(a - 3b) - 1(-3a - 2b)] \qquad \text{Insert 1 before second pair of parentheses.}$$

$$= -4[2a - 6b + 3a + 2b] \qquad \text{Remove parentheses (distributive property).}$$

$$= -4[2a + 3a - 6b + 2b] \qquad \text{Rearrange terms in brackets.}$$

$$= -4[5a - 4b] \qquad \text{Combine like terms within brackets.}$$

$$= -20a + 16b \qquad \text{Remove brackets (distributive property).}$$

⫸ **Trial Run**

Simplify each expression.

———— **1.** $-(3x - 5y)$ ———— **2.** $-(-2a - 11b + 2)$

———— **3.** $(3x + 6) - (2x + 3)$ ———— **4.** $(8a - 5b) - (-a - 3b)$

———— **5.** $4(2m - 3) - 2(3m + 1)$ ———— **6.** $3[2x - (x - 2)]$

———— **7.** $3[-2(3x - 4) - (x + 1)]$

———— **8.** $5[3x - (x + 2)] - 2[2(x - 1) - 4]$

Answers are on page 160.

▶ Examples You Completed _____

Example 2. Multiply $-2(-8a)$.

Solution

$$-2(-8a) = [-2(-8)]a$$
$$= 16a$$

Example 4. Multiply $7(-4)(-x)$.

Solution

$$7(-4)(-x) = 7(-4)(-1x)$$
$$= [7(-4)](-1x)$$
$$= (-28)(-1x)$$
$$= [-28(-1)]x$$
$$= 28x$$

Example 6. Find $2(5a) + 8a$.

Solution

$$2(5a) + 8a = (2 \cdot 5)a + 8a$$
$$= 10a + 8a$$
$$= 18a$$

Example 8. Find $\dfrac{-28x}{-4}$.

Solution

$$\frac{-28x}{-4} = \frac{-28}{-4} \cdot x$$
$$= 7x$$

Example 10. Find $\dfrac{7a}{7}$.

Solution

$$\frac{7a}{7} = \frac{7}{7} \cdot a$$
$$= 1a$$
$$= a$$

Example 14. Find $5(6x - 3y)$.

Solution

$$5(6x - 3y) = 5(6x - 3y)$$
$$= 30x - 15y$$

Example 15. Simplify $(8x + 7) - (2x + 3)$.

Solution

$$(8x + 7) - (2x + 3)$$
$$= 1(8x + 7) - 1(2x + 3)$$
$$= 8x + 7 - 2x - 3$$
$$= 8x - 2x + 7 - 3$$
$$= 6x + 4$$

Example 18. Simplify

$$3(x - 5y) + 4(-x + 3y)$$

Solution

$$3(x - 5y) + 4(-x + 3y)$$
$$= 3x - 15y - 4x + 12y$$
$$= 3x - 4x - 15y + 12y$$
$$= -x - 3y$$

Answers to Trial Runs _____

page 156 **1.** $20x$ **2.** $24x$ **3.** $-33x$ **4.** $-70x$ **5.** $4x$ **6.** $8x$ **7.** $-x$ **8.** $-a$

page 157 **1.** $3x + 3y$ **2.** $-2x - 6y$ **3.** $3x - y$ **4.** $-6a + 4b$ **5.** $-6m - 3n$
6. $5x - 30y$

page 159 **1.** $-3x + 5y$ **2.** $2a + 11b - 2$ **3.** $x + 3$ **4.** $9a - 2b$ **5.** $2m - 14$ **6.** $3x + 6$
7. $-21x + 21$ **8.** $6x + 2$

EXERCISE SET 3.3

Find each product or quotient.

_____ 1. $2(4x)$ _____ 2. $3(7x)$

_____ 3. $3(-0.5a)$ _____ 4. $4(-0.2a)$

_____ 5. $-12\left(\dfrac{3}{4}\,m\right)$ _____ 6. $-9\left(\dfrac{2}{3}\,m\right)$

_____ 7. $-3(-2x)$ _____ 8. $-2(-4x)$

_____ 9. $-2\left(\dfrac{1}{2}\right)(3y)$ _____ 10. $-3\left(\dfrac{1}{3}\right)(2y)$

_____ 11. $-3(2)(-5k)$ _____ 12. $-4(3)(-3k)$

_____ 13. $\dfrac{20x}{-4}$ _____ 14. $\dfrac{27x}{-3}$

_____ 15. $\dfrac{-45x}{-9}$ _____ 16. $\dfrac{-34x}{-2}$

_____ 17. $\dfrac{5a}{5}$ _____ 18. $\dfrac{7a}{7}$

_____ 19. $\dfrac{-m}{-1}$ _____ 20. $\dfrac{m}{-1}$

_____ 21. $\dfrac{-10y}{2}$ _____ 22. $\dfrac{-15y}{3}$

_____ 23. $\dfrac{-7n}{7}$ _____ 24. $\dfrac{-15n}{15}$

Simplify each expression and combine like terms.

_____ 25. $5x + 3(2x)$ _____ 26. $4x + 3(3x)$

_____ 27. $2(5x) - 3x$ _____ 28. $3(4x) - 7x$

_____ 29. $\dfrac{1}{4}\,(a - b)$ _____ 30. $\dfrac{1}{5}\,(a - b)$

_____ 31. $-3(x + 2y)$ _____ 32. $-4(x + 5y)$

_____ 33. $1.5(6x - 2y)$ _____ 34. $0.5(2x - 4y)$

_____ 35. $-4(5a - 7b)$ _____ 36. $-6(a - 2b)$

_____ 37. $9(-2m - 3n)$

_____ 38. $8(-3m - 5n)$

_____ 39. $-3(-x + 4y)$

_____ 40. $-7(-x + 2y)$

_____ 41. $-\left(\dfrac{1}{2}x - 5y\right)$

_____ 42. $-\left(\dfrac{1}{3}x - 2y\right)$

_____ 43. $-(-3a - 7b + 5.4)$

_____ 44. $-(-4a - 6b + 3.2)$

_____ 45. $(3x + 5) - (2x + 4)$

_____ 46. $(5x + 3) - (3x + 1)$

_____ 47. $(12a + 3b) - (5a - 2b)$

_____ 48. $(9a + 4b) - (7a - b)$

_____ 49. $(-4m - 3n) - (2m - 3n)$

_____ 50. $(-5m - 2n) - (m - 2n)$

_____ 51. $(7x - 3y) - (-2x + y)$

_____ 52. $(8x - 5y) - (-3x + 2y)$

_____ 53. $3(5m - 4) - 4(m + 2)$

_____ 54. $5(2m - 1) - 2(m + 5)$

_____ 55. $-3(x - 2y) + 4(-x + y)$

_____ 56. $-2(2x - y) + 3(-x + 2y)$

_____ 57. $4[2a - (a - 3.5)]$

_____ 58. $5[a - (2a - 1.4)]$

_____ 59. $5[(3x - 2y) - (x + y)]$

_____ 60. $7[(2x - 3y) - (x + 2y)]$

_____ 61. $-2[5(a - 2b) + 7(2a + b)]$

_____ 62. $-3[4(2a - b) + 5(a + 3b)]$

_____ 63. $4[-3(x - 4) - 2(x + 6)]$

_____ 64. $5[-6(x - 1) - 3(x + 2)]$

☆ Stretching the Topics

Simplify each expression and combine like terms.

_____ 1. $0.4(2x - 3y) - 0.5(-y - 3) + 1.2(2x - 1)$

_____ 2. $-3[(2a - b) - 5(4a + b)] + 7[-2(3a + b) + (b - 3a)]$

_____ 3. $9 - 3\{4x - 7[5x - (x -2) + 6] - [2x + (4x - 5)] + 9\}$

Check your answers in the back of your book.

If you can simplify the expressions in **Checkup 3.3**, you are ready to do the **Review Exercises** for Chapter 3.

CHECKUP 3.3

Simplify each expression and combine like terms.

_____ **1.** $4(-8x)$

_____ **2.** $-2(4)(-5y)$

_____ **3.** $\dfrac{48x}{-12}$

_____ **4.** $\dfrac{10a}{-10}$

_____ **5.** $7x - 3(4x)$

_____ **6.** $5(-2x + 3y)$

_____ **7.** $(4x - 5) - (2x + 3)$

_____ **8.** $2(3a - 6b) - 3(-2a + b)$

_____ **9.** $-6[4a - (3a + 1)]$

_____ **10.** $7[3(2m - n) - 2(3m - 5n)]$

Check your answers in the back of your book.

If You Missed Problem:	You Should Review Examples:
1	1, 2
2	3, 4
3	7, 8
4	7, 10
5	5, 6
6	11–14
7	15, 16
8	17, 18
9	19
10	20

Summary

In this chapter we learned to recognize, evaluate, and simplify **algebraic expressions.**

In Order to	We Must	Example
Evaluate an algebraic expression for some value of the variable	Substitute the given value for the variable in the expression.	If $x = 2$, then $1 - 3x$ $= 1 - 3(2)$ $= 1 - 6$ $= -5$
Combine like terms in an expression	Add the numerical coefficients and keep the same variable part.	$3x + 2y - 7x$ $= 3x - 7x + 2y$ $= -4x + 2y$
Multiply a constant times a variable term	Multiply the constant times the numerical coefficient and keep the same variable part.	$-5(2x)$ $= (-5 \cdot 2)x$ $= -10x$
Divide a variable term by a constant	Divide the numerical coefficient by the constant and keep the same variable part.	$\dfrac{-21a}{-7}$ $= \dfrac{-21}{-7} \cdot a$ $= 3a$
Multiply a constant times a sum of terms	Use the distributive property and simplify.	$3(-4x + y)$ $= 3(-4x) + 3y$ $= -12x + 3y$

☐ Speaking the Language of Algebra

Complete each statement with the appropriate word or phrase.

1. In the expression $2x - 3y + 5$, 2 is the _____ _____ of x, -3 is the _____ _____ of y, and 5 is the _____ term.

2. To combine like terms, we add the _____ _____ and keep the same variable part.

3. In the expression $x + y + 5x - 6y$, we say that x and $5x$ are _____ _____ because their variable parts are exactly the same.

4. When we rewrite $2(x + 5)$ as $2x + 10$, we are using the _____ property.

5. When we begin to remove symbols of grouping, we always work with the _____ symbols of grouping first.

Check your answers in the back of your book.

△ Writing About Mathematics

Write your response to each question in complete sentences.

1. Write three different word expressions that could be translated into the algebraic expression $x + 2(x + 3)$.

2. Write an algebraic expression that contains two unlike variable terms and a constant term. Then identify the variables, their numerical coefficients, and the constant term.

3. Explain the steps you would use to simplify the algebraic expression $2(x + 3) - (2x - 1)$.

REVIEW EXERCISES for Chapter 3

Write an algebraic expression for each word expression.

_____ 1. Multiply 6 times x.

_____ 2. The sum of y and 9.

_____ 3. Ten more than 3 times m.

_____ 4. Four times k is subtracted from 23.7.

_____ 5. Nine times the sum of x and 7.

Evaluate each expression.

_____ 6. $x - 9$ when $x = 2$; when $x = -12$

_____ 7. $\dfrac{56}{n}$ when $n = -7$; when $n = 0$

_____ 8. $3(9 - x)$ when $x = 11$; when $x = 9$

_____ 9. $2x - 7y$ when $x = 5$, $y = -1$

_____ 10. $\dfrac{4a - 12}{3}$ when $a = 3$; when $a = 0$

_____ 11. A hiker left Horseshoe Cave walking along the trail at 4 miles per hour. Write an algebraic expression for the distance traveled after h hours. Find the distance traveled after 3.5 hours.

_____ 12. Aunt Vina left her entire estate to her three nephews. If the estate is valued at x dollars, write an algebraic expression for each nephew's share. Find the value of each nephew's share if the estate is worth $210,000.

_____ 13. The cost of renting a cabin at the lake is $200 for the first week and $19.50 for each additional day. Write an algebraic expression for the total cost of the cabin after 1 week and d additional days. Find the total cost for staying 1 week and 3 days.

Identify the constant terms and the variable terms.

_____ 14. $3x - 12$ _____ 15. $-2x + 7y$

_____ 16. $4a - 3b + 2$ _____ 17. $3m - 2 + 4n - 9 - p$

Identify the numerical coefficient in each variable term.

_____ 18. $3x - 9$ _____ 19. $7x - 2y$

_____ 20. $-3a + b - 1$ _____ 21. $m - n + 2p$

Simplify each expression and combine like terms.

_____ 22. $2x - 9x + 5x$ _____ 23. $3m - 4 + 2m + 6$

_____ 24. $2a - 13b - 9a + 3$ _____ 25. $-4x - 3y + 7z + 8x + 3y - 2z$

_____ 26. $4(-5x)$ _____ 27. $-2(-3)(4y)$

_____ 28. $\dfrac{-54x}{9}$ _____ 29. $\dfrac{-a}{-1}$

_____ 30. $-3(5x) + 12x$ _____ 31. $7(4a - 5b)$

_____ 32. $-(7x - 2y)$ _____ 33. $(-7x + 2) + (3x + 5)$

_____ 34. $(-9m - 3n) - (7m - 3n)$ _____ 35. $3[4a - (7a + 2)]$

_____ 36. $5[-3(x + y) + 7(x - y)]$

Check your answers in the back of your book.

If You Missed Exercises:	You Should Review Examples:	
1–5	Section 3.1	1–4
6–10		7–10
11		11
12		13
13		14
14–21	Section 3.2	1–4
22–25		7, 8
26, 27	Section 3.3	1–4
28, 29		7–10
30		5, 6
31, 32		11–14
33, 34		17, 18
35, 36		19, 20

If you have completed the **Review Exercises** and have corrected your errors, you are ready to take the **Practice Test** for Chapter 3.

PRACTICE TEST for Chapter 3

Write an algebraic expression for each word expression.   SECTION   EXAMPLES

_____ **1.** Decrease m by 5.   3.1   2

_____ **2.** Nine more than the product of x and 3.   3.1   3

_____ **3.** Ten times the sum of b and 4.6.   3.1   6

Evaluate each expression.

_____ **4.** $6x - 4$ when $x = 1$; when $x = -2$ 3.1   7, 8

_____ **5.** $-2(7 - x)$ when $x = 0$; when $x = -1$ 3.1   10

_____ **6.** $\dfrac{2r - 4s}{6}$ when $r = 5$, $s = -7$ 3.1   9

_____ **7.** If a hotel room costs \$33.79 per night (including tax), write an algebraic expression for the cost of staying n nights. Find the cost of staying 5 nights.   3.1   11

_____ **8.** A cable TV subscription salesperson is paid \$200 per week plus \$4.10 for every new subscriber. Write an algebraic expression for the weekly pay of a salesperson enrolling x new subscribers. Find the weekly pay if 30 new subscribers are enrolled.   3.1   14

_____ **9.** The \$180 rental of a banquet room is to be shared equally by the English Club members attending the banquet. Write an algebraic expression for each member's share if m members attend the banquet. Find each member's share if 45 members attend.   3.1   13

Simplify each expression and combine like terms.

_____ **10.** $5n - 6n - 3n$ 3.2   7

_____ **11.** $4x - 10 - 7x + 3$ 3.2   8

_____ **12.** $-8(-2)(-y)$ 3.3   3, 4

_____ **13.** $-5x - 2y + 3z + 5x - 2y - z$ 3.2   8

_____ **14.** $\dfrac{-18x}{-3}$ 3.3   8

_____ **15.** $-5(4n) + 8n$ 3.3   6

_____ **16.** $\dfrac{2}{3}(9x + 3)$ 3.3   12

_____ **17.** $-6(3x - 2y)$ 3.3   13

169

	SECTION	EXAMPLE

_____ **18.** $y - (7x - 3y)$ ⟶ 3.3 ⟶ 16

_____ **19.** $4[2a - (5a - 3)]$ ⟶ 3.3 ⟶ 19

_____ **20.** $7[-2(x - y) - 4(x + y)]$ ⟶ 3.3 ⟶ 20

Check your answers in the back of your book.

SHARPENING YOUR SKILLS after Chapters 1–3

		SECTION

_____ **1.** Change 0.875 to a fraction and reduce. — 1.4

_____ **2.** Change $\dfrac{5}{6}$ to a decimal number and round to the hundredths place. — 1.4

_____ **3.** Find the value of $\left(\dfrac{2}{3}\right)^3$. — 1.3

_____ **4.** Find 9% of 35. — 1.4

_____ **5.** Complete the statement $2 \cdot 3 - 7$ _____ -8 using $<$, $>$, or $=$. — 2.1

Perform the indicated operations.

_____ **6.** $\dfrac{13(9-2)}{5-\dfrac{15}{3}}$ — 1.2

_____ **7.** $\dfrac{5}{8} + \dfrac{9}{10}$ — 1.3

_____ **8.** $(2.36)(0.24)$ — 2.4

_____ **9.** $(23-9)-(11-6)$ — 1.2

_____ **10.** $3\left(2 \cdot 3 + \dfrac{15-3}{4}\right)$ — 1.2

_____ **11.** $-4(7-10)$ — 2.3

_____ **12.** $\dfrac{-2[3(-1)-(-9)]}{-13+10}$ — 2.3

_____ **13.** $(-2)^4$ — 2.3

_____ **14.** $|-9| - |3|$ — 2.1

_____ **15.** Write a number expression for the word expression and simplify. The sum of -13 and 12 is divided by 7. — 2.5

Check your answers in the back of your book.

Solving
First-Degree Equations

Charlotte now earns $6 per hour as a grounds keeper working 35 hours per week. She has been offered another job paying $280 per week for the same work. How much of an increase in her hourly wage must her present employer give Charlotte to meet the better offer?

Before you finish this chapter, you will be able to solve this problem using an **equation**. Equations are mathematical sentences that say that one quantity represents the same number as another quantity.

In this chapter we learn how to

1. Solve equations using one property.
2. Solve equations using several properties.
3. Switch from word statements to equations.

4.1 Solving Equations Using One Property

Equations may or may not contain variables. Using the symbol $=$ to mean "is equal to" or "equals," we may write

$$2 + 3 = 6 - 1 \qquad x + 3 = 3 + x$$
$$2(5 - 7) = -4 \qquad 2(x + 5) = 2x + 10$$

and we know that each of these equations is *always* true. Such statements are called **identities**, but they are not our primary concern. Instead, we shall concentrate on statements that are sometimes, but not always, true.

Satisfying Equations

If an equation containing variables is true only for certain values of the variable, we call it a **conditional equation**.

> The **solution** for any equation is the value of the variable that makes the equation a true statement.

For example,

Equation	Solution	Reason
$x + 3 = 5$	$x = 2$	because $2 + 3 = 5$
$17 - y = 11$	$y = 6$	because $17 - 6 = 11$
$5x = 20$	$x = 4$	because $5 \cdot 4 = 20$
$\dfrac{x}{3} = 4$	$x = 12$	because $\dfrac{12}{3} = 4$

If we substitute a number for the variable in an equation and obtain a true statement, we say that the number *satisfies* the equation and that the number is a solution for the equation. In this chapter we are concerned with equations containing one variable with no exponent except 1. Such equations are called **first-degree equations**.

Example 1. Is -1 a solution for $1 - 3x = 4$?

Solution. Substituting -1 for x we have

$$1 - 3(-1) \overset{?}{=} 4$$
$$1 + 3 \overset{?}{=} 4$$
$$4 = 4$$

So -1 is a solution.

You complete Example 2.

Example 2. Is 3 a solution for $2x + 1 = 10 - x$?

Solution. Substituting 3 for x, we have

$$2(\underline{}) + 1 \overset{?}{=} 10 - \underline{}$$
$$\underline{} \overset{?}{=} 7$$
$$\underline{} = 7$$

So $\underline{}$.

Check your work on page 183. ▶

Example 3. Is -3 a solution for $4(5 - x) = 8$?

Solution. Substituting -3 for x, we have

$$4[5 - (-3)] \overset{?}{=} 8$$

$$4[5 + 3] \overset{?}{=} 8$$

$$4(8) \overset{?}{=} 8$$

$$32 \neq 8$$

So -3 is *not* a solution.

You try Example 4.

Example 4. Is -5 a solution for

$$\frac{2(y + 2)}{3} = -2?$$

Solution. Substituting -5 for y, we have

$$\frac{2[(\underline{}) + 2]}{3} \overset{?}{=} -2$$

So _____ .

Check your work on page 183. ▶

⫸ Trial Run

_____ **1.** Is 3 a solution for $3x - 1 = 8$?

_____ **2.** Is -5 a solution for $6 - 2y = 16$?

_____ **3.** Is 9 a solution for $\dfrac{m}{3} + 2 = -7$?

_____ **4.** Is 0 a solution for $2(a - 5) = -10$?

_____ **5.** Is -1 a solution for $2x + 8 = x + 7$?

Answers are on page 184.

Finding Solutions by Addition and Subtraction

As we have already discovered, it is sometimes possible to find a solution just by looking at an equation. However, there are many equations that are too complicated to be solved by inspection. We need some steps that will help us find solutions in an orderly and accurate way.

To solve a first-degree equation we would like to have the variable (with a numerical coefficient of 1) all by itself on one side of the equation, and some number (the solution) on the other side. In other words, we must **isolate the variable**.

To isolate x in the equation

$$x + 6 = 10$$

we must get rid of the 6 being added to x. By the property of opposites we know that $6 + (-6) = 0$, so we can get rid of the 6 by *subtracting* 6 (or adding -6) on the left-hand side. But *we cannot do something to one side of an equation without doing the same thing to the other side*. If we subtract 6 (or add -6) on the left, we *must* also subtract 6 (or add -6) on the right.

$$x + 6 = 10$$

$$x + 6 - 6 = 10 - 6 \qquad \text{Subtract 6 from both sides.}$$

$$x + 0 = 4 \qquad \text{Simplify the sums.}$$

$$x = 4 \qquad \text{Simplify (addition property of 0).}$$

To solve the equation

$$x - 9 = 3$$

we can get rid of the 9 being subtracted from x by *adding* 9 on the left. But we must also add 9 on the right at the same time. Here are the steps.

$$x - 9 = 3$$

$$x - 9 + 9 = 3 + 9 \qquad \text{Add 9 to both sides.}$$

$$x + 0 = 12 \qquad \text{Simplify the sums.}$$

$$x = 12 \qquad \text{Simplify (addition property of 0).}$$

> We may add the same quantity to both sides of an equation. We may subtract the same quantity from both sides of an equation. In each case the resulting equation has the same solution as the original equation.

This property of equations can be stated in symbols.

Addition Property of Equality

If $\qquad A = B$

then $\quad A + C = B + C$

and $\quad A - C = B - C$

Remember that we can always test our solution by substituting it into the original equation to see if it makes the original statement true. This is called **checking the solution**.

Example 5. Solve $x + 9 = 32$ and check.

Solution

$$x + 9 = 32$$

$$x + 9 - 9 = 32 - 9 \qquad \text{Subtract 9 from both sides.}$$

$$x + 0 = 23 \qquad \text{Simplify the sums.}$$

$$x = 23 \qquad \text{Simplify (addition property of 0).}$$

CHECK

$$x + 9 = 32 \qquad \text{Original equation.}$$

$$23 + 9 \overset{?}{=} 32 \qquad \text{Substitute 23 for } x.$$

$$32 = 32 \qquad \text{Simplify.}$$

Now you provide the reason for each step in Example 6.

Example 6. Solve $x - 15.2 = 100$ and check.

Solution

$$x - 15.2 = 100$$
$$x - 15.2 + 15.2 = 100 + 15.2 \qquad \underline{\hspace{5cm}}$$
$$x + 0 = 115.2 \qquad \underline{\hspace{5cm}}$$
$$x = 115.2 \qquad \underline{\hspace{5cm}}$$

CHECK

$$x - 15.2 = 100 \qquad \underline{\hspace{5cm}}$$
$$115.2 - 15.2 \stackrel{?}{=} 100 \qquad \underline{\hspace{5cm}}$$
$$100 = 100 \qquad \underline{\hspace{5cm}}$$

Check your reasons on page 183. ▶

Example 7. Solve $-5 + y = -3$ and check.

Solution

$$-5 + y = -3$$
$$5 - 5 + y = 5 - 3$$
$$0 + y = 2$$
$$y = 2$$

CHECK

$$-5 + y = -3$$
$$-5 + 2 \stackrel{?}{=} -3$$
$$-3 = -3$$

You try Example 8.

Example 8. Solve $y + 7 = -2$ and check.

Solution

$$y + 7 = -2$$

CHECK

$$y + 7 = -2$$

Check your work on page 183. ▶

⇒ Trial Run

Solve each equation and check.

_____ **1.** $x + 3 = 8$ _____ **2.** $6 = y - 4$

_____ **3.** $3 + m = 3$ _____ **4.** $-5 = a + 4$

_____ **5.** $-8 + x = -4$ _____ **6.** $9 = -2 + x$

Answers are on page 184.

Finding Solutions by Multiplication and Division

Consider the equation

$$3x = 15$$

We must decide how to *isolate the variable* by getting rid of the 3 that is being multiplied times the variable. We would like to operate on both sides of the equation so that we are left with $1x$ or x by itself.

Look at what happens if we *divide* both sides of this equation by 3.

$$3x = 15$$
$$\frac{3x}{3} = \frac{15}{3}$$
$$\frac{3}{3} \cdot x = \frac{15}{3}$$
$$1x = 5$$
$$x = 5$$

Let's check this solution in the original equation.

$$3x = 15$$
$$3(5) \stackrel{?}{=} 15$$
$$15 = 15$$

Look at what happens if we *multiply* both sides of this equation by $\frac{1}{3}$.

$$3x = 15$$
$$\frac{1}{3}(3x) = \frac{1}{3}(15)$$
$$\left(\frac{1}{3} \cdot 3\right)x = \frac{1}{3}(15)$$
$$1x = 5$$
$$x = 5$$

Notice that both methods give the same solution for the equation.

To solve an equation in which the variable is being multiplied by a constant, we may

1. Divide both sides by the coefficient of the variable; *or*
2. Multiply both sides by the *reciprocal* of the coefficient of the variable.

This property of equations may be stated in symbols.

Multiplication Property of Equality

If $A = B$

then $AC = BC$

and $\frac{A}{C} = \frac{B}{C}$ $(C \neq 0)$

Example 9. Solve $-6x = 24$ and check.

Solution

$$-6x = 24$$

$$\frac{-6x}{-6} = \frac{24}{-6}$$ Divide both sides by -6.

$$1x = -4$$ Find quotients. (Remember: $\frac{A}{A} = 1$.)

$$x = -4$$ Simplify (multiplication property of 1).

CHECK

$$-6x = 24$$ Original equation.

$$-6(-4) \overset{?}{=} 24$$ Substitute -4 for x.

$$24 = 24$$ Find the product.

Now you provide the reason for each step in Example 10.

Example 10. Solve $-24 = -\dfrac{12}{5}x$ and check.

Solution

$$-24 = -\frac{12}{5}x$$

$$-\frac{5}{12}\left(-\frac{24}{1}\right) = -\frac{5}{12}\left(-\frac{12}{5}x\right)$$ _____

$$\frac{-5}{\cancel{12}}\cdot\frac{\overset{-2}{\cancel{-24}}}{1} = \left(\frac{\overset{-1}{\cancel{-5}}}{\cancel{12}}\cdot\frac{\overset{-1}{\cancel{-12}}}{\cancel{5}}\right)x$$ Rewrite and reduce fractions.

$$10 = 1x$$ _____

$$10 = x$$ _____

CHECK

$$-24 = -\frac{12}{5}x$$ _____

$$-24 \overset{?}{=} -\frac{12}{5}(10)$$ _____

$$-24 = -24$$ _____

Check your reasons on page 183. ▶

Example 11. Solve $-x = 13$ and check.

Solution

$$-x = 13$$
$$-1x = 13$$
$$\frac{-1x}{-1} = \frac{13}{-1}$$
$$x = -13$$

CHECK

$$-x = 13$$
$$-(-13) \stackrel{?}{=} 13$$
$$13 = 13$$

▌▌▊➡ Trial Run

Solve each equation and check.

_____ 1. $4x = 24$ _____ 2. $25 = -5x$

_____ 3. $3x = 0$ _____ 4. $-x = 7$

_____ 5. $\frac{1}{3}a = 6$ _____ 6. $-\frac{7}{2}x = 21$

Answers are on page 184.

We now know how to solve an equation in which the variable is being *multiplied* by a constant. But how can we get rid of a constant that is *dividing* the variable? Consider the equation

$$\frac{x}{2} = 9$$

$$\frac{1}{2}x = 9 \qquad \text{Rewrite the quotient as a product (definition of division).}$$

$$\frac{2}{1}\left(\frac{1}{2}x\right) = \frac{2}{1} \cdot 9 \qquad \text{Multiply both sides by the } \textit{reciprocal} \text{ of the coefficient of } x.$$

$$\left(\frac{2}{1} \cdot \frac{1}{2}\right)x = 2 \cdot 9 \qquad \text{Use the associative property.}$$

$$1 \cdot x = 18 \qquad \text{Find the products.}$$

$$x = 18 \qquad \text{Simplify.}$$

This method can be shortened if we realize that we have used multiplication to "un-do" division.

$$\frac{x}{2} = 9$$

$$\frac{2}{1}\left(\frac{x}{2}\right) = 2 \cdot 9 \qquad \text{Multiply both sides by the denominator, 2}$$

$$x = 18 \qquad \text{Find the products.}$$

In each method we solved the equation by the *multiplication property of equality*. From the following examples you will see that we can choose whichever method seems more appropriate.

Example 12. Solve $\dfrac{-x}{3} = 21$ and check.

Solution. Since the variable is being multiplied (by -1) *and* divided (by 3), we choose to rewrite the left side before solving the equation.

$$\frac{-x}{3} = 21$$

$$-\frac{1}{3}x = 21 \qquad \text{Rewrite the quotient as a product.}$$

$$-\frac{3}{1}\left(-\frac{1}{3}\right)x = -3(21) \qquad \begin{array}{l}\text{Multiply both sides by } -3, \text{ the reciprocal} \\ \text{of } -\frac{1}{3} \text{ (the coefficient of } x\text{).}\end{array}$$

$$\left(\frac{-3}{1} \cdot \frac{-1}{3}\right)x = -3(21) \qquad \text{Use the associative property.}$$

$$x = -63 \qquad \text{Find the products.}$$

CHECK

$$\frac{-x}{3} = 21 \qquad \text{Original equation.}$$

$$\frac{-(-63)}{3} \overset{?}{=} 21 \qquad \text{Substitute } -63 \text{ for } x.$$

$$\frac{63}{3} \overset{?}{=} 21 \qquad \text{Remove parentheses.}$$

$$21 = 21 \qquad \text{Find the quotient.}$$

Example 13. Solve $\dfrac{a}{2.5} = -8$ and check.

Solution. Since the variable a is only being divided (by 2.5), we choose to use the shorter method.

$$\frac{a}{2.5} = -8$$

$$\frac{2.5}{1}\left(\frac{a}{2.5}\right) = 2.5\,(-8) \qquad \text{Multiply both sides by the denominator 2.5.}$$

$$a = -20 \qquad \text{Find the products.}$$

CHECK

$$\frac{a}{2.5} = -8 \qquad \text{Original equation.}$$

$$\frac{-20}{2.5} \overset{?}{=} -8 \qquad \text{Substitute } -20 \text{ for } a.$$

$$-8 = -8 \qquad \text{Find the quotient.}$$

⮕ Trial Run

Solve each equation and check.

_____ 1. $\dfrac{x}{3} = -2$ _____ 2. $4 = \dfrac{2x}{7}$

_____ 3. $\dfrac{-x}{2.4} = 5$ _____ 4. $\dfrac{x}{6} = 0$

Answers are on page 184.

We have discovered that in finding the solution of an equation we are permitted to

1. Add the same quantity to both sides.
2. Subtract the same quantity from both sides.
3. Multiply both sides by the same quantity (not zero).
4. Divide both sides by the same quantity (not zero).

We can use these properties to "un-do" an operation that is being performed on the variable in order to isolate that variable.

To "Un-do"	We Use	Example	Method
Addition	Subtraction	$x + 3 = 6$	Subtract 3
Subtraction	Addition	$x - 3 = 6$	Add 3
Multiplication	Division	$3x = 6$	Divide by 3
Division	Multiplication	$\dfrac{x}{3} = 6$	Multiply by 3

▶ **Examples You Completed** ────────────────────

Example 2. Is 3 a solution for
$2x + 1 = 10 - x$?

Solution. Substituting 3 for x, we have

$$2(3) + 1 \overset{?}{=} 10 - 3$$
$$6 + 1 \overset{?}{=} 7$$
$$7 = 7$$

So 3 is a solution.

Example 4. Is -5 a solution for
$\dfrac{2(y + 2)}{3} = -2$?

Solution. Substituting -5 for y, we have

$$\frac{2[(-5) + 2]}{3} \overset{?}{=} -2$$
$$\frac{2(-3)}{3} \overset{?}{=} -2$$
$$2(-1) \overset{?}{=} -2$$
$$-2 = -2$$

So -5 is a solution.

Example 6. Solve $x - 15.2 = 100$ and check.

Solution

$x - 15.2 = 100$	
$x - 15.2 + 15.2 = 100 + 15.2$	Add 15.2 to both sides.
$x + 0 = 115.2$	Simplify the sums.
$x = 115.2$	Simplify (addition property of 0).

CHECK

$x - 15.2 = 100$	Original equation.
$115.2 - 15.2 \overset{?}{=} 100$	Substitute 115.2 for x.
$100 = 100$	Simplify.

Example 8. Solve $y + 7 = -2$ and check.

Solution

$$y + 7 = -2$$
$$y + 7 - 7 = -2 - 7$$
$$y + 0 = -9$$
$$y = -9$$

CHECK
$$y + 7 = -2$$
$$-9 + 7 \overset{?}{=} -2$$
$$-2 = -2$$

Example 10. Solve $-24 = -\dfrac{12}{5}x$ and check.

Solution

$$-24 = -\frac{12}{5}x$$

$$-\frac{5}{12}\left(-\frac{24}{1}\right) = -\frac{5}{12}\left(-\frac{12}{5}x\right)$$
Multiply both sides by $-\dfrac{5}{12}$.

$$\frac{-5}{\cancel{12}_1}\cdot\frac{\cancel{-24}^{-2}}{1} = \left(\frac{\cancel{-5}^{-1}}{\cancel{12}_1}\cdot\frac{\cancel{-12}^{-1}}{\cancel{5}_1}\right)x$$
Rewrite right side (associative property)

$$10 = 1x$$
Simplify the products.

$$10 = x$$
Simplify (multiplication property of 1).

CHECK

$$-24 = -\frac{12}{5}x \qquad \text{Original equation.}$$

$$-24 \stackrel{?}{=} -\frac{12}{5}(10) \qquad \text{Substitute 10 for } x.$$

$$-24 = -24 \qquad \text{Find the product.}$$

Answers to Trial Runs

page 175   1. Yes   2. Yes   3. No   4. Yes   5. Yes

page 178   1. $x = 5$ 2. $y = 10$ 3. $m = 0$ 4. $a = -9$ 5. $x = 4$ 6. $x = 11$

page 180   1. $x = 6$ 2. $x = -5$ 3. $x = 0$ 4. $x = -7$ 5. $a = 18$ 6. $x = -6$

page 182   1. $x = -6$ 2. $x = 14$ 3. $x = -12$ 4. $x = 0$

CHECKUP 4.1

_____ **1.** Is -2 a solution for $7 - 3x = 13$?

Solve the equations.

_____ **2.** $x + 9 = 8$ _____ **3.** $x - 9 = -2$ _____ **4.** $1.4 = x + 0.2$

_____ **5.** $x + 11 = 11$ _____ **6.** $6x = 54$ _____ **7.** $42 = -\dfrac{7}{2}x$

_____ **8.** $4x = 0$ _____ **9.** $\dfrac{1}{8}x = 3$ _____ **10.** $-10 = \dfrac{x}{6}$

Check your answers in the back of your book.

If You Missed Problems:	You Should Review Examples:
1	1–4
2–5	5–8
6–8	9–11
9, 10	12–14

4.2 Solving Equations Using Several Properties

Many equations involve more than one operation. Equations such as

$$3x + 1 = 7 \quad \text{and} \quad -7 - \frac{x}{6} = -2$$

require more than one step to isolate the variable.

Undoing Two Operations

In the equation $3x + 1 = 7$, we must undo the operations of addition (get rid of the 1) and multiplication (get rid of the 3). Let's start by subtracting 1 from both sides of the equation.

$$3x + 1 = 7$$
$$3x + 1 - 1 = 7 - 1$$
$$3x = 6$$

Now we can divide both sides by 3.

$$\frac{3x}{3} = \frac{6}{3}$$
$$x = 2$$

Again we can check this solution.

$$3x + 1 = 7$$
$$3(2) + 1 \overset{?}{=} 7$$
$$6 + 1 \overset{?}{=} 7$$
$$7 = 7$$

When solving equations with more than one operation, we follow this general procedure.

> If there are no parentheses in the equation, first perform the necessary addition or subtraction, and then perform the necessary multiplication or division.

Example 1. Solve $-5x + 1.3 = 7.8$ and check.

Solution

$$-5x + 1.3 = 7.8$$

$$-5x + 1.3 - 1.3 = 7.8 - 1.3 \qquad \text{Subtract } 1.3 \text{ (to undo addition).}$$

$$-5x = 6.5 \qquad \text{Find the sums.}$$

$$\frac{-5x}{-5} = \frac{6.5}{-5} \qquad \text{Divide by } -5 \text{ (to undo multiplication).}$$

$$x = -1.3 \qquad \text{Find the quotients.}$$

CHECK

$$-5x + 1.3 = 7.8 \qquad \text{Original equation.}$$

$$-5(-1.3) + 1.3 \overset{?}{=} 7.8 \qquad \text{Substitute } -1.3 \text{ for } x.$$

$$6.5 + 1.3 \overset{?}{=} 7.8 \qquad \text{Remove parentheses.}$$

$$7.8 = 7.8 \qquad \text{Find the sum.}$$

Example 2. Solve $-7 = \dfrac{x}{6} - 2$.

Solution

$$-7 = \frac{x}{6} - 2$$

$$-7 + 2 = \frac{x}{6} - 2 + 2 \qquad\qquad \text{Add } 2 \text{ (to undo subtraction).}$$

$$-5 = \frac{x}{6} \qquad\qquad\qquad\qquad \text{Find the sums.}$$

$$6(-5) = \frac{6}{1}\left(\frac{x}{6}\right) \qquad\qquad \text{Multiply by } 6 \text{ (to undo division).}$$

$$-30 = x \qquad\qquad\qquad\qquad \text{Find the products.}$$

Example 3. Solve $\dfrac{2x}{3} - 1 = 5$.

Solution. We show two methods for solving this equation.

METHOD 1

$$\frac{2x}{3} - 1 = 5$$

$$\frac{2x}{3} - 1 + 1 = 5 + 1 \qquad\qquad \text{Add } 1 \text{ (to undo subtraction).}$$

$$\frac{2x}{3} = 6 \qquad\qquad\qquad\qquad \text{Find the sums.}$$

$$\frac{2}{3}x = 6 \qquad\qquad\qquad\qquad \text{Rewrite the variable term.}$$

$$\frac{3}{2}\left(\frac{2}{3}x\right) = \frac{3}{2}\left(\frac{6}{1}\right) \qquad \text{Multiply by } \frac{3}{2} \text{ (reciprocal of variable coefficient).}$$

$$x = 9 \qquad\qquad\qquad\qquad \text{Find the products.}$$

METHOD 2

$$\frac{2x}{3} - 1 = 5$$

$$\frac{2x}{3} - 1 + 1 = 5 + 1 \qquad\qquad \text{Add } 1.$$

$$\frac{2x}{3} = 6 \qquad\qquad\qquad\qquad \text{Find the sums.}$$

$$\frac{3}{1}\left(\frac{2x}{3}\right) = 3\,(6) \qquad\qquad \text{Multiply by } 3.$$

$$2x = 18 \qquad\qquad\qquad\qquad \text{Find the products.}$$

$$\frac{2x}{2} = \frac{18}{2} \qquad\qquad\qquad\qquad \text{Divide by } 2.$$

$$x = 9 \qquad\qquad\qquad\qquad \text{Find the quotients.}$$

Notice that both methods result in the same solution. Either method is acceptable. In our examples we will usually use Method 1.

⫸ Trial Run

Solve each equation and check.

_____ 1. $2x - 0.4 = 1.2$ _____ 2. $-3y + 9 = 27$

_____ 3. $\dfrac{x}{5} + 6 = 8$ _____ 4. $5 - \dfrac{a}{2} = -1$

_____ 5. $8 = -3m + 2$ _____ 6. $3 + \dfrac{3x}{4} = 14$

Answers are on page 198.

Working with Symbols of Grouping in Equations

Parentheses and other symbols of grouping provide us with important directions. As before, we shall deal with symbols of grouping first, then combine like terms, and then solve the equation.

$$2(x + 3) + (x + 5) = 17$$

$2x + 6 + x + 5 = 17$	Remove parentheses (distributive property).
$3x + 11 = 17$	Combine like terms.
$3x + 11 - 11 = 17 - 11$	Subtract 11.
$3x = 6$	Simplify sums.
$\dfrac{3x}{3} = \dfrac{6}{3}$	Divide by 3.
$x = 2$	Simplify quotients.

We shall check our solution by substituting 2 for x in the *original equation*.

$$2(x + 3) + (x + 5) = 17$$
$$2(2 + 3) + (2 + 5) \stackrel{?}{=} 17$$
$$2(5) + (7) \stackrel{?}{=} 17$$
$$10 + 7 \stackrel{?}{=} 17$$
$$17 = 17$$

When solving equations involving more than one operation and symbols of grouping, we should follow these steps.

1. Remove symbols of grouping.
2. Combine like terms.
3. Perform necessary addition or subtraction to move constant terms to one side, keeping the variable term on the other side.
4. Perform necessary multiplication or division to isolate the variable.
5. Check the solution in the original equation.

Example 4. Solve $2(x + 7) - 9 = 31$ and check.

Solution

		CHECK
$2(x + 7) - 9 = 31$		$2(x + 7) - 9 = 31$
$2x + 14 - 9 = 31$	Remove parentheses.	$2(13 + 7) - 9 \stackrel{?}{=} 31$
$2x + 5 = 31$	Simplify sum.	$2(20) - 9 \stackrel{?}{=} 31$
$2x + 5 - 5 = 31 - 5$	Subtract 5.	$40 - 9 \stackrel{?}{=} 31$
$2x = 26$	Simplify sums.	$31 = 31$
$\dfrac{2x}{2} = \dfrac{26}{2}$	Divide by 2.	
$x = 13$	Simplify quotients.	

Complete the work in Examples 5 and 6.

Example 5. Solve
$1 = 3(2x - 7) - 5(3x + 1)$

Solution

$1 = 3(2x - 7) - 5(3x + 1)$

$1 = 6x - 21 - 15x - 5$

Check your work on page 197. ▶

Example 6. Solve $3[x + 2(x - 7)] = -60$.

Solution

$3[x + 2(x - 7)] = -60$

$3[x + 2x - 14] = -60$

$3[3x - 14] = -60$

Check your work on page 197. ▶

⫸ Trial Run

Solve each equation.

_____ **1.** $3(x - 2) + 7 = 22$ _____ **2.** $2(x - 6) - (x + 8) = -12$

_____ **3.** $11 = 3(3x - 2) - 5(x - 1)$ _____ **4.** $4[2x + 3(x - 2)] = -4$

Answers are on page 198.

Solving Equations with Variables on Both Sides

In an equation such as

$$3x + 1 = x + 5$$

we must find some way to *get the variable terms together on one side of the equation and the constant terms together on the other side*. We shall do this by addition and subtraction.

$3x + 1 = x + 5$	
$3x - x + 1 = x - x + 5$	Subtract x from both sides.
$2x + 1 = 5$	Combine like terms.
$2x + 1 - 1 = 5 - 1$	Subtract 1.
$2x = 4$	Simplify sums.
$\dfrac{2x}{2} = \dfrac{4}{2}$	Divide by 2.
$x = 2$	Simplify.

We check our solution by substituting 2 for x in the original equation.

$$3x + 1 = x + 5$$
$$3(2) + 1 \stackrel{?}{=} 2 + 5$$
$$6 + 1 \stackrel{?}{=} 7$$
$$7 = 7$$

Example 7. Solve $0.7x - 0.3 = 3.3 - 0.2x$.

Solution

$$0.7x - 0.3 = 3.3 - 0.2x$$
$$0.7x + 0.2x - 0.3 = 3.3 - 0.2x + 0.2x$$
$$0.9x - 0.3 = 3.3$$
$$0.9x - 0.3 + 0.3 = 3.3 + 0.3$$
$$0.9x = 3.6$$
$$\frac{0.9x}{0.9} = \frac{3.6}{0.9}$$
$$x = 4$$

Example 8. Solve $5x + 6 = 12x + 41$.

Solution

$$5x + 6 = 12x + 41$$
$$5x - 5x + 6 = 12x - 5x + 41$$
$$6 = 7x + 41$$
$$6 - 41 = 7x + 41 - 41$$
$$-35 = 7x$$
$$\frac{-35}{7} = \frac{7x}{7}$$
$$-5 = x$$

From Examples 7 and 8, we notice that we may move the variable terms to *either* side of the equation. *You* must decide on which side you wish the variable terms to appear, and then move the constants to the *other* side.

⟹ Trial Run

Solve each equation.

_____ **1.** $4x - 3 = x + 6$ _____ **2.** $3x + 9 = 8x + 6$

_____ **3.** $-3x + 7 = 2x - 18$ _____ **4.** $9 + 2x = -x - 3$

Answers are on page 198.

Equations with variables on both sides that also contain symbols of grouping are solved in the same basic way.

Example 9. Solve $2(x + 3) + 5(x + 1) = 3x - 9$ and check.

Solution

$$2(x + 3) + 5(x + 1) = 3x - 9$$

$2x + 6 + 5x + 5 = 3x - 9$	Remove parentheses.
$7x + 11 = 3x - 9$	Combine like terms.
$7x + 11 - 3x = 3x - 9 - 3x$	Subtract $3x$ from both sides.
$4x + 11 = -9$	Combine like terms.
$4x + 11 - 11 = -9 - 11$	Subtract 11 from both sides.
$4x = -20$	Simplify sums.
$\dfrac{4x}{4} = \dfrac{-20}{4}$	Divide both sides by 4.
$x = -5$	

CHECK

$$2(x + 3) + 5(x + 1) = 3x - 9$$
$$2(-5 + 3) + 5(-5 + 1) \overset{?}{=} 3(-5) - 9$$
$$2(-2) + 5(-4) \overset{?}{=} -15 - 9$$
$$-4 - 20 \overset{?}{=} -24$$
$$-24 = -24$$

Now you complete the work for Examples 10 and 11.

Example 10. Solve
$5(3 - 2x) + 4 = 3(4 - x)$

Solution

$$5(3 - 2x) + 4 = 3(4 - x)$$
$$15 - 10x + 4 = 12 - 3x$$
$$19 - 10x = 12 - 3x$$

Check your work on page 198. ▶

Example 11. Solve
$2[x + 3(5 - x)] = 7(x + 1) - 21$

Solution .

$$2[x + 3(5 - x)] = 7(x + 1) - 21$$
$$2[x + 15 - 3x] = 7x + 7 - 21$$
$$2[15 - 2x] = 7x - 14$$

Check your work on page 198. ▶

⮕ Trial Run

Solve each equation.

_____ **1.** $2(x + 3) + 5(x - 1) = 3x - 7$

_____ **2.** $4(2x - 3) + 2 = 2(5 - x)$

_____ **3.** $3 - 2(4 - 5x) = 4(2x - 1) + 7$

_____ **4.** $2[y + 5(y - 3)] = 4(y - 3) - 10$

Answers are on page 198.

Let's summarize the general method for solving first-degree equations.

> **Steps for Solving First-Degree Equations**
> **1.** Remove symbols of grouping.
> **2.** Combine like terms on each side.
> **3.** Move all variable terms to one side.
> **4.** Move all constant terms to the other side.
> **5.** Isolate the variable to obtain the solution.
> **6.** Check your solution in the original equation.

Solving an Equation for a Certain Variable

Often an equation or formula contains more than one variable. To ''solve'' such an equation, we must be told which variable is to be isolated. Then we can solve for that specified variable by the methods we have already learned.

Example 12. The perimeter P of a rectangle with length l and width w is given by the formula $P = 2l + 2w$. Solve this formula for l.

Solution. We wish to isolate l.

$$P = 2l + 2w$$

$$P - 2w = 2l + 2w - 2w \qquad \text{Subtract } 2w \text{ from both sides (to isolate } l\text{-term).}$$

$$P - 2w = 2l \qquad \text{Combine like terms.}$$

$$\frac{P - 2w}{2} = \frac{2l}{2} \qquad \text{Divide both sides by 2 (to isolate } l\text{).}$$

$$\frac{P - 2w}{2} = l \qquad \text{Simplify the quotient on the right.}$$

Example 13. If P dollars are deposited in a savings account paying a simple annual interest rate of r percent, the total amount A after t years is given by the formula $A = P(1 + rt)$. Solve this formula for t.

Solution. We wish to isolate t.

$$A = P(1 + rt)$$

$$A = P + Prt \qquad \text{Remove parentheses.}$$

$$A - P = P - P + Prt \qquad \text{Subtract } P \text{ from both sides (to isolate } t\text{-term).}$$

$$A - P = Prt \qquad \text{Combine like terms.}$$

$$\frac{A - P}{Pr} = \frac{Prt}{Pr} \qquad \text{Divide both sides by } Pr \text{ (to isolate } t\text{).}$$

$$\frac{A - P}{Pr} = t \qquad \text{Simplify the quotient on the right.}$$

Example 14. Solve the equation $3x + \frac{1}{4}y = 2$ for y.

Solution. We wish to isolate y.

$$3x + \frac{1}{4}y = 2$$

$$-3x + 3x + \frac{1}{4}y = -3x + 2 \qquad \text{Subtract } 3x \text{ from both sides (to isolate } y\text{-term).}$$

$$\frac{1}{4}y = -3x + 2 \qquad \text{Simplify.}$$

$$\frac{4}{1} \cdot \frac{1}{4}y = 4(-3x + 2) \qquad \begin{array}{l}\text{Multiply both sides by } 4 \\ \text{(the reciprocal of the coefficient of } y\text{).}\end{array}$$

$$y = -12x + 8 \qquad \text{Find the products.}$$

▌▐▶ Trial Run

Solve each equation for the specified variable.

_____ 1. $A = \dfrac{bh}{2}$ for h _____ 2. $xy = 3$ for y

_____ 3. $C = 2\pi r$ for r _____ 4. $2A = h(B + b)$ for B

Answers are on page 198.

▶ Examples You Completed

Example 5. Solve
$1 = 3(2x - 7) - 5(3x + 1)$

Solution

$$1 = 3(2x - 7) - 5(3x + 1)$$
$$1 = 6x - 21 - 15x - 5$$
$$1 = -9x - 26$$
$$1 + 26 = -9x - 26 + 26$$
$$27 = -9x$$
$$\frac{27}{-9} = \frac{-9x}{-9}$$
$$-3 = x$$

Example 6. Solve $3[x + 2(x - 7)] = -60$.

Solution

$$3[x + 2(x - 7)] = -60$$
$$3[x + 2x - 14] = -60$$
$$3[3x - 14] = -60$$
$$9x - 42 = -60$$
$$9x - 42 + 42 = -60 + 42$$
$$9x = -18$$
$$\frac{9x}{9} = \frac{-18}{9}$$
$$x = -2$$

Example 10. Solve
$$5(3 - 2x) + 4 = 3(4 - x)$$

Solution

$$5(3 - 2x) + 4 = 3(4 - x)$$
$$15 - 10x + 4 = 12 - 3x$$
$$19 - 10x = 12 - 3x$$
$$19 - 10x + 10x = 12 - 3x + 10x$$
$$19 = 12 + 7x$$
$$19 - 12 = 12 + 7x - 12$$
$$7 = 7x$$
$$\frac{7}{7} = \frac{7x}{7}$$
$$1 = x$$

Example 11. Solve
$$2[x + 3(5 - x)] = 7(x + 1) - 21$$

Solution

$$2[x + 3(5 - x)] = 7(x + 1) - 21$$
$$2[x + 15 - 3x] = 7x + 7 - 21$$
$$2[15 - 2x] = 7x - 14$$
$$30 - 4x = 7x - 14$$
$$30 - 4x + 4x = 7x - 14 + 4x$$
$$30 = 11x - 14$$
$$30 + 14 = 11x - 14 + 14$$
$$44 = 11x$$
$$\frac{44}{11} = \frac{11x}{11}$$
$$4 = x$$

Answers to Trial Runs

page 191 **1.** $x = 0.8$ **2.** $y = -6$ **3.** $x = 10$ **4.** $a = 12$ **5.** $m = -2$ **6.** $x = \frac{44}{3}$

page 193 **1.** $x = 7$ **2.** $x = 8$ **3.** $x = 3$ **4.** $x = 1$

page 194 **1.** $x = 3$ **2.** $x = \frac{3}{5}$ **3.** $x = 5$ **4.** $x = -4$

page 195 **1.** $x = -2$ **2.** $x = 2$ **3.** $x = 4$ **4.** $y = 1$

page 197 **1.** $h = \frac{2A}{b}$ **2.** $y = \frac{3}{x}$ **3.** $r = \frac{C}{2\pi}$ **4.** $B = \frac{2A - hb}{h}$

EXERCISE SET 4.2

Solve each equation.

_____ 1. $3x + 5 = 26$

_____ 2. $4x + 6 = 26$

_____ 3. $2x - 9 = 11$

_____ 4. $7x - 9 = 5$

_____ 5. $5y + 3 = -22$

_____ 6. $2y + 8 = -12$

_____ 7. $0.8x - 1.5 = -0.7$

_____ 8. $0.6x + 2.3 = 1.1$

_____ 9. $-4a + 9 = 45$

_____ 10. $-5a + 10 = 55$

_____ 11. $13 + 2m = 0$

_____ 12. $20 + 3m = 0$

_____ 13. $-x - 7 = 6$

_____ 14. $-x - 8 = 9$

_____ 15. $7\frac{1}{2} - 2x = 13\frac{1}{2}$

_____ 16. $6\frac{3}{4} - 5x = 41\frac{3}{4}$

_____ 17. $24 = 3x - 15$

_____ 18. $17 = 6x - 7$

_____ 19. $4y + \frac{2}{5} = \frac{2}{5}$

_____ 20. $7y + \frac{1}{9} = \frac{1}{9}$

_____ 21. $-23 = 9 - 4a$

_____ 22. $-21 = 15 - 9a$

_____ 23. $-1.7x + 17 = 0$

_____ 24. $-0.4x + 3.2 = 0$

_____ 25. $5 = \frac{3}{2}x - 4$

_____ 26. $2 = \frac{5}{3}x - 8$

_____ 27. $-9 = \frac{1}{5}x - 9$

_____ 28. $-7 = \frac{1}{8}x - 7$

_____ 29. $8 - \frac{2}{5}x = 12$

_____ 30. $9 - \frac{3}{2}x = 15$

_____ 31. $\frac{x}{6} + 3 = 5$

_____ 32. $\frac{x}{7} + 4 = 9$

_____ 33. $0.7 + \frac{a}{5} = -0.1$

_____ 34. $0.6 + \frac{a}{3} = -0.2$

_____ 35. $13 = \frac{y}{4} + 9$

_____ 36. $17 = \frac{y}{5} + 8$

_____ 37. $6 - \frac{x}{4} = 7$

_____ 38. $9 - \frac{x}{2} = 12$

_____ 39. $(x + 3) + (x - 5) = 8$

_____ 40. $(x + 5) + (x - 6) = 9$

_____ 41. $12 = 4\left(\frac{1}{2}x - 7\right) + 8$

_____ 42. $12 = 6\left(\frac{1}{2}x - 6\right) + 9$

_____ 43. $2(x - 8) - (x + 3) = -22$

_____ 44. $3(x - 4) - (2x + 7) = -26$

_____ 45. $3 = 4(x - 5) - 2(x + 6)$

_____ 46. $2 = 5(x - 5) - 3(x - 2)$

_____ 47. $2[3x + 3(x - 1)] = 6$

_____ 48. $3[4x + 3(x - 2)] = 24$

_____ 49. $5[2(x - 4) - 2] = 0$

_____ 50. $6[3(x - 2) + 12] = 0$

_____ 51. $3x + 2 = -x + 10$

_____ 52. $5x + 3 = -3x + 15$

_____ 53. $6x - 0.5 = 3x + 1.6$

_____ 54. $7x - 0.5 = 4x + 0.7$

_____ 55. $11x - 3 = 5x - 15$

_____ 56. $12x - 9 = 7x - 14$

_____ 57. $3x - 9 = 6 - 4x$

_____ 58. $3x - 9 = 12 - 7x$

_____ 59. $5x + 7 = 9x + 23$

_____ 60. $5x + 5 = 11x + 32$

_____ 61. $x + 3 = -5x - 9$

_____ 62. $7x + 6 = -9x - 4$

_____ 63. $3x - 8 = 5x$

_____ 64. $7x - 15 = 12x$

_____ 65. $7.2x + 8 = 8 - 3.8x$

_____ 66. $9.6x + 12 = 12 - 7.4x$

_____ 67. $2(x + 1) + 3(x - 2) = 2x - 37$

_____ 68. $4(x + 1) + (x - 5) = 3x - 19$

_____ 69. $5(2x - 4) - 2 = 2(7 - x)$

_____ 70. $3(2x - 7) - 7 = 5(1 - x)$

_____ 71. $x - (-3x + 16) = 3(x - 12) + 18$

_____ 72. $-4 - 6x = 2(-3 + 4x) + 4(3 - x)$

_____ 73. $3[2x - 5(x - 1)] = 6 - [3 - (x + 2)]$

_____ 74. $4[x - 3(x - 2)] = 9 - [5 - (x - 1)]$

_____ 75. $2[x + 3(x - 2)] = 5(x - 4) - 1$

_____ 76. $3[2x + (x - 5)] = 2(x - 7) + 6$

Solve each equation for the specified variable.

_____ 77. $C = \pi d$ for d

_____ 78. $A = lw$ for w

_____ 79. $x + 2y = 3$ for y

_____ 80. $2x + 3y = 5$ for x

_____ 81. $I = prt$ for t

_____ 82. $V = lwh$ for h

_____ 83. $P = a + b + c$ for a

_____ 84. $M = x + y + z$ for z

_____ 85. $K = c(1 + 2p)$ for p

_____ 86. $Q = m(1 + 3l)$ for l

_____ 87. $2x - 2b = -5(b - x)$ for x

_____ 88. $6x - 8b = -2(b + x)$ for x

_____ **89.** $l = a + d(n - 1)$ for n _____ **90.** $E = I(R + r)$ for r

_____ **91.** $A = P + Prt$ for r _____ **92.** $P = 2l + 2w$ for w

_____ **93.** $V = \dfrac{1}{3} Bh$ for h _____ **94.** $A = \dfrac{1}{2} bh$ for b

_____ **95.** $3x - 2y = 1$ for x _____ **96.** $2x - 5y = 3$ for y

_____ **97.** $l = a + (n - 1)d$ for d _____ **98.** $m = p + (t - 2)r$ for t

_____ **99.** $2(4x - y) = 3(x + y - 2)$ for x

_____ **100.** $3(7x - y) = 4(x - y + 1)$ for x

☆ Stretching the Topics _____

Solve each equation.

_____ **1.** $2.3 + 0.2x = -0.9 - (9.4x + 1.6)$

_____ **2.** $3(x - 2) = 2(-x - 6) + 3x - (-x - 4)$

_____ **3.** $V = \dfrac{1}{6} h (B + b + 4M)$, for M

Check your answers in the back of your book.

If you can solve the equations in **Checkup 4.2**, you are ready to go on to Section 4.3.

Name _____   Date _____

✓ CHECKUP 4.2

Solve each equation.

_____ **1.** $3x - 7 = 8$

_____ **2.** $0.7 = 0.8x + 1.5$

_____ **3.** $-2y - 5 = 7$

_____ **4.** $8 - \dfrac{x}{3} = 9$

_____ **5.** $\dfrac{3}{2}x - 1 = 5$

_____ **6.** $2(x - 3) - 4(x + 2) = 2$

_____ **7.** $3[2x - (x + 4)] = 6$

_____ **8.** $y + 5 = 7y - 8$

_____ **9.** $3(3y - 1) - 9 = 5(4 + y)$

_____ **10.** $A = P(1 + rt)$ for r

Check your answers in the back of your book.

If You Missed Problems:	You Should Review Examples:
1–4	1, 2
5	3
6, 7	4–6
8	8
9	10
10	14

4.3 Switching from Word Statements to Equations

In Chapter 3 we learned to use variables and constants to switch from words to algebraic expressions. Now we must learn to switch from word statements to equations.

If 1 credit hour at a university costs $25, we know the cost of h hours can be written as $25 \cdot h$ dollars. Suppose that a student has saved $150 for part-time courses. How can we figure out how many credit hours that student can take?

We must solve this equation.

$$25h = 150$$

$$\frac{25h}{25} = \frac{150}{25}$$

$$h = 6$$

The student can take 6 credit hours.

Switching problems from words to equations should be done in an orderly way, using these steps.

1. Read the problem carefully to be sure you understand what you are being asked to find.
2. Use a drawing if it helps you visualize the problem.
3. Write down what you want the variable to stand for.
4. Write down any expressions containing the variable.
5. Write an equation from the information in the problem.
6. Solve the equation. Answer the question in a sentence.
7. Check your solution in the original problem, being sure that your solution makes sense.

Example 1. Gary has $1760 in his savings account now. If he deposits $40 per week in this account, for *how many weeks* must he continue to save if he wants a total of $5000 in his account?

Solution

We Think	*We Write*
We do not know the number of weeks.	Let w = number of weeks
The amount saved in w weeks is found by *multiplying* the weekly savings times the number of weeks.	$40w$ = dollars saved in w weeks
Gary's total amount is the *sum* of his current balance and the amount he will save in w weeks.	$1760 + 40w$ = total dollars saved
Gary wishes his total to be $5000.	$1760 + 40w = 5000$

Now we have an equation to solve by the usual methods.

$$1760 + 40w = 5000$$

$$1760 - 1760 + 40w = 5000 - 1760$$

$$40w = 3240$$

$$\frac{40w}{40} = \frac{3240}{40}$$

$$w = 81$$

Gary must save for 81 weeks. Let's check. If Gary saves $40 each week for 81 weeks, he will save $40(81) = $3240. Since he has already saved $1760, he will have saved a total of $1760 + $3240 = $5000. Our solution is correct.

Example 2. Beverly has 36 yards of fencing to use for enclosing a rectangular garden. If she wishes the garden to be twice as long as it is wide, *what should the garden's dimensions be?*

   Solution

We Think	*We Write*

The amount of fencing is the *perimeter* of the rectangle.

$$P = 2 \text{ lengths} + 2 \text{ widths}$$

We do not know the garden's dimensions, but we do know the length is *twice* the width.

Let x = width

$2x$ = length

Perimeter = 2 lengths + 2 widths.

$$P = 2(2x) + 2 \cdot x$$

Beverly has 36 yards of fencing.

$$36 = 2(2x) + 2x$$

Now we have an equation to solve by the usual methods.

$$2(2x) + 2x = 36$$

$$4x + 2x = 36$$

$$6x = 36$$

$$\frac{6x}{6} = \frac{36}{6}$$

$$x = 6$$

In this problem we are asked to find both dimensions of the garden.

Width: $x = 6$

Length: $2x = 12$

The width is 6 yards and the length is 12 yards.

CHECK

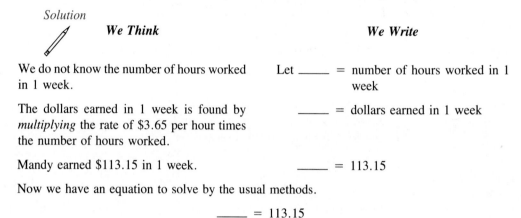

12 yards

6 yards   6 yards

12 yards

FENCING: 6 + 12 + 6 + 12

= 36 yards

Now you try to complete Example 3.

Example 3. If Mandy earns $3.65 per hour, *for how many hours* did she work in 1 week if she earned $113.15?

Solution

We Think	*We Write*
We do not know the number of hours worked in 1 week.	Let _____ = number of hours worked in 1 week
The dollars earned in 1 week is found by *multiplying* the rate of $3.65 per hour times the number of hours worked.	_____ = dollars earned in 1 week
Mandy earned $113.15 in 1 week.	_____ = 113.15

Now we have an equation to solve by the usual methods.

$$\underline{\hspace{1cm}} = 113.15$$

Mandy worked _____ hours.

Check your work on page 208. ▶

Now let's return to the problem stated at the beginning of this chapter and see if we can solve it.

Example 4. Charlotte now earns $6 per hour as a grounds keeper working 35 hours per week. She has been offered a similar job paying $280 per week. *How much of an increase* in her hourly wage must her present employer give Charlotte to meet the better offer?

Solution

We Think	*We Write*
We do not know the increase in Charlotte's hourly wage.	Let x = increase in hourly wage (dollars)
Her new hourly wage is the *sum* of her old wage and the increase.	$6 + x$ = new hourly wage (dollars)
Her new weekly earnings are found by *multiplying* her new wage per hour times the number of hours.	$35(6 + x)$ = new weekly earnings
Her new weekly earnings must *equal* the amount of the better offer.	$35(6 + x) = 280$

Now we have an equation to solve by the usual methods.

$$35(6 + x) = 280$$

$$210 + 35x = 280$$

$$210 - 210 + 35x = 280 - 210$$

$$35x = 70$$

$$\frac{35x}{35} = \frac{70}{35}$$

$$x = 2$$

Charlotte must receive an hourly increase of $2.

CHECK: Charlotte now earns $6 per hour. If her hourly wage is increased by $2, she will earn $8 per hour. If she works 35 hours at $8 per hour, she will earn

$$35(8) = 280 \text{ dollars}$$

Example 5. Three roommates share monthly rent and food expenses equally. If each person's share of the rent is $65, *how much can they spend altogether on food* so that each roommate pays $150 for total monthly expenses?

Solution

We Think	*We Write*
We don't know the total spent on food.	Let F = total spent on food (dollars)
To find each roommate's share of the food cost, we *divide* the total by 3.	$\dfrac{F}{3}$ = each roommate's share of food cost (dollars)
Each roommate also pays $65 for rent.	$\dfrac{F}{3} + 65$ = each roommate's share of expenses (dollars)
Each roommate wishes to pay $150 for total monthly expenses.	$\dfrac{F}{3} + 65 = 150$

Now we have an equation to solve by the usual methods.

$$\frac{F}{3} + 65 = 150$$

$$\frac{F}{3} + 65 - 65 = 150 - 65$$

$$\frac{F}{3} = 85$$

$$3 \cdot \frac{F}{3} = 3(85)$$

$$F = 255$$

They can spend $255 on food.

CHECK: If they spend $255 each month on food, each roommate's share of food expenses is $255 ÷ 3 = $85. Since each roommate also pays $65 for rent, each pays a total each month of $85 + $65 = $150.

Complete Example 6.

Example 6. After a session of Weight Worriers, John announced "I lost 3 pounds more than twice as much as Carla lost this month. I lost 17 pounds." *How many pounds* did Carla lose?

Solution

We Think	**We Write**
We don't know Carla's loss.	Let x = number of pounds Carla lost
John lost 3 pounds *more* than *twice* Carla's loss.	_____ = twice Carla's loss (pounds)
	_____ + _____ = John's loss (pounds)
We know John lost 17 pounds.	_____ + _____ = 17
And we have an equation to solve by the usual methods.	

$$\text{_____} = 17$$

Carla lost _____ pounds.

CHECK

Check your solution on page 208. ▶

Example 7. In 1990, the enrollment at State University increased by 3 percent to 12,360 students. *What was the enrollment* in 1989?

Solution

We Think	**We Write**
We don't know the 1989 enrollment.	Let x = enrollment in 1989
The increase was 3 percent of the 1989 enrollment.	3% of x = increase in enrollment
	$0.03x$ = increase in enrollment
The 1990 enrollment is the *sum* of the 1989 enrollment and the increase.	$x + 0.03x$ = enrollment in 1990
We know the 1990 enrollment was 12,360 students.	$x + 0.03x = 12{,}360$

Now we have an equation to solve by the usual methods.

$$x + 0.03x = 12,360$$

$$1x + 0.03x = 12,360 \qquad \text{Rewrite } x \text{ as } 1x.$$

$$1.03x = 12,360 \qquad \text{Combine like terms.}$$

$$\frac{1.03x}{1.03} = \frac{12,360}{1.03} \qquad \text{Divide both sides by } 1.03.$$

$$x = 12,000$$

The 1989 enrollment was 12,000 students.

▶ Examples You Completed

Example 3. If Mandy earns $3.65 per hour, *for how many hours* did she work in 1 week if she earned $113.15?

Solution

We Think	We Write
We do not know the number of hours worked in 1 week.	Let h = number of hours worked in 1 week
The dollars earned in 1 week is found by *multiplying* the rate of $3.65 per hour times the number of hours worked.	$3.65h$ = dollars earned in 1 week
Mandy earned $113.15 in 1 week.	$3.65h = 113.15$

Now we have an equation to solve by the usual methods.

$$3.65h = 113.15$$

$$\frac{3.65h}{3.65} = \frac{113.15}{3.65}$$

$$h = 31$$

Mandy worked 31 hours.

Example 6. John lost 3 pounds more than twice as much as Carla lost. John lost 17 pounds. How many pounds did Carla lose?

Solution

We Think	We Write
We don't know Carla's loss.	Let x = number of pounds Carla lost
John lost 3 pounds *more* than *twice* Carla's loss.	$2x$ = twice Carla's loss (pounds)
	$2x + 3$ = John's loss (pounds)
We know John lost 17 pounds.	$2x + 3 = 17$

And we have an equation to solve by the usual methods.

$$2x + 3 = 17$$

$$2x + 3 - 3 = 17 - 3$$

$$2x = 14$$

$$\frac{2x}{2} = \frac{14}{2}$$

$$x = 7$$

Carla lost 7 pounds.

CHECK: Carla lost 7 pounds. Twice what Carla lost is 14 pounds, and 3 pounds more than twice Carla's loss is 14 + 3 or 17 pounds. John lost 17 pounds.

EXERCISE SET 4.3

Use the given information to write an equation. Then solve and check.

_____ 1. The Waller twins' combined weight at birth was 12 pounds. If Lyle weighed 2 more pounds than Lydia, how much did each weigh?

_____ 2. If Carl finds he can save $12 a week, how many weeks will it take him to save $252?

_____ 3. If Henry drives at a rate of 55 miles per hour, how long will it take him to travel 495 miles?

_____ 4. Cheri bought a shirt and a blazer. The cost of the blazer was $15 less than 3 times the cost of the shirt. If the blazer cost $47.97, what was the cost of the shirt?

_____ 5. Julia and Sara are selling tickets for a concert to benefit the Special Olympics. Sara has sold 4 times as many tickets as Julia and together they have sold 75 tickets. How many tickets has each sold?

_____ 6. Carol is building a bookcase that will have to hold 162 books. How many shelves will the bookcase have if each shelf will hold 18 books?

_____ 7. The area of a rectangle is the product of the length and the width. Find the width of a rectangle with a length of 15 meters and an area of 195 square meters.

_____ 8. The restaurant manager tells the chairperson of the banquet committee that she will charge $200 for the use of the room plus $8.75 per person for the dinner. If the total cost of the banquet is $900, how many persons will be attending?

_____ 9. At the Bargain Barn Restaurant, a hamburger costs $0.15 less than twice the cost of french fries. If a hamburger costs $0.95, what is the cost of the fries?

_____ 10. The width of a room is 5 feet more than one-half the length. If a paperhanger needs 70 feet of ceiling border, what are the dimensions of the room?

_____ 11. Four neighbors share equally in the expenses for maintaining a nearby empty lot. If each neighbor's share of the mowing is $17 per month, how much can they spend on shrubs and flowers so that each neighbor's share is $26 per month?

_____ 12. Mickey sells garden tillers. He earns $75 for each tiller he sells. How many must he sell each month so that he will have a monthly salary of at least $1400?

_____ 13. If Ella loses 16 pounds, 9 pounds, and 11 pounds during the first 3 months of her diet, how much will she have to lose during the fourth month so that she will have an average weight loss of 12 pounds per month?

_____ 14. Tyrone is buying a bedroom suite on the installment plan. He paid $150 as a down payment and is paying $55.75 a month until it is paid for. If the bedroom suite costs $819 (including finance charges), for how many months will he be making payments?

_____ 15. If Maxine has 134 inches of framing, what will be the dimensions of the largest picture frame she can make if she wants the length to be 5 inches more than the width?

_____ 16. Bernard is saving money for a vacation during spring break. He deposits $50 and then plans to save a certain amount each week for the next 8 weeks. If he estimates he will need $290 for his vacation, how much must he save each week?

_____ 17. Sidney spends 30 percent of his monthly salary for rent and 25 percent for food. If the total for these two expenditures is $660, find Sidney's monthly salary.

_____ 18. Mr. McDonald has 120 feet of fencing to build a loading pen for cattle. If the pen must be 3 times as long as it is wide, find the width of the pen.

_____ 19. Lucindy is using her savings to attend an exercise salon twice a week to improve her figure. She must pay $6 each time she attends. How many weeks can she attend if she has only $792 in her savings account?

_____ 20. If Ira has scored 80, 62, 75, and 69 on his first 4 mathematics tests, what score must he earn on his fifth test to have a test average of 70?

_____ 21. The Miller family decided to save $247.50 monthly. This represented $12\frac{1}{2}$ percent of the family's income per month. What was the family's monthly income?

_____ 22. Each consumer receives a discount of 5 percent if gas bills are paid within 10 days after billing. If Mr. Bryant saved $3.29 by paying promptly, what was the original amount of his gas bill?

_____ 23. A business firm increased its capital by 25 percent in 1 year. If the firm's capital this year was $525,000, what was its capital last year?

_____ 24. A manufacturing company has decreased its inventory by 8 percent. If they now have 2070 items in stock, how many items did they have before the decrease?

_____ 25. Lynn is selling a china cabinet he built in a shop class. He wishes to make a profit of 40 percent of the cost. If he sells the china cabinet for $581, what was the cost of building the cabinet?

☆ Stretching the Topics _____

_____ 1. A board 50 inches in length is to be sawed into 4 pieces so that each piece will be 1 inch shorter than the preceding piece. Find the length of each piece.

_____ 2. A business firm increased its capital by 25 percent and then invested 40 percent of the total capital in real estate. The amount of capital remaining on hand was $32,250. What was the original capital?

Check your answers in the back of your book.
If you can do the problems in **Checkup 4.3**, you are ready to do the **Review Exercises** for Chapter 4.

✓ CHECKUP 4.3

Use the given information to write an equation. Then solve and check.

_____ 1. Maurice buys a stereo set for $500. He makes a $150 down payment and pays the balance by making payments of $25 a week. For how many weeks will he be making payments?

_____ 2. Jo Ann is buying framing for a rectangular picture that is 3 times as long as it is wide. The salesperson tells her that she needs 72 inches of framing. What are the dimensions of the frame?

_____ 3. Tickets to Opryland are $14.25 each. How many members of Pi Mu Epsilon can go if the organization has $342?

_____ 4. In physical fitness class, Ramon did 25 more than one-half as many push-ups as his father. If Ramon did 65 push-ups, how many did his father do?

_____ 5. In 1990, the corn production on Mr. Kurtz's farm increased by 7 percent, to 192.6 bushels per acre. What was the production per acre in 1989?

Check your answers in the back of your book.

If You Missed Problem:	You Should Review Example:
1	1
2	2
3	3
4	6
5	7

Summary

In this chapter we learned that **equations** are mathematical sentences which state that one quantity represents the same number as another quantity. The **solution** for an equation containing a variable is the value of the variable that makes the original equation a true statement.

To solve a first-degree equation we learned to **isolate the variable** using the following steps.

1. Remove symbols of grouping.
2. Combine like terms on each side.
3. Perform additions and/or subtractions needed to move variable terms to one side and constant terms to the other side.
4. Perform multiplications and/or divisions needed to isolate the variable.
5. Check the solution by substituting it into the original equation.

We discovered that these same steps can be used to solve for a specified variable in an equation containing several variables. Then we practiced switching from word statements to equations containing variables, using an orderly approach.

❑ Speaking the Language of Algebra

Complete each statement with the appropriate word or phrase.

1. An equation is a _____ which states that one quantity represents the same number as another quantity.

2. A value of the variable that makes an equation true is called a _____ for the equation, and we say that the value _____ the equation.

3. In solving a first-degree equation, our goal is to _____ the variable.

4. To check a solution, we must _____ it for the variable in the _____ equation.

Check your answers in the back of your book.

△ Writing About Mathematics

Write your response to each question in complete sentences.

1. Explain how you can decide whether a given number is a solution for a given equation.

2. Show and explain the steps you would use to solve the equation

$$3(x - 1) - (2x + 3) = 2x + 1$$

3. Make up a word problem that would lead you to write the equation

$$\$6x + \$7(2x + 1) = \$127$$

REVIEW EXERCISES for Chapter 4

_____ 1. Is 6 a solution for $19 - 2x = 7$?

_____ 2. Is 0.5 a solution for $\dfrac{y}{5} - 3 = -4.5$?

_____ 3. Is 1 a solution for $7(2a - 3) = 7$?

_____ 4. Is 10 a solution for $3m - 9 = m + 11$?

Solve each equation and check.

_____ 5. $x - 7 = 3$ _____ 6. $5 + y = -9$

_____ 7. $-3 = m - 7$ _____ 8. $a + \dfrac{3}{2} = \dfrac{7}{2}$

_____ 9. $0.9x = 7.2$ _____ 10. $2x = -36$

_____ 11. $-4x = 28$ _____ 12. $5x = 0$

_____ 13. $\dfrac{x}{3} = -1$ _____ 14. $2 = \dfrac{y}{-7}$

_____ 15. $4x - 7 = 31$ _____ 16. $8 - 2x = 6$

_____ 17. $-15 = 12 + 9x$ _____ 18. $\dfrac{x}{4} - 2 = -1$

_____ 19. $-\dfrac{7}{3}x = 14$ _____ 20. $9 = \dfrac{5}{2}x - 1$

_____ 21. $8 - \dfrac{3x}{4} = 5$ _____ 22. $0.2(x - 4) - (0.5x + 0.2) = 2$

_____ 23. $8 = -2[3(y - 4) - 7]$ _____ 24. $7x - 5 = 3x + 12$

_____ 25. $4a - 7 = 3 - a$ _____ 26. $-4x - 9 = 10 - 3x$

_____ 27. $18 + 9x = -2x - 15$ _____ 28. $2(x + 2) + 6 = 5(2 + x)$

_____ 29. $3(x - 7) - (x + 1) = 0$ _____ 30. $P = 2a + c$, for a

Use an equation to solve each problem.

_____ 31. Dawn has 5 times as many cassettes as Eddie. Together they have 54 cassettes. How many cassettes does Dawn have?

_____ 32. The length of a rectangle is 3 times the width. If the perimeter of the rectangle is 32 feet, find its dimensions.

_____ 33. Raymond earns $40 a day plus $7.35 an hour for each hour he works overtime. If Raymond earned $69.40 last Tuesday, how many hours of overtime did he work?

Check your answers in the back of your book.

If You Missed Exercises:	You Should Review Examples:	
1–4	SECTION 4.1	1–4
5–8		5–8
9–12		9
13, 14		12, 13
15–21	SECTION 4.2	1–3
22, 23		4–6
24–27		7, 8
28, 29		9–11
30		13
31–33	SECTION 4.3	1–6

If you have completed the **Review Exercises** and corrected your errors, you are ready to take the **Practice Test** for Chapter 4.

PRACTICE TEST for Chapter 4

		SECTION	EXAMPLE
_____	**1.** Is 2 a solution for $7a - 2 = 2(a + 4)$?	4.1	3
_____	**2.** Is 1 a solution for $2 - 7x = 6x - 1$?	4.1	2

Solve each equation.

		SECTION	EXAMPLE
_____	**3.** $y - 8 = 5$	4.1	6
_____	**4.** $0.3x = 3.9$	4.1	9
_____	**5.** $\dfrac{x}{-5} = 2$	4.1	13
_____	**6.** $6x + 7 = 31$	4.2	1
_____	**7.** $\dfrac{5}{4}x - 3 = 7$	4.2	3
_____	**8.** $4(x - 3) - (6x + 1) = 1$	4.2	5
_____	**9.** $11 - x = 3x - 7$	4.2	8
_____	**10.** $5(x - 1) + 6 = 3(4 - 2x)$	4.2	10
_____	**11.** $8\left(\dfrac{1}{2}x - 2\right) - (5x - 1) = 0$	4.2	9
_____	**12.** Solve $2x + \dfrac{1}{5}y = 3$ for y.	4.2	15

Use an equation to solve each problem.

		SECTION	EXAMPLE
_____	**13.** Martha sold 4 times as many tickets to the Chili Supper as Harvey sold. Together they sold 185 tickets. How many tickets did each person sell?	4.3	6
_____	**14.** The width of a rectangular photograph is 1 inch more than twice the length. If the perimeter of the photograph is 14 inches, find the dimensions of the photograph.	4.3	2
_____	**15.** A limousine rental agency charges customers $50 per day plus $2 per mile driven. If Adolph was charged $216 for 1 day's rental of a limousine, how far did he drive?	4.3	1

Check your answers in the back of your book.

SHARPENING YOUR SKILLS after Chapters 1–4

Perform the indicated operations.

SECTION

_____ **1.** $(-25.8) \div (0.3)$ 2.4

_____ **2.** $\dfrac{28}{15} \cdot \dfrac{6}{21}$ 1.3

_____ **3.** $-8\left[\dfrac{-15}{3} + 2(7 - 11)\right]$ 2.3

_____ **4.** $\dfrac{5}{8} - \left(-\dfrac{1}{2} - \dfrac{3}{4}\right)$ 2.4

_____ **5.** $-3.92 - 7.45$ 2.4

_____ **6.** $|3 - 6| + |-15 + 2|$ 2.1

_____ **7.** $(-3)^3 - 4(-5)$ 2.3

_____ **8.** $3x - 7y - 5x + 4y$ 3.2

_____ **9.** $-5(2x) - (-4x)$ 3.3

_____ **10.** $4[-2(x - y) + 5(x + y)]$ 3.3

_____ **11.** Evaluate $7(3 - a)$ when $a = -4$.   3.1

_____ **12.** Identify the numerical coefficient of y in the expression $5x - 3y + z$.   3.2

_____ **13.** Compare $\dfrac{2}{3}$ and $\dfrac{3}{4}$ using $<$ or $>$.   2.1

_____ **14.** Evaluate $\dfrac{3m - 2n}{4}$ when $m = -8$ and $n = 8$.   3.1

_____ **15.** A building contractor must pay a penalty of \$500 a day for each day over the deadline for completion of a building. Write an algebraic expression for the amount of penalty after n days. Find the amount of penalty if the building is completed 12 days late.   3.1

Check your answers in the back of your book.

5

Working with Exponents

Mary has a square mirror measuring *s* inches on each side. After framing, each side is $1\frac{1}{4}$ times as long as it was. Write an algebraic expression for the *area* of the framed mirror.

In Chapter 1 we learned to use exponents to show that a number was repeated as a factor in a product. Now we turn our attention to variable expressions involving exponents. In this chapter we

1. Define whole-number exponents.
2. Use the laws of exponents.
3. Work with negative exponents.

5.1 Multiplying with Whole-Number Powers

Using Exponents

Sometimes it is necessary to deal with products in which the same factor is repeated. Mathematicians have invented notation to handle such products.

Product	Notation	Read as
$3 \cdot 3$	3^2	3 squared
$x \cdot x$	x^2	x squared
$a \cdot a \cdot a$	a^3	a cubed
$b \cdot b \cdot b \cdot b \cdot b$	b^5	b to the fifth
$x \cdot x \cdot y \cdot y \cdot y \cdot y$	x^2y^4	x squared times y to the fourth

Such expressions (sometimes called **powers**) always contain two important pieces of information. The small raised positive integer is called the **exponent** and tells us how many times the other number, called the **base**, is to be used as a factor in a product.

$$\overset{\text{exponent}}{x^5} = \underbrace{x \cdot x \cdot x \cdot x \cdot x}_{\text{5 factors}}$$

base

If n is a positive integer

$$\overset{\text{exponent}}{a^n} = \underbrace{a \cdot a \cdot \ldots \cdot a}_{n \text{ factors}}$$

base

Example 1. Find $\left(\dfrac{1}{4}\right)^3$.

Solution

$$\left(\frac{1}{4}\right)^3 = \frac{1}{4} \cdot \frac{1}{4} \cdot \frac{1}{4}$$

$$= \frac{1}{64}$$

You complete Example 2.

Example 3. Use exponents to rewrite $a \cdot a \cdot a \cdot x \cdot x$.

Solution

$$a \cdot a \cdot a \cdot x \cdot x$$

$$= a^3 x^2$$

Now complete Example 4.

Example 2. Find $(-2)^5$.

Solution

$$(-2)^5$$

$$= (-2)(-2)(-2)(\underline{\quad})(\underline{\quad})$$

$$= \underline{\quad}$$

Check your work on page 229. ▶

Example 4. Write x^4y without exponents.

Solution

$$x^4y$$

$$= \underline{\ } \cdot \underline{\ } \cdot \underline{\ } \cdot \underline{\ } \cdot \underline{\ }$$

Check your work on page 229. ▶

Unless parentheses indicate otherwise, an exponent belongs *only* with the number to which it is attached.

Complete Example 5.

Example 5. Write $2x^3$ without exponents.

Solution

$2x^3 = 2 \cdot \underline{\quad} \cdot \underline{\quad} \cdot \underline{\quad}$

Check your work on page 229. ▶

Example 6. Write $(2x)^3$ without parentheses.

Solution

$$(2x)^3 = (2x)(2x)(2x)$$
$$= 2 \cdot x \cdot 2 \cdot x \cdot 2 \cdot x$$
$$= 2 \cdot 2 \cdot 2 \cdot x \cdot x \cdot x$$
$$= 8x^3$$

In Example 6, notice that we regrouped the factors using the commutative and associative properties for multiplication. Notice also that $2x^3$ and $(2x)^3$ do *not* represent the same number.

Example 7. Write $-x^4$ without exponents.

Solution

$$-x^4$$
$$= -1 \cdot x^4$$
$$= -1 \cdot x \cdot x \cdot x \cdot x$$

Example 8. Write $(-x)^4$ without exponents.

Solution

$$(-x)^4$$
$$= (-1x)^4$$
$$= (-1x)(-1x)(-1x)(-1x)$$
$$= (-1)(-1)(-1)(-1) \cdot x \cdot x \cdot x \cdot x$$
$$= 1 \cdot x \cdot x \cdot x \cdot x$$
$$= x \cdot x \cdot x \cdot x$$

Example 9. Evaluate a^3b^5 when $a = -1$ and $b = 2$.

Solution

$$a^3b^5$$
$$= (-1)^3(2)^5$$
$$= (-1)(-1)(-1)(2)(2)(2)(2)(2)$$
$$= (-1)(32)$$
$$= -32$$

Try to complete Example 10.

Example 10. Evaluate $\left(\frac{1}{2}x^2y\right)^3$ when $x = -1$ and $y = 4$.

Solution

$$\left(\frac{1}{2}x^2y\right)^3 = \left[\frac{1}{2}(-1)^2(4)\right]^3$$
$$= \left[\frac{1}{2} \cdot 1 \cdot 4\right]^3$$
$$= [\underline{\quad}]^3$$
$$= \underline{\quad}$$

Check your work on page 229. ▶

We note here that zero and one can be used as exponents if we follow these rules.

One as an Exponent	**Zero as an Exponent**
$a^1 = a$	$a^0 = 1$ provided that $a \neq 0$

Example 11. Find $(3y)^0$.

Solution

$$(3y)^0 = 1$$

Example 12. Find $3y^0$.

Solution

$$3y^0 = 3 \cdot 1$$
$$= 3$$

Example 13. Find $5x^0y^1$.

Solution

$$5x^0y^1 = 5 \cdot 1 \cdot y$$
$$= 5y$$

Now you try Example 14.

Example 14. Find $4(7xyz)^0$.

Solution

 $$4(7xyz)^0 =$$

Check your work on page 229. ▶

⫸ Trial Run

 _____ 1. Use exponents to rewrite $2 \cdot a \cdot a \cdot a \cdot b \cdot b$.

_____ 2. Write $-y^2$ without exponents.

_____ 3. Write $(3x)^3$ without parentheses.

_____ 4. Evaluate $(-2)^4$.

_____ 5. Evaluate $(-3x)^2$ when $x = 2$.

_____ 6. Evaluate $-3x^2$ when $x = 2$.

_____ 7. Evaluate $(2x^5y)^3$ when $x = -1$ and $y = 3$.

_____ 8. Find $3x^1y^0$.

Answers are on page 230.

Multiplying Powers

Suppose that we consider some *products* of powers of the same base.

$$3^2 \cdot 3^4 = 3 \cdot 3 \cdot 3 \cdot 3 \cdot 3 \cdot 3 \quad = 3^6$$

$$x^3 \cdot x = x \cdot x \cdot x \cdot x \quad\quad\quad = x^4$$

$$y^4 \cdot y^3 = y \cdot y \cdot y \cdot y \cdot y \cdot y \cdot y = y^7$$

$$a^2 \cdot a^2 = a \cdot a \cdot a \cdot a \quad\quad\quad = a^4$$

Do you see a rule that we could use to multiply powers? The base in each answer is the original base, and the exponent in each answer is the *sum* of the original exponents.

$$3^2 \cdot 3^4 = 3^{2+4} = 3^6$$

$$x^3 \cdot x = x^{3+1} = x^4$$

$$y^3 \cdot y^4 = y^{3+4} = y^7$$

$$a^2 \cdot a^2 = a^{2+2} = a^4$$

To multiply powers of the *same* base, we keep the same base and *add* the exponents.

This rule is called the **First Law of Exponents**, and we may state it in general, using *a* to represent any base and *m* and *n* to represent any whole numbers.

First Law of Exponents

$$a^m \cdot a^n = a^{m+n}$$

Example 15. Find $x^6 \cdot x^9$.

Solution

$$x^6 \cdot x^9 = x^{6+9}$$

$$= x^{15}$$

You try Example 16.

Example 16. Find $y \cdot y^3 \cdot y^5$.

Solution

$$y \cdot y^3 \cdot y^5 =$$

Check your work on page 229. ▶

Example 17. Find $x^2 \cdot y$.

Solution

$$x^2 \cdot y = x^2y$$

The bases are *different*, so we cannot add exponents.

Example 18. Find $a^{11} \cdot a^0$.

Solution

$$a^{11} \cdot a^0 = a^{11+0}$$

$$= a^{11}$$

Example 18 should help you see why a^0 must be defined to be 1. The only number that leaves another number unchanged in multiplication is 1, the identity for multiplication. Remember

$$a^0 = 1 \quad \text{(provided that } a \neq 0\text{)}$$

If a product involves powers of several bases, we may use the associative and commutative properties of multiplication to rearrange the factors and then use the First Law of Exponents to multiply powers of the same base.

Example 19. Find $2x^2y \cdot 3xy^3$.

Solution

$$2x^2y \cdot 3xy^3$$
$$= 2 \cdot 3 \cdot x^2 \cdot x \cdot y \cdot y^3$$
$$= 6x^{2+1} \cdot y^{1+3}$$
$$= 6x^3y^4$$

You complete Example 20.

Example 20. Find $abc \cdot a^2bc^4$.

Solution

$$abc \cdot a^2bc^4$$
$$= a \cdot a^2 \cdot b \cdot b \cdot c \cdot c^4$$

Check your work on page 230. ▶

⫸ Trial Run

Find the products.

_____ 1. $x^3 \cdot x^6$

_____ 2. $x^3 \cdot y^2$

_____ 3. $a^0 \cdot a^5$

_____ 4. $y^2 \cdot y \cdot y^3$

_____ 5. $-2x^2 \cdot x^5$

_____ 6. $3a \cdot a^2 \cdot a^6$

_____ 7. $-5x^2 \cdot 2x^3y^4$

_____ 8. $xy^5z^3 \cdot x^3yz^2$

Answers are on page 230.

Raising a Power to a Power

How would you simplify an expression such as $(x^2)^3$? We know that the exponent 3 tells us to use the base x^2 as a factor in a product 3 times. Using what we know about the First Law of Exponents, we can simplify this expression and others.

$$(x^2)^3 = x^2 \cdot x^2 \cdot x^2 \qquad = x^6$$
$$(y^5)^2 = y^5 \cdot y^5 \qquad = y^{10}$$
$$(a^4)^5 = a^4 \cdot a^4 \cdot a^4 \cdot a^4 \cdot a^4 = a^{20}$$
$$(x^3)^3 = x^3 \cdot x^3 \cdot x^3 \qquad = x^9$$

This is called "raising a power to a power." You should see that it can be done by *multiplying* exponents.

$$(x^2)^3 = x^{2 \cdot 3} = x^6$$
$$(y^5)^2 = y^{5 \cdot 2} = y^{10}$$
$$(a^4)^5 = a^{4 \cdot 5} = a^{20}$$
$$(x^3)^3 = x^{3 \cdot 3} = x^9$$

To raise a power to a power, we keep the base and *multiply* the exponents.

This is called the **Second Law of Exponents** and we state it in general, using a to represent any base and m and n to represent any whole numbers.

Second Law of Exponents

$$(a^m)^n = a^{m \cdot n}$$

Example 21. Find $(x^7)^2$.

Solution

$$(x^7)^2 = x^{7 \cdot 2}$$
$$= x^{14}$$

Now you do Example 22.

Example 22. Find $(a^9)^3$.

Solution

 $$(a^9)^3 =$$

Check your work on page 230. ▶

Example 23. Find $(y^2)^3 \cdot (y^4)^2$.

Solution

$$(y^2)^3 \cdot (y^4)^2 = y^{2 \cdot 3} \cdot y^{4 \cdot 2}$$
$$= y^6 \cdot y^8$$
$$= y^{6+8}$$
$$= y^{14}$$

Now you try Example 24.

Example 24. Find $(a^5)^2 \cdot (a^3)^4$.

Solution

 $$(a^5)^2 \cdot (a^3)^4 =$$

Check your work on page 230. ▶

Raising a Product to a Power

To "raise a product to a power," we again use what we know about exponents.

$$(xy)^3 = (xy)(xy)(xy) \qquad = x \cdot x \cdot x \cdot y \cdot y \cdot y \qquad = x^3y^3$$
$$(2b)^5 = (2b)(2b)(2b)(2b)(2b) = 2 \cdot 2 \cdot 2 \cdot 2 \cdot 2 \cdot b \cdot b \cdot b \cdot b \cdot b = 2^5b^5$$

Do you see how to raise a product to a power?

> To raise a product to a power, we *raise each factor* to that power.

This rule is called the **Third Law of Exponents** and can be written in general as

---

Third Law of Exponents

$$(ab)^n = a^n b^n$$

Example 25. Find $(2x)^7$.

Solution

$$(2x)^7 = 2^7 x^7$$
$$= 128x^7$$

Now you try Example 26.

Example 26. Find $\left(\dfrac{1}{4}\,y\right)^3$.

Solution

 $$\left(\frac{1}{4}\,y\right)^3 = \left(\frac{1}{4}\right) \underline{\quad}\; y\underline{\quad}$$
$$= \underline{\quad\quad}$$

Check your work on page 230. ▶

Students often make errors in raising products of *negative constants* and variables to a certain power. The safest way to handle such powers is to be sure to write the step in which each factor is raised to the power. Look at this power.

$$(-x)^6 = (-1 \cdot x)^6 \qquad \text{Rewrite } -x \text{ as } -1x.$$
$$= (-1)^6 x^6 \qquad \text{Raise each factor to the power.}$$
$$= 1 \cdot x^6 \qquad \text{Find the power of the constant.}$$
$$= x^6 \qquad \text{Use the multiplication property of 1.}$$

Writing these steps may seem tedious, but they will assure accuracy, especially when problems become more involved later in this chapter.

Example 27. Simplify $(-3y)^5$.

Solution

$$(-3y)^5 = (-3)^5 y^5$$
$$= -243y^5$$

You complete Example 28.

Example 28. Simplify $(-3y)^4$.

Solution

$$(-3y)^4 = (-3) \underline{\quad}\; y\underline{\quad}$$
$$= \underline{\quad\quad}$$

Check your work on page 230. ▶

Example 29. Simplify $(5x^3 y^4)^2$.

Solution

$$(5x^3 y^4)^2$$
$$= 5^2 (x^3)^2 (y^4)^2 \quad \text{Raise each factor to the power.}$$
$$= 5^2 x^6 y^8 \qquad \text{Raise each power to the power.}$$
$$= 25x^6 y^8 \qquad \text{Simplify the constant.}$$

Now you complete Example 30.

Example 30. Simplify $x(x^2 z)^5$.

Solution

$$x(x^2 z)^5$$
$$= x \cdot (x^2)^5 \cdot z^5 \quad \text{Raise each factor to the power.}$$
$$= x \cdot x\underline{\quad} \cdot z^5 \quad \text{Raise the power to the power.}$$
$$= \underline{\quad\quad\quad} \qquad \text{Multiply powers of the same base.}$$

Check your work on page 230. ▶

⫸ Trial Run

Simplify each expression.

_____ **1.** $(a^3)^4$

_____ **2.** $(x^{10})^8 (x^2)^5$

———— 3. $(xy)^5$ ———— 4. $(4y)^3$

———— 5. $(-2y)^4$ ———— 6. $(a^2b)^3$

———— 7. $(3x^2y^4)^2$ ———— 8. $x(x^2y^3)^5$

Answers are on page 230.

▶ Examples You Completed

Example 2. Find $(-2)^5$.

Solution

$$(-2)^5 = (-2)(-2)(-2)(-2)(-2)$$
$$= -32$$

Example 4. Write x^4y without exponents.

Solution

$$x^4y = x \cdot x \cdot x \cdot x \cdot y$$

Example 5. Write $2x^3$ without exponents.

Solution

$$2x^3 = 2 \cdot x \cdot x \cdot x$$

Example 10. Evaluate $\left(\dfrac{1}{2}x^2y\right)^3$ when $x = -1$ and $y = 4$.

Solution

$$\left(\frac{1}{2}x^2y\right)^3 = \left[\frac{1}{2}(-1)^2(4)\right]^3$$
$$= \left[\frac{1}{2} \cdot 1 \cdot 4\right]^3$$
$$= [2]^3$$
$$= 2 \cdot 2 \cdot 2$$
$$= 8$$

Example 14. Find $4(7xyz)^0$.

Solution

$$4(7xyz)^0 = 4(1)$$
$$= 4$$

Example 16. Find $y \cdot y^3 \cdot y^5$.

Solution

$$y \cdot y^3 \cdot y^5 = y^{1+3+5}$$
$$= y^9$$

Example 20. Find $abc \cdot a^2bc^4$.

Solution

$$abc \cdot a^2bc^4$$
$$= a \cdot a^2 \cdot b \cdot b \cdot c \cdot c^4$$
$$= a^{1+2}b^{1+1}c^{1+4}$$
$$= a^3b^2c^5$$

Example 22. Find $(a^9)^3$.

Solution

$$(a^9)^3 = a^{9 \cdot 3}$$
$$= a^{27}$$

Example 24. Find $(a^5)^2 \cdot (a^3)^4$

Solution

$$(a^5)^2 \cdot (a^3)^4 = a^{5 \cdot 2} \cdot a^{3 \cdot 4}$$
$$= a^{10} \cdot a^{12}$$
$$= a^{10+12}$$
$$= a^{22}$$

Example 26. Find $\left(\dfrac{1}{4}\, y\right)^3$.

Solution

$$\left(\frac{1}{4}\, y\right)^3 = \left(\frac{1}{4}\right)^3 y^3$$
$$= \frac{1}{64}\, y^3$$

Example 28. Simplify $(-3y)^4$.

Solution

$$(-3y)^4 = (-3)^4 y^4$$
$$= 81y^4$$

Example 30. Simplify $x(x^2z)^5$.

Solution

$$x(x^2z)^5 = x \cdot (x^2)^5 \cdot z^5$$
$$= x \cdot x^{10} \cdot z^5$$
$$= x^{11}z^5$$

Answers to Trial Runs

page 224 **1.** $2a^3b^2$ **2.** $-1 \cdot y \cdot y$ **3.** $27x^3$ **4.** 16 **5.** 36 **6.** -12 **7.** -216 **8.** $3x$

page 226 **1.** x^9 **2.** x^3y^2 **3.** a^5 **4.** y^6 **5.** $-2x^7$ **6.** $3a^9$ **7.** $-10x^5y^4$ **8.** $x^4y^6z^5$

page 228 **1.** a^{12} **2.** x^{90} **3.** x^5y^5 **4.** $64y^3$ **5.** $16y^4$ **6.** a^6b^3 **7.** $9x^4y^8$ **8.** $x^{11}y^{15}$

EXERCISE SET 5.1

Use exponents to rewrite each expression.

_____ 1. $3xxyy$

_____ 2. $-2xxxxyy$

_____ 3. $-a \cdot a \cdot a \cdot a$

_____ 4. $-b \cdot b \cdot b \cdot b \cdot b$

_____ 5. $(-2x)(-2x)(-2x)$

_____ 6. $(3x)(3x)(3x)(3x)$

_____ 7. $3x \cdot x \cdot x - 4y \cdot y \cdot y$

_____ 8. $x \cdot x \cdot x \cdot x + 2y \cdot y$

Write each expression without exponents.

_____ 9. $8a^3$

_____ 10. $5a^4$

_____ 11. $(3a)^3$

_____ 12. $(2a)^5$

_____ 13. $-x^4$

_____ 14. $-3x^3$

_____ 15. $\frac{2}{3}x^3y^3$

_____ 16. $\frac{4}{7}x^4y^2$

Evaluate each expression.

_____ 17. $(-3)^3$

_____ 18. $(-2)^5$

_____ 19. -2^4

_____ 20. -3^4

_____ 21. $(2x)^2$ when $x = -1$

_____ 22. $(-3x)^2$ when $x = -3$

_____ 23. $(xy^2)^3$ when $x = 1$ and $y = -2$

_____ 24. $(3x^5y)^2$ when $x = 1$ and $y = 2$

_____ 25. $(0.2xy^3)^0$ when $x = 5$ and $y = 3$

_____ 26. $(0.3xy^5)^0$ when $x = 2$ and $y = -1$

Perform the indicated operations.

_____ 27. $x^2 \cdot x^3$

_____ 28. $a^5 \cdot a^2$

_____ 29. $a^0 \cdot a^7$

_____ 30. $a^9 \cdot a^0$

_____ 31. $y^3 \cdot y \cdot y^5$

_____ 32. $y^2 \cdot y \cdot y^7$

_____ 33. $-2a^5 \cdot a^7$

_____ 34. $-3a^4 \cdot a^6$

_____ 35. $x^2y^3 \cdot x^5y^2$

_____ 36. $x^3y^4 \cdot x^2y^5$

_____ 37. $2m^2 \cdot m \cdot m^5$

_____ 38. $3m^4 \cdot m^3 \cdot m$

_____ 39. $\frac{1}{2}x^4y^3 \cdot x^2y \cdot x^3$

_____ 40. $-\frac{1}{5}x^5y^2 \cdot xy^2 \cdot y$

_____ 41. $0.5x^2y^3z \cdot x^2yz^4$

_____ 42. $0.9x^2y^3z^5 \cdot xyz^2$

_____ 43. $3x^2y \cdot x^7z^2$

_____ 44. $5xy^3 \cdot x^5z^3$

_____ 45. $(a^2)^3$

_____ 46. $(a^3)^2$

_____ 47. $(xy)^5$

_____ 48. $(xy)^7$

_____ 49. $(7x)^2$

_____ 50. $(8y)^2$

_____ 51. $\left(-\dfrac{1}{3}y^2\right)^2$

_____ 52. $\left(-\dfrac{1}{5}y^2\right)^2$

_____ 53. $(-4x)^3$

_____ 54. $(-3x)^3$

_____ 55. $(x^2)^3 \cdot (x^3)^4$

_____ 56. $(x^4)^2 \cdot (x^2)^5$

_____ 57. $(9x^2y^3)^2$

_____ 58. $(8x^2y^4)^2$

_____ 59. $(-2xy^3)^4$

_____ 60. $(-3x^2y)^5$

_____ 61. $\dfrac{1}{4}x^2 \cdot (xy^2)^3$

_____ 62. $\dfrac{1}{3}x^2 \cdot (x^2y)^4$

☆ Stretching the Topics

Perform the indicated operations.

_____ 1. $\left(-\dfrac{1}{2}x^3y^2z\right)^3 \cdot (xy^2z^3)^5$

_____ 2. $7xy(-2xy)(-3x^2y)$

_____ 3. $9x^2y^3z\left(-\dfrac{1}{3}x^5yz^4\right)^2$

_____ 4. $(2ab^2)^3 \cdot (-a^3b^2c)^4 \cdot (3a^2c^0)^2$

Check your answers in the back of your book.

If you can complete **Checkup 5.1**, you are ready to go on to Section 5.2.

Name _____ Date _____

 CHECKUP 5.1

_____ **1.** Evaluate $(-2)^4$.

_____ **2.** Use exponents to write $-\dfrac{1}{3} x \cdot x \cdot x \cdot x$.

_____ **3.** Write $0.5x^3y^2$ without exponents.

_____ **4.** Evaluate $(-3x^2y)^3$ when $x = -1$ and $y = 2$.

Perform the indicated operations.

_____ **5.** $x^6 \cdot x^2$ _____ **6.** $-y^3 \cdot y^2 \cdot y$

_____ **7.** $-5x^2y^3 \cdot xy^4$ _____ **8.** $(a^3)^2 \cdot (a^4)^3$

_____ **9.** $\left(\dfrac{2}{3}x\right)^2$ _____ **10.** $(-2xy)^3$

Check your answers in the back of your book.

If You Missed Problems:	You Should Review Examples:
1	1, 2
2	3, 4
3	5–8
4	9, 10
5, 6	15, 16
7	19, 20
8	23, 24
9, 10	25–29

5.2 Dividing with Whole-Number Powers

To learn how to simplify quotients of powers, we recall our earlier work with fractions.

Dividing Powers of the Same Base

Let's look at some quotients of powers of the same base. We shall rewrite each power using the definition of exponents.

$$\frac{6^5}{6^2} = \frac{6 \cdot 6 \cdot 6 \cdot 6 \cdot 6}{6 \cdot 6} = \frac{6 \cdot 6}{6 \cdot 6} \cdot 6 \cdot 6 \cdot 6 = 1 \cdot 6 \cdot 6 \cdot 6 = 6^3$$

$$\frac{x^7}{x^3} = \frac{x \cdot x \cdot x \cdot x \cdot x \cdot x \cdot x}{x \cdot x \cdot x} = \frac{x \cdot x \cdot x}{x \cdot x \cdot x} \cdot x \cdot x \cdot x \cdot x = 1 \cdot x \cdot x \cdot x \cdot x = x^4$$

$$\frac{a^3}{a^2} = \frac{a \cdot a \cdot a}{a \cdot a} = \frac{a \cdot a}{a \cdot a} \cdot a = 1 \cdot a = a$$

$$\frac{x^3}{x^6} = \frac{x \cdot x \cdot x}{x \cdot x \cdot x \cdot x \cdot x \cdot x} = \frac{x \cdot x \cdot x}{x \cdot x \cdot x} \cdot \frac{1}{x \cdot x \cdot x} = 1 \cdot \frac{1}{x \cdot x \cdot x} = \frac{1}{x^3}$$

Can you spot a pattern in these quotients? In each case, the exponent on the base in the answer was the *difference* between the original exponents. If the exponent in the numerator was larger than the exponent in the denominator, the power ended up in the numerator. If the exponent in the denominator was larger than the exponent in the numerator, the power ended up in the denominator.

This is the **Fourth Law of Exponents**, and it can be stated in general as follows:

Fourth Law of Exponents

$$\frac{a^m}{a^n} = a^{m-n} \qquad \text{if } m \text{ is larger than or equal to } n \ (m \geq n)$$

$$\frac{a^m}{a^n} = \frac{1}{a^{n-m}} \qquad \text{if } n \text{ is larger than } m \ (n > m)$$

provided that $a \neq 0$

The statement $a \neq 0$ arises from the fact that *division by zero is undefined*. Recall from Chapter 2 that a divisor can never be zero.

Example 1. Find $\dfrac{x^{17}}{x^{10}}$.

Solution

$$\frac{x^{17}}{x^{10}} = x^{17-10}$$

$$= x^7$$

You try Example 2.

Example 2. Find $\dfrac{y^7}{y}$.

Solution

$$\frac{y^7}{y} =$$

Check your work on page 240. ▶

Example 3. Find $\dfrac{x^{13}}{x^{13}}$.

Solution

$$\frac{x^{13}}{x^{13}} = x^{13-13}$$

$$= x^0$$

$$= 1$$

Remember that $x^0 = 1$.

You complete Example 4.

Example 4. Find $\dfrac{x^7}{x^{12}}$.

Solution

$$\frac{x^7}{x^{12}} = \frac{1}{x^{12-7}}$$

$$= \underline{}$$

Check your answers on page 240. ▶

If a quotient contains constant factors and variable factors, we reduce the constant parts as we reduce a numerical fraction and use the Fourth Law of Exponents with the variable parts.

Example 5. Find $\dfrac{-42x^{10}}{7x}$.

Solution

$$\frac{-42x^{10}}{7x} = \frac{-42}{7} \cdot \frac{x^{10}}{x}$$

$$= -6x^{10-1}$$

$$= -6x^9$$

Example 6. Find $\dfrac{3x^2}{12x^4}$.

Solution

$$\frac{3x^2}{12x^4} = \frac{3}{12} \cdot \frac{x^2}{x^4}$$

$$= \frac{1}{4} \cdot \frac{1}{x^{4-2}}$$

$$= \frac{1}{4} \cdot \frac{1}{x^2}$$

$$= \frac{1}{4x^2}$$

If a quotient contains more than one variable, we reduce the constant parts first, then simplify each of the variable parts by subtracting exponents.

Example 7. Find $\dfrac{-2x^5y^2}{4x^4y^7}$.

Solution

$$\frac{-2x^5y^2}{4x^4y^7} = \frac{-2}{4} \cdot \frac{x^5}{x^4} \cdot \frac{y^2}{y^7}$$

$$= \frac{-1}{2} \cdot x^{5-4} \cdot \frac{1}{x^{7-2}}$$

$$= \frac{-1}{2} \cdot \frac{x}{1} \cdot \frac{1}{y^5}$$

$$= -\frac{x}{2y^5}$$

You complete Example 8.

Example 8. Find $\dfrac{-x^7y^3z}{x^5y^8z}$.

Solution

$$\frac{-x^7y^3z}{x^5y^8z}$$

$$= -1 \cdot \frac{x^7}{x^5} \cdot \frac{y^3}{y^8} \cdot \frac{z}{z}$$

$$= -1 \cdot x^{7-5} \cdot \frac{1}{y^{8-3}} \cdot z^{1-1}$$

Check your work on page 240. ▶

Example 9. Find $\dfrac{(x^2y)^3}{(-xy)^2}$.

Solution

$$\frac{(x^2y)^3}{(-xy)^2} = \frac{(x^2)^3\,y^3}{(-1)^2\,x^2y^2} \qquad \text{Use Third Law of Exponents in numerator and in denominator.}$$

$$= \frac{x^6y^3}{1\cdot x^2y^2} \qquad \text{Use Second Law of Exponents in numerator.}$$

$$= x^{6-2}y^{3-2} \qquad \text{Use Fourth Law of Exponents.}$$

$$= x^4y \qquad \text{Simplify.}$$

⮕ Trial Run

Find the quotients.

_____ 1. $\dfrac{x^9}{x^2}$ _____ 2. $\dfrac{x^4}{x^5}$

_____ 3. $\dfrac{3y^3}{9y^7}$ _____ 4. $\dfrac{-24a^2b^5}{8a^3b^4}$

_____ 5. $\dfrac{64a^7b^2c^3}{56a^2bc}$ _____ 6. $\dfrac{(2xy^2)^4}{(-4xy^3)^3}$

Answers are on page 241.

Raising a Quotient to a Power

To "raise a quotient to a power," we again use what we know about exponents and fractions.

$$\left(\frac{x}{y}\right)^2 = \frac{x}{y}\cdot\frac{x}{y} = \frac{x\cdot x}{y\cdot y} = \frac{x^2}{y^2}$$

$$\left(\frac{b}{2}\right)^3 = \frac{b}{2}\cdot\frac{b}{2}\cdot\frac{b}{2} = \frac{b\cdot b\cdot b}{2\cdot 2\cdot 2} = \frac{b^3}{2^3} = \frac{b^3}{8}$$

These examples should help you agree that

To raise a quotient to a power, we raise the dividend (numerator) to that power and raise the divisor (denominator) to that power.

This rule is called the **Fifth Law of Exponents** and can be written in general as

> **Fifth Law of Exponents**
>
> $$\left(\frac{a}{b}\right)^n = \frac{a^n}{b^n} \qquad \text{provided that } b \neq 0$$

Example 10. Find $\left(\dfrac{x}{3}\right)^4$.

Solution

$$\left(\frac{x}{3}\right)^4 = \frac{x^4}{3^4}$$

$$= \frac{x^4}{81}$$

You complete Example 11.

Example 12. Find $\left(-\dfrac{2a^3}{3c}\right)^4$.

Solution

$$\left(-\frac{2a^3}{3c}\right)^4 = \left(\frac{-2a^3}{3c}\right)^4$$

$$= \frac{(-2a^3)^4}{(3c)^4} \qquad \text{Raise numerator and denominator to power.}$$

$$= \frac{(-2)^4(a^3)^4}{3^4 \cdot c^4} \qquad \text{Raise products to power.}$$

$$= \frac{16a^{12}}{81c^4} \qquad \text{Raise power to power and simplify constants.}$$

Example 11. Find $\left(-\dfrac{2}{y}\right)^5$.

Solution

$$\left(-\frac{2}{y}\right)^5 = \left(\frac{-2}{y}\right)^5$$

$$= \frac{(-2)^5}{y^5}$$

$$=$$

Check your answer on page 240. ▶

To find a power of a quotient containing the same variables in the numerator and denominator, we should *simplify the quotient first* before we raise it to the power.

Example 13. Find $\left(\dfrac{14a^2b^3}{7ab^9}\right)^4$.

Solution

$$\left(\frac{14a^2b^3}{7ab^9}\right)^4 = \left(\frac{14}{7} \cdot \frac{a^2}{a} \cdot \frac{b^3}{b^9}\right)^4 \qquad \text{Rewrite the quotient.}$$

$$= \left(2 \cdot a^{2-1} \cdot \frac{1}{b^{9-3}}\right)^4 \qquad \text{Use the Fourth Law of Exponents inside parentheses.}$$

$$= \left(\frac{2a}{b^6}\right)^4 \qquad \text{Simplify inside parentheses.}$$

$$= \frac{(2a)^4}{(b^6)^4} \qquad \text{Raise numerator and denominator to power.}$$

$$= \frac{2^4a^4}{b^{24}} \qquad \text{Raise product to power and raise power to power.}$$

$$= \frac{16a^4}{b^{24}} \qquad \text{Simplify constant.}$$

Let's try the mirror problem stated at the beginning of this chapter.

Example 14. Mary has a square mirror measuring s inches on each side. After framing, each side is $1\frac{1}{4}$ times as long as it was. Write an algebraic expression for the *area* of the framed mirror.

Solution. Each side of the framed mirror is $1\frac{1}{4}s$ or $\frac{5}{4}s$, and we illustrate the mirror here. The *area* of a square is found by *squaring* its side.

$$\text{Area} = \left(\frac{5}{4}s\right)^2$$

$$= \left(\frac{5s}{4}\right)^2 \qquad \text{Rewrite within parentheses.}$$

$$= \frac{(5s)^2}{4^2} \qquad \text{Raise numerator and denominator to the power.}$$

$$= \frac{25s^2}{16} \qquad \text{Find the powers.}$$

The area of the framed mirror is $\frac{25}{16}s^2$. Note that since the unframed mirror had an area of s^2, the area of the framed mirror is $\frac{25}{16}$ (or $1\frac{9}{16}$) times the area of the unframed mirror.

⫸ Trial Run

Find the powers.

_____ 1. $\left(\dfrac{x}{5}\right)^3$ _____ 2. $\left(\dfrac{-a}{b}\right)^5$

_____ 3. $\left(\dfrac{2a}{5b}\right)^2$ _____ 4. $\left(\dfrac{-3a^2}{4b}\right)^3$

_____ 5. $\left(\dfrac{abc}{2ac}\right)^5$ _____ 6. $\left(\dfrac{81x^2y^3}{-9xy^5}\right)^2$

Answers are on page 241.

▶ Examples You Completed

Example 2. Find $\dfrac{y^7}{y}$.

Solution

$$\frac{y^7}{y} = \frac{y^7}{y^1}$$
$$= y^{7-1}$$
$$= y^6$$

Example 8. Find $\dfrac{-x^7y^3z}{x^5y^8z}$.

Solution

$$\frac{-x^7y^3z}{x^5y^8z} = -1 \cdot \frac{x^7}{x^5} \cdot \frac{y^3}{y^8} \cdot \frac{z}{z}$$
$$= -1 \cdot x^{7-5} \cdot \frac{1}{y^{8-3}} \cdot z^{1-1}$$
$$= -1 \cdot x^2 \cdot \frac{1}{y^5} \cdot z^0$$
$$= -\frac{x^2}{y^5} \qquad \text{Remember, } z^0 = 1.$$

Example 4. Find $\dfrac{x^7}{x^{12}}$.

Solution

$$\frac{x^7}{x^{12}} = \frac{1}{x^{12-7}}$$
$$= \frac{1}{x^5}$$

Example 11. Find $\left(-\dfrac{2}{y}\right)^5$.

Solution

$$\left(-\frac{2}{y}\right)^5 = \left(\frac{-2}{y}\right)^5$$
$$= \frac{(-2)^5}{y^5}$$
$$= \frac{-32}{y^5}$$
$$= -\frac{32}{y^5}$$

Answers to Trial Runs

page 237   1. x^7 2. $\dfrac{1}{x}$ 3. $\dfrac{1}{3y^4}$ 4. $-\dfrac{3b}{a}$ 5. $\dfrac{8a^5bc^2}{7}$ 6. $-\dfrac{x}{4y}$

page 240   1. $\dfrac{x^3}{125}$ 2. $-\dfrac{a^5}{b^5}$ 3. $\dfrac{4a^2}{25b^2}$ 4. $-\dfrac{27a^6}{64b^3}$ 5. $\dfrac{b^5}{32}$ 6. $\dfrac{81x^2}{y^4}$

For handy reference, we summarize the definitions and laws that we use in working with exponents. Remember that a and b stand for any bases, and m and n represent whole numbers.

	Symbols
Definition of a^n	$a^n = \underbrace{a \cdot a \cdot \ldots \cdot a}_{n \text{ factors}}$
Definition of a^0	$a^0 = 1 \quad (a \neq 0)$
First Law of Exponents	$a^m \cdot a^n = a^{m+n}$
Second Law of Exponents	$(a^m)^n = a^{m \cdot n}$
Third Law of Exponents	$(a \cdot b)^n = a^n \cdot b^n$
Fourth Law of Exponents	$\dfrac{a^m}{a^n} = a^{m-n} \quad \text{if } m \geq n \quad (a \neq 0)$
	$\dfrac{a^m}{a^n} = \dfrac{1}{a^{n-m}} \quad \text{if } n > m \quad (a \neq 0)$
Fifth Law of Exponents	$\left(\dfrac{a}{b}\right)^n = \dfrac{a^n}{b^n} \quad (b \neq 0)$

EXERCISE SET 5.2

Find the quotients. (Assume that variables in the denominators are not zero.)

_____ 1. $\dfrac{x^{12}}{x^5}$

_____ 2. $\dfrac{x^{13}}{x^7}$

_____ 3. $\dfrac{a^5}{a}$

_____ 4. $\dfrac{a^9}{a}$

_____ 5. $\dfrac{m^6}{m^6}$

_____ 6. $\dfrac{m^{12}}{m^{12}}$

_____ 7. $\dfrac{y^9}{y^{11}}$

_____ 8. $\dfrac{y^7}{y^{10}}$

_____ 9. $\dfrac{32x^5}{8x^2}$

_____ 10. $\dfrac{35x^7}{5x^5}$

_____ 11. $\dfrac{-9y^7}{3y^2}$

_____ 12. $\dfrac{-12y^8}{3y^3}$

_____ 13. $\dfrac{5x^2}{15x^4}$

_____ 14. $\dfrac{6x^3}{18x^7}$

_____ 15. $\dfrac{-21a^3}{7a^5}$

_____ 16. $\dfrac{-60a^5}{12a^{10}}$

_____ 17. $\dfrac{-5x^3y^2}{25xy^5}$

_____ 18. $\dfrac{-6x^5y^3}{36xy^4}$

_____ 19. $\dfrac{49a^3b^5}{7a^4b^7}$

_____ 20. $\dfrac{64a^4b^7}{8a^5b^9}$

_____ 21. $\dfrac{-x^5y^3z}{x^4y^9z}$

_____ 22. $\dfrac{-x^7yz^3}{x^5yz^7}$

_____ 23. $\dfrac{-12x^2yz^3}{6x^2y^3z}$

_____ 24. $\dfrac{-15x^4y^3z^5}{5x^4yz^7}$

_____ 25. $\dfrac{63a^5b^3c^3}{54a^2bc^5}$

_____ 26. $\dfrac{56a^{10}b^5c^4}{42a^6b^9c^7}$

_____ 27. $\dfrac{(xy^2)^5}{(-xy)^6}$

_____ 28. $\dfrac{(x^3y^5)^4}{(-x^2y^2)^7}$

_____ 29. $\dfrac{(-3xy^3)^3}{(9x^2y)^2}$

_____ 30. $\dfrac{(-4xy^2)^2}{(2x^3y)^3}$

Find the powers. (Assume that variables in the denominators are not zero.)

_____ 31. $\left(\dfrac{x}{2}\right)^3$ _____ 32. $\left(\dfrac{x}{3}\right)^2$

_____ 33. $\left(-\dfrac{3}{y}\right)^3$ _____ 34. $\left(-\dfrac{1}{y}\right)^7$

_____ 35. $\left(\dfrac{x^2}{y}\right)^4$ _____ 36. $\left(\dfrac{x}{y^3}\right)^2$

_____ 37. $\left(\dfrac{x^2}{-y^3}\right)^5$ _____ 38. $\left(\dfrac{-x}{y^2}\right)^3$

_____ 39. $\left(\dfrac{2x^2}{y}\right)^4$ _____ 40. $\left(\dfrac{5x^3}{y^2}\right)^2$

_____ 41. $\left(\dfrac{-5a^3}{2b^2}\right)^2$ _____ 42. $\left(\dfrac{-4a^5}{3b^3}\right)^2$

_____ 43. $\left(-\dfrac{2x^2y}{3z}\right)^5$ _____ 44. $\left(-\dfrac{3xy^3}{2z}\right)^3$

_____ 45. $\left(\dfrac{15ab^5}{5a^3b^2}\right)^2$ _____ 46. $\left(\dfrac{12a^3b}{4ab^5}\right)^3$

_____ 47. $\left(\dfrac{-9xy^2z^5}{27x^3y^2z^2}\right)^3$ _____ 48. $\left(\dfrac{-5x^5y^3z}{45x^3y^3z^4}\right)^2$

_____ 49. Mr. Jones has a square garden measuring x feet on each side. He wishes to enlarge the garden by doubling each side. Write an expression for the area of the enlarged garden.

_____ 50. Hillary has a square tablecloth measuring x inches on each side. After bordering the tablecloth with lace, each side is $1\frac{1}{5}$ times as long as it was. Write an algebraic expression for the area of the bordered tablecloth.

☆ Stretching the Topics _____

Simplify each expression. (Assume variables in the denominators are not zero.)

_____ 1. $\left(\dfrac{-x^3y^2z}{4x^4yz^3}\right)^2 \cdot \left(\dfrac{-2x^5y^2z^3}{x^4y^2z^6}\right)^3$ _____ 2. $\dfrac{(a^2b^3)^2}{c^9} \cdot \dfrac{(ac^5)^3}{(-b^3)^5} \cdot \left(\dfrac{-b^5}{a^3c^3}\right)^2$

_____ 3. Felicia and Holly each have a picture measuring x inches on each side. Felicia framed her picture after cutting off the border. After framing, each side was $\frac{2}{3}$ as long as it was before framing. Holly framed her picture using a mat. After framing, each side was $1\frac{1}{6}$ as long as it was before framing. Find the difference between the areas of the framed pictures.

Check your answers in the back of your book.

If you can complete **Checkup 5.2**, you are ready to go on to Section 5.3.

Name _____   Date _____

 CHECKUP 5.2

Find the quotients. (Assume variables in the denominators are not zero.)

_____ **1.** $\dfrac{x^{15}}{x^{10}}$

_____ **2.** $\dfrac{-a^7}{a^{10}}$

_____ **3.** $\dfrac{21x^9}{7x^{11}}$

_____ **4.** $\dfrac{-8a^5b^3}{16a^3b^7}$

_____ **5.** $\dfrac{x^5y^3z^2}{x^4y^3z^5}$

_____ **6.** $\dfrac{(xy^2)^5}{(-x^2y^3)^2}$

Find the powers. (Assume variables in the denominators are not zero.)

_____ **7.** $\left(\dfrac{x}{2}\right)^5$

_____ **8.** $\left(-\dfrac{2}{a}\right)^4$

_____ **9.** $\left(\dfrac{-x^4}{2y}\right)^3$

_____ **10.** $\left(\dfrac{20a^4b^2}{5ab^4}\right)^3$

Check your answers in the back of your book.

If You Missed Problems:	You Should Review Examples:
1, 2	1–4
3–5	5–8
6	9
7, 8	10, 11
9	12
10	13

5.3 Working with Negative Exponents

So far we have used only whole numbers $\{0, 1, 2, 3, 4, \ldots\}$ as exponents, and you may have wondered if negative integers $\{\ldots, -4, -3, -2, -1\}$ could ever appear as exponents. The answer is yes, but let's see how to give meaning to expressions such as

$$2^{-1} \quad \text{or} \quad x^{-2} \quad \text{or} \quad y^{-3}$$

However we decide to define negative exponents, we must agree that we do not wish to lose all the laws of exponents that we have already learned. We would like negative exponents to continue to obey *all* those laws.

Let's see if we can find some meaning for negative exponents by considering the expression x^{-2}.

By the First Law of Exponents $x^2 \cdot x^{-2} = x^{2+(-2)} = x^0 = 1$

By the property of reciprocals $x^2 \cdot \dfrac{1}{x^2} = \dfrac{x^2}{1} \cdot \dfrac{1}{x^2} \quad = 1$

Since both products yield the same result, it makes sense to conclude that

$$x^{-2} = \frac{1}{x^2}$$

Thus, x^{-2} *is the reciprocal of* x^2. We can use similar reasoning with other negative integer exponents.

Negative Exponent Form	Positive Exponent Form
y^{-3}	$\dfrac{1}{y^3}$
x^{-5}	$\dfrac{1}{x^5}$
2^{-1}	$\dfrac{1}{2^1} = \dfrac{1}{2}$

Now we are ready to state the general definition for a base raised to a negative power.

Definition of Negative Exponent

$$a^{-n} = \frac{1}{a^n} \quad \text{provided that } a \neq 0$$

We emphasize that a^n and a^{-n} are *reciprocals* for each other. In particular, a^{-1} is the reciprocal of a; in other words, $a^{-1} = \dfrac{1}{a}$.

See if you can complete Example 1.

Example 1. Write 2^{-5}, y^{-4}, x^{-1}, and $(-3)^{-2}$ without negative exponents.

Solution

$$2^{-5} = \frac{1}{2^5} = \frac{1}{32}$$

$$y^{-4} = \underline{\hspace{1cm}}$$

$$x^{-1} = \underline{\hspace{1cm}}$$

$$(-3)^{-2} = \frac{1}{(-3)^2} = \underline{\hspace{0.7cm}}$$

Check your work on page 254. ▶

Example 2. Write $\left(\frac{2}{3}\right)^{-1}$ without negative exponents.

Solution. $\left(\frac{2}{3}\right)^{-1}$ is the *reciprocal* of $\frac{2}{3}$.

Therefore,

$$\left(\frac{2}{3}\right)^{-1} = \frac{3}{2}$$

Example 3. Write $(2x)^{-1}$ without negative exponents.

Solution

$$(2x)^{-1} = \frac{1}{(2x)^1}$$

$$= \frac{1}{2x}$$

Example 4. Write $2x^{-1}$ without negative exponents.

Solution

$$2x^{-1} = 2 \cdot \frac{1}{x^1}$$

$$= \frac{2}{1} \cdot \frac{1}{x}$$

$$= \frac{2}{x}$$

We should notice two important facts from these examples.

1. *A negative exponent has no effect on the sign of the answer.* The negative exponent merely says "rewrite the expression as 1 divided by the base raised to the opposite (positive) power." The sign of the answer is determined *after* the expression is rewritten with a positive exponent. So

$$(-2)^{-4} = \frac{1}{(-2)^4} = \frac{1}{16}$$

$$(-2)^{-5} = \frac{1}{(-2)^5} = \frac{1}{-32} = -\frac{1}{32}$$

2. *A negative exponent applies only to the base to which it is attached, unless parentheses indicate otherwise.* Recall that this fact was true for positive exponents also. So

$$3 \cdot 5^2 = 3 \cdot 25 = 75 \qquad \text{but} \qquad (3 \cdot 5)^2 = 15^2 = 225$$

$$2x^{-3} = 2 \cdot \frac{1}{x^3} = \frac{2}{x^3} \qquad \text{but} \qquad (2x)^{-3} = \frac{1}{(2x)^3} = \frac{1}{2^3 x^3} = \frac{1}{8x^3}$$

⫸ Trial Run ▬▬▬▬▬▬▬▬▬▬▬▬▬▬▬▬▬▬▬▬▬▬

Write each expression without negative exponents.

————— 1. 5^{-2} ————— 2. $(-2)^{-3}$

_____ 3. $\left(\dfrac{3}{4}\right)^{-1}$ _____ 4. $(3a)^{-1}$

_____ 5. $5x^{-2}$ _____ 6. $(5x)^{-2}$

Answers are on page 255.

Negative exponents obey *all* the laws of exponents. Perhaps you should look back at those laws stated earlier in this chapter before trying to simplify some expressions. Let's agree to *use positive exponents in our answers.*

The **First Law of Exponents** helps us to multiply factors that are powers of the same base.

Example 5. Find $x^2 \cdot x^{-7}$.

Solution

$$x^2 \cdot x^{-7} = x^{2+(-7)}$$
$$= x^{-5}$$
$$= \dfrac{1}{x^5}$$

Let's rework Example 5 using the definition of a negative exponent, to verify that the First Law of Exponents does work with negative exponents.

$$x^2 \cdot x^{-7} = \dfrac{x^2}{1} \cdot \dfrac{1}{x^7}$$
$$= \dfrac{x^2}{x^7}$$
$$= \dfrac{1}{x^5}$$

Now complete Example 6.

Example 6. Find $a^{-3} \cdot a^{-4}$.

Solution

$$a^{-3} \cdot a^{-4} = a^{-3+(-4)}$$
$$= a^{---}$$
$$=$$

Check your work on page 254. ▶

Example 7. Find $(2x^{-3})(x^{-1})$.

Solution

$$(2x^{-3})(x^{-1}) = 2 \cdot x^{-3} \cdot x^{-1}$$
$$= 2x^{-3+(-1)}$$
$$= 2x^{-4}$$
$$= \dfrac{2}{1} \cdot \dfrac{1}{x^4}$$
$$= \dfrac{2}{x^4}$$

Example 8. Find $x^{-3} \cdot x^5 \cdot x^{-1}$.

Solution

$$x^{-3} \cdot x^5 \cdot x^{-1} = x^{-3+5+(-1)}$$
$$= x^1$$
$$= x$$

▶ Trial Run

Find the products. Write answers without negative exponents.

_____ **1.** $a^4 \cdot a^{-1}$ _____ **2.** $x^{-3} \cdot x^{-5} \cdot x$

_____ **3.** $(3x^{-6})(x^2)$ _____ **4.** $(-5x^3)(x^{-3})$

Answers are on page 255.

We can use the **Second Law of Exponents** to raise a power to a power.

Example 9. Find $(x^{-3})^2$.

Solution

$$(x^{-3})^2 = x^{-3(2)}$$
$$= x^{-6}$$
$$= \frac{1}{x^6}$$

Now you try Example 10.

Example 10. Find $(x^{-5})^{-2}$.

Solution

$$(x^{-5})^{-2} =$$

Check your work on page 254. ▶

Let's rework Example 9 using the definition of a negative exponent, to verify that the Second Law of Exponents does work with negative exponents.

$$(x^{-3})^2 = \left(\frac{1}{x^3}\right)^2$$
$$= \frac{1^2}{(x^3)^2}$$
$$= \frac{1}{x^6}$$

Example 11. Find $(x^{10})^{-1} \cdot (y^{-4})^{-2}$.

Solution

$$(x^{10})^{-1}(y^{-4})^{-2} = x^{10(-1)} \cdot y^{-4(-2)}$$
$$= x^{-10} \cdot y^8$$
$$= \frac{1}{x^{10}} \cdot \frac{y^8}{1}$$
$$= \frac{y^8}{x^{10}}$$

Example 12. Find $(x^{-1})^9 \cdot (x^2)^{-3}$.

Solution

$$(x^{-1})^9 \cdot (x^2)^{-3} = x^{-1(9)} \cdot x^{2(-3)}$$
$$= x^{-9} \cdot x^{-6}$$
$$= x^{-9+(-6)}$$
$$= x^{-15}$$
$$= \frac{1}{x^{15}}$$

Now you try Example 13.

Example 13. Find $x^5 \cdot (x^{-3})^2$.

Solution

$$x^5 \cdot (x^{-3})^2 =$$

Check your work on page 254. ▶

The **Third Law of Exponents** gives us a method for raising a product to a power.

Example 14. Find $(2a)^{-3}$.

Solution

$$(2a)^{-3} = 2^{-3}a^{-3}$$

$$= \frac{1}{2^3} \cdot \frac{1}{a^3}$$

$$= \frac{1}{8} \cdot \frac{1}{a^3}$$

$$= \frac{1}{8a^3}$$

Let's rework Example 14 using the definition of a negative exponent to verify that the Third Law of Exponents works for negative exponents.

$$(2a)^{-3} = \frac{1}{(2a)^3}$$

$$= \frac{1}{2^3 a^3}$$

$$= \frac{1}{8a^3}$$

Example 15. Find $(-5x)^{-2}$.

Solution

$$(-5x)^{-2} = (-5)^{-2} \cdot x^{-2}$$

$$= \frac{1}{(-5)^2} \cdot \frac{1}{x^2}$$

$$= \frac{1}{25} \cdot \frac{1}{x^2}$$

$$= \frac{1}{25x^2}$$

Now you complete Example 16.

Example 16. Find $(3x^{-1})^4$.

Solution

$$(3x^{-1})^4 = 3^4(x^{-1})^4$$

$$= 3^4 x^{-\!\!-}$$

$$= 81 \cdot \frac{1}{x^{-\!\!-}}$$

$$= \underline{\hspace{1cm}}$$

Check your work on page 254. ▶

Example 17. Find $(2^{-1}x^{-3}y^{-5})^{-2}$.

Solution

$$(2^{-1}x^{-3}y^{-5})^{-2} = (2^{-1})^{-2}(x^{-3})^{-2}(y^{-5})^{-2}$$

$$= 2^2 x^6 y^{10}$$

$$= 4x^6 y^{10}$$

From this example, notice that a step can be skipped if you realize that you must multiply the exponent on each factor within the parentheses times the exponent on the entire quantity. You may be able to do this computation in your head if you are careful.

Example 18. Find $(2^{-2}x^4z^{-1})^{-3}$.

Solution

$$(2^{-2}x^4z^{-1})^{-3} = 2^6 x^{-12} z^3$$

$$= 64 \cdot \frac{1}{x^{12}} \cdot z^3$$

$$= \frac{64z^3}{x^{12}}$$

Now complete Example 19.

Example 19. Find $(2^{-1}x^2)^{-4}(2x^{-2})^{-3}$.

Solution

$$(2^{-1}x^2)^{-4}(2x^{-2})^{-3}$$

$$= 2^4 x^{-8} \cdot 2^{-3} x^6$$

$$= 2^4 \cdot 2^{-3} \cdot x^{-8} \cdot x^6$$

Check your work on page 254. ▶

⫸ Trial Run

Simplify. Write answers without negative exponents.

_____ **1.** $(x^{-4})^2$ _____ **2.** $(a^{-3})^{-4}$

_____ **3.** $(y^5)^{-1}(y^{-3})^{-2}$ _____ **4.** $3x^{-2}(x^{-1})^3$

_____ **5.** $(3x)^{-2}$ _____ **6.** $(2a^{-2})^3$

_____ **7.** $(3^{-1}y^4)^{-3}$ _____ **8.** $(2a^2b^{-3})^2$

Answers are on page 255.

The **Fourth Law of Exponents** involves division. Now that we are allowed to use negative exponents, we need only one version of that law.

> **Fourth Law of Exponents**
>
> $$\frac{a^m}{a^n} = a^{m-n} \qquad \text{provided that } a \neq 0$$

Let's see why our new version for this law works whether the exponent in the denominator is less than or greater than the exponent in the numerator.

Using Old Version	Using New Version
$\dfrac{x^5}{x^2} = x^{5-2}$	$\dfrac{x^5}{x^2} = x^{5-2}$
$= x^3$	$= x^3$
$\dfrac{a^6}{a^{10}} = \dfrac{1}{a^{10-6}}$	$\dfrac{a^6}{a^{10}} = a^{6-10}$
$= \dfrac{1}{a^4}$	$= a^{-4}$
	$= \dfrac{1}{a^4}$

Since the results are the same, we agree to use the new version of the Fourth Law of Exponents.

Example 20. Find $\dfrac{x^9}{x^{-6}}$.

Solution

$$\dfrac{x^9}{x^{-6}} = x^{9-(-6)}$$

$$= x^{9+6}$$

$$= x^{15}$$

You try Example 21.

Example 21. Find $\dfrac{10x^{-8}}{2x^{-2}}$.

Solution

$$\dfrac{10x^{-8}}{2x^{-2}} = \dfrac{10}{2} \cdot x^{-8-(-2)}$$

Check your work on page 254. ▶

Example 22. Find $\dfrac{2x^{-3}y^{-4}}{6x^8y^{-5}}$.

Solution

$$\dfrac{2x^{-3}y^{-4}}{6x^8y^{-5}} = \dfrac{2}{6} \cdot x^{-3-8} \cdot y^{-4-(-5)}$$

$$= \dfrac{1}{3} \cdot x^{-11} \cdot y^{-4+5}$$

$$= \dfrac{1}{3} \cdot x^{-11} \cdot y$$

$$= \dfrac{1}{3} \cdot \dfrac{1}{x^{11}} \cdot \dfrac{y}{1}$$

$$= \dfrac{y}{3x^{11}}$$

You complete Example 23.

Example 23. Find $\dfrac{(-9x^{-3})(x^{-5})}{15x^{-10}}$.

Solution

$$\dfrac{(-9x^{-3})(x^{-5})}{15x^{-10}} = \dfrac{-9}{15} \cdot \dfrac{x\text{—}}{x^{-10}}$$

Check your work on page 255. ▶

⫸ **Trial Run**

Simplify. Write answers without negative exponents.

_____ **1.** $\dfrac{x^{10}}{x^{-6}}$

_____ **2.** $\dfrac{a^{-4}}{a^4}$

_____ **3.** $\dfrac{12y^{-7}}{6y^{-3}}$

_____ **4.** $\dfrac{3a^{-2}b^{-3}}{9a^3b^{-6}}$

_____ **5.** $\dfrac{(-3x^{-2})(x^{-3})}{9x^{-11}}$

_____ **6.** $\dfrac{(2x^3)^{-2}}{x^{-10}}$

Answers are on page 255.

The **Fifth Law of Exponents** provides the rule for raising a quotient (or fraction) to a power.

Example 24. Find $\left(\dfrac{x}{3}\right)^{-2}$.

Solution

$$\left(\dfrac{x}{3}\right)^{-2} = \dfrac{x^{-2}}{3^{-2}}$$

$$= x^{-2} \div 3^{-2}$$

$$= \dfrac{1}{x^2} \div \dfrac{1}{3^2}$$

$$= \dfrac{1}{x^2} \cdot \dfrac{3^2}{1}$$

$$= \dfrac{9}{x^2}$$

Example 25. Find $\left(\dfrac{2a^2}{b^{-5}}\right)^{-4}$.

Solution

$$\left(\dfrac{2a^2}{b^{-5}}\right)^{-4} = \dfrac{2^{-4}a^{2(-4)}}{b^{-5(-4)}}$$

$$= \dfrac{2^{-4}a^{-8}}{b^{20}}$$

$$= \dfrac{1}{2^4} \cdot \dfrac{1}{a^8} \div \dfrac{b^{20}}{1}$$

$$= \dfrac{1}{2^4} \cdot \dfrac{1}{a^8} \cdot \dfrac{1}{b^{20}}$$

$$= \dfrac{1}{16a^8b^{20}}$$

Notice from Examples 24 and 25 that we first use the laws of exponents to simplify an expression. After simplifying, we may write our answers using positive exponents if we observe the following:

Observation	Examples
A factor in the numerator with a negative exponent should be moved to the denominator and the exponent will be positive.	$\dfrac{5x^{-3}}{4} = \dfrac{5}{4x^3}$
A factor in the denominator with a negative exponent should be moved to the numerator and the exponent will be positive.	$\dfrac{7}{x^{-1}} = 7x$ $\dfrac{x^{-2}}{3^{-2}} = \dfrac{3^2}{x^2} = \dfrac{9}{x^2}$
A factor with a positive exponent stays where it is.	$\dfrac{2^{-4}a^{-8}}{b^{20}} = \dfrac{1}{2^4 a^8 b^{20}}$ $= \dfrac{1}{16a^8b^{20}}$

These observations make it much easier to rewrite a simplified fraction using positive exponents. Just be sure you have used the laws of exponents to simplify the expression first.

Example 26. Simplify $\left(\dfrac{2y^{-5}}{3y^3}\right)^{-2}$.

Solution

$$\left(\frac{2y^{-5}}{3y^3}\right)^{-2} = \left(\frac{2y^{-5-3}}{3}\right)^{-2}$$ Simplify the quotient *inside* parentheses by subtracting exponents.

$$= \left(\frac{2y^{-8}}{3}\right)^{-2}$$ Simplify the exponent.

$$= \frac{2^{-2}y^{16}}{3^{-2}}$$ Raise the numerator and denominator to the power.

$$= \frac{3^2 y^{16}}{2^2}$$ Move factors to have positive exponents.

$$= \frac{9y^{16}}{4}$$ Simplify the constants.

Example 27. Simplify $\left(\dfrac{-x^3}{y^{-2}}\right)^{-3}$.

Solution

$$\left(\frac{-x^3}{y^{-2}}\right)^{-3} = \left(\frac{-1 \cdot x^3}{y^{-2}}\right)^{-3}$$ Rewrite the numerator.

$$= \frac{(-1)^{-3}x^{-9}}{y^6}$$ Raise the numerator and denominator to the power.

$$= \frac{1}{(-1)^3 \cdot x^9 \cdot y^6}$$ Move factors to have positive exponents.

$$= \frac{1}{-1 \cdot x^9 y^6}$$ Simplify $(-1)^3$.

$$= -\frac{1}{x^9 y^6}$$

▐▶ Trial Run

Simplify. Write answers without negative exponents.

_____ 1. $\left(\dfrac{x}{y}\right)^{-1}$ _____ 2. $\left(\dfrac{a^{-2}}{b^3}\right)^{-3}$

_____ 3. $\left(\dfrac{x^{-4}}{y^{-2}}\right)^{-1}$ _____ 4. $\left(\dfrac{3x^2}{y^{-3}}\right)^{-2}$

Answers are on page 255.

▶ Examples You Completed ────────────────

Example 1. Write 2^{-5}, y^{-4}, x^{-1}, and $(-3)^{-2}$ without negative exponents.

Solution

$$2^{-5} = \frac{1}{2^5} = \frac{1}{32}$$

$$y^{-4} = \frac{1}{y^4}$$

$$x^{-1} = \frac{1}{x}$$

$$(-3)^{-2} = \frac{1}{(-3)^2} = \frac{1}{9}$$

Example 6. Find $a^{-3} \cdot a^{-4}$.

Solution

$$a^{-3} \cdot a^{-4} = a^{-3+(-4)}$$

$$= a^{-7}$$

$$= \frac{1}{a^7}$$

Example 10. Find $(x^{-5})^{-2}$.

Solution

$$(x^{-5})^{-2} = x^{-5(-2)}$$

$$= x^{10}$$

Example 13. Find $x^5 \cdot (x^{-3})^2$.

Solution

$$x^5 \cdot (x^{-3})^2 = x^5 \cdot x^{-3(2)}$$

$$= x^5 \cdot x^{-6}$$

$$= x^{5+(-6)}$$

$$= x^{-1}$$

$$= \frac{1}{x}$$

Example 16. Find $(3x^{-1})^4$.

Solution

$$(3x^{-1})^4 = 3^4(x^{-1})^4$$

$$= 3^4 x^{-4}$$

$$= 81 \cdot \frac{1}{x^4}$$

$$= \frac{81}{x^4}$$

Example 19. Find $(2^{-1}x^2)^{-4}(2x^{-2})^{-3}$.

Solution

$$(2^{-1}x^2)^{-4}(2x^{-2})^{-3}$$

$$= 2^4 x^{-8} \cdot 2^{-3} x^6$$

$$= 2^4 \cdot 2^{-3} \cdot x^{-8} \cdot x^6$$

$$= 2^1 \cdot x^{-2}$$

$$= 2 \cdot \frac{1}{x^2}$$

$$= \frac{2}{x^2}$$

Example 21. Find $\dfrac{10x^{-8}}{2x^{-2}}$.

Solution

$$\frac{10x^{-8}}{2x^{-2}} = \frac{10}{2} \cdot x^{-8-(-2)}$$

$$= 5 \cdot x^{-8+2}$$

$$= 5 \cdot x^{-6}$$

$$= 5 \cdot \frac{1}{x^6}$$

$$= \frac{5}{x^6}$$

Example 23. Find $\dfrac{(-9x^{-3})(x^{-5})}{15x^{-10}}$.

Solution

$$\frac{-9 \cdot x^{-3} \cdot x^{-5}}{15x^{-10}} = \frac{-9}{15} \cdot \frac{x^{-8}}{x^{-10}}$$

$$= \frac{-3}{5} \cdot x^{-8-(-10)}$$

$$= \frac{-3}{5} \cdot x^{-8+10}$$

$$= \frac{-3}{5} \cdot \frac{x^2}{1}$$

$$= -\frac{3x^2}{5}$$

Answers to Trial Runs _____

page 246   1. $\dfrac{1}{25}$ 2. $-\dfrac{1}{8}$ 3. $\dfrac{4}{3}$ 4. $\dfrac{1}{3a}$ 5. $\dfrac{5}{x^2}$ 6. $\dfrac{1}{25x^2}$

page 248   1. a^3 2. $\dfrac{1}{x^7}$ 3. $\dfrac{3}{x^4}$ 4. -5

page 250   1. $\dfrac{1}{x^8}$ 2. a^{12} 3. y 4. $\dfrac{3}{x^5}$ 5. $\dfrac{1}{9x^2}$ 6. $\dfrac{8}{a^6}$ 7. $\dfrac{27}{y^{12}}$ 8. $\dfrac{4a^4}{b^6}$

page 251   1. x^{16} 2. $\dfrac{1}{a^8}$ 3. $\dfrac{2}{y^4}$ 4. $\dfrac{b^3}{3a^5}$ 5. $-\dfrac{x^6}{3}$ 6. $\dfrac{x^4}{4}$

page 253   1. $\dfrac{y}{x}$ 2. a^6b^9 3. $\dfrac{x^4}{y^2}$ 4. $\dfrac{1}{9x^4y^6}$

EXERCISE SET 5.3

Simplify using the laws of exponents. Write answers without negative exponents. (Assume variables in denominators are not zero.)

_____ 1. $(2x^2)^3$

_____ 2. $(-3x^4)^2$

_____ 3. $\dfrac{a^5}{a^2}$

_____ 4. $\dfrac{a^9}{a^3}$

_____ 5. $\dfrac{5y^4}{y^7}$

_____ 6. $\dfrac{4y^5}{y^9}$

_____ 7. $\left(\dfrac{-3a^2}{b}\right)^3$

_____ 8. $\left(\dfrac{-2a^3}{b}\right)^5$

_____ 9. $(3x)^{-2}$

_____ 10. $(2x)^{-4}$

_____ 11. $(-5)^{-3}$

_____ 12. $(-4)^{-2}$

_____ 13. $\left(\dfrac{4}{5}\right)^{-2}$

_____ 14. $\left(\dfrac{2}{3}\right)^{-3}$

_____ 15. $(6x)^{-2}$

_____ 16. $(2x)^{-3}$

_____ 17. $6x^{-2}$

_____ 18. $2x^{-3}$

_____ 19. $4y^7 \cdot y^{-7}$

_____ 20. $5y^9 y^{-9}$

_____ 21. $a^8 \cdot a^{-11}$

_____ 22. $a^{-9} \cdot a^6$

_____ 23. $2x^{-3}(x^7)$

_____ 24. $4x^{-2}(x^5)$

_____ 25. $-5x^{-1}(x^{-2})$

_____ 26. $-8x^{-2}(x^{-3})$

_____ 27. $(a^{-6})^2$

_____ 28. $(a^{-7})^3$

_____ 29. $(y^9)^{-1} \cdot (y^{-5})^{-2}$

_____ 30. $(y^{-3})^{-2} \cdot (y^5)^{-1}$

_____ 31. $9x^{-3}(x^{-7})^0$

_____ 32. $8x^{-7}(x^{-11})^0$

_____ 33. $(4x)^{-3}$

_____ 34. $(9x)^{-2}$

_____ 35. $(3a^{-4})^3$

_____ 36. $(6a^{-5})^2$

_____ 37. $(2^3 y^{-5})^{-2}$

_____ 38. $(3y^{-7})^{-3}$

_____ 39. $(5a^3 b^{-5})^2$

_____ 40. $(6a^{-3} b^7)^2$

_____ 41. $(9x^{-5} y^{-3})^{-2}$

_____ 42. $(11x^{-4} y^{-5})^{-2}$

_____ 43. $(7^{-1} x^3)^{-2} \cdot (7^{-1} x^2)^3$

_____ 44. $(5^{-3} x^4)^{-1} \cdot (5^{-1} x^2)^2$

_____ 45. $\dfrac{x^{13}}{x^{-2}}$ _____ 46. $\dfrac{x^9}{x^{-3}}$

_____ 47. $\dfrac{9a^{-5}}{15a^5}$ _____ 48. $\dfrac{8a^{-6}}{12a^6}$

_____ 49. $\dfrac{7a^{-4}b^{-5}}{21a^3b^7}$ _____ 50. $\dfrac{8a^{-9}b^{-2}}{32a^6b^9}$

_____ 51. $\dfrac{(-4x^{-3})(x^{-7})}{6x^{-12}}$ _____ 52. $\dfrac{(-9x^{-7})(x^{-4})}{-6x^{-8}}$

_____ 53. $\left(\dfrac{2x}{3}\right)^{-3}$ _____ 54. $\left(\dfrac{3x}{4}\right)^{-2}$

_____ 55. $\left(\dfrac{a^{-1}}{x^2}\right)^{-2}$ _____ 56. $\left(\dfrac{a^{-2}}{b^3}\right)^{-1}$

_____ 57. $\left(\dfrac{x^{-4}}{y^{-2}}\right)^{-3}$ _____ 58. $\left(\dfrac{x^{-5}}{y^{-3}}\right)^{-2}$

_____ 59. $\left(\dfrac{-x^4}{3y^{-1}}\right)^{-2}$ _____ 60. $\left(\dfrac{-x^2}{2y^{-3}}\right)^{-3}$

☆ Stretching the Topics

Simplify. Write answers without negative exponents.

_____ 1. $(3^{-1}x^5y^{-2}z^3)^{-2}(6x^{-2}y^{-4}z^{-6})^{-1}$ _____ 2. $\left(\dfrac{2a^{-2}b}{b^{-3}}\right)^2 \cdot \left(\dfrac{3a^4b^{-1}}{b^3}\right)^{-3}$

_____ 3. $\dfrac{(-x^{-3}y^{-2})^{-2}}{(x^{-2}y^2)^{-3}}$

Check your answers in the back of your book.

If you can simplify the expressions in **Checkup 5.3**, you are ready to do the **Review Exercises** for Chapter 5.

Name _____   **Date** _____

✓ CHECKUP 5.3

Simplify and write answers without negative exponents. (Assume that variables in denominators are not zero.)

_____ **1.** $(2x)^{-5}$

_____ **2.** $a^7 \cdot a^{-3}$

_____ **3.** $3y^{-2}$

_____ **4.** $2x^{-8}(x^{-4})^{-2}$

_____ **5.** $\dfrac{-12x^5}{4x^{-2}}$

_____ **6.** $\dfrac{9a^{-3}}{18a^{-5}}$

_____ **7.** $\dfrac{(-2x^{-4})(x^6)}{6x^{-2}}$

_____ **8.** $\dfrac{9x^{-4}y^{-5}}{3x^4y^{-5}}$

_____ **9.** $\left(\dfrac{x^{-2}}{3}\right)^{-2}$

_____ **10.** $\left(\dfrac{3a^3}{b^{-2}}\right)^{-3}$

Check your answers in the back of your book.

If You Missed Problems:	You Should Review Examples:
1	3
2	5, 6
3	4
4	7–10
5–8	21–23
9, 10	24, 25

258

Summary

In this chapter we discussed integer exponents, beginning with whole-number exponents. We again state the definitions and laws that we used in working with exponents. Here a and b stand for any bases, and m and n represent whole numbers.

	Symbols	Examples
Definition of a^n	$a^n = \underbrace{a \cdot a \cdot \ldots \cdot a}_{n \text{ factors}}$	$2^5 = 2 \cdot 2 \cdot 2 \cdot 2 \cdot 2 = 32$ $x^3 = x \cdot x \cdot x$
Definition of a^0	$a^0 = 1 \ (a \neq 0)$	$17^0 = 1$ $x^0 = 1$
First Law of Exponents	$a^m \cdot a^n = a^{m+n}$	$2^5 \cdot 2^3 = 2^{5+3} = 2^8$ $x^2 \cdot x = x^{2+1} = x^3$
Second Law of Exponents	$(a^m)^n = a^{m \cdot n}$	$(3^2)^4 = 3^{2 \cdot 4} = 3^8$ $(x^3)^6 = x^{3 \cdot 6} = x^{18}$
Third Law of Exponents	$(a \cdot b)^n = a^n \cdot b^n$	$(2 \cdot 5)^3 = 2^3 \cdot 5^3$ $(3xy^2)^4 = 3^4 \cdot x^4 \cdot (y^2)^4 = 81x^4y^8$
Fourth Law of Exponents $(a \neq 0)$	$\dfrac{a^m}{a^n} = a^{m-n}$ if $m \geq n$ $\dfrac{a^m}{a^n} = \dfrac{1}{a^{n-m}}$ if $n > m$	$\dfrac{2^9}{2^3} = 2^{9-3} = 2^6$ $\dfrac{y}{y^5} = \dfrac{1}{y^{5-1}} = \dfrac{1}{y^4}$
Fifth Law of Exponents	$\left(\dfrac{a}{b}\right)^n = \dfrac{a^n}{b^n} \ (b \neq 0)$	$\left(\dfrac{2}{3}\right)^4 = \dfrac{2^4}{3^4} = \dfrac{16}{81}$ $\left(\dfrac{x}{y}\right)^3 = \dfrac{x^3}{y^3}$ $\left(\dfrac{3x^3}{z}\right)^2 = \dfrac{(3x^3)^2}{z^2} = \dfrac{3^2(x^3)^2}{z^2} = \dfrac{9x^6}{z^2}$

We also learned that *negative integers* can be used as exponents and continue to obey all the laws of exponents.

	Symbols	Examples
Definition	$a^{-n} = \dfrac{1}{a^n}$	$x^{-3} = \dfrac{1}{x^3}$
First Law of Exponents	$a^m \cdot a^n = a^{m+n}$	$x^{-3} \cdot x^{-2} = x^{-5}\dfrac{1}{x^5}$
		$x^2 \cdot x \cdot x^{-3} = x^0 = 1$
Second Law of Exponents	$(a^m)^n = a^{mn}$	$(x^{-3})^{-4} = x^{12}$
		$(x^{-1})^3 = x^{-3} = \dfrac{1}{x^3}$
Third Law of Exponents	$(ab)^n = a^n \cdot b^n$	$(2x)^{-3} = 2^{-3}x^{-3} = \dfrac{1}{8x^3}$
		$(x^{-1}y^2)^{-2} = x^2y^{-4} = \dfrac{x^2}{y^4}$
Fourth Law of Exponents	$\dfrac{a^m}{a^n} = a^{m-n}$ $(a \neq 0)$	$\dfrac{x^3}{x^5} = x^{-2} = \dfrac{1}{x^2}$
		$\dfrac{y^{-2}}{y^{-6}} = y^{-2-(-6)} = y^4$
Fifth Law of Exponents	$\left(\dfrac{a}{b}\right)^n = \dfrac{a^n}{b^n}$ $(b \neq 0)$	$\left(\dfrac{2}{x}\right)^{-3} = \dfrac{2^{-3}}{x^{-3}} = \dfrac{x^3}{8}$
		$\left(\dfrac{x^2}{2y^{-1}}\right)^{-2} = \dfrac{x^{-4}}{2^{-2}y^2} = \dfrac{2^2}{x^4y^2} = \dfrac{4}{x^4y^2}$

☐ Speaking the Language of Algebra

Complete each statement with the appropriate word or phrase.

1. In the expression a^n, we call a the _____ and we call n the _____ . If n is a positive integer, the expression a^n tells us to use _____ as a factor in a product _____ times.

2. An exponent belongs only to the base to which it is attached unless _____ indicate otherwise.

3. To multiply powers of the same base, we keep the same base and _____ the exponents.

4. To raise a power to a power, we keep the same base and _____ the exponents.

5. To raise a product to a power, we raise each _____ to that power.

6. To divide powers of the same base, we keep the same base and _____ the exponent in the _____ from the exponent in the _____ .

7. To raise a fraction to a power, we raise the _____ and the _____ to that power.

8. The expression a^{-1} tells us to find the _____ of a.

9. If a factor with a negative exponent appears in the denominator of a fraction, that factor should be moved to the _____ , where its exponent will be positive.

10. For any base a (except 0), $a^n \cdot a^{-n} =$ _____ .

Check your answers in the back of your book.

△ Writing About Mathematics

Write your response to each question in complete sentences.

1. Explain the steps that you would use to find the power $(-2x^2y)^5$.

2. Explain the difference between $-x^4$ and $(-x)^4$.

3. By substituting nonzero constants for a and b, decide whether $(a + b)^3 = a^3 + b^3$.

4. Explain why $3x^{-4}$ and $(3x)^{-4}$ do not mean the same thing.

REVIEW EXERCISES for Chapter 5

Use exponents to rewrite each expression.

_____ 1. $-2 \cdot x \cdot x \cdot y \cdot y \cdot y$

_____ 2. $\left(\dfrac{1}{3}a\right)\left(\dfrac{1}{3}a\right)\left(\dfrac{1}{3}a\right)$

Write each expression without exponents.

_____ 3. $-3m^3n^2$

_____ 4. $(0.5a)^3$

Evaluate each expression.

_____ 5. $(-3)^3$

_____ 6. $\left(-\dfrac{1}{2}\right)^4$

_____ 7. $-2a^2$ when $a = -1$

_____ 8. $(5m^2n)^0$ when $m = 3$ and $n = -1$

Perform the indicated operations and simplify. Write answers without negative exponents.
(Assume that variables in the denominators are not zero.)

_____ 9. $x^7 \cdot x^3$

_____ 10. $-2a^3(a^2)(a^4)$

_____ 11. $9xy^6 \cdot x^4y^2$

_____ 12. $(m^2)^3$

_____ 13. $(-3y)^3$

_____ 14. $(x^3y^2z)^2$

_____ 15. $3a(a^4b^2)^3$

_____ 16. $\left(\dfrac{1}{3}x\right)^3$

_____ 17. $\dfrac{x^5}{x^8}$

_____ 18. $\dfrac{-40x^{12}}{8x^3}$

_____ 19. $\dfrac{-12x^6y^3}{4x^3y^9}$

_____ 20. $\dfrac{-x^9y^5z}{x^{12}y^3z}$

_____ 21. $\dfrac{(2x^3y)^3}{(-xy^2)^2}$

_____ 22. $\left(\dfrac{-3a^2}{2b}\right)^3$

_____ 23. $\left(\dfrac{22a^4b^2}{11ab^7}\right)^4$

_____ 24. $(5x)^{-2}$

_____ 25. $4y^{-3}$

_____ 26. $x^5 \cdot x^{-9}$

_____ 27. $4a^{-5} \cdot a^{-3} \cdot a^{10}$

_____ 28. $(2x^{-1})^3 \cdot (x^4)^{-2}$

_____ 29. $(-3x^{-1})^{-2} \cdot (x^{-2})^2$

_____ 30. $\dfrac{-35x^6}{5x^{-2}}$

————— 31. $\dfrac{13a^{-2}}{52a^{-5}}$ ————— 32. $\left(\dfrac{3x}{2}\right)^{-4}$

————— 33. $\left(\dfrac{-2y^{-3}}{y^{-4}}\right)^{2}$ ————— 34. $\dfrac{(-2x^{-3})^{2}(x^{7})}{16x^{-3}}$

Check your answers in the back of your book.

If You Missed Exercises:	You Should Review Examples:	
1, 2	Section 5.1	1–3
3, 4		4–8
5–8		1, 9–11
9, 10		15, 16
11		19
12		25
13		27
14–16		29, 30
17	Section 5.2	4
18–20		5–8
21		9
22		12
23		13
24, 25	Section 5.3	1–4
26, 27		5–8
28, 29		12–16
30, 31		21, 22
32, 33		24, 25
34		23

If you have completed the **Review Exercises** and corrected your errors, you are ready to take the **Practice Test** for Chapter 5.

PRACTICE TEST for Chapter 5

Evaluate each expression.

		Section	Examples
_____	**1.** $(-4)^3$	5.1	2
_____	**2.** $\frac{1}{3}a^2b^3$ when $a = 3$ and $b = -1$	5.1	10
_____	**3.** $4(2x - 3y)^0$ when $x = 1$ and $y = 1$	5.1	14

Perform the indicated operation and simplify. Write answers without negative exponents.
(Assume that variables in the denominators are not zero.)

		Section	Examples
_____	**4.** $x^6 \cdot x^3$	5.1	15
_____	**5.** $(-2y^3)(y^2)(y^3)$	5.1	16
_____	**6.** $4x^5y^2z \cdot x^2yz^3$	5.1	19, 20
_____	**7.** $(a^4)^3$	5.1	22
_____	**8.** $(x^3)^4 \cdot (y^2)^3$	5.1	24
_____	**9.** $(-3x)^3$	5.1	28
_____	**10.** $(-2a^4b^3)^4$	5.1	29
_____	**11.** $5a(a^2b)^3$	5.1	30
_____	**12.** $\dfrac{x^7}{x^{15}}$	5.2	4
_____	**13.** $\dfrac{-36x^{11}}{18x^6}$	5.2	5
_____	**14.** $\dfrac{-24x^9y^3}{8x^4y^7}$	5.2	7
_____	**15.** $\dfrac{-a^{12}b^3c^2}{a^{15}bc^2}$	5.2	8
_____	**16.** $\dfrac{(-2x^2y^3)^5}{(x^2y)^4}$	5.2	9
_____	**17.** $\left(\dfrac{-3a^5}{4b^2}\right)^3$	5.2	12
_____	**18.** $\left(\dfrac{54x^4y^3}{9xy^7}\right)^4$	5.2	13

		SECTION	EXAMPLE
_____ 19. $7x^{-4}$		5.3	4
_____ 20. $-5w^{-3}(w^{-2})$		5.3	7
_____ 21. $(3a)^{-4}$		5.3	14
_____ 22. $(-5x^{-1}y^4)^{-2}(x^3y^{-2})^2$		5.3	19
_____ 23. $\dfrac{-12x^{-6}}{4x^{-2}}$		5.3	21
_____ 24. $\dfrac{28x^2y^{-4}}{7x^{-3}y}$		5.3	22
_____ 25. $\left(\dfrac{-4a^{-5}}{3b^5}\right)^{-1}$		5.3	26

Check your answers in the back of your book.

Name _____ Date _____

SHARPENING YOUR SKILLS after Chapters 1–5

Perform the indicated operations.

SECTION

_____ 1. $\dfrac{4(-9) + (-5)(-2)}{13(-1)}$

2.3

_____ 2. $-4[2(3a - 2b) - 3(2a - 3b)]$

3.3

_____ 3. $9x - 5y + 3z - 7x + 4y - 2z$

3.2

_____ 4. $19.24 - 3(6.13)$

1.3

_____ 5. $\dfrac{(-0.4)^2 - 2^2}{(-8)^2 - 5(6)(2)}$

2.4

_____ 6. $\dfrac{13}{28} - \dfrac{20}{21} + \dfrac{7}{12}$

2.4

_____ 7. Is -9 a solution for $7x - 4 = 6x + 5$?

4.1

_____ 8. Write an algebraic expression for 8 times the sum of a variable and 13.

3.1

Solve each equation.

_____ 9. $\dfrac{3}{11} x = 6$

4.1

_____ 10. $5 - 0.3x = -7$

4.2

_____ 11. $4[2(x - 2) - (x + 4)] = 0$

4.2

_____ 12. $3[x - (2x - 1)] = -2(x - 3) - 6$

4.2

_____ 13. Solve $P = 2l + 2w$ for w.

4.2

Use an equation to solve each problem.

_____ 14. Sam and Ethel are collecting aluminum cans for recycling. Sam has collected 50 more cans than twice the number Ethel has collected. If the total number of cans both collected is 290, how many did each collect?

4.3

_____ 15. The length of a rectangle is 4 times the width. If the perimeter is 40 feet, find its dimensions.

4.3

Check your answers in the back of your book.

6

Working with Polynomials

Mr. Abell's garden is in the shape of a square. Next year he plans to enlarge his garden by increasing the length by 6 feet and the width by 4 feet. Write an expression for the area of the garden after it has been enlarged.

Before you finish this chapter, you will be able to write an expression for the area of Mr. Abell's garden. To do this, we must learn how to

1. Identify polynomials.
2. Add and subtract polynomials.
3. Multiply polynomials.
4. Divide polynomials.
5. Switch from word expressions to polynomials.

6.1 Simplifying Polynomials

Identifying Polynomials

Now that we have discussed the meaning of exponents, we can state the definition of a **polynomial**.

Definition of a Polynomial. A polynomial in x is a *sum* of one or more terms of the form

$$ax^n$$

where a is a constant and n is a *whole number*.

It is important to understand that the exponent on the variable in each term of a polynomial must be a *whole number* and that a polynomial must be expressible as a *sum* of such terms.

Learning to recognize algebraic expressions that are *not* polynomials will help us understand what polynomials must look like.

Polynomials	Not Polynomials
x^9	$\dfrac{1}{x^9}$
$x^2 - 3x + 8$	$\dfrac{x^2 - 3x + 8}{x + 2}$
$\dfrac{1}{2}x^3 + 5x$	$\sqrt{x} + 5$

Mathematicians give different names to polynomials depending on *how many terms* the polynomial contains.

Number of Terms	Name of Polynomial	Examples
One	Monomial	$7x$ $-y^3$
Two	Binomial	$2x + 1$ $x^2 - 5y$
Three	Trinomial	$3a + 2b - 7$ $6x^2 + x - 2$

If a polynomial contains *more* than 3 terms, it is simply referred to as a polynomial of so-many terms.

$$2x + 3y - 4z + 8 \quad \text{is a polynomial of 4 terms}$$

$$a + b - c + d + e \quad \text{is a polynomial of 5 terms}$$

Example 1. Name each polynomial.

$-16x$ is a monomial.
$x^2 + x + 2$ is a trinomial.
$ab - c$ is a binomial.

Complete Example 2.

Example 2. Name each polynomial.

$x^2 - y^2$ is a _____ .
xyz is a _____ .
$6a^3 - a^2 + 3$ is a _____ .

Check your work on page 274. ▶

We should combine like terms, if possible, before naming a polynomial. Remember that terms are called "like terms" if their variable parts contain exactly the same variable raised to the same power.

Example 3. Name the polynomial
$x^2 + x + 3x^2 + 2$.

Solution. We must combine like terms first.

$x^2 + x + 3x^2 + 2$ Look for like terms.

$= x^2 + 3x^2 + x + 2$ Rearrange terms.

$= 4x^2 + x + 2$ Combine like terms.

This is a trinomial.

Now you try Example 4.

Example 4. Name the polynomial
$ab + 2ab + ac$.

Solution. We must combine like terms first.

$ab + 2ab + ac$

$= \underline{\quad} + \underline{\quad}$

This is a _____ .

Check your work on page 274. ▶

⇒ Trial Run

Name each polynomial.

_____ 1. $-3x^2$

_____ 2. $-3x + 5$

_____ 3. $3ab - 2a + b$

_____ 4. $-2x + x - 2y^3$

_____ 5. $-3m + 4m + 2m + m$

_____ 6. $xy - 3xz + 4yz$

Answers are on page 274.

Simplifying Polynomials

Adding polynomials is simply a matter of **combining like terms**. Remember that *to combine like terms, we combine the numerical coefficients and keep the variable part.*

Example 5. Simplify

$$(2x^2 + 7x - 3) + (x^2 - 5x - 1)$$

Solution

$$(2x^2 + 7x - 3) + (x^2 - 5x - 1)$$

$$= 2x^2 + 7x - 3 + x^2 - 5x - 1$$

Remove parentheses.

$$= 2x^2 + x^2 + 7x - 5x - 3 - 1$$

Rearrange terms.

$$= 3x^2 + 2x - 4 \qquad \text{Combine like terms.}$$

Now you try Example 6.

Example 6. Simplify

$$(-x^2 - 5x + 9) + (x^2 + x - 2)$$

Solution

Check your work on page 274. ▶

To multiply a constant times a monomial, we use the method learned in Chapter 3. We multiply the constant times the numerical coefficient and keep the variable part.

Example 7. Simplify $5(-x)$.

Solution

$$5(-x) = 5(-1x)$$

$$= 5(-1)x$$

$$= -5x$$

Now complete Example 8.

Example 8. Simplify $-\dfrac{3}{4}(4y^3)$.

Solution

$$-\frac{3}{4}(4y^3) = \left(\frac{-3}{4} \cdot \underline{\qquad}\right)y^3$$

$$= \underline{\qquad}y^3$$

Check your work on page 274. ▶

The **distributive property** allows us to multiply constants times polynomials, just as we did in Chapter 3.

Example 9. Simplify $2(x^2 + 3x + 1)$.

Solution

$$2(x^2 + 3x + 1)$$

$$= 2x^2 + 6x + 2$$

You complete Example 10.

Example 10. Simplify $-(5a^2 - 3a + 2)$.

Solution

$$-(5a^2 - 3a + 2)$$

$$= -1(5a^2 - 3a + 2)$$

$$= \underline{\qquad}$$

Check your work on page 274. ▶

We must use all these ideas to simplify polynomial sums or differences.

Example 11. Simplify $3(2a - a^2) - (a^2 - a + 3)$.

Solution

$$3(2a - a^2) - (a^2 - a + 3)$$

$= 3(2a - a^2) - 1(a^2 - a + 3)$ Insert 1 before second parentheses.

$= 6a - 3a^2 - a^2 + a - 3$ Use the distributive property.

$= -4a^2 + 7a - 3$ Combine like terms.

Now you complete Example 12.

Example 12. Simplify $2(x^2 - x) + 5(2x^2 - 1)$.

Solution

$$2(x^2 - x) + 5(2x^2 - 1)$$

$= \underline{\quad}x^2 - \underline{\quad}x + \underline{\quad}x^2 - \underline{\quad}$ Use the distributive property.

$= \underline{\quad\quad}$ Combine like terms.

Check your work on page 274. ▶

Notice that it is customary to write polynomials with the exponents in decreasing order from left to right. This is called **writing a polynomial in descending powers of the variable.**

⫸ Trial Run

Perform the indicated operations.

_____ **1.** $5(-2xy)$ _____ **2.** $2(x + y) - 3(2x - y)$

_____ **3.** $5(x^2 - 1) + 2(2x^2 - 3x)$ _____ **4.** $(a^2 - 2a + 1) - (3a^2 + a + 5)$

_____ **5.** $-2(x^2 - 3) + (2x^2 - 5)$ _____ **6.** $(y^2 + 2y + 1) - (y^2 - 2y + 1)$

Answers are on page 274.

▶ **Examples You Completed** _____

Example 2. Name each polynomial.

$$x^2 - y^2 \text{ is a binomial.}$$

$$xyz \text{ is a monomial}$$

$$6a^3 - a^2 + 3 \text{ is a trinomial.}$$

Example 4. Name the polynomial $ab + 2ab + ac$.

Solution. We must combine like terms first.

$$ab + 2ab + ac$$

$$= 3ab + ac$$

This is a binomial.

Example 6. Simplify
$(-x^2 - 5x + 9) + (x^2 + x - 2)$

Solution

$$(-x^2 - 5x + 9) + (x^2 + x - 2)$$

$$= -x^2 - 5x + 9 + x^2 + x - 2$$

$$= -x^2 + x^2 - 5x + x + 9 - 2$$

$$= -4x + 7$$

Example 8. Simplify $-\dfrac{3}{4}(4y^3)$.

Solution

$$-\frac{3}{4}(4y^3) = \left(\frac{-3}{4} \cdot \frac{4}{1}\right)y^3$$

$$= -3y^3$$

Example 10. Simplify $-(5a^2 - 3a + 2)$.

Solution

$$-(5a^2 - 3a + 2)$$

$$= -1(5a^2 - 3a + 2)$$

$$= -5a^2 + 3a - 2$$

Example 12. Simplify $2(x^2 - x) + 5(2x^2 - 1)$.

Solution

$$2(x^2 - x) + 5(2x^2 - 1)$$

$$= 2x^2 - 2x + 10x^2 - 5$$

$$= 12x^2 - 2x - 5$$

Answers to Trial Runs _____

page 271   **1.** Monomial   **2.** Binomial   **3.** Trinomial   **4.** Binomial   **5.** Monomial   **6.** Trinomial

page 273   **1.** $-10xy$ **2.** $-4x + 5y$ **3.** $9x^2 - 6x - 5$ **4.** $-2a^2 - 3a - 4$ **5.** 1   **6.** $4y$

EXERCISE SET 6.1

Name each expression.

_____ 1. $-3x^2$

_____ 2. $-5y^4$

_____ 3. $7x - \dfrac{1}{4}y$

_____ 4. $-3x + \dfrac{2}{5}y$

_____ 5. $4ab - 2a + b$

_____ 6. $a - b + 3ab$

_____ 7. $-7m^2 + 2n^2 + 10m^2$

_____ 8. $8n^2 - 6m^2 + n^2$

_____ 9. $0.3a - 0.5b - 0.7a + 0.6b$

_____ 10. $0.9a - 0.6b - 0.4a + 0.9b$

_____ 11. $ab - ac + bc$

_____ 12. $3ab + 2ac - 5bc$

_____ 13. $4a^3 - 2a^2 + a - 5$

_____ 14. $-5a^3 - 4a^2 + 3a - 9$

Perform the indicated operations.

_____ 15. $(x^2 - 3x + 5) + (3x^2 + x - 1)$

_____ 16. $(3x^2 - 4x + 2) + (x^2 + x - 5)$

_____ 17. $(a - 3b + 2c) + (4a + 3b - 5c)$

_____ 18. $(2a - 5b + c) + (7a + 5b - 2c)$

_____ 19. $(-5x^2 + x - 4) + (x^2 - 3x - 5)$

_____ 20. $(-4x^2 + 3x - 6) + (x^2 - x - 7)$

_____ 21. $-3(4x^2)$

_____ 22. $-2(5x^4)$

_____ 23. $4(-2xy)$

_____ 24. $7(-5xy)$

_____ 25. $\dfrac{2}{3}(9a)$

_____ 26. $\dfrac{3}{5}(15a)$

_____ 27. $0.4(-2ab^3)$

_____ 28. $0.5(-4ab^3)$

_____ 29. $-8\left(\dfrac{3}{4}x^5y^4\right)$

_____ 30. $-12\left(\dfrac{5}{6}x^4y^3\right)$

_____ 31. $5(x^2 - 3x + 2)$

_____ 32. $3(x^2 - 5x - 6)$

_____ 33. $-(6a - 4b + c)$

_____ 34. $-(3a - 2b + 5c)$

_____ 35. $-4(2a^2 - 2a + 1)$

_____ 36. $-3(5a^2 - a - 1)$

_____ 37. $\dfrac{1}{3}(9x^2 - 6x + 3)$

_____ 38. $\dfrac{1}{4}(8x^2 - 4x + 16)$

_____ 39. $(x + y) - (2x - y)$

_____ 40. $(2x + y) - (3x - y)$

_____ **41.** $-2(x + y) + 3(2x - 5y)$ _____ **42.** $-3(2x - y) + 4(x + y)$

_____ **43.** $0.4(x^2 - 3x) - 0.5(3x - 2)$ _____ **44.** $0.3(2x^2 - 5x) - 0.7(-2x + 5)$

_____ **45.** $-3(y^2 + 3y - 5) - (2y^2 + 3y - 6)$ _____ **46.** $-4(2y^2 - y - 4) - (y^2 - 2y + 2)$

_____ **47.** $-3(a^2 + 2) + (3a^2 + 7)$ _____ **48.** $(3x^2 - 6) - 3(x^2 - 2)$

_____ **49.** $5(y^2 - 2y + 1) - 2(3y^2 - 5y + 4)$ _____ **50.** $3(x^2 - 2x + 1) - 5(x^2 + x - 1)$

☆ Stretching the Topics ────────────────

Perform the indicated operations.

_____ **1.** $2(h + 3k) - 4(2h - 5k) - (h + 2k) + 6k$

_____ **2.** $2x(y + z) - 3x(2y - z) + x(y - 5z) - 3xz$

_____ **3.** The length of a rectangle is $2a - b$ inches. The width is $2b$ inches. What is its perimeter?

Check your answers in the back of your book.

If you can complete **Checkup 6.1**, you are ready to go on to Section 6.2.

CHECKUP 6.1

Name each expression.

_____1. $3x - y + 4xy$ _____2. $0.2a - 0.5b - 0.3a + 0.5b$

Perform the indicated operations.

_____3. $(-x^2 + 3x - 5) + (4x^2 - 3x - 4)$ _____4. $-4(-3xy)$

_____5. $\dfrac{4}{7}(21a^3b^2)$ _____6. $0.5(4x^2 - 8x + 2)$

_____7. $-(3a - 4b - 7c)$ _____8. $4(3x - x^2) - (x^2 + x - 20)$

_____9. $5(2x^2 - x) - 3(4x^2 - 2)$ _____10. $(3y^2 - 2y + 5) - (y^2 + 2y - 5)$

Check your answers in the back of your book.

If You Missed Problems:	You Should Review Examples:
1, 2	1–4
3	5, 6
4, 5	7, 8
6, 7	9, 10
8–10	11, 12

6.2 Multiplying with Monomials

Multiplying Monomials

In Chapter 5 we learned to use the First Law of Exponents to multiply powers of the same base. Recall that $a^m \cdot a^n = a^{m+n}$. We use the commutative and associative properties of multiplication when multiplying monomials containing constants *and* variables.

Example 1. Find $2x^3(4x^2)$.

Solution

$$2x^3(4x^2) = 2 \cdot x^3 \cdot 4 \cdot x^2$$
$$= 2 \cdot 4 \cdot x^3 \cdot x^2$$
$$= 8 \cdot x^{3+2}$$
$$= 8x^5$$

You try to complete Example 2.

Example 2. Find $-5y^8(6y^7)$.

Solution

$$-5y^8(6y^7) = -5 \cdot y^8 \cdot 6 \cdot y^7$$
$$= -5 \cdot 6 \cdot y^8 \cdot y^7$$

Check your work on page 281. ▶

To multiply monomials, first multiply the numerical coefficients, then multiply the variables using the First Law of Exponents.

Example 3. Find $4a^5\left(\dfrac{1}{3}a^7\right)$.

Solution

$$4a^5\left(\frac{1}{3}a^7\right) = 4 \cdot \frac{1}{3} \cdot a^5 \cdot a^7$$
$$= \frac{4}{3} \cdot a^{5+7}$$
$$= \frac{4}{3}a^{12}$$

You complete Example 4.

Example 4. Find $3x^2(-2x^4)(-7x)$.

Solution

$$3x^2(-2x^4)(-7x)$$
$$= 3(-2)(-7) \cdot x^2 \cdot x^4 \cdot x$$

Check your work on page 281. ▶

Example 5. Find $0.8x(3x^2y)$.

Solution

$$0.8x(3x^2y)$$
$$= 0.8 \cdot 3 \cdot x \cdot x^2 \cdot y$$
$$= 2.4x^3y$$

Try to complete Example 6.

Example 6. Find $(-2xy^2z)(-9x^2y^5z^7)$.

Solution

$$(-2xy^2z)(-9x^2y^5z^7)$$
$$= (-2)(-9) \cdot x \cdot x^2 \cdot y^2 \cdot y^5 \cdot z \cdot z^7$$
$$= \underline{\hspace{2cm}}$$

Check your work on page 282. ▶

⫸ Trial Run ▬▬▬▬▬▬▬▬▬▬▬▬▬▬▬▬▬▬▬▬▬▬▬

Find the products.

_____ 1. $(-4x^5)(7x^3)$

_____ 2. $(0.2a^3)(-5a^2)(6a^6)$

_____ 3. $(3x^2)(2x^5y^7)$

_____ 4. $(4x^2yz^5)\left(-\dfrac{1}{2}xy^3z^7\right)$

Answers are on page 282.

Multiplying Monomials Times Polynomials

The distributive property, the rules for signs, and the laws of exponents are all we need to multiply a monomial times a polynomial. The distributive property tells us to multiply each term inside the parentheses by the monomial outside the parentheses and the First Law of Exponents tells us how to simplify each of those multiplications.

Example 7. Find $x(x^2 + 3x + 1)$.

Solution

$$x(x^2 + 3x - 1)$$
$$= x \cdot x^2 + x \cdot 3x - x \cdot 1$$
$$= x^3 + 3x^2 - x$$

You complete Example 8.

Example 8. Find $2a(4a^3 - 5a - 3)$.

Solution

$$2a(4a^3 - 5a - 3)$$
$$= 2a \cdot 4a^3 - 2a \cdot 5a - 2a \cdot 3$$
$$= \text{_____}$$

Check your work on page 282. ▶

Remember that we learned to do such products without writing down each step, using arrows to keep our work organized.

Example 9. Find $-4xy(2x + y - 7)$.

Solution

$$-4xy(2x + y - 7)$$
$$= -8x^2y - 4xy^2 + 28xy$$

Complete Example 10.

Example 10. Find

$$x^3y(x^2y^2 - 4xy + 3xy^5)$$

Solution

$$x^3y(x^2y^2 - 4xy + 3xy^5)$$
$$= \text{_____}$$

Check your work on page 282. ▶

Example 11. Find

$$x^2(2x - y) + x(xy - 3x^2)$$

Solution

$$x^2(2x - y) + x(xy - 3x^2)$$

$$= 2x^3 - x^2y + x^2y - 3x^3$$

   Remove parentheses.

$$= 2x^3 - 3x^3 - x^2y + x^2y$$

   Rearrange terms.

$$= -x^3 \qquad \text{Combine like terms}$$

Now you try Example 12.

Example 12. Find

$$2a(x - 3xy) - 5x(a - 4ay)$$

Solution

Check your work on page 282. ▶

⫸ Trial Run

Perform the indicated operations.

_____ **1.** $x(x^2 - 5x + 2)$

_____ **2.** $-2a^2(a - 6)$

_____ **3.** $3x(4x^2 - 2x + 7)$

_____ **4.** $a^2b(a^2b^2 - 3ab + 4ab^4)$

_____ **5.** $x^2(3x + y) - x(xy - 2x^2)$

_____ **6.** $2a(x - 2xy) - 3x(a + 3ay)$

Answers are on page 282.

▶ Examples You Completed

Example 2. Find $-5y^8(6y^7)$.

   Solution

$$-5y^8(6y^7) = -5 \cdot y^8 \cdot 6 \cdot y^7$$

$$= -5 \cdot 6 \cdot y^8 \cdot y^7$$

$$= -30 \cdot y^{8+7}$$

$$= -30y^{15}$$

Example 4. Find $3x^2(-2x^4)(-7x)$.

   Solution

$$3x^2(-2x^4)(-7x)$$

$$= 3(-2)(-7) \cdot x^2 \cdot x^4 \cdot x$$

$$= 3(14) \cdot x^{2+4+1}$$

$$= 42x^7$$

Example 6. Find $(-2xy^2z)(-9x^2y^5z^7)$.

Solution

$$(-2xy^2z)(-9x^2y^5z^7)$$
$$= (-2)(-9) \cdot x \cdot x^2 \cdot y^2 \cdot y^5 \cdot z \cdot z^7$$
$$= 18x^3y^7z^8$$

Example 8. Find $2a(4a^3 - 5a - 3)$.

Solution

$$2a(4a^3 - 5a - 3)$$
$$= 2a \cdot 4a^3 - 2a \cdot 5a - 2a \cdot 3$$
$$= 8a^4 - 10a^2 - 6a$$

Example 10. Find $x^3y(x^2y^2 - 4xy + 3xy^5)$.

Solution

$$x^3y(x^2y^2 - 4xy + 3xy^5)$$
$$= x^5y^3 - 4x^4y^2 + 3x^4y^6$$

Example 12. Find $2a(x - 3xy) - 5x(a - 4ay)$.

Solution

$$2a(x - 3xy) - 5x(a - 4ay)$$
$$= 2ax - 6axy - 5ax + 20axy$$
$$= 2ax - 5ax - 6axy + 20axy$$
$$= -3ax + 14axy$$

Answers to Trial Runs

page 280   **1.** $-28x^8$ **2.** $-6a^{11}$ **3.** $6x^7y^7$ **4.** $-2x^3y^4z^{12}$

page 281   **1.** $x^3 - 5x^2 + 2x$ **2.** $-2a^3 + 12a^2$ **3.** $12x^3 - 6x^2 + 21x$ **4.** $a^4b^3 - 3a^3b^2 + 4a^3b^5$
5. $5x^3$ **6.** $-ax - 13axy$

EXERCISE SET 6.2

Perform the indicated operations.

_____ **1.** $2x^2(5x^3)$ _____ **2.** $3x^3(7x^2)$

_____ **3.** $\frac{1}{2}x^7(4x)$ _____ **4.** $\frac{2}{3}x^5(3x)$

_____ **5.** $-2a^5(-5a^7)$ _____ **6.** $-3a^4(-5a^6)$

_____ **7.** $0.9y^3(-y)$ _____ **8.** $0.7y^4(-y)$

_____ **9.** $3m^2(-2m)(m^5)$ _____ **10.** $4m^3(-3m)(m^2)$

_____ **11.** $4x^2y^3(-3x^5y^2)$ _____ **12.** $8x^3y^4(-2x^2y^5)$

_____ **13.** $\frac{1}{4}x(-x^2y)(-8x^4y^3)$ _____ **14.** $-\frac{1}{5}x(-x^3y)(15x^4y^3)$

_____ **15.** $7xy(-2xy)(-3x^2y)$ _____ **16.** $-2xy^2(8x^2y^3)(-3xy)$

_____ **17.** $5x^2y^3z(7xyz^4)$ _____ **18.** $9x^2y^3z^5(3xyz^2)$

_____ **19.** $-3x^2y(8x^3z^2)$ _____ **20.** $-4x^3y^2(9x^5z^3)$

_____ **21.** $5x(2x^3)^2$ _____ **22.** $6x(3x^4)^2$

_____ **23.** $(5y^4)^2(-2y^5)^3$ _____ **24.** $(7y^2)^2(-y^4)^5$

_____ **25.** $(-6a^2b)(2ab^3)^2$ _____ **26.** $(2a^2b)(-3a^4b^2)^3$

_____ **27.** $5(2x+3)$ _____ **28.** $4(x+5)$

_____ **29.** $-7(2x^2-4)$ _____ **30.** $-3(4x^2-7)$

_____ **31.** $3x(2x-4)$ _____ **32.** $5x(3x-2)$

_____ **33.** $y^2(2y-5)$ _____ **34.** $y^2(3y-8)$

_____ **35.** $-a^2(a^2-2)$ _____ **36.** $-a^2(2a^2-3)$

_____ **37.** $4x^2(9x^2-4)$ _____ **38.** $5x^2(8x^2-1)$

_____ **39.** $-3a^3(7a^2-3a)$ _____ **40.** $-4a^3(6a^2-5a)$

_____ **41.** $2xy(5x+2y)$ _____ **42.** $-5xy(3x+5y)$

_____ **43.** $-x^2y^2(11x-4y)$ _____ **44.** $-2x^2y^2(x-7y)$

_____ **45.** $2a^3b^2(7a^2-b^2)$ _____ **46.** $6a^3b^2(a^2-2b^2)$

_____ **47.** $2m(m^2-5m+6)$ _____ **48.** $9m(m^2-m-6)$

_____ 49. $5x^2y^2(3x^2y^2 - xy - 4)$

_____ 50. $4x^2y^2(5x^2y^2 - xy - 6)$

_____ 51. $-a^2b(3 - 2ab + 4a^2b^2)$

_____ 52. $-a^2b(5 - 3ab + 2a^2b^2)$

_____ 53. $4mn(8m^3 - 2m^2n + mn^2 - 3n^3)$

_____ 54. $5mn(6m^3 - m^2 + 2mn^2 - 7n^3)$

_____ 55. $x^2(2xy - 6) + 2x^2(xy + 3)$

_____ 56. $x(3xy - x^2) - x^2(y - x)$

_____ 57. $x^2(x^2 + 5) - 3(x^2 - 8)$

_____ 58. $x^2(x^2 - 4) + 2(x^2 - 5)$

_____ 59. $3m(3m + n) - n(3m + n)$

_____ 60. $7m(2m - n) - 2n(4m + n)$

_____ 61. $3ax(2 - 3y) - 2ax(1 - 2y)$

_____ 62. $5ax(3 - y) - 7ax(4 - 3y)$

_____ 63. $5xy(3x - 4y) - 2xy(4x - 5y)$

_____ 64. $8xy(2x - y) - xy(x - 2y)$

☆ Stretching the Topics

Perform the indicated operations.

_____ 1. $-m^3n^2(4m^2 + 5mn + n^2) + m^3n^2(m^2 - 2mn + 3n^2)$

_____ 2. $xy^2(4x^2 - 3xy + 4xy^2) - x^2y(xy + 4y^2 - 3y^3)$

_____ 3. $2ab^2(-a^3b^{-2} + 3a^2b^{-1} + a - a^{-1}b) + 6a^2b(a^2b^{-1} - a - b + 4a^{-2}b^2)$

Check your answers in the back of your book.

If you can complete **Checkup 6.2**, you are ready to go on to Section 6.3.

✓ **CHECKUP 6.2**

Perform the indicated operations.

_____ **1.** $-3x^4(7x^7)$ _____ **2.** $2a^3(-4a^5)(-a^4)$

_____ **3.** $(0.5xy^3z)(-4x^3yz^2)$ _____ **4.** $-2a^2(a^2 - 7a + 5)$

_____ **5.** $3xy(5x - 2y + 4)$ _____ **6.** $6x^2y^2(2x^2y^2 - 3xy + 4)$

_____ **7.** $-2mn(4m^3 - m^2n + 2mn^2 - n^3)$ _____ **8.** $2x^2(3xy - 2) - x^2(7xy - 4)$

_____ **9.** $5x(3x + y) - y(4x - y)$ _____ **10.** $-3ax(1 - 2y) + 2ax(5 - 3y)$

Check your answers in the back of your book.

If You Missed Problems:	You Should Review Examples:
1	1–3
2, 3	4–6
4–7	7–10
8–10	11, 12

6.3 Multiplying Binomials

Multiplying Binomials

Recall how we used the distributive property to multiply monomials times binomials. Now suppose that we wish to find the product of *two binomials*, such as

$$(2x + 5)(x + 3)$$

According to the distributive property, each term in the first binomial $2x + 5$ must be multiplied times the second binomial, $x + 3$. Then we combine like terms.

Multiply first term, $2x$, times $x + 3$.

$$(2x + 5)(x + 3) = 2x(x + 3) + 5(x + 3)$$

Multiply second term, 5, times $x + 3$.

$$= 2x^2 + 6x + 5x + 15 \qquad \text{Use the distributive property again.}$$
$$= 2x^2 + 11x + 15 \qquad \text{Combine like terms.}$$

Example 1. Find $(x - 2)(3x + 1)$.

Solution

$$(x - 2)(3x + 1)$$
$$= x(3x + 1) - 2(3x + 1)$$
$$= 3x^2 + x - 6x - 2$$
$$= 3x^2 - 5x - 2$$

Now you try Example 2.

Example 2. Find $(5 - x)(4 + 7x)$.

Solution

$$(5 - x)(4 + 7x)$$
$$= 5(4 + 7x) - x(4 + 7x)$$

Check your work on page 292. ▶

Let us now consider some "special products."

Example 3. Find $(x + 5y)^2$.

Solution

$$(x + 5y)^2$$
$$= (x + 5y)(x + 5y)$$
$$= x(x + 5y) + 5y(x + 5y)$$
$$= x^2 + 5xy + 5xy + 25y^2$$
$$= x^2 + 10xy + 25y^2$$

You try Example 4.

Example 4. Find $(7 - a)^2$.

Solution

$$(7 - a)^2$$
$$= (7 - a)(7 - a)$$
$$= 7(7 - a) - a(7 - a)$$

Check your work on page 292. ▶

Examples 3 and 4 illustrate a very interesting "special product." In each case, we were asked to *square a binomial*. Notice how the terms in the answer are related to the terms in the original binomial.

$$\underset{\text{first term}}{\downarrow} \quad \underset{\substack{\text{square of} \\ \text{first term}}}{\downarrow} \qquad \underset{\substack{\text{square of} \\ \text{second term}}}{\downarrow}$$

$$(x + 5y)^2 = x^2 + 10xy + 25y^2$$

$$\underset{\text{second term}}{\uparrow} \qquad \underset{\substack{\textit{twice} \text{ the product} \\ \text{of first and} \\ \text{second terms}}}{\uparrow}$$

Indeed, this will always occur when we square a binomial.

> **Squaring a Binomial**
>
> $(A + B)^2 = A^2 + 2AB + B^2$
>
> $(A - B)^2 = A^2 - 2AB + B^2$

The trinomial obtained by squaring a binomial is sometimes called a **perfect square trinomial**.

Example 5. Find $(x + 3)(x - 3)$.

Solution

$$(x + 3)(x - 3)$$
$$= x(x - 3) + 3(x - 3)$$
$$= x^2 - 3x + 3x - 9$$
$$= x^2 - 9$$

Now try Example 6.

Example 6. Find $(8 - y)(8 + y)$.

Solution

$$(8 - y)(8 + y)$$
$$= 8(8 + y) - y(8 + y)$$

Check your work on page 292. ▶

Examples 5 and 6 illustrate another interesting "special product." In each case we multiplied the *sum* of two terms times the *difference* of the same two terms and our answer was just the **difference of the squares** of the original two terms.

> **Multiplying the Sum and Difference of Two Terms**
>
> $(A + B)(A - B) = A^2 - B^2$

We practice the special product formulas in the next two examples.

Example 7. Use the formula for squaring a binomial to find $(3x - 4)^2$.

Solution. Here, $A = 3x$ and $B = 4$.

$$(3x - 4)^2$$
$$= (3x)^2 - 2(3x)(4) + 4^2$$
$$= 9x^2 - 24x + 16$$

Example 8. Use the formula for the difference of two squares to find

$$\left(y + \frac{6}{5}\right)\left(y - \frac{6}{5}\right)$$

Solution. Here, $A = y$ and $B = \frac{6}{5}$.

$$\left(y + \frac{6}{5}\right)\left(y - \frac{6}{5}\right) = y^2 - \left(\frac{6}{5}\right)^2$$
$$= y^2 - \frac{36}{25}$$

These formulas can come in handy, but remember that you can always multiply binomials using the distributive property.

Example 9. Find $(x^2 - 0.3)(x - 0.5)$.

Solution

$$(x^2 - 0.3)(x - 0.5)$$
$$= x^2(x - 0.5) - 0.3(x - 0.5)$$
$$= x^3 - 0.5x^2 - 0.3x + 0.15$$

Note that no like terms appeared here after we multiplied the binomials.

Now try Example 10.

Example 10. Find $(x^3 - 4)(7x^3 - 10)$.

Solution

$$(x^3 - 4)(7x^3 - 10)$$
$$= x^3(7x^3 - 10) - 4(7x^3 - 10)$$

Check your work on page 292. ▶

⟾ Trial Run

Find the products.

_____ **1.** $(2x + 3)(x + 4)$

_____ **2.** $(x - 5)(3x + 2)$

_____ **3.** $(3 - x)(5 + 2x)$

_____ **4.** $(-2a - 1)(3b + 2)$

_____ **5.** $(2x - y)^2$

_____ **6.** $\left(y + \dfrac{1}{5}\right)\left(y - \dfrac{1}{5}\right)$

Answers are on page 292.

Multiplying Binomials by FOIL

We shall now discover a shortcut that we may use in multiplying two binomials. This shortcut actually uses the distributive property with less writing. Notice what happens if we multiply our binomials using the following steps.

1. Multiply the **F**irst terms in the binomials.
2. Multiply the **O**uter terms in the binomials.
3. Multiply the **I**nner terms in the binomials.
4. Multiply the **L**ast terms in the binomials.
5. Combine the results of those multiplications.

We abbreviate First by **F**, Outer by **O**, Inner by **I**, and Last by **L**.

$$\text{(x + 7)(2x - 3)} = 2x^2 - 3x + 14x - 21$$
$$= 2x^2 + 11x - 21$$

This is called the **FOIL method**.

Example 11. Find $(5x - 1)(3x + 3)$ using FOIL.

Solution

$$(5x - 1)(3x + 3)$$

$$= (5x - 1)(3x + 3)$$

$$= 15x^2 + 15x - 3x - 3$$

$$= 15x^2 + 12x - 3$$

Now you use FOIL for Example 12.

Example 12. Find $(2x + 3y)(2x - 8y)$ using FOIL.

Solution

$$(2x + 3y)(2x - 8y)$$

$$= (2x + 3y)(2x - 8y)$$

$$= 4x^2 - \underline{\quad} + \underline{\quad} - \underline{\quad}$$

$$= 4x^2 - \underline{\quad} - \underline{\quad}$$

Check your work on page 292. ▶

Remember that the FOIL method is merely a shortcut for writing down the terms of the product of two binomials. The process still uses the *distributive property*.

⫸ Trial Run

Find the products.

_____ 1. $(3x - 7)(x + 2)$

_____ 2. $(x^2 - 0.9)(x^2 + 0.7)$

_____ 3. $(7y - x)(2y + x)$

_____ 4. $(a - 2)(2b + 5)$

_____ 5. $(3x + 2)^2$

_____ 6. $(2x^3 - 7)(x^3 - 4)$

Answers are on page 292.

Multiplying More Polynomials

To find a product such as $(x + 2)(x^2 + 5x + 3)$ we must use the distributive property again. In this problem, we must multiply x times $x^2 + 5x + 3$ and then multiply 2 times $x^2 + 5x + 3$. Let's perform this multiplication.

$$(x + 2)(x^2 + 5x + 3) = x(x^2 + 5x + 3) + 2(x^2 + 5x + 3)$$

$$= x^3 + 5x^2 + 3x + 2x^2 + 10x + 6$$

$$= x^3 + 7x^2 + 13x + 6$$

To multiply polynomials, we use the distributive property and then combine like terms.

Try to complete Example 13.

Example 13. Find the product

$$(x - 3)(x^2 + 2x - 4)$$

Solution

$$(x - 3)(x^2 + 2x - 4)$$
$$= x(x^2 + 2x - 4) - 3(x^2 + 2x - 4)$$

Check your work on page 292. ▶

Example 14. Find the product

$$(2x^2 + 1)(x^2 - x - 1)$$

Solution

$$(2x^2 + 1)(x^2 - x - 1)$$
$$= 2x^2(x^2 - x - 1) + 1(x^2 - x - 1)$$
$$= 2x^4 - 2x^3 - 2x^2 + x^2 - x - 1$$
$$= 2x^4 - 2x^3 - x^2 - x - 1$$

To multiply a monomial times two binomials it is easier to multiply the binomials together first and then multiply that product by the monomial.

Example 15. Find $2x(x - 3)(5x + 1)$.

Solution

$$2x(x - 3)(5x + 1)$$
$$= 2x(5x^2 + x - 15x - 3)$$
$$= 2x(5x^2 - 14x - 3)$$
$$= 10x^3 - 28x^2 - 6x$$

You try Example 16.

Example 16. Find $-3y^2(6 - y)(5 + y)$.

Solution

$$-3y^2(6 - y)(5 + y)$$

Check your work on page 292. ▶

Unfortunately, there are no shortcuts to use here and lots of places to make errors. You should take your time and be careful when finding products of polynomials.

◗ **Trial Run**

Perform the indicated operations.

_____ **1.** $(x - 2)(x^2 + 2x + 4)$

_____ **2.** $-0.4(x + 1)^2$

_____ **3.** $3x(x - 2)(4x - 1)$

_____ **4.** $-y(y + 5)(y - 5)$

Answers are on page 292.

▶ Examples You Completed

Example 2. Find $(5 - x)(4 + 7x)$.

Solution

$$(5 - x)(4 + 7x)$$
$$= 5(4 + 7x) - x(4 + 7x)$$
$$= 20 + 35x - 4x - 7x^2$$
$$= 20 + 31x - 7x^2$$

Example 4. Find $(7 - a)^2$.

Solution

$$(7 - a)^2$$
$$= (7 - a)(7 - a)$$
$$= 7(7 - a) - a(7 - a)$$
$$= 49 - 7a - 7a + a^2$$
$$= 49 - 14a + a^2$$

Example 6. Find $(8 - y)(8 + y)$

Solution

$$(8 - y)(8 + y)$$
$$= 8(8 + y) - y(8 + y)$$
$$= 64 + 8y - 8y - y^2$$
$$= 64 - y^2$$

Example 10. Find $(x^3 - 4)(7x^3 - 10)$.

Solution

$$(x^3 - 4)(7x^3 - 10)$$
$$= x^3(7x^3 - 10) - 4(7x^3 - 10)$$
$$= 7x^6 - 10x^3 - 28x^3 + 40$$
$$= 7x^6 - 38x^3 + 40$$

Example 12. Find $(2x + 3y)(2x - 8y)$ using FOIL.

Solution

$$(2x + 3y)(2x - 8y)$$

$$= (2x + 3y)(2x - 8y)$$
F, L, I, O

$$= 4x^2 - 16xy + 6xy - 24y^2$$
$$= 4x^2 - 10xy - 24y^2$$

Example 13. Find $(x - 3)(x^2 + 2x - 4)$.

Solution

$$(x - 3)(x^2 + 2x - 4)$$
$$= x(x^2 + 2x - 4) - 3(x^2 + 2x - 4)$$
$$= x^3 + 2x^2 - 4x - 3x^2 - 6x + 12$$
$$= x^3 - x^2 - 10x + 12$$

Example 16. Find $-3y^2(6 - y)(5 + y)$.

Solution

$$-3y^2(6 - y)(5 + y)$$
$$= -3y^2(30 + 6y - 5y - y^2)$$
$$= -3y^2(30 + y - y^2)$$
$$= -90y^2 - 3y^3 + 3y^4$$

Answers to Trial Runs

page 289   **1.** $2x^2 + 11x + 12$ **2.** $3x^2 - 13x - 10$ **3.** $15 + x - 2x^2$ **4.** $-6ab - 4a - 3b - 2$ **5.** $4x^2 - 4xy + y^2$ **6.** $y^2 - \frac{1}{25}$

page 290   **1.** $3x^2 - x - 14$ **2.** $x^4 - 0.2x^2 - 0.63$ **3.** $14y^2 + 5xy - x^2$ **4.** $2ab + 5a - 4b - 10$ **5.** $9x^2 + 12x + 4$ **6.** $2x^6 - 15x^3 + 28$

page 291   **1.** $x^3 - 8$ **2.** $-0.4x^2 - 0.8x - 0.4$ **3.** $12x^3 - 27x^2 + 6x$ **4.** $-y^3 + 25y$

EXERCISE SET 6.3

Perform the indicated operations.

_____ **1.** $(x - 2)(x + 3)$ _____ **2.** $(x + 4)(x - 1)$

_____ **3.** $(x + 9)(x + 4)$ _____ **4.** $(x + 5)(x + 7)$

_____ **5.** $(x - 0.5)(x - 0.3)$ _____ **6.** $(x - 0.2)(x - 0.8)$

_____ **7.** $(y - 9)^2$ _____ **8.** $(y - 6)^2$

_____ **9.** $(a + 0.2)^2$ _____ **10.** $(a + 0.3)^2$

_____ **11.** $(x - 4)(x + 4)$ _____ **12.** $(x - 5)(x + 5)$

_____ **13.** $\left(y - \frac{1}{3}\right)\left(y + \frac{1}{3}\right)$ _____ **14.** $\left(y - \frac{2}{5}\right)\left(y + \frac{2}{5}\right)$

_____ **15.** $(4y - 1)(3y - 5)$ _____ **16.** $(5y - 2)(3y - 1)$

_____ **17.** $(2a + 7)(-3a + 5)$ _____ **18.** $(3a + 1)(-2a + 7)$

_____ **19.** $(6x - 5)(6x + 5)$ _____ **20.** $(5x - 3)(5x + 3)$

_____ **21.** $(9 - 2y)(2 - y)$ _____ **22.** $(8 - 3y)(3 - y)$

_____ **23.** $(2a - b)(a - 5b)$ _____ **24.** $(3a - b)(a - 4b)$

_____ **25.** $(x - 3y)(2x + y)$ _____ **26.** $(x - 2y)(3x + y)$

_____ **27.** $(-x + y)(2x - 3y)$ _____ **28.** $(-x + y)(3x - 5y)$

_____ **29.** $(a + 3b)^2$ _____ **30.** $(a + 5b)^2$

_____ **31.** $(0.2 + 0.5x)(0.2 - 0.5x)$ _____ **32.** $(0.3 - 0.2x)(0.3 + 0.2x)$

_____ **33.** $\left(\frac{1}{2}x - y\right)^2$ _____ **34.** $\left(\frac{1}{3}x - 6y\right)^2$

_____ **35.** $(a - 2b)(c + d)$ _____ **36.** $(x - y)(z + 2w)$

_____ **37.** $(a^2 - 4)(a - 3)$ _____ **38.** $(x^2 - 5)(x - 2)$

_____ **39.** $(6a^2 - 1)(a^2 - 5)$ _____ **40.** $(5a^2 - 2)(a^2 - 1)$

_____ **41.** $(5a^2 - 2)^2$ _____ **42.** $(4a^2 - 1)^2$

_____ **43.** $(a^3 - 3)(a^3 + 1)$ _____ **44.** $(x^3 - 4)(x^3 + 1)$

_____ **45.** $(xy - 10)(xy + 5)$ _____ **46.** $(xy - 7)(2xy + 1)$

_____ **47.** $(ab - 2c)(ab + 2c)$ _____ **48.** $(xy - 3z)(xy + 3z)$

_____ **49.** $(3y - 2z)^2$ _____ **50.** $(4a - 3b)^2$

_____ **51.** $(6a^2 + b)(6a^2 - b)$ _____ **52.** $(5x^2 - y)(5x^2 + y)$

_____ **53.** $\left(10 + \dfrac{1}{2}x^2\right)^2$ _____ **54.** $\left(6 + \dfrac{1}{3}x^2\right)^2$

_____ **55.** $(x - 3)(x^2 - 2x + 1)$ _____ **56.** $(x - 5)(x^2 - 6x + 9)$

_____ **57.** $(b - 5)(b^2 + 5b + 25)$ _____ **58.** $(x - 2)(x^2 + 2x + 4)$

_____ **59.** $(x^2 - 4)(x^2 + 3x + 2)$ _____ **60.** $(x^2 - 1)(x^2 - x - 2)$

_____ **61.** $(1 - x + 2x^2)(2 + 3x)$ _____ **62.** $(1 - 3x + x^2)(2 + x)$

_____ **63.** $0.3x(x - 2)(x - 5)$ _____ **64.** $0.5x(x - 3)(x + 1)$

_____ **65.** $-y^2(7 - y)(7 + y)$ _____ **66.** $-y^2(4 - y)(4 + y)$

☆ Stretching the Topics _____

Perform the indicated operations.

_____ **1.** $(x - 2y)(2x - y)(3x - 2y)$

_____ **2.** $(x - 1)^3$

_____ **3.** Find the area of a rectangle with a width of $2x - 3y$ inches and length of $x + y$ inches.

Check your answers in the back of your book.

If you can complete **Checkup 6.3**, you are ready to go on to Section 6.4.

CHECKUP 6.3

Perform the indicated operations.

_____ **1.** $(x - 2)(x + 5)$ _____ **2.** $(2y - 3)(y - 6)$

_____ **3.** $(5 - 2y)^2$ _____ **4.** $(2a + 5b)^2$

_____ **5.** $(5 - 2a)(5 + 2a)$ _____ **6.** $\left(y - \dfrac{3}{5}\right)\left(y + \dfrac{3}{5}\right)$

_____ **7.** $(a - 6)(a^2 + 7)$ _____ **8.** $(x^2 - 4)(3x^2 + 5)$

_____ **9.** $(x + 2)(x^2 - 2x + 4)$ _____ **10.** $-0.2x(3 - x)(5 + x)$

Check your answers in the back of your book.

If You Missed Problems:	You Should Review Examples:
1, 2	1
3, 4	3, 4, 7
5, 6	5, 6, 8
7	9
8	10
9	13
10	15

6.4 Dividing Polynomials

Now that we have learned to add, subtract, and multiply polynomials, we shall consider the operation of division.

Dividing Polynomials by Monomials

To discover the method for dividing a polynomial by a monomial, we must use the definition of division $\left(\dfrac{A}{B} = A \cdot \dfrac{1}{B}\right)$ and the distributive property. Consider the following quotient.

$$\frac{3x^3 + 12x^2 + 3x}{3x}$$

$$= \frac{1}{3x}(3x^3 + 12x^2 + 3x) \qquad \text{Rewrite the quotient, using the definition of division.}$$

$$= \frac{1}{3x} \cdot 3x^3 + \frac{1}{3x} \cdot 12x^2 + \frac{1}{3x} \cdot 3x \qquad \text{Use the distributive property.}$$

$$= \frac{3x^3}{3x} + \frac{12x^2}{3x} + \frac{3x}{3x} \qquad \text{Rewrite each term, using the definition of division.}$$

$$= x^2 + 4x + 1 \qquad \text{Simplify each term.}$$

We may skip the first two steps in this procedure provided we understand that the definition of division and the distributive property are being used mentally.

> To divide a polynomial by a monomial, we may divide each term of the polynomial by the monomial.

As usual, we can describe this method using symbols.

> **Dividing a Polynomial by a Monomial**
> $$\frac{A + B}{C} = \frac{A}{C} + \frac{B}{C} \qquad (C \neq 0)$$

Example 1. Find $\dfrac{2x^3 - 5x^2}{2x^2}$.

Solution

$$\frac{2x^3 - 5x^2}{2x^2}$$

$$= \frac{2x^3}{2x^2} - \frac{5x^2}{2x^2} \quad \text{Divide each term by } 2x^2.$$

$$= x - \frac{5}{2} \quad \text{Reduce each fraction.}$$

Now try Example 2.

Example 2. Find $\dfrac{x^3y + x^2y^2 - 8xy^2}{xy}$.

Solution

$$\frac{x^3y + x^2y^2 - 8xy^2}{xy}$$

$$= \frac{x^3y}{xy} + \frac{x^2y^2}{xy} - \frac{8xy^2}{xy}$$

$$= \underline{\qquad}$$

Check your answer on page 301. ▶

Example 3. Find $\dfrac{x^2 + 3x - 1}{-x}$.

Solution

$$\dfrac{x^2 + 3x - 1}{-x}$$

$$= \dfrac{x^2}{-x} + \dfrac{3x}{-x} - \dfrac{1}{-x} \qquad \text{Divide each term by } -x.$$

$$= -x - 3 + \dfrac{1}{x} \qquad \text{Reduce each fraction.}$$

Example 4. Find $\dfrac{3a^2b - 6ab - 4b^2}{-3a^2b}$.

Solution

$$\dfrac{3a^2b - 6ab - 4b^2}{-3a^2b}$$

$$= \dfrac{3a^2b}{-3a^2b} - \dfrac{6ab}{-3a^2b} - \dfrac{4b^2}{-3a^2b}$$

$$= -1 + \dfrac{2}{a} + \dfrac{4b}{3a^2}$$

Be sure to watch the signs as you reduce each fraction.

⫸ Trial Run

Find the quotients.

_____ **1.** $\dfrac{10x + 5}{5}$

_____ **2.** $\dfrac{x^2 + 3x + 1}{x}$

_____ **3.** $\dfrac{2 - 3y + y^2}{-y}$

_____ **4.** $\dfrac{x^3 - 5x^2 + x}{x^2}$

_____ **5.** $\dfrac{4x^2y - x^2y^2 + 3xy^2}{xy}$

_____ **6.** $\dfrac{-14a^2 - 7ab + 3b^2}{-7ab}$

Answers are on page 301.

Dividing Polynomials by Polynomials

To divide a polynomial by a polynomial, we use a method that is very similar to the method of **long division** in arithmetic.

$$\begin{array}{r} 23 \\ 17\overline{)391} \\ \underline{34} \\ 51 \\ \underline{51} \\ 0 \end{array}$$

CHECK

$17(23) \stackrel{?}{=} 391$

$391 = 391$

Keeping the steps of this long division process in mind, let's see how to divide polynomials, starting with the quotient $(x^2 + x - 12) \div (x - 3)$.

$$x - 3 \overline{)x^2 + x - 12}$$

Set up quotient as long-division problem.

$$\begin{array}{r} x \\ x - 3 \overline{)x^2 + x - 12} \end{array}$$

Divide first term (x) of divisor into first term (x^2) of dividend to get first term (x) of quotient.

$$\begin{array}{r} x \\ x - 3 \overline{)x^2 + x - 12} \\ \underline{x^2 - 3x} \end{array}$$

Multiply quotient term (x) times *entire* divisor $(x - 3)$ and write product beneath like terms in dividend.

$$\begin{array}{r} x \\ x - 3 \overline{)x^2 + x - 12} \\ \underline{{}^\ominus x^2 {}^\oplus 3x} \\ 4x - 12 \end{array}$$

Subtract carefully (change the sign and add) and bring down the next term of the dividend (-12).

$$\begin{array}{r} x + 4 \\ x - 3 \overline{)x^2 + x - 12} \\ \underline{x^2 - 3x} \\ 4x - 12 \\ \underline{{}^\ominus 4x {}^\oplus 12} \\ 0 \end{array}$$

Divide first term (x) of divisor into first term $(4x)$ of new dividend; then multiply new term (4) of quotient times divisor $(x - 3)$ and subtract again.

Notice that we have circled the changed signs to distinguish them from the original signs. To check this problem, we use the same process as in arithmetic.

$$(x - 3)(x + 4) \overset{?}{=} x^2 + x - 12$$

$$x^2 + 4x - 3x - 12 \overset{?}{=} x^2 + x - 12$$

$$x^2 + x - 12 = x^2 + x - 12$$

Example 5. Find $(y^2 + 4y - 12) \div (y - 2)$ and check.

Solution

$$\begin{array}{r} y + 6 \\ y - 2 \overline{)y^2 + 4y - 12} \\ \underline{{}^\ominus y^2 {}^\oplus 2y} \\ 6y - 12 \\ \underline{{}^\ominus 6y {}^\oplus 12} \\ 0 \end{array}$$

CHECK

$$(y - 2)(y + 6) \overset{?}{=} y^2 + 4y - 12$$

Now you complete the check.

Check your work on page 301. ▶

Example 6. Find $(x^3 - 2x^2 + 3x - 1) \div (x + 2)$ and check.

Solution

$$\begin{array}{r} x^2 - 4x + 11 \\ x + 2 \overline{)x^3 - 2x^2 + 3x - 1} \\ \underline{{}^\ominus x^3 {}^\ominus 2x^2} \\ - 4x^2 + 3x \\ \underline{{}^\oplus 4x^2 {}^\oplus 8x} \\ 11x - 1 \\ \underline{{}^\ominus 11x {}^\ominus 22} \\ - 23 \end{array}$$

Here we have a *remainder* of -23, so we may write our answer in either of two ways.

$$(x^3 - 2x^2 + 3x - 1) \div (x + 2) = x^2 - 4x + 11 \quad R -23$$

or

$$(x^3 - 2x^2 + 3x - 1) \div (x + 2) = x^2 - 4x + 11 + \frac{-23}{x + 2}$$

As in arithmetic, we check a division problem with a remainder by multiplying the quotient times the divisor and adding the remainder. The result should be the original dividend. Let's check Example 6.

$$(x + 2)(x^2 - 4x + 11) + (-23) \overset{?}{=} x^3 - 2x^2 + 3x - 1$$

$$x(x^2 - 4x + 11) + 2(x^2 - 4x + 11) + (-23) \overset{?}{=} x^3 - 2x^2 + 3x - 1$$

$$x^3 - 4x^2 + 11x + 2x^2 - 8x + 22 - 23 \overset{?}{=} x^3 - 2x^2 + 3x - 1$$

$$x^3 - 2x^2 + 3x - 1 = x^3 - 2x^2 + 3x - 1$$

In performing long division of polynomials, your work will be simplified if you remember to write the terms of the dividend and divisor in *descending powers* of the same variable and to use a 0 coefficient for any missing terms in the dividend.

Example 7. Find $(11x^2 + 4x^4 - 3) \div (2x - 1)$.

Solution. We arrange the terms in descending powers of x and note the missing x^3-term and x-term in the dividend.

$$
\begin{array}{r}
2x^3 + x^2 + 6x + 3 \\
2x - 1 \overline{)4x^4 + 0x^3 + 11x^2 + 0x - 3} \\
\ominus\underline{4x^4 \oplus 2x^3} \\
2x^3 + 11x^2 \\
\ominus\underline{2x^3 \oplus x^2} \\
12x^2 + 0x \\
\ominus\underline{12x^2 \oplus 6x} \\
6x - 3 \\
\ominus\underline{6x \oplus 3} \\
0
\end{array}
$$

$$(4x^4 + 11x^2 - 3) \div (2x - 1) = 2x^3 + x^2 + 6x + 3$$

Example 8. Find $(a^3 - 8) \div (a^2 + 2a + 4)$ and check.

Solution. We note the missing a^2-term and a-term in the dividend.

$$
\begin{array}{r}
a - 2 \\
a^2 + 2a + 4 \overline{)a^3 + 0a^2 + 0a - 8} \\
\ominus\underline{a^3 \ominus 2a^2 \ominus 4a} \\
- 2a^2 - 4a - 8 \\
\oplus\underline{\ominus 2a^2 \ominus 4a \oplus 8} \\
0
\end{array}
$$

CHECK

$$(a^2 + 2a + 4)(a - 2) \overset{?}{=} a^3 - 8$$

$$a^2(a - 2) + 2a(a - 2) + 4(a - 2) \overset{?}{=} a^3 - 8 \qquad \text{Use distributive property.}$$

$$a^3 - 2a^2 + 2a^2 - 4a + 4a - 8 \overset{?}{=} a^3 - 8 \qquad \text{Remove parentheses.}$$

$$a^3 - 8 = a^3 - 8 \qquad \text{Combine like terms.}$$

▌▶ Trial Run

Divide and check.

_____ **1.** $(x^2 + 3x + 2) \div (x + 1)$ _____ **2.** $(3 + x^2 - x) \div (x - 2)$

_____ **3.** $(9x^3 + 45x^2 - x - 5) \div (3x + 1)$ _____ **4.** $(y^4 - 1) \div (y - 1)$

Answers are below.

▶ Examples You Completed

Example 2. Find $\dfrac{x^3y + x^2y^2 - 8xy^2}{xy}$.

Solution

$$\frac{x^3y + x^2y^2 - 8xy^2}{xy}$$

$$= \frac{x^3y}{xy} + \frac{x^2y^2}{xy} - \frac{8xy^2}{xy}$$

$$= x^2 + xy - 8y$$

Example 5. Find $(y^2 + 4y - 12) \div (y - 2)$ and check.

CHECK

$$(y - 2)(y + 6) \overset{?}{=} y^2 + 4y - 12$$

$$y^2 + 6y - 2y - 12 \overset{?}{=} y^2 + 4y - 12$$

$$y^2 + 4y - 12 = y^2 + 4y - 12$$

Answers to Trial Runs

page 298   **1.** $2x + 1$ **2.** $x + 3 + \dfrac{1}{x}$ **3.** $-\dfrac{2}{y} + 3 - y$ **4.** $x - 5 + \dfrac{1}{x}$ **5.** $4x - xy + 3y$

6. $\dfrac{2a}{b} + 1 - \dfrac{3b}{7a}$

page 301   **1.** $x + 2$ **2.** $x + 1 + \dfrac{5}{x - 2}$ **3.** $3x^2 + 14x - 5$ **4.** $y^3 + y^2 + y + 1$

EXERCISE SET 6.4

Find the quotients.

_____ 1. $\dfrac{4x + 20}{4}$

_____ 2. $\dfrac{16x - 32}{16}$

_____ 3. $\dfrac{8x^3 - 7x^2}{x^2}$

_____ 4. $\dfrac{9x^5 - 5x^4}{x^3}$

_____ 5. $\dfrac{-8x + 6x^2}{-2x}$

_____ 6. $\dfrac{-12x + 8x^2}{-4x}$

_____ 7. $\dfrac{20x^3y^2 - 8xy}{4xy}$

_____ 8. $\dfrac{15x^5y^3 - 9xy}{3xy}$

_____ 9. $\dfrac{8d^3 - 12d^2 - d}{4d^2}$

_____ 10. $\dfrac{16d^3 - 24d^2 - d}{8d^2}$

_____ 11. $\dfrac{y^3 - ay^2 - by}{-y}$

_____ 12. $\dfrac{x^3 - mx^2 - nx}{-x}$

_____ 13. $\dfrac{-m^4n - 3m^3n^2 + 5m^2n^3}{m^2n}$

_____ 14. $\dfrac{-a^3b^2 - 5a^2b^3 + 4ab^4}{ab^2}$

_____ 15. $\dfrac{6ab + 8ac - ad}{2a}$

_____ 16. $\dfrac{15xw + 9xy - xz}{3x}$

_____ 17. $(x^2 + 10x + 21) \div (x + 3)$

_____ 18. $(x^2 - 6x - 16) \div (x + 2)$

_____ 19. $(2a^2 - 5a - 12) \div (2a + 3)$

_____ 20. $(10a^2 + 11a - 6) \div (5a - 2)$

_____ 21. $(z^3 - 8 + 14z - 7z^2) \div (z - 4)$

_____ 22. $(x^3 + 6 - 11x + 2x^2) \div (x - 2)$

_____ 23. $(x^3 - 27) \div (x + 3)$

_____ 24. $(x^3 - 64) \div (x + 4)$

_____ 25. $(4x^3 - 39x^2 + 92x - 63) \div (x^2 - 8x + 9)$

_____ 26. $(3x^3 - 22x^2 + 47x - 28) \div (x^2 - 5x + 4)$

_____ 27. $(6x^4 - 7x^3 + 5x^2 - 10x - 3) \div (2x - 3)$

_____ 28. $(8x^4 + 2x^3 + 3x^2 + 12x + 5) \div (2x + 1)$

_____ 29. $(3x^3 - 4x + 2) \div (x - 1)$

_____ 30. $(4x^3 - 5x + 11) \div (x + 2)$

☆ Stretching the Topics

Find the quotients.

_____ **1.** $(x^5 - 32y^5) \div (x - 2y)$

_____ **2.** $(6a^4 - a^3b - 2a^2b^2 - 11ab^3 - 6b^4) \div (2a - 3b)$

_____ **3.** The area of a rectangle is $12x^2y^3$. The length of the rectangle is $3xy^3$. Write expressions for the width and the perimeter of the rectangle.

Check your answers in the back of your book.

If you can complete **Checkup 6.4**, you are ready to go on to Section 6.5.

CHECKUP 6.4

Find the quotients.

_____ 1. $\dfrac{6x^4 - x^2}{3x^2}$

_____ 2. $\dfrac{a^3b - 4a^2b^2 - 3ab^3}{-a^2b}$

_____ 3. $(2y^2 + 11y + 9) \div (y + 4)$

_____ 4. $(x^3 - x^2 - 7x + 3) \div (x - 3)$

_____ 5. $(9x^4 + 8x^2 - 1) \div (3x - 1)$

_____ 6. $(a^3 - 27) \div (a^2 + 3a + 9)$

Check your answers in the back of your book.

If You Missed Problems:	You Should Review Examples:
1, 2	1–4
3, 4	5, 6
5	7
6	8

6.5 Switching from Word Expressions to Polynomials

Let's continue our practice of switching word expressions to variable expressions, making use of some of the skills learned in this chapter. As before, we always begin by deciding what the variable represents. Then we translate the word expressions in the problem into variable expressions. An illustration will sometimes help us get started.

Using Expressions in Geometry Problems

Example 1. If one side of a square is s, write an expression for the perimeter. Write an expression for the area.

Solution

Perimeter	Area
$s + s + s + s$	$s \cdot s$
$= 4s$	$= s^2$

Now you complete Example 2.

Example 2. If one side of a square is $2x$, write an expression for the perimeter. Write an expression for the area.

Solution

Perimeter	Area
$4(\underline{\quad})$	$(\underline{\quad})^2$
$= \underline{\quad}$	$= \underline{\quad}$

Check your work on page 309. ▶

Example 3. If the length of a rectangle is l and its width is w, write an expression for the perimeter. Write an expression for the area.

Solution

Perimeter	Area
$l + w + l + w$	$l \cdot w$
$= 2l + 2w$	

You try Example 4.

Example 4. If the length of a rectangular room is 5 feet more than its width, write an expression for the perimeter. Write an expression for the area.

   Solution

We Think	**We Write**
We need an expression for the width.	Let x = width (feet)
The length is 5 feet *more than* the width.	$x + 5$ = length (feet)

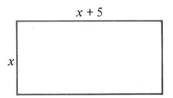

The *perimeter* of a rectangle is the *sum* of 2 lengths and 2 widths.	$P = 2(\underline{\hspace{1cm}}) + 2(\underline{\hspace{1cm}})$
	$= 2x + 10 + 2x$
	$= \underline{\hspace{1cm}} + \underline{\hspace{1cm}}$
The *area* of a rectangle is the *product* of the length and width.	$A = (\underline{\hspace{1cm}}) \cdot x$
	$= \underline{\hspace{1cm}} + \underline{\hspace{1cm}}$

The expression _____ represents the perimeter of the rectangle and the expression _____ represents its area.

Check your work on page 309. ▶

Let's return to the problem stated at the beginning of this chapter.

Example 5. Mr. Abell's garden is in the shape of a square. Next year he plans to enlarge his garden by increasing the length by 6 feet and the width by 4 feet. Write an expression for the area of the garden after it has been enlarged.

   Solution. Let's illustrate Mr. Abell's old and new gardens.

Old garden

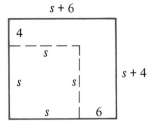

New garden

The area of the new garden is found by multiplying the new length times the new width.

$$A = (s + 6)(s + 4)$$
$$= s^2 + 6s + 4s + 24$$
$$= s^2 + 10s + 24$$

A triangle in which 1 side is perpendicular to another side is called a **right triangle**. The perpendicular sides form a **right angle** (90°). The side opposite the right angle is called the **hypotenuse** of the right triangle. The other sides are called **legs**.

Example 6. If a and b are legs of a right triangle with hypotenuse c, write an expression for the perimeter of the triangle. Write an expression for the area.

Solution

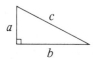

Perimeter	Area
$a + b + c$	$\dfrac{a \cdot b}{2}$

Now you try Example 7.

Example 7. One leg of a right triangle is twice the other leg and the hypotenuse is 3 inches more than the longer leg. Write an expression for the perimeter. Write an expression for the area.

Solution

We Think	**We Write**
We know the least about the shorter leg.	Let $x =$ length of shorter leg (inches)
The longer leg is *twice* as long as the shorter leg.	$2x =$ length of longer leg (inches)
The hypotenuse is 3 inches *more than* the longer leg.	$2x + 3 =$ length of hypotenuse (inches)

The perimeter is the *sum* of the lengths of the 3 sides.	$P = x + 2x + (2x + 3)$ $\quad = $ _____ (inches)
The area is the *product* of the two legs, *divided* by 2.	$A = \dfrac{x \cdot 2x}{2}$ $\quad = \dfrac{2x^2}{2}$ $\quad = $ _____ (square inches)

The expression _____ represents the perimeter of the triangle and the expression _____ represents its area.

Check your work on page 309. ▶

A Greek mathematician named **Pythagoras** discovered that if the lengths of the legs of a right triangle are squared and then added together, the sum is equal to the square of the length of the hypotenuse.

Pythagorean Theorem

$$a^2 + b^2 = c^2$$

For instance, if the legs of a right triangle are 3 feet and 4 feet long, the hypotenuse must be 5 feet long, because

$$3^2 + 4^2 = 5^2$$
$$9 + 16 = 25$$
$$25 = 25$$

Example 8. If one leg of a right triangle is 5 inches longer than the other leg, write an algebraic expression for the square of the hypotenuse.

Solution

We Think	We Write
We know the least about the shorter leg.	Let a = length of shorter leg
The longer leg is 5 inches *longer than* the shorter leg.	$a + 5$ = length of longer leg
The square of the hypotenuse is the *sum* of the *squares* of the 2 legs.	The square of the hypotenuse is $$a^2 + (a + 5)^2$$ $$= a^2 + a^2 + 10a + 25$$ $$= 2a^2 + 10a + 25$$

The square of the hypotenuse can be expressed as $2a^2 + 10a + 25$.

Using Expressions in Integer Problems

Consecutive integers are integers that are next to each other when the list of integers is written in order. For instance

$$22, 23, 24 \quad \text{are consecutive integers}$$
$$-2, -1, 0, 1 \quad \text{are consecutive integers}$$

To get from one integer to the next consecutive integer, do you see that we must add 1? So if a first integer is x, the next integer must be $x + 1$. Sometimes we must work with *sums* or *products* of two consecutive integers. Let

$$x = \text{first integer}$$
$$x + 1 = \text{next consecutive integer}$$

Sum	Product
$x + (x + 1)$	$x(x + 1)$
$= 2x + 1$	$= x^2 + x$

Try to complete Example 9.

Example 9. Write an expression for the square of the sum of two consecutive integers.

Solution

We Think	We Write
We need an expression for each of two consecutive integers.	Let x = first consecutive integer
	$x + 1$ = next consecutive integer
The *sum* of two consecutive integers is found by adding them.	$x + (x + 1)$
	$= \underline{\hspace{1cm}}$

Now we must find the *square* of this sum.   (———)²

                                                        = ———

The square of the sum of two consecutive integers is expressed as ——————————— .

Check your work below. ▶

▶ Examples You Completed ————————————————————

Example 2. If one side of a square is $2x$, write an expression for the perimeter. Write an expression for the area.

    Solution

Perimeter	Area
$4(2x)$	$(2x)^2$
$= 8x$	$= 4x^2$

Example 4 (*Solution*)

Perimeter	Area
$2(x + 5) + 2(x)$	$(x + 5) \cdot x$
$= 2x + 10 + 2x$	$= x^2 + 5x$
$= 4x + 10$	

Example 7 (*Solution*)

We Think	**We Write**
The perimeter is the *sum* of the lengths of the 3 sides.	$P = x + 2x + (2x + 3)$
	$= 5x + 3$ (inches)
The area is the *product* of the two legs, divided by 2.	$A = \dfrac{x \cdot 2x}{2}$
	$= \dfrac{2x^2}{2}$
	$= x^2$ (square inches)

The expression $5x + 3$ represents the perimeter of the triangle and the expression x^2 represents its area.

Example 9. Write an expression for the square of the sum of two consecutive integers.

    Solution

We Think	**We Write**
We need an expression for each of two consecutive integers.	Let x = first consecutive integer
	$x + 1$ = next consecutive integer
The *sum* of two consecutive integers is found by adding them.	$x + (x + 1)$
	$= 2x + 1$
Now we must find the *square* of this sum.	$(2x + 1)^2$
	$= 4x^2 + 4x + 1$

The square of the sum of two consecutive integers is expressed as $4x^2 + 4x + 1$.

EXERCISE SET 6.5

Write the indicated expression and simplify.

_____ 1. If the width of a rectangle is 7 less than the length, write an expression for the perimeter. Write an expression for the area.

_____ 2. If the length of a rectangle is 10 more than the width, write an expression for the perimeter. Write an expression for the area.

_____ 3. If one side of a square is $3x$, write an expression for the perimeter. Write an expression for the area.

_____ 4. If one side of a square is $5x$, write an expression for the perimeter. Write an expression for the area.

_____ 5. One leg of a right triangle is 7 more than the other leg and the hypotenuse is 8 more than the shorter leg. Write an expression for the perimeter. Write an expression for the area.

_____ 6. One leg of a right triangle is 2 more than the other leg and the hypotenuse is 4 more than the shorter leg. Write an expression for the perimeter. Write an expression for the area.

_____ 7. If the length of one leg of a right triangle is twice the length of the other leg, write an expression for the square of the hypotenuse.

_____ 8. If the length of one leg of a right triangle is 4 times the length of the other leg, write an expression for the square of the hypotenuse.

_____ 9. Write an expression for the product of two consecutive integers.

_____ 10. Write an expression for the sum of the squares of two consecutive integers.

_____ 11. One number is 12 more than the other. Write an expression for their product.

_____ 12. One integer is 3 less than twice another. Write an expression for their product.

_____ 13. One number is 9 less than another. Write an expression for the square of the larger added to 5 times the smaller.

_____ 14. One number is 2 more than another. Write an expression for the square of the smaller added to 7 times the larger.

_____ 15. Write an expression for the product of two consecutive integers increased by 3 times the smaller.

_____ 16. Write an expression for the product of two consecutive integers decreased by twice the larger.

_____ 17. Wick plans to plow a rectangular field that is 15 rods longer than it is wide. Write an expression for the area.

_____ 18. Kaaryn is building a room on her house that is 5 feet longer than it is wide. Write an expression for the area of the room.

_____ 19. Lamar is building a rectangular pen for his pony that is 15 feet longer than it is wide. Write an expression for the number of feet of fence he will need.

_____ 20. Carol is buying lace to border a tablecloth that is 20 inches longer than it is wide. Write an expression for the number of inches of lace she will need.

_____ 21. Maxine has a garden that is a square. She plans to enlarge the garden by increasing the length of one side by 10 feet and decreasing the other by 2 feet. Write an expression for the area of the larger garden.

_____ 22. Robert has a square family room that he plans to enlarge by increasing the length of one side by 6 feet and increasing the other by 2 feet. Write an expression for the area of the enlarged family room.

_____ 23. The length of a room is 6 feet longer than twice its width. Write an expression for the number of square feet of carpet needed to cover the floor.

_____ 24. A yard is 6 feet longer than it is wide. Write an expression for the number of square feet of sodding it would take to cover the yard.

_____ 25. A flower bed is in the shape of a right triangle with one side 9 feet shorter than the other. Write an expression for the area.

_____ 26. Chris is building a corner table for her kitchen. The top will be a right triangle with equal sides. Write an expression for the area.

_____ 27. One leg of a right triangle is 5 more than twice the other leg. Write an expression for the square of the hypotenuse.

_____ 28. One leg of a right triangle is 3 more than 4 times the other leg. Write an expression for the square of the hypotenuse.

_____ 29. If one number is 5 times the other, write an expression for the product of the numbers increased by twice their sum.

_____ 30. If one number is 2 smaller than twice the other, write an expression for the product of the two numbers decreased by 3 times their sum.

☆ Stretching the Topics _____

_____ 1. The width of a rectangle is 3 less than one-half its length. Write an expression for the area of a square with a side equal to the rectangle's width.

_____ 2. If a train travels $2t^2 - 1$ miles in t hours, write an expression for the rate of speed of the train.

_____ 3. If n represents an odd number, write an expression for the product of the next two consecutive higher even numbers.

Check your answers in the back of your book.

If you can complete **Checkup 6.5**, you are ready to do the **Review Exercises** for Chapter 6.

CHECKUP 6.5

_____ **1.** If one side of a square is $3x$, write an expression for the perimeter. Write an expression for the area.

_____ **2.** If the length of a rectangle is 3 less than 7 times the width, write an expression for the perimeter. Write an expression for the area.

_____ **3.** One leg of a right triangle is 7 more than the other and the hypotenuse is 3 more than twice the shorter leg. Write an expression for the perimeter. Write an expression for the area.

_____ **4.** If one leg of a right triangle is 7 more than the other, write an expression for the square of the hypotenuse.

_____ **5.** Write an expression for the square of the sum of two consecutive integers, decreased by twice the smaller.

Check your answers in the back of your book.

If You Missed Problem:	You Should Review Example:
1	2
2	4
3	7
4	8
5	9

312

Summary

In this chapter we learned to work with **polynomials**, which are sums of terms that are constants times whole number powers of variables. We identify polynomials according to the number of terms they contain.

Name	Number of Terms	Examples
Monomial	One	$3x$ $-x^2y$
Binomial	Two	$3x + 1$ $x^2 - y^2$
Trinomial	Three	$x^2 + x - 2$ $a + b + c$

Then we discussed the methods for performing the basic operations of addition, subtraction, multiplication, and division with polynomials.

In Order to	We Must	Example
Add polynomials	Combine like terms.	$(x^2 + 2x - 1) + (2x^2 - 5)$ $= 3x^2 + 2x - 6$
Subtract polynomials	Remove parentheses and combine like terms.	$(3x^2 - 5x + 2) - (x^2 - x + 1)$ $= 3x^2 - 5x + 2 - x^2 + x - 1$ $= 2x^2 - 4x + 1$
Multiply monomials	Multiply coefficients, multiply variables using the First Law of Exponents.	$-3x^2y(2x^3y)$ $= -6x^5y^2$
Multiply a monomial times a polynomial	Use the distributive property.	$-3x^2(x^2 - 5x + 1)$ $= -3x^4 + 15x^3 - 3x^2$
Multiply binomials	Use FOIL as a shortcut for the distributive property.	$(2x + 3)(x - 1)$ $= 2x^2 - 2x + 3x - 3$ $= 2x^2 + x - 3$
Multiply polynomials	Use the distributive property.	$(x + 3)(x^2 - 5x - 2)$ $= x(x^2 - 5x - 2) + 3(x^2 - 5x - 2)$ $= x^3 - 5x^2 - 2x + 3x^2 - 15x - 6$ $= x^3 - 2x^2 - 17x - 6$
Divide a polynomial by a monomial	Divide each term of the polynomial by the monomial.	$\dfrac{5x^2 - 3x - 6}{3x}$ $= \dfrac{5x^2}{3x} - \dfrac{3x}{3x} - \dfrac{6}{3x}$ $= \dfrac{5x}{3} - 1 - \dfrac{2}{x}$
Divide polynomials	Use long division.	$\begin{array}{r} x + 3 \\ x + 2\overline{)x^2 + 5x + 6} \\ \ominus\underline{x^2 + 2x} \\ 3x + 6 \\ \ominus\underline{3x + 6} \\ 0 \end{array}$

In multiplying binomials, we discovered two *special products* that can be found quickly using formulas.

Special Product	Formula	Examples
Square of a binomial	$(A + B)^2 = A^2 + 2AB + B^2$ $(A - B)^2 = A^2 - 2AB + B^2$	$(x + 5)^2 = x^2 + 10x + 25$ $(2y - 3)^2 = 4y^2 - 12y + 9$
Product of sum and difference of two terms	$(A + B)(A - B) = A^2 - B^2$	$(2x + 7)(2x - 7)$ $= 4x^2 - 49$

Finally, we practiced switching from word expressions to variable expressions that were polynomials.

❑ Speaking the Language of Algebra ⸻⸻⸻⸻

Complete each statement with the appropriate word or phrase.

1. The polynomial $7y$ is called a ⸻⸻ because it contains ⸻⸻ term. The polynomial $5x - y$ is a ⸻⸻ because it contains ⸻⸻ terms. The polynomial $6 - x + 3x^2$ is a ⸻⸻ because it contains ⸻⸻ terms.

2. To add polynomials, we must ⸻⸻ ⸻⸻ ⸻⸻ .

3. To multiply a monomial times a polynomial, we must use the ⸻⸻ property.

4. The FOIL method can be used only when multiplying two ⸻⸻ .

5. To square a binomial, we use the special product formula $(A + B)^2 = $ ⸻⸻ .

6. To multiply the sum and difference of the same two terms, we use the special product formula $(A + B)(A - B) = $ ⸻⸻ .

7. To divide a polynomial by a monomial, we divide each term of the polynomial by the ⸻⸻ .

8. To divide two polynomials, we use the method of ⸻⸻ ⸻⸻ .

Check your answers in the back of your book.

△ Writing About Mathematics ⸻⸻⸻⸻

Write your response to each question in complete sentences.

1. Explain how to use the *distributive property* to find the product $(3x + 1)(2x - 3)$.

2. Show and explain your steps in using long division to find the quotient $(x^3 - 7x - 6) \div (x - 3)$.

3. Explain, in your own words, the meaning of the Pythagorean Theorem.

REVIEW EXERCISES for Chapter 6

Name each expression.

_____ **1.** $3a + 2b$

_____ **2.** $7x - 2y + 5x$

_____ **3.** $-3m^2$

_____ **4.** $x^2 - 3x - 1$

Perform the indicated operations.

_____ **5.** $(3x^2 - 2x) + (5x - 3)$

_____ **6.** $(-3a^2 + 2a - 4) + (a^2 - 5a + 4)$

_____ **7.** $-2(7xy)$

_____ **8.** $\dfrac{2}{3}(9x^2)$

_____ **9.** $-4(0.8a^2b^3)$

_____ **10.** $5(-x^2 - 3x + 2)$

_____ **11.** $-(x^2 - 2x + 1)$

_____ **12.** $0.2(8x^2 - 3x + 2)$

_____ **13.** $(4x - 3y) - (2x + 3y)$

_____ **14.** $-5(2a + b) + 3(a + b)$

_____ **15.** $2(x^2 - x - 9) - (5x^2 + 2x - 4)$

_____ **16.** $x^7 \cdot x^3$

_____ **17.** $2a^3(-3a)(a^4)$

_____ **18.** $9xy^6(-2x^4y^2)$

_____ **19.** $(m^2)^3$

_____ **20.** $(-3y)^4$

_____ **21.** $(x^3y^2z)^2$

_____ **22.** $3a(a^4b^2)^3$

_____ **23.** $\left(\dfrac{1}{2}x\right)^3$

_____ **24.** $0.8(5x - 3)$

_____ **25.** $7y(3y + 2)$

_____ **26.** $3m^2(m^2 - 4mn - 5n^2)$

_____ **27.** $5xy(7x - 11y)$

_____ **28.** $-a^2b(6 - 3ab + 4a^2b^2)$

_____ **29.** $4xy(3x - y) - 5xy(2x - y)$

_____ **30.** $(x + 9)(x - 8)$

_____ **31.** $(3y - 2)(4y + 1)$

_____ **32.** $(2a - 3b)(2a + 3b)$

_____ **33.** $(x - 5y)^2$

_____ **34.** $(2xy - 1)^2$

_____ **35.** $(m^2 - 3)(m^2 + 5)$

_____ **36.** $(3x^2 - 5y^2)(2x^2 + 4y^2)$

_____ **37.** $3x(3x - 2)^2$

_____ **38.** $2x(x - 3)(x + 3)$

_____ **39.** $-4a^2(3 - 2a)(3 + 2a)$

_____ **40.** $(y - 1)(y^2 - 3y - 4)$

_____ 41. $\dfrac{3x^3 - 6x^2 - 2x}{3x}$ _____ 42. $\dfrac{x^3y + 15x^2y^2 - 5xy^2}{-xy}$

_____ 43. $(5x^2 + 33x - 14) \div (5x - 2)$ _____ 44. $(x^2 - 2x + 5) \div (x + 4)$

_____ 45. $(x^3 - 3x^2 + x + 4) \div (x + 1)$ _____ 46. $(8x^4 - 2x^2 + 1) \div (2x - 1)$

_____ 47. $(y^3 + 1) \div (y^2 + y + 1)$

_____ 48. Joyce is adding a family room to her house. If the width of the room is to be 10 feet less than the length, write an expression for the area of the room, and simplify.

_____ 49. If one leg of a right triangle is 2 inches longer than twice the other leg, write an expression for the square of the hypotenuse, and simplify.

_____ 50. Write an expression for the product of two consecutive integers, decreased by 3 times the square of the smaller one.

Check your answers in the back of your book.

If You Missed Exercises:	You Should Review Examples:	
1–4	Section 6.1	1–4
5, 6		5, 6
7–9		7, 8
12–15		9–12
16–23	Section 6.2	1–6
24–28		7–10
29		11, 12
30, 31	Section 6.3	1, 2
32		8
33, 34		7
35, 36		9, 10
37–39		13–16
41, 42	Section 6.4	1–4
43, 44		5
45		6
46		7
47		8
48	Section 6.5	4
49		8
50		9

If you have completed the **Review Exercises** and corrected your errors, you are ready to take the **Practice Test** for Chapter 6.

PRACTICE TEST for Chapter 6

Name each expression.

		SECTION	EXAMPLES
_____	**1.** $-4a - 3b + 4a - b$	6.1	1
_____	**2.** $3x^2 - 5x + 2$	6.1	3

Perform the indicated operations.

		SECTION	EXAMPLES
_____	**3.** $(2x + 7y - 3z) + (4x - y + 2z)$	6.1	6
_____	**4.** $(-0.3)(7a^2)$	6.1	8
_____	**5.** $5(2x^2 - x + 3) - (x^2 - 3x - 1)$	6.1	11
_____	**6.** $y^5(-3y)(5y^4)$	6.2	4
_____	**7.** $(-4x^3y^2)(5x^2y)$	6.2	5, 6
_____	**8.** $-5c(a^2 - 3ac - c^2)$	6.2	8
_____	**9.** $8xy(2x - 3y) - 4xy(x - 6xy)$	6.2	11
_____	**10.** $(x + 11)(x - 10)$	6.3	1
_____	**11.** $(4x - 5)(3x - 2)$	6.3	2
_____	**12.** $(a - 7)^2$	6.3	3, 4, 7
_____	**13.** $\left(x - \dfrac{2}{3}\right)\left(x + \dfrac{2}{3}\right)$	6.3	5, 6, 8
_____	**14.** $(5a + 3c)(4a - c)$	6.3	11, 12
_____	**15.** $(y^2 - 0.5)(y^2 + 0.7)$	6.3	9, 10
_____	**16.** $(4x - z)^2$	6.3	7
_____	**17.** $(x + 2)(x^2 - 2x + 4)$	6.3	13
_____	**18.** $-2y(y + 9)(y - 9)$	6.3	15, 16
_____	**19.** $-3(4x - y)^2$	6.3	15, 16
_____	**20.** $\dfrac{9x^3 - 15x^2 - 12x}{3x}$	6.4	1, 2

		SECTION	EXAMPLES

_____ **21.** $\dfrac{6x^2 - 12xy - 5y}{-6x^2y}$ 6.4   3, 4

_____ **22.** $(6x^2 - 17x - 14) \div (3x + 2)$ 6.4   5

_____ **23.** $(x^3 - 5x - 9) \div (x - 3)$ 6.4   6

_____ **24.** If one leg of a right triangle is 2 inches longer than 3 times the other leg, write an expression for the square of the hypotenuse, and simplify.   6.5   8

_____ **25.** If all four sides of a square patio are increased by 3 feet, write an expression for the area of the new patio, and simplify.   6.5   5

_____ **26.** Write an expression for the product of two consecutive integers increased by 5 times the larger integer.   6.5   9

Check your answers in the back of your book.

SHARPENING YOUR SKILLS after Chapters 1–6

_____ 1. Change $\dfrac{5}{8}$ to a decimal number.   1.3

_____ 2. Use the associative property for addition to rewrite $(-3 + 4) + (-10)$.   2.3

_____ 3. Evaluate $\dfrac{2a - 5b}{a - b}$ when $a = 4$ and $b = -2$.   3.1

_____ 4. Is -2 a solution for $3x - 2(x + 5) = -12$?   4.1

_____ 5. Evaluate $-2(x^2 - xy)$ when $x = -1$ and $y = 2$.   5.1

Perform the indicated operations.

_____ 6. $\left(\dfrac{2}{3}\right)^2 - \dfrac{5}{12} + \dfrac{5}{6}$ 1.2

_____ 7. $\dfrac{-3(8 - 4) - 7(-4)}{-6 + [(-2)^2 \cdot 5]}$ 2.4

_____ 8. $-3a + 7b - 4c + 9a - 3b + 5c$ 3.2

_____ 9. $6[-4(x - y) - (2x - 3y)]$ 3.3

_____ 10. $(-3x)^2(xy)(x^3y^4)$ 5.1

_____ 11. $\dfrac{-72x^3y^2z}{81xy^5z^4}$ 5.2

_____ 12. $(-7y^{-5})(y^3)(-y)^2$ 5.3

Solve.

_____ 13. $4(x - 5) - 7 = 3(2x - 4) - 5x$ 4.2

_____ 14. $4[2(x - 2) - (x + 4)] = 0$ 4.2

_____ 15. The length of a rectangle is 3 less than twice the width. If the perimeter of the rectangle is 30 feet, find the dimensions of the rectangle.   4.3

Check your answers in the back of your book.

7

Factoring Polynomials and Solving Quadratic Equations

After a tree was broken over by a storm, the part that was bending over was 26 feet long. The part left standing was 14 feet less than the distance from the foot of the tree to the point where the broken part touched the ground. Find the height of the tree before the storm.

Before you finish this chapter, you will be able to solve this problem using a **second-degree equation**. A second-degree equation is a polynomial equation in which the largest exponent on the variable is 2.

One way to begin solving a second-degree equation is to rewrite the polynomial as a product of **factors**. Our purpose in this chapter is to learn procedures for figuring out what factors were multiplied together to give a certain polynomial.

In this chapter we learn to

1. Find the common factor of a polynomial.
2. Factor a difference of two squares.
3. Factor a general trinomial.
4. Solve quadratic equations by factoring.
5. Switch from word statements to quadratic equations.

7.1 Looking for Common Factors

Factoring is a process by which we rewrite an expression as a **product**. In order to figure out what quantities would multiply together to give a certain product, you must be very familiar with the multiplication of polynomials.

Looking for Common Monomial Factors

Recall that when we multiplied monomials times polynomials the products were found using the **distributive property.** For example,

$$-7(2x - y) = -14x + 7y$$

$$8(x^2 + 3x - 5) = 8x^2 + 24x - 40$$

Can you write $2x + 2y$ as a product of factors?

$$2x + 2y = 2(x + y)$$

In factoring, you can *always* decide whether your "guess" is correct. All you must do is multiply your factors together and see if your answer is the same as the original polynomial.

Let's try factoring

$$9y^2 + 27y - 18$$

We notice that 3 is a factor of each term, but 9 is also a factor of each term. Which do we choose? We must agree to choose the *largest possible* common factor. Here we choose 9 and write

$$9y^2 + 27y - 18 = 9(y^2 + 3y - 2)$$

Example 1. Factor $-5x^2 + 10$.

Solution

$$-5x^2 + 10 = -5(x^2 - 2)$$

Notice that if the first numerical coefficient in a polynomial is negative, we usually agree to factor out a negative common factor.

You try Examples 2 and 3.

Example 2. Factor $14x^2 - 70x - 7$.

Solution

$$14x^2 - 70x - 7 = 7(\underline{\qquad})$$

Example 3. Factor $-19x + 19$.

Solution

$$-19x + 19 = \underline{\qquad}$$

Did you remember to check your factors by multiplying?

Check your work on page 326. ▶

If you always remember to check your factors by multiplying, you will not make factoring errors such as these.

Polynomial	Wrong	Right
$-7x + 7$	$-7(x + 1)$	$-7(x - 1)$
$2x^2 + 2$	$2(x^2)$	$2(x^2 + 1)$
$3x + 6y$	$3(x + 6y)$	$3(x + 2y)$

The process used to factor a polynomial such as $3x^2 + 6x$ is similar to the process used in factoring out common numerical factors. Once again we are looking for factors common to every term in the polynomial. Those factors may contain a numerical part and/or a variable part.

In the binomial $3x^2 + 6x$, we notice that both terms contain the factor 3 *and* the factor x. Therefore we write

$$3x^2 + 6x = 3x(x + 2)$$

Let's factor the expression $98a^2 - 28a$.

$98a^2 - 28a$	Notice that 2 and a are factors of both terms.
$= 2a(49a - 14)$	Notice that 7 is a factor of $49a - 14$.
$= 2a \cdot 7(7a - 2)$	It is not too late to factor out the 7.
$= 14a(7a - 2)$	Simplify the monomial factor.

Indeed, $14a$ could have been factored out at the start, but you were not wrong if you did not spot it right away. *It is never too late to factor out a common factor.*

Example 4. Factor $10xy - 2x$.

Solution

$$10xy - 2x = 2x(5y - 1)$$

Now try Example 5.

Example 5. Factor $3x^2 - 15x$.

Solution

$$3x^2 - 15x = \text{_____}$$

Check your factors on page 326. ▶

In the polynomial $x^5 + 2x^3 + x^2$ we see that each term contains a power of x. The highest power of x that can be factored out is actually the *lowest* power contained in any one of the terms. In this case, the lowest power is x^2, so

$$x^5 + 2x^3 + x^2 = x^2(x^3 + 2x + 1)$$

Example 6. Factor

$$8x^7 - 2x^6 + 10x^5 - 2x^3$$

Solution

$$8x^7 - 2x^6 + 10x^5 - 2x^3$$
$$= 2x^3(4x^4 - x^3 + 5x^2 - 1)$$

Try Example 7.

Example 7. Factor $-x^3 + 3x^2 - 4x$.

Solution

$$-x^3 + 3x^2 - 4x$$
$$= -x(\text{_____})$$

Remember to check by multiplying.

Check your factors on page 326. ▶

Example 8. Factor $3a^2x^2 - 5a^2x$.

Solution

$$3a^2x^2 - 5a^2x = a^2x(3x - 5)$$

Now try Example 9.

Example 9. Factor $x^7y^8 - 2x^5y^6$.

Solution

$$x^7y^8 - 2x^5y^6$$
$$= \text{_____}$$

Check your factors on page 326. ▶

To factor out the largest common factor of a polynomial, we follow these steps.

1. Look for the largest monomial factor common to all terms of the polynomial.
2. Use the distributive property to rewrite the polynomial as a product of that monomial factor and a new polynomial factor.
3. Check by multiplying your factors.

⇒ Trial Run

Rewrite each expression as a product of factors.

_____ **1.** $15x^2 - 10x - 50$ _____ **2.** $13y - 13$

_____ **3.** $-5x - 5$ _____ **4.** $-x^3 + 2x^2 - 10x$

_____ **5.** $x^5y^4 + 5x^4y^3$ _____ **6.** $2x^2y^2z^3 - 6x^2y^3z^2 + 4x^3y^2z^2$

Answers are on page 327.

Looking for Common Binomial Factors (Factoring by Grouping)

In an expression such as $3(x + 1) + y(x + 1)$ notice that there are just *two terms* to consider. The first term is $3(x + 1)$ and the second term is $y(x + 1)$. You should see that both terms contain the common factor $(x + 1)$. The distributive property still allows us to write

$$3(x + 1) + y(x + 1) = (x + 1)(3 + y)$$

Let's try factoring another expression of this type. Factor $a(x + y) + (x + y)$.

$$a(x + y) + (x + y)$$ Note that there are two terms.

$$= a(x + y) + 1(x + y)$$ Rewrite the expression.

$$= (x + y)(a + 1)$$ The binomial factor $(x + y)$ is common to both terms.

Example 10. Factor $(x + 5) + 2a(x + 5)$.

Solution

$$(x + 5) + 2a(x + 5)$$

$$= 1(x + 5) + 2a(x + 5)$$

$$= (x + 5)(1 + 2a)$$

Now finish Example 11.

Now try Examples 12 and 13.

Example 11. Factor $x(a - b) + (a - b)$.

Solution

$$x(a - b) + (a - b)$$

$$= x(a - b) + 1(a - b)$$

$$= (a - b)(\underline{\quad\quad})$$

Check your work on page 326. ▶

Example 12. Factor

$$y(x^2 + 3) - 2(x^2 + 3)$$

Solution

$$y(x^2 + 3) - 2(x^2 + 3)$$

$$= (\underline{\quad\quad})(\underline{\quad\quad})$$

Example 13. Factor

$$3x(x^2 + 1) - (x^2 + 1)$$

Solution

Check your factors on page 326. ▶   Check your factors on page 326. ▶

Suppose that we wish to factor the four-term polynomial $5x + 5y + bx + by$. We must try to *group* the terms into *two pairs of terms*, each pair containing a common factor. We notice that the first two terms contain the common factor 5 and the second two terms contain the common factor b. We can rewrite our four terms as

$$5x + 5y + bx + by = 5(x + y) + b(x + y)$$
$$= (x + y)(5 + b)$$

and we have written our four-term polynomial as a product of factors. This technique is called **factoring by grouping**.

Let's try factoring another polynomial of four terms by grouping.

$$6y^2 + 3y - 2ay - a$$

The first two terms contain the common factor $3y$ and the second two terms contain a common factor a or $-a$. Which should we remove? Look at the choices.

$$6y^2 + 3y - 2ay - a \qquad\qquad \text{or} \qquad\qquad 6y^2 + 3y - 2ay - a$$
$$= 3y(2y + 1) + a(-2y - 1) \qquad\qquad = 3y(2y + 1) - a(2y + 1)$$

Both versions are correct, but which one will allow us to continue factoring? The second version leaves us with the common binomial factor $(2y + 1)$ in both terms, so we choose that version.

$$6y^2 + 3y - 2ay - a = 3y\,(2y + 1) - a\,(2y + 1)$$
$$= (2y + 1)(3y - a)$$

Example 14. Factor

$$4ax - 12ay + x - 3y$$

Solution

$$4ax - 12ay + x - 3y$$
$$= 4a(x - 3y) + 1(x - 3y)$$
$$= (x - 3y)(4a + 1)$$

Now complete Example 15.

Example 15. Factor $8rx - 4r - 2x + 1$.

Solution

$$8rx - 4r - 2x + 1$$
$$= 4r(\underline{\hspace{1cm}}) - 1(\underline{\hspace{1cm}})$$
$$= (\underline{\hspace{1cm}})(\underline{\hspace{1cm}})$$

Check your work on page 326. ▶

Example 16. Factor $x^3 + 5x^2 + 2x + 10$.

Solution

$$x^3 + 5x^2 + 2x + 10$$
$$= x^2(x + 5) + 2(x + 5)$$
$$= (x + 5)(x^2 + 2)$$

You complete Example 17.

Example 17. Factor
$-2x^2 - 2y^2 + 9x^2y + 9y^3$

Solution

$$-2x^2 - 2y^2 + 9x^2y + 9y^3$$
$$= -2(\underline{\hspace{1cm}}) + 9y(\underline{\hspace{1cm}})$$
$$= (\underline{\hspace{1cm}})(\underline{\hspace{1cm}})$$

Check your work on page 326. ▶

⫸ Trial Run ━━━━━━━━━━━━━━━━━━━━━━━━━━━━

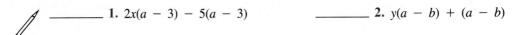

Rewrite each expression as a product of factors.

_____ **1.** $2x(a - 3) - 5(a - 3)$ _____ **2.** $y(a - b) + (a - b)$

_____ 3. $4x + 4y + bx + by$ _____ 4. $x^3 - 7x^2 + 3x - 21$

_____ 5. $3ax - 15ay + 2x - 10y$ _____ 6. $8a^2 + 4a - 2ab - b$

Answers are on page 327.

▶ **Examples You Completed** _____

Example 2. Factor $14x^2 - 70x - 7$.

Solution

$$14x^2 - 70x - 7 = 7(2x^2 - 10x - 1)$$

Example 5. Factor $3x^2 - 15x$.

Solution

$$3x^2 - 15x = 3x(x - 5)$$

Example 9. Factor $x^7y^8 - 2x^5y^6$.

Solution

$$x^7y^8 - 2x^5y^6 = x^5y^6(x^2y^2 - 2)$$

Example 12. Factor $y(x^2 + 3) - 2(x^2 + 3)$.

Solution

$$y(x^2 + 3) - 2(x^2 + 3)$$
$$= (x^2 + 3)(y - 2)$$

Example 15. Factor $8rx - 4r - 2x + 1$.

Solution

$$8rx - 4r - 2x + 1$$
$$= 4r(2x - 1) - 1(2x - 1)$$
$$= (2x - 1)(4r - 1)$$

Example 3. Factor $-19x + 19$.

Solution

$$-19x + 19 = -19(x - 1)$$

Example 7. Factor $-x^3 + 3x^2 - 4x$.

Solution

$$-x^3 + 3x^2 - 4x$$
$$= -x(x^2 - 3x + 4)$$

Example 11. Factor $x(a - b) + (a - b)$.

Solution

$$x(a - b) + (a - b)$$
$$= x(a - b) + 1\,(a - b)$$
$$= (a - b)(x + 1)$$

Example 13. Factor $3x(x^2 + 1) - (x^2 + 1)$.

Solution

$$3x(x^2 + 1) - (x^2 + 1)$$
$$= 3x(x^2 + 1) - 1(x^2 + 1)$$
$$= (x^2 + 1)(3x - 1)$$

Example 17. Factor

$$-2x^2 - 2y^2 + 9x^2y + 9y^3$$

Solution

$$-2x^2 - 2y^2 + 9x^2y + 9y^3$$
$$= -2(x^2 + y^2) + 9y(x^2 + y^2)$$
$$= (x^2 + y^2)(-2 + 9y)$$

Answers to Trial Runs

page 324   **1.** $5(3x^2 - 2x - 10)$ **2.** $13(y - 1)$ **3.** $-5(x + 1)$ **4.** $-x(x^2 - 2x + 10)$
5. $x^4y^3(xy + 5)$ **6.** $2x^2y^2z^2(z - 3y + 2x)$

page 325   **1.** $(a - 3)(2x - 5)$ **2.** $(a - b)(y + 1)$ **3.** $(x + y)(4 + b)$ **4.** $(x - 7)(x^2 + 3)$
5. $(x - 5y)(3a + 2)$ **6.** $(2a + 1)(4a - b)$

EXERCISE SET 7.1

Rewrite each expression as a product of factors.

_____ 1. $7x + 14$ _____ 2. $5x - 15$

_____ 3. $21x - 28$ _____ 4. $9x + 21$

_____ 5. $42 - 6y$ _____ 6. $45 - 9y$

_____ 7. $6a^2 - 36$ _____ 8. $8a^2 - 40$

_____ 9. $-12x + 18y$ _____ 10. $-15x + 20y$

_____ 11. $30x^2 + 15y^2$ _____ 12. $60x^2 - 12y^2$

_____ 13. $7x^2 - 21x - 7$ _____ 14. $9x^2 + 27x + 9$

_____ 15. $10y^2 + 15y - 35$ _____ 16. $12y^2 - 18y - 33$

_____ 17. $28 - 7a + 35a^2$ _____ 18. $33 - 22a + 55a^2$

_____ 19. $50x^2 - 10xy + 20y^2$ _____ 20. $36x^2 - 24xy + 48y^2$

_____ 21. $2x^3 - 6x^2 + 4x + 8$ _____ 22. $8x^3 - 16x^2 + 24x + 8$

_____ 23. $x^2 + x$ _____ 24. $2x^2 + x$

_____ 25. $-y^3 + y^2 - 2y$ _____ 26. $-y^3 - y^2 + 4y$

_____ 27. $2b^3 - 6b^2 - 4b$ _____ 28. $5b^3 - 10b^2 + 25b$

_____ 29. $7b^3 + 2b^5 - 3b^7$ _____ 30. $9b^4 + 4b^6 - 2b^8$

_____ 31. $3x^3 - 6x^2y + 12xy^2$ _____ 32. $5x^3 + 15x^2y + 55xy^2$

_____ 33. $-a^4b + 8a^3b^2 - 5a^2b^3$ _____ 34. $-a^5b - 4a^4b^2 - 5a^3b^3$

_____ 35. $4x^3y^3 - 20x^2y^2 + 2xy$ _____ 36. $9x^4y^4 + 18x^3y^3 - 3x^2y^2$

_____ 37. $x^3y^2z - 2x^4y^3z - x^5y^4z$ _____ 38. $x^6y^4z^2 + 5x^4y^2z^2 - x^2z^2$

_____ 39. $63x^2 + 72x$ _____ 40. $36x^3 + 45x^2$

_____ 41. $100a^3 - 60b^3$ _____ 42. $75a^3 - 100b^3$

_____ 43. $16b^3 + 64b^4$ _____ 44. $12b^5 + 60b^7$

_____ 45. $6a^3 - 42a^2 - 78a$ _____ 46. $9a^3 - 72a^2 + 81a$

_____ 47. $5x^3y + 80xy^3$ _____ 48. $7x^4y^2 - 56x^3y^3$

_____ 49. $-7b^2 - 42b^3 + 35b^4$ _____ 50. $-12b^4 + 36b^5 - 48b^6$

_____ 51. $9x^3y - 12x^2y^2 + 20xy^3$ _____ 52. $27x^3y + 45x^2y^2 - 63xy^3$

_____ 53. $72a^4 - 48a^3 - 8a^2$ _____ 54. $72a^5 - 54a^4 - 6a^3$

_____ 55. $12x^3y - 8x^2y^2 + 4xy^3$ _____ 56. $21x^3y - 15x^2y^2 - 3xy^3$

_____ 57. $9x^4yz + 18x^3y^2z - 27x^2y^3z - 45xy^4z$

_____ 58. $35xy^4a - 28xy^3a - 14xy^2a + 7xya$

_____ 59. $5a + ac + 5b + bc$ _____ 60. $7x + 7y + xz + yz$

_____ 61. $x^3 - 5x^2 + 6x - 30$ _____ 62. $x^3 - 9x^2 + 9x - 81$

_____ 63. $70a^2 - 40ab + 21a - 12b$ _____ 64. $30a^2 - 75ab + 4a - 10b$

_____ 65. $3xy - 6x - y + 2$ _____ 66. $2ab - 10a - b + 5$

_____ 67. $2xy^2 + 8x + 3y^2 + 12$ _____ 68. $3ab^2 - 2a + 12b^3 - 8b$

_____ 69. $3x^3 - 6x^2 + x - 2$ _____ 70. $2y^3 - 6y^2 + y - 3$

☆ Stretching the Topics _____

Rewrite each expression as a product of factors.

_____ 1. $2(m - n)^2 - (m - n)$

_____ 2. $3x^{2m+5} + 9x^{2m}$

_____ 3. $\frac{1}{6}Bh + \frac{1}{6}bh + \frac{2}{3}Mh$

Check your answers in the back of your book.

If you can complete **Checkup 7.1**, you are ready to go on to Section 7.2.

CHECKUP 7.1

Rewrite each expression as a product of factors.

_____ **1.** $3x + 15$

_____ **2.** $35 - 14y$

_____ **3.** $-x^2 + x$

_____ **4.** $15x^2 - 10xy + 25y^2$

_____ **5.** $8a^3 - 16a^2 - 40a$

_____ **6.** $9a^3b^3 - 27a^2b^2 + 3ab$

_____ **7.** $-8m^3 - 24m$

_____ **8.** $x^4yz - x^3y^2z + x^2y^3z$

_____ **9.** $5m + 5n + pm + pn$

_____ **10.** $3x^2 + 6xy - x - 2y$

Check your answers in the back of your book.

If You Missed Problems:	You Should Review Examples:
1–3	1–5
4–8	6–9
9–10	14–17

7.2 Factoring the Difference of Two Squares

In Chapter 5 we noted that whenever we multiplied the sum of two terms times the difference of those same two terms, the product was the **difference of the squares** of the two terms. For example, remember that

$$(x + 7)(x - 7) = x^2 - 49$$

In this chapter we are concerned with *reversing* the operation of multiplication. Factoring is a process by which we rewrite a polynomial as a product of factors.

Let's try to factor $x^2 - 16$. Is this a difference of two squares? Is x^2 a square? Of course. Is 16 a square? Yes, 16 is 4^2. You should agree that

$$x^2 - 16 = (x + 4)(x - 4)$$

As usual, we can check our factors by multiplication.

We see that the difference of two squares can be factored in the following way.

Difference of Two Squares

$$A^2 - B^2 = (A + B)(A - B)$$

Let's factor the expression $25y^2 - 121$.

Is this a difference of 2 terms?	Yes.
Is the first term a square?	Yes, $25y^2$ is the square of $5y$.
What is A?	$A = 5y$.
Is the second term a square?	Yes, 121 is the square of 11.
What is B?	$B = 11$.

$$25y^2 - 121 = (5y + 11)(5y - 11)$$

The difference of two squares is not difficult to spot. You must look for two terms with a minus sign in between and decide whether each term is the square of some expression. If so, one factor will be the *sum* of those expressions and the other factor will be the *difference* of those expressions. *Remember to remove any common monomial factors first.*

Example 1. Factor $36 - x^2$.

Solution

$$36 - x^2$$
$$= (6 + x)(6 - x)$$

You try Example 2.

Example 2. Factor $\frac{1}{9} a^2 - y^2$.

Solution

$$\frac{1}{9} a^2 - y^2$$

$$= (\underline{\hspace{1.5cm}})(\underline{\hspace{1.5cm}})$$

Check your factors on page 334. ▶

Example 3. Factor $5x^2 - 20$.

Solution

$$5x^2 - 20$$
$$= 5(x^2 - 4)$$
$$= 5(x + 2)(x - 2)$$

Try Example 4 and watch for a *common factor*.

Example 4. Factor $3x^3 - 3x$.

Solution

$$3x^3 - 3x$$
$$= 3x(\underline{\qquad})$$
$$= 3x(\underline{\qquad})(\underline{\qquad})$$

Check your factors on page 334. ▶

Note that we removed the common monomial factors in Examples 3 and 4. Be sure that you do not skip the step in which you factor out the common monomial factor. Even mathematicians who have been factoring for years would not dream of jumping from the original polynomial to its completely factored form.

Now try Example 5.

Example 5. Factor $99x^2y^2 - 44x^2z^2$.

Solution

Check your factors on page 334. ▶

Example 6. Factor $x^2 + 4y^2$.

Solution. $x^2 + 4y^2$ is a *sum* of two squares. It is *not* a *difference* of two squares. It cannot be factored.

We note that a *sum of two squares* cannot be factored in the context of this course. When a polynomial cannot be factored, we say that the polynomial is **prime** (or **irreducible**).

Example 7. Factor $x^4 - 1$.

Solution

$$x^4 - 1$$
$$= (x^2 + 1)(x^2 - 1)$$
$$= (x^2 + 1)(x + 1)(x - 1)$$

Now try Example 8.

Example 8. Factor $16y^4 - 81z^4$.

Solution

$$16y^4 - 81z^4$$
$$= (4y^2 + 9z^2)(\underline{\qquad})$$
$$= (4y^2 + 9z^2)(\underline{\qquad})(\underline{\qquad})$$

Check your factors on page 334. ▶

Notice in Examples 7 and 8 that after factoring the first time, one of our factors was the difference of two squares. Therefore, we continued factoring because an expression should always be factored completely.

Since it is important that we be able to recognize numbers that are squares, let's complete the following table of squares of the first 20 natural numbers.

Table of Squares

n	n^2	n	n^2
1	1	11	121
2	____	12	____
3	____	13	____
4	____	14	196
5	____	15	____
6	____	16	256
7	____	17	289
8	64	18	____
9	81	19	361
10	____	20	____

Check your table on page 334. ▶

⫸ Trial Run

Rewrite each expression as a product of factors.

_____ **1.** $x^2 - 196$

_____ **2.** $9x^2 - 49$

_____ **3.** $32 - 2x^2$

_____ **4.** $x^2 + y^2$

_____ **5.** $12a^2 - 75b^2$

_____ **6.** $45x^3 - 5x$

Answers are on page 334.

▶ **Examples You Completed** —————————————

Example 2. Factor $\frac{1}{9}a^2 - y^2$.

Solution

$$\frac{1}{9}a^2 - y^2$$

$$= \left(\frac{1}{3}a + y\right)\left(\frac{1}{3}a - y\right)$$

Example 4. Factor $3x^3 - 3x$.

Solution

$$3x^3 - 3x$$

$$= 3x(x^2 - 1)$$

$$= 3x(x + 1)(x - 1)$$

Example 5. Factor $99x^2y^2 - 44x^2z^2$.

Solution

$$99x^2y^2 - 44x^2z^2$$

$$= 11x^2(9y^2 - 4z^2)$$

$$= 11x^2(3y + 2z)(3y - 2z)$$

Example 8. Factor $16y^4 - 81z^4$.

Solution

$$16y^4 - 81z^4$$

$$= (4y^2 + 9z^2)(4y^2 - 9z^2)$$

$$= (4y^2 + 9z^2)(2y + 3z)(2y - 3z)$$

Table of Squares

n	n^2	n	n^2
1	1	11	121
2	4	12	144
3	9	13	169
4	16	14	196
5	25	15	225
6	36	16	256
7	49	17	289
8	64	18	324
9	81	19	361
10	100	20	400

Answers to Trial Run —————————————

page 333 **1.** $(x - 14)(x + 14)$ **2.** $(3x + 7)(3x - 7)$ **3.** $2(4 + x)(4 - x)$ **4.** $x^2 + y^2$
5. $3(2a - 5b)(2a + 5b)$ **6.** $5x(3x + 1)(3x - 1)$

Name _____ **Date** _____

EXERCISE SET 7.2

Rewrite each expression as a product of factors.

_____ 1. $x^2 - 25$

_____ 2. $x^2 - 16$

_____ 3. $9y^2 - 1$

_____ 4. $144y^2 - 1$

_____ 5. $49 - a^2$

_____ 6. $36 - a^2$

_____ 7. $4a^2 - 121$

_____ 8. $9a^2 - 100$

_____ 9. $25b^2 - \dfrac{4}{9}$

_____ 10. $49b^2 - \dfrac{25}{64}$

_____ 11. $x^2 - y^2$

_____ 12. $a^2 - b^2$

_____ 13. $a^2 - 0.36b^2$

_____ 14. $a^2 - 0.81b^2$

_____ 15. $4x^2 - 225y^2$

_____ 16. $9x^2 - 25y^2$

_____ 17. $m^2n^2 - 144$

_____ 18. $x^2y^2 - 169$

_____ 19. $9a^2b^2 - c^2$

_____ 20. $16a^2b^2 - c^2$

_____ 21. $0.64x^2y^2 - 0.25z^2$

_____ 22. $0.16x^2y^2 - 0.81z^2$

_____ 23. $x^4 - 81$

_____ 24. $x^4 - 16$

_____ 25. $25x^4 - 121$

_____ 26. $9x^4 - 64$

_____ 27. $225 - a^4b^4$

_____ 28. $289 - a^4b^4$

_____ 29. $3x^2 - 75$

_____ 30. $2x^2 - 32$

_____ 31. $11x^2 - 44y^2$

_____ 32. $5x^2 - 45y^2$

_____ 33. $16y - 9x^2y$

_____ 34. $25y - 4x^2y$

_____ 35. $6m^3 - 150m$

_____ 36. $7m^3 - 112m$

_____ 37. $49a^2b - \dfrac{1}{4}b^3$

_____ 38. $64a^2b - \dfrac{1}{9}b^3$

_____ 39. $2x^2y^2 - 72y^4$

_____ 40. $3x^2y^2 - 48y^4$

_____ 41. $3m^4 - 243$

_____ 42. $5m^4 - 80$

_____ 43. $2x^3y - 8xy^3$

_____ 44. $3x^3y - 27xy^3$

_____ 45. $27x^3y^3 - 12xy$

_____ 46. $50x^3y^3 - 2xy$

_____ **47.** $a^6 - 25a^2$ _____ **48.** $a^6 - 81a^2$

_____ **49.** $289x^2y^2z^2 - z^2$ _____ **50.** $256x^2y^2z^2 - z^2$

☆ Stretching the Topics

Rewrite each expression as a product of factors.

_____ **1.** $(a + b)^2 - 9c^2$

_____ **2.** $(2x - y)^2 - (y + 3z)^2$

_____ **3.** $x^{2a} - 25y^{2b}$

Check your answers in the back of your book.

If you can do **Checkup 7.2,** you are ready to go to Section 7.3.

✓ **CHECKUP 7.2**

Rewrite each expression as a product of factors.

_____ 1. $x^2 - 64$ _____ 2. $25 - y^2$

_____ 3. $9a^2 - 100$ _____ 4. $x^2 - \dfrac{25}{49}y^2$

_____ 5. $x^4 - 16$ _____ 6. $3x^2 - 0.75y^2$

_____ 7. $7x^2y^2 - 28y^4$ _____ 8. $5m^4 - 405n^4$

_____ 9. $4x^3y - 400xy^3$ _____ 10. $3a - 243a^3$

Check your answers in the back of the book.

If You Missed Problems:	You Should Review Examples:
1–4	1, 2
5	7
6, 7	3, 4
8	3, 8
9	5
10	1, 4

7.3 Factoring Trinomials

Recall from Chapter 6 the FOIL method that we used to multiply binomials.

$$
\begin{array}{ccccc}
\text{F} & \text{L} & \text{F} & \text{O} \qquad \text{I} & \text{L} \\
\end{array}
$$

$$(x + 5)(x - 2) = x^2 - 2x + 5x - 10 = x^2 + 3x - 10$$

$$
\begin{array}{c}
\text{I} \\
\text{O}
\end{array}
$$

When we multiply two binomials, our product often turns out to be a trinomial. In this section we must learn how to decide what binomials were multiplied together to give trinomials; we must learn how to *factor* trinomials.

Factoring Trinomials (Leading Coefficient of 1)

Notice in our FOIL example that the first term in our trinomial (the F term) came from multiplying the first terms of the binomials. The last term in the trinomial (the L term) came from multiplying the second terms of the binomials. The middle term in the trinomial was the sum of the O and I terms.

Keeping that in mind, let's try to factor:	$x^2 + 4x + 3$
We are looking for two *binomials*.	(_____)(_____)
Since the first term of our trinomial is x^2, we can write:	$(x \quad)(x \quad)$
Since the last term of our trinomial is 3, we know the second terms of our	$(x + 1)(x + 3)$
binomials must be 1 and 3 or -1 and -3.	$(x - 1)(x - 3)$

Do both pairs of factors give the correct product? We check by multiplication.

$$
\begin{array}{ll}
(x + 1)(x + 3) & (x - 1)(x - 3) \\
= x^2 + 3x + x + 3 & = x^2 - 3x - x + 3 \\
= x^2 + 4x + 3 & = x^2 - 4x + 3 \\
\qquad \text{Yes} & \qquad \text{No}
\end{array}
$$

Therefore, the factored form of our trinomial is

$$x^2 + 4x + 3 = (x + 1)(x + 3)$$

Checking your factors by multiplying is a *must*. Suppose that we consider another problem where a choice must be made from more possibilities.

Let's factor this trinomial:	$x^2 - 7x + 12$
We are looking for two binomials.	$(x \quad)(x \quad)$

Since the last term of our trinomial is 12, there are several possibilities for the second terms of the binomials.	$(x + 1)(x + 12)$	$(x - 1)(x - 12)$
	$(x + 6)(x + 2)$	$(x - 6)(x - 2)$
	$(x + 3)(x + 4)$	$(x - 3)(x - 4)$

Since the middle term of the trinomial is *negative* $(-7x)$, we may reject the first three possibilities because they will yield a *positive* middle term. We check the last three possibilities.

$$(x - 1)(x - 12) = x^2 - 12x - x + 12$$
$$= x^2 - 13x + 12$$
$$(x - 6)(x - 2) = x^2 - 2x - 6x + 12$$
$$= x^2 - 8x + 12$$
$$(x - 3)(x - 4) = x^2 - 4x - 3x + 12$$
$$= x^2 - 7x + 12$$

We find that only the last pair of factors gives the correct middle term and we conclude that

$$x^2 - 7x + 12 = (x - 3)(x - 4)$$

After some practice you will learn to leave out the possibilities that do not seem likely. You need not list *all* the possibilities, but be sure to check your factors by multiplying before deciding that they are correct.

Example 1. Factor $x^2 - 9xy + 20y^2$.

Solution

$$x^2 - 9xy + 20y^2$$
$$= (x - 5y)(x - 4y)$$

You try Example 2.

Example 2. Factor $x^2 + 12xy + 20y^2$.

Solution

$$x^2 + 12xy + 20y^2$$
$$= (\text{_____})(\text{_____})$$

Remember to check your factors by multiplying. Check your factors on page 348. ▶

Example 3. Factor $8 + 6x + x^2$.

Solution

$$8 + 6x + x^2$$
$$= (4 + x)(2 + x)$$

You factor Example 4.

Example 4. Factor $15 - 8y + y^2$.

Solution

$$15 - 8y + y^2$$
$$= (\text{_____})(\text{_____})$$

Check your factors on page 348. ▶

⫸ Trial Run

Rewrite each trinomial as a product of factors.

_____ **1.** $x^2 + 8x + 12$

_____ **2.** $12 + 13a + a^2$

_____ **3.** $y^2 - 9y + 18$

_____ **4.** $x^2 - 6xy + 9y^2$

Answers are on page 349.

What happens when the last term of our trinomial is *negative*?

Let's try to factor: $x^2 - 5x - 6$

We are looking for two binomials. $(x \quad)(x \quad)$

Since the last term of the trinomial is -6, there are several possibilities for the second terms of the binomials.

$(x + 1)(x - 6)$ $\qquad$ $(x + 2)(x - 3)$

$(x - 1)(x + 6)$ $\qquad$ $(x - 2)(x + 3)$

Which of these possibilities gives the correct product?

$(x + 1)(x - 6)$	$(x - 1)(x + 6)$	$(x + 2)(x - 3)$	$(x - 2)(x + 3)$
$= x^2 - 6x + x - 6$	$= x^2 + 6x - x - 6$	$= x^2 - 3x + 2x - 6$	$= x^2 + 3x - 2x - 6$
$= x^2 - 5x - 6$	$= x^2 + 5x - 6$	$= x^2 - x - 6$	$= x^2 + x - 6$

Our trinomial can be factored as

$$x^2 - 5x - 6 = (x + 1)(x - 6)$$

Example 5. Factor $x^2 - x - 12$.

Solution

$$x^2 - x - 12$$
$$= (x - 4)(x + 3)$$

Try Example 6.

Example 6. Factor $x^2 + 4x - 12$.

Solution

$$x^2 + 4x - 12$$
$$= (\text{_____})(\text{_____})$$

Check your factors on page 348. ▶

Example 7. Factor $3x^2 - 33xy - 36y^2$.

Solution

$$3x^2 - 33xy - 36y^2$$
$$= 3(x^2 - 11xy - 12y^2)$$
$$= 3(x - 12y)(x + y)$$

Notice that we remembered to look first for a *common monomial factor*.

You try Example 8.

Example 8. Factor $x^3 - 13x^2 - 30x$.

Solution

$$x^3 - 13x^2 - 30x$$
$$= x(\text{_____})$$
$$= x(\text{_____})(\text{_____})$$

Check your factors on page 348. ▶

The order in which we write the factors is not important, but the factors themselves are **unique**, which means that if a polynomial can be factored, there is *one and only one* correct set of factors. After some practice you will become skillful at finding the correct factors.

⫸ Trial Run

Rewrite each trinomial as a product of factors.

_____ **1.** $x^2 + x - 6$ $\qquad\qquad$ _____ **2.** $10 - 3a - a^2$

_____ **3.** $2m^2 - 10m - 48$ $\qquad\qquad$ _____ **4.** $a^2 - 14ab - 15b^2$

Answers are on page 349.

Factoring Trinomials (Leading Coefficient Not 1)

We have learned to factor trinomials in which the coefficient of the squared term is 1. There are many trinomials in which the coefficient of the squared term is *not* 1. The techniques used to factor such trinomials are exactly the same as the techniques already learned. We continue to look for a common monomial factor first and then for two binomials that multiply together to give the trinomial.

Let's try to factor: $3x^2 + 7x + 2$

Since there is no common monomial factor, we know the first terms in the binomials must multiply to give $3x^2$. $(3x\quad)(x\quad)$

Since the last term of the trinomial is 2, there are two possibilities for the second terms of the binomials. $(3x + 2)(x + 1)$

$(3x + 1)(x + 2)$

We check both possibilities by multiplying. $(3x + 2)(x + 1) = 3x^2 + 3x + 2x + 2$

$$= 3x^2 + 5x + 2$$

$$(3x + 1)(x + 2) = 3x^2 + 6x + x + 2$$

$$= 3x^2 + 7x + 2$$

Therefore, our trinomial factors as

$$3x^2 + 7x + 2 = (3x + 1)(x + 2)$$

If the first term and/or the third term of a trinomial can be obtained in several ways, there may be many possible pairs of binomial factors to try.

Let's look at: $2x^2 - x - 10$

Noting that the terms contain *no common monomial factor*, we list the possible pairs of binomial factors.

$(2x + 2)(x - 5)$ $(2x + 1)(x - 10)$

$(2x - 2)(x + 5)$ $(2x - 1)(x + 10)$

$(2x + 10)(x - 1)$ $(2x + 5)(x - 2)$

$(2x - 10)(x + 1)$ $(2x - 5)(x + 2)$

We can reject the first four pairs of factors because in each pair, the first factor contains a common monomial factor of 2. If these were the correct binomials, a common factor of 2 could have been removed from the *original* trinomial. Now we can check the remaining four pairs of factors.

$(2x + 1)(x - 10)$	$(2x - 1)(x + 10)$	$(2x + 5)(x - 2)$	$(2x - 5)(x + 2)$
$= 2x^2 - 20x + x - 10$	$= 2x^2 + 20x - x - 10$	$= 2x^2 - 4x + 5x - 10$	$= 2x^2 + 4x - 5x - 10$
$= 2x^2 - 19x - 10$	$= 2x^2 + 19x - 10$	$= 2x^2 + x - 10$	$= 2x^2 - x - 10$

Since the last pair gives the product we were seeking,

$$2x^2 - x - 10 = (2x - 5)(x + 2)$$

Writing all the possible pairs of factors can sometimes be a major task. It is often easier to jump in and try some, hoping to hit the correct pair before too long. You should *not* erase pairs of factors after rejecting them or you are likely to lose track of which pairs you have tried.

Example 9. Factor $8x^2 + 2x - 15$.

Solution

$$8x^2 + 2x - 15$$
$$= (2x + 3)(4x - 5)$$

You try Example 10.

Example 10. Factor $7x^2 - x - 6$.

Solution

$$7x^2 - x - 6$$
$$= (\underline{\hspace{1.5cm}})(\underline{\hspace{1.5cm}})$$

Check your factors on page 348. ▶

Example 11. Factor $-21y - 5y^2 + 6y^3$.

Solution

$$-21y - 5y^2 + 6y^3$$
$$= -y(21 + 5y - 6y^2)$$
$$= -y(3 + 2y)(7 - 3y)$$

Now try Example 12.

Example 12. Factor $3x^3y + 6x^2y^2 + 3xy^3$.

Solution

Check your factors on page 348. ▶

⇒ Trial Run

Rewrite each trinomial as a product of factors.

_____ **1.** $2x^2 + 7x - 15$

_____ **2.** $2 - 11m + 12m^2$

_____ **3.** $2x^2 - 9xy + 10y^2$

_____ **4.** $4ax^3 - 12ax^2 + 9ax$

Answers are on page 349.

Factoring Perfect Square Trinomials

In Chapter 6 we learned that the square of a binomial yielded a perfect square trinomial found by the following formulas.

$$(A + B)^2 = A^2 + 2AB + B^2$$
$$(A - B)^2 = A^2 - 2AB + B^2$$

If we recognize that a trinomial that we wish to factor is indeed a perfect square trinomial, we may reverse these formulas to find the factors. In a perfect square trinomial, the first term must be the square of some quantity A, the last term must be the square of some quantity B, and the middle term must be twice the product of A times B. Then we can use one of our formulas. Notice that the sign of the middle term tells us which formula to use.

> **Perfect Square Trinomials**
>
> $A^2 + 2AB + B^2 = (A + B)^2$
>
> $A^2 - 2AB + B^2 = (A - B)^2$

Consider the trinomial $x^2 + 6x + 9$.

Is the first term a *square*?	Yes, x^2 is the square of x.
What is A?	$A = x$.
Is the third term a *square*?	Yes, 9 is the square of 3.
What is B?	$B = 3$.
Is the second term *twice* the product of A times B?	Yes, $6x = 2 \cdot 3 \cdot x$.
What is the *sign* of the middle term?	The middle term is *positive*.

$$x^2 + 6x + 9 = (x + 3)^2$$

Example 13. Factor $x^2 - 10xy + 25y^2$.

Solution. Is $x^2 - 10xy + 25y^2$ a perfect square trinomial? Yes, because

$$A = x$$
$$B = 5y$$
$$2AB = 2 \cdot x \cdot 5y = 10xy$$

The middle term is *negative*.

$$x^2 - 10xy + 25y^2 = (x - 5y)^2$$

You try Example 14.

Example 14. Factor $4a^2 + 28a + 49$.

Solution. Is $4a^2 + 28a + 49$ a perfect square trinomial? Yes, because

$$A = \underline{\quad}$$
$$B = \underline{\quad}$$
$$2AB = 2 \cdot \underline{\quad} \cdot \underline{\quad} = \underline{\quad}$$

The middle term is $\underline{\quad}$.

$$4a^2 + 28a + 49 = (\underline{\quad} + \underline{\quad})^2$$

Check your work on page 348. ▶

Example 15. Factor $y^2 + \dfrac{2}{3}y + \dfrac{1}{9}$.

Solution. Is $y^2 + \dfrac{2}{3}y + \dfrac{1}{9}$ a perfect square trinomial? Yes, because

$$A = y$$
$$B = \frac{1}{3}$$
$$2AB = 2 \cdot y \cdot \frac{1}{3} = \frac{2}{3}y$$

The middle term is *positive*.

$$y^2 + \frac{2}{3}y + \frac{1}{9} = \left(y + \frac{1}{3}\right)^2$$

You try Example 16.

Example 16. Factor $x^2 - 0.8x + 0.16$.

Solution. Is $x^2 - 0.8x + 0.16$ a perfect square trinomial? Yes, because

$$A = \underline{\quad}$$
$$B = 0.4$$
$$2AB = 2 \cdot \underline{\quad} (0.4) = \underline{\quad}$$

The middle term is $\underline{\quad}$.

$$x^2 - 0.8x + 0.16 = (x - \underline{\quad})^2$$

Check your work on page 348. ▶

Although perfect square trinomials can always be factored by the trial-and-error process discussed earlier, you should try to recognize them and use the formulas whenever possible.

⫸ Trial Run

Rewrite each trinomial as a product of factors.

———— **1.** $x^2 + 12x + 36$ ———— **2.** $100y^2 - 20xy + x^2$

———— **3.** $a^2 - \dfrac{2}{7}a + \dfrac{1}{49}$ ———— **4.** $0.01 + 0.2x + x^2$

Answers are on page 349.

Factoring Trinomials by the ac-Method (Optional)

Perhaps you would like to know if there is another way to factor a trinomial of the form

$$ax^2 + bx + c$$

where a, b, and c are integers. We have found that trial and error is a quick method to use when the leading coefficient, a, is 1, but trial and error can sometimes become frustrating when a is different from 1.

Suppose that we consider factoring the trinomial $3x^2 + 14x + 8$.

1. First we *multiply* the leading coefficient (3) times the constant term (8). $3 \cdot 8 = 24$

2. Now we list all possible pairs of integer factors whose products are 24.

$$\begin{array}{ll} (1)(24) & (-1)(-24) \\ (2)(12) & (-2)(-12) \\ (3)(8) & (-3)(-8) \\ (4)(6) & (-4)(-6) \end{array}$$

3. See which (if any) of the listed pairs of factors will *add* together to give the numerical coefficient of the *middle* term of the original trinomial (14). $2 + 12 = 14$

4. Now split the middle term (14x) into the sum of two separate terms with coefficients from the chosen pair. $14x = 2x + 12x$

5. Rewrite the original polynomial with the split middle term and factor this four-term polynomial by grouping.

$$3x^2 + 14x + 8$$
$$= 3x^2 + 2x + 12x + 8$$
$$= x(3x + 2) + 4(3x + 2)$$
$$= (3x + 2)(x + 4)$$

Finally, check these factors by multiplying.

$$(3x + 2)(x + 4)$$
$$= 3x^2 + 12x + 2x + 8$$
$$= 3x^2 + 14x + 8$$

We have found the correct factors, and we write

$$3x^2 + 14x + 8 = (3x + 2)(x + 4)$$

Because we use the leading coefficient, a, and the constant term, c, to begin our search, this method is sometimes called the **ac-method** for factoring trinomials.

You complete Example 17.

Example 17. Use the ac-method to factor $2x^2 - x - 15$.

Solution. Here $a = 2$ and $c = -15$.

Find $a \cdot c$.

$$a \cdot c = (\underline{\qquad})(\underline{\qquad}) = -30$$

List pairs of factors of -30.

$(1)(-30)$	$(-1)(30)$
$(2)(-15)$	$(\underline{\quad})(\underline{\quad})$
$(\underline{\quad})(\underline{\quad})$	$(-3)(10)$
$(5)(-6)$	$(\underline{\quad})(\underline{\quad})$

Find the pair whose sum is the coefficient of the middle term, -1.

5 and $\underline{\qquad}$

Split the original middle term.

$$2x^2 - x - 15$$
$$= 2x^2 + 5x - 6x - 15$$

Factor the four-term polynomial.

$$= x(\underline{\qquad}) - 3(\underline{\qquad})$$
$$= (\underline{\qquad})(x - 3)$$

Check your factors on page 348. ▶

Example 18. Use the ac-method to factor $2x^2 + 4x + 5$.

Solution. Here $a = 2$ and $c = 5$, so $a \cdot c = 2 \cdot 5 = 10$. Pairs of factors of 10 are

$(1)(10)$	$(-1)(-10)$
$(2)(5)$	$(-2)(-5)$

None of the paired factors add to give the coefficient of the middle term (4). Therefore, we conclude that the trinomial $2x^2 + 4x + 5$ cannot be factored using integers. It is *prime*.

Our last example illustrates that the ac-method is very useful in deciding whether a trinomial is prime. If none of the paired factors of $a \cdot c$ will add to give the coefficient of the middle term, we can conclude that the trinomial cannot be factored using integers.

⫸ **Trial Run**

Factor by the ac-method.

_____ **1.** $6x^2 - 11x + 3$ _____ **2.** $5x^2 - 13x - 6$

_____ **3.** $8x^2 - 26x + 15$ _____ **4.** $6x^2 + 5x - 6$

Answers are on page 349.

Factoring Trinomials of the Form $ax^4 + bx^2 + c$ (Optional)

Let's try to factor $x^4 - 5x^2 - 6$. From our earlier experience with factoring other trinomials, we might guess that this trinomial is the product of two binomials.

We consider: $\qquad\qquad\qquad\qquad\qquad\qquad x^4 - 5x^2 - 6$

We look for two binomials like: $\qquad\qquad\qquad (x^2 \quad)(x^2 \quad)$

Since the constant term is -6, we use our former $\quad (x^2 - 6)(x^2 + 1)$
methods to decide on the correct second terms.

Neither of our binomial factors can be factored further, so we conclude that

$$x^4 - 5x^2 - 6 = (x^2 - 6)(x^2 + 1)$$

Example 19. Factor $x^4 - 8x^2 - 9$.

Solution

$$x^4 - 8x^2 - 9$$
$$= (x^2 + 1)(x^2 - 9)$$
$$= (x^2 + 1)(x + 3)(x - 3)$$

Notice that we continue factoring when possible.

You try Example 20.

Example 20. Factor $x^4 - 5x^2 + 4$.

Solution

$$x^4 - 5x^2 + 4$$
$$= (x^2 - 4)(x^2 - \underline{})$$
$$=$$

Check your factors on page 348. ▶

Example 21. Factor $5 - 10x^2 + 5x^4$.

Solution

$$5 - 10x^2 + 5x^4$$
$$= 5(1 - 2x^2 + x^4)$$
$$= 5(1 - x^2)(1 - x^2)$$
$$= 5(1 + x)(1 - x)(1 + x)(1 - x)$$
$$= 5(1 + x)^2(1 - x)^2$$

Now try Example 22.

Example 22. Factor $3x^4 + 3x^2 - 60$.

Solution

Check your factors on page 348. ▶

IIII▶ Trial Run

Rewrite each trinomial as a product of factors.

_____ 1. $x^4 + 8x^2 + 12$

_____ 2. $2x^4 + 8x^2 + 6$

_____ 3. $x^4 + 2x^2 - 3$

_____ 4. $x^4 - 13x^2 + 36$

Answers are on page 349.

▶ Examples You Completed _____

Example 2. Factor $x^2 + 12xy + 20y^2$.

Solution

$$x^2 + 12xy + 20y^2$$
$$= (x + 10y)(x + 2y)$$

Example 4. Factor $15 - 8y + y^2$.

Solution

$$15 - 8y + y^2$$
$$= (5 - y)(3 - y)$$

Example 6. Factor $x^2 + 4x - 12$.

Solution

$$x^2 + 4x - 12$$
$$= (x + 6)(x - 2)$$

Example 8. Factor $x^3 - 13x^2 - 30x$.

Solution

$$x^3 - 13x^2 - 30x$$
$$= x(x^2 - 13x - 30)$$
$$= x(x - 15)(x + 2)$$

Example 10. Factor $7x^2 - x - 6$.

Solution

$$7x^2 - x - 6$$
$$= (7x + 6)(x - 1)$$

Example 12. Factor $3x^3y + 6x^2y^2 + 3xy^3$.

Solution

$$3x^3y + 6x^2y^2 + 3xy^3$$
$$= 3xy(x^2 + 2xy + y^2)$$
$$= 3xy(x + y)(x + y)$$

Example 14. Factor $4a^2 + 28a + 49$.

Solution. Is $4a^2 + 28a + 49$ a perfect square trinomial? Yes, because

$$A = 2a$$
$$B = 7$$
$$2AB = 2 \cdot 2a \cdot 7 = 28a$$

The middle term is positive.

$$4a^2 + 28a + 49 = (2a + 7)^2$$

Example 16. Factor $x^2 - 0.8x + 0.16$.

Solution. Is $x^2 - 0.8x + 0.16$ a perfect square trinomial? Yes, because

$$A = x$$
$$B = 0.4$$
$$2AB = 2 \cdot x \,(0.4) = 0.8x$$

The middle term is negative.

$$x^2 - 0.8x + 0.16 = (x - 0.4)^2$$

Example 17. Use the *ac*-method to factor $2x^2 - x - 15$.

Solution

$$a \cdot c = (2)(-15) = -30$$

Pairs of factors:

$$(1)(-30) \qquad (-1)(30)$$
$$(2)(-15) \qquad (-2)(15)$$
$$(3)(-10) \qquad (-3)(10)$$
$$(5)(-6) \qquad (-5)(6)$$

Correct pair: 5 and -6.

$$2x^2 - x - 15$$
$$= 2x^2 + 5x - 6x - 15$$
$$= x(2x + 5) - 3(2x + 5)$$
$$= (2x + 5)(x - 3)$$

Example 20. Factor $x^4 - 5x^2 + 4$.

Solution

$$x^4 - 5x^2 + 4$$
$$= (x^2 - 4)(x^2 - 1)$$
$$= (x + 2)(x - 2)(x + 1)(x - 1)$$

Example 22. Factor $3x^4 + 3x^2 - 60$.

Solution

$$3x^4 + 3x^2 - 60$$
$$= 3(x^4 + x^2 - 20)$$
$$= 3(x^2 + 5)(x^2 - 4)$$
$$= 3(x^2 + 5)(x + 2)(x - 2)$$

Answers to Trial Runs

page 340   **1.** $(x + 6)(x + 2)$ **2.** $(12 + a)(1 + a)$ **3.** $(y - 6)(y - 3)$ **4.** $(x - 3y)(x - 3y)$

page 341   **1.** $(x + 3)(x - 2)$ **2.** $(5 + a)(2 - a)$ **3.** $2(m - 8)(m + 3)$ **4.** $(a - 15b)(a + b)$

page 343   **1.** $(2x - 3)(x + 5)$ **2.** $(2 - 3m)(1 - 4m)$ **3.** $(2x - 5y)(x - 2y)$
4. $ax(2x - 3)(2x - 3)$

page 345   **1.** $(x + 6)^2$ **2.** $(10y - x)^2$ **3.** $\left(a - \dfrac{1}{7}\right)^2$ **4.** $(0.1 + x)^2$

page 346   **1.** $(3x - 1)(2x - 3)$ **2.** $(5x + 2)(x - 3)$ **3.** $(4x - 3)(2x - 5)$ **4.** $(3x - 2)(2x + 3)$

page 347   **1.** $(x^2 + 6)(x^2 + 2)$ **2.** $2(x^2 + 1)(x^2 + 3)$ **3.** $(x^2 + 3)(x + 1)(x - 1)$
4. $(x + 3)(x - 3)(x + 2)(x - 2)$

EXERCISE SET 7.3

Rewrite each trinomial as a product of factors.

_____ 1. $x^2 - 7x + 12$

_____ 2. $x^2 - 7x + 10$

_____ 3. $x^2 + 0.9x + 0.20$

_____ 4. $x^2 + 0.8x + 0.15$

_____ 5. $x^2 - 24x + 144$

_____ 6. $x^2 - 18x + 81$

_____ 7. $a^2 - 5a - 14$

_____ 8. $a^2 + 6a - 27$

_____ 9. $x^2 + 8x + 16$

_____ 10. $x^2 + 14x + 49$

_____ 11. $x^2 - 0.2x - 0.35$

_____ 12. $x^2 + 1.1x - 0.26$

_____ 13. $x^2 + x - 72$

_____ 14. $x^2 + x - 30$

_____ 15. $x^2 - 4xy + 4y^2$

_____ 16. $x^2 - 8xy + 16y^2$

_____ 17. $x^2 - \frac{2}{5}xy + \frac{1}{25}y^2$

_____ 18. $x^2 - \frac{2}{7}xy + \frac{1}{49}y^2$

_____ 19. $m^2 + mn - 6n^2$

_____ 20. $m^2 - 4mn - 45n^2$

_____ 21. $39 - 16m + m^2$

_____ 22. $72 - 17m + m^2$

_____ 23. $x^2y^2 + 0.2xy - 0.15$

_____ 24. $x^2y^2 + 0.5xy - 0.24$

_____ 25. $x^4 - 5x^2 + 6$

_____ 26. $x^4 - 11x^2 + 24$

_____ 27. $3x^2 - 18x + 27$

_____ 28. $5x^2 - 40x + 80$

_____ 29. $-2y^3 + 6y^2 + 20y$

_____ 30. $-3y^3 - 24y^2 + 27y$

_____ 31. $4m^4 + 4m^2n^2 - 24n^4$

_____ 32. $5m^4 - 25m^2n^2 - 30n^4$

_____ 33. $x^3y - 18x^2y^2 + 80xy^3$

_____ 34. $x^3y - 15x^2y^2 + 26xy^3$

_____ 35. $2a^4b - 28a^3b^2 - 64a^2b^3$

_____ 36. $3a^3b^2 - 15a^2b^3 - 450ab^4$

_____ 37. $2x^2 + 7x + 3$

_____ 38. $3x^2 + 16x + 5$

_____ 39. $9a^2 - 30a + 25$

_____ 40. $25a^2 - 20a + 4$

_____ 41. $10y^2 - 7y - 12$

_____ 42. $14y^2 + 17y - 6$

_____ 43. $8m^2 + 14m - 15$

_____ 44. $8m^2 + 18m - 35$

_____ 45. $6x^2 - 5xy - 56y^2$

_____ 46. $16x^2 + 6xy - 27y^2$

_____ 47. $81x^2 - 36xy + 4y^2$

_____ 48. $25x^2 - 30xy + 9y^2$

_____ 49. $121 + 132a + 36a^2$

_____ 50. $144 + 168a + 49a^2$

_____ **51.** $30c^2 - 145cd + 45d^2$ _____ **52.** $20c^2 - 68cd + 24d^2$

_____ **53.** $-8y^2 + 56y - 98$ _____ **54.** $-48y^2 + 72y - 27$

_____ **55.** $2x^2 - 7x - 3$ _____ **56.** $5x^2 - x - 8$

_____ **57.** $4a^2b^4 - 25a^2b^3 + 6a^2b^2$ _____ **58.** $4a^4b^2 - 23a^3b^2 - 6a^2b^2$

_____ **59.** $24x^3y + 26x^2y - 70xy$ _____ **60.** $105x^3y + 57x^2y - 72xy$

_____ **61.** $0.09k^2 - 3hk + 25h^2$ _____ **62.** $0.04h^2 - 1.2hk + 9k^2$

_____ **63.** $x^4 - 9x^2 + 20$ _____ **64.** $x^4 - 8x^2 - 9$

☆ Stretching the Topics

Rewrite each expression as a product of factors.

_____ **1.** $(a + b)^2 - 5(a + b) - 14$

_____ **2.** $x^5y - 53x^3y^3 + 196xy^5$

_____ **3.** $5a^{2x} - 8a^xb^y - 21b^{2y}$

Check your answers in the back of your book.

If you can complete **Checkup 7.3**, you are ready to go on to Section 7.4.

✓ **CHECKUP 7.3**

Rewrite each trinomial as a product of factors.

_____ 1. $x^2 - 3xy - 40y^2$

_____ 2. $21 - 10x + x^2$

_____ 3. $x^2 + 0.9x - 0.22$

_____ 4. $5x^2 - 10xy - 15y^2$

_____ 5. $x^2 - 12x + 36$

_____ 6. $x^3 - \dfrac{2}{3}x^2 + \dfrac{1}{9}x$

_____ 7. $2x^3y + 14x^2y^2 - 16xy^3$

_____ 8. $5x^2 - 17x + 6$

_____ 9. $-8y + 2y^2 + 3y^3$

_____ 10. $12a^3b + 26a^2b^2 - 10ab^3$

Check your answers in the back of your book.

If You Missed Problems:	You Should Review Examples:
1–4	1–6
5, 6	13–16
7	7, 8
8	9, 10
9, 10	11, 12

7.4 Using All Types of Factoring

Factoring is not a guessing game. If you are systematic in your approach, you will become successful after lots of practice. Get into the habit of asking yourself these questions.

1. Do all terms contain a common factor?
   If so, use the distributive property to rewrite the expression.

$$AB + AC = A(B + C)$$

2. Does the expression contain exactly *two* terms that are perfect squares separated by a *minus* sign?
   If so, use the formula for factoring the difference of two squares.

$$A^2 - B^2 = (A + B)(A - B)$$

3. Does the expression contain exactly *three* terms?
   If so, is it the square of a binomial that can be factored using a formula?

$$A^2 + 2AB + B^2 = (A + B)^2 \quad \text{or} \quad A^2 - 2AB + B^2 = (A - B)^2$$

   Or is it the product of two different binomials that can be found by trial and error?

4. Does the expression contain exactly *four* terms?
   If so, try to group the terms two by two.

In some cases you will use more than one factoring technique. In other cases you will discover that the expression cannot be factored.

Example 1. Factor completely $3x^3 - 48x$.

   Solution

$$3x^3 - 48x$$
$$= 3x(x^2 - 16) \qquad \text{Remove common factor } 3x.$$
$$= 3x(x + 4)(x - 4) \qquad \text{Factor the difference of two squares.}$$

Example 2. Factor completely $x^2 + xy + 6x + 6y$.

   Solution

$$x^2 + xy + 6x + 6y$$
$$= \underbrace{x^2 + xy} + \underbrace{6x + 6y} \qquad \text{Mentally group the terms two by two.}$$
$$= x(x + y) + 6(x + y) \qquad \text{Remove the common factor from each pair.}$$
$$= (x + y)(x + 6) \qquad \text{Remove the common binomial factor, } (x + y).$$

Example 3. Factor completely $x^4 + 8x^2 - 9$.

   Solution

$$x^4 + 8x^2 - 9$$
$$= (x^2 + 9)(x^2 - 1) \qquad \text{Factor the trinomial as a product of two binomials.}$$
$$= (x^2 + 9)(x + 1)(x - 1) \qquad \text{Factor the difference of two squares.}$$
$$\qquad\qquad\qquad\qquad\qquad \text{The sum of two squares will not factor.}$$

⟶ Trial Run

Rewrite each expression as a product of factors.

_____ **1.** $10x^2 - 90y^2$ _____ **2.** $-45x^2 + 30x - 5$

_____ **3.** $x^4 - 3x^2 - 4$ _____ **4.** $ac + bc - ad - bd$

Answers are given below.

Answers to Trial Run

page 354   **1.** $10(x + 3y)(x - 3y)$ **2.** $-5(3x - 1)^2$ **3.** $(x + 2)(x - 2)(x^2 + 1)$ **4.** $(a + b)(c - d)$

EXERCISE SET 7.4

Rewrite each expression as a product of factors.

_____ 1. $x^2 - 11x + 30$ _____ 2. $x^2 - 14x + 48$

_____ 3. $42c^2 - 66cd + 24d^2$ _____ 4. $35c^2 - 91cd + 56d^2$

_____ 5. $20ax + 4ay - 15x - 3y$ _____ 6. $10ax - 15ay - 8x + 12y$

_____ 7. $49x^2 - 42xy + 9y^2$ _____ 8. $81x^2 - 36xy + 4y^2$

_____ 9. $16b^2 - 121$ _____ 10. $49b^2 - 169$

_____ 11. $12a^3b - 16a^2 + 20ab^3$ _____ 12. $35a^3b - 25a^2b^2 + 20b^2$

_____ 13. $2a^3 - 200a$ _____ 14. $3a^3 - 75a$

_____ 15. $a^2b^2 - 11abc - 12c^2$ _____ 16. $a^2b^2 - 13abc - 14c^2$

_____ 17. $9x^6 + 30x^4y^2 + 25x^2y^4$ _____ 18. $16x^6 + 24x^4y^2 + 9x^2y^4$

_____ 19. $169 + 52a + 4a^2$ _____ 20. $144 - 120a + 25a^2$

_____ 21. $-24x^2 + 8x - 56$ _____ 22. $-28x^2 - 21x + 35$

_____ 23. $a^3 - 3a^2 + 9a - 27$ _____ 24. $a^3 - 4a^2 + 16a - 64$

_____ 25. $x^4 - 2x^2 - 63$ _____ 26. $x^4 - 22x^2 - 75$

_____ 27. $14a^3 - 28a^2 + 14a$ _____ 28. $6a^3 - 36a^2 + 54a$

_____ 29. $x^3yz - 8x^2y^2z + 15xy^3z$ _____ 30. $x^3yz - 4x^2y^2z - 21xy^3z$

_____ 31. $-x^4 + 19x^3 - 34x^2$ _____ 32. $-x^4 + 21x^3 - 38x^2$

_____ 33. $21x^2 - 9x + 3$ _____ 34. $18x^2 - 7x + 5$

_____ 35. $x^2 - 18xy + 81y^2$ _____ 36. $x^2 - 14xy + 49y^2$

_____ 37. $10x^2 - 3xy - 27y^2$ _____ 38. $28x^2 + xy - 15y^2$

_____ 39. $-30y^2 - 225y + 40$ _____ 40. $-28y^2 - 56y + 105$

_____ 41. $4a^2b^4 - 28a^2b^3 + 49a^2b^2$ _____ 42. $9a^2b^4 - 30a^2b^3 + 25a^2b^2$

_____ 43. $x^3 - 4x^2 - 9x + 36$ _____ 44. $2x^3 - 3x^2 - 8x + 12$

_____ 45. $4x^3 - 20x^2y + 16xy^2$ _____ 46. $6x^3 - 42x^2y - 48xy^2$

_____ 47. $a^6 - 16a^2$ _____ 48. $a^6 - 81a^2$

_____ 49. $-121y^2 + 16x^2y^2$ _____ 50. $-144y^2 + 25x^2y^2$

☆ Stretching the Topics ————————————

Rewrite each expression as a product of factors.

———— **1.** $-12a^5b^2 + 19a^3b^2 + 21ab^2$

———— **2.** $(a + b)^2 - 10c(a + b) + 25c^2$

———— **3.** $4(x + y)^2 + 12(x + y)(x - y) + 9(x - y)^2$

Check your answers in the back of your book.

If you can complete **Checkup 7.4**, you are ready to go on to Section 7.5.

✔ **CHECKUP 7.4**

Rewrite each expression as a product of factors.

_____ 1. $9x^2 - 42x + 49$

_____ 2. $18x^2 - 63x - 9$

_____ 3. $4x^2 - 36$

_____ 4. $81x^4 - y^4$

_____ 5. $x^3 - x^2 - 9x + 9$

_____ 6. $y^2 - 3y - 54$

_____ 7. $9m^2 - 18m - 16$

_____ 8. $64a^2b - 25b^3$

_____ 9. $30x^3y - 18x^2y - 99xy$

_____ 10. $10x^4y^2 + 80x^2y^4 - 480y^6$

Check your answers in the back of your book.

If You Missed Problem:	You Should Review Example:	
1	Section 7.3	14
2	Section 7.1	2
3	Section 7.2	4
4		8
5	Section 7.1	15
6	Section 7.3	5
7		10
8	Section 7.2	4
9	Section 7.3	7
10		21

7.5 Using Factoring to Solve Quadratic Equations

In Chapter 4 we learned to solve equations such as

$$3x + 1 = 6 \qquad 5 - 2x = 25 \qquad 2(y + 3) = y - 1$$

by finding the value of the variable that satisfied the equation. In each of these equations, the variable appears with an exponent of 1. Such equations are called **first-degree equations** or **linear equations**.

A polynomial equation containing one variable in which the largest exponent on the variable is a 2 is called a **second-degree equation** or a **quadratic equation**. Before we learn to solve quadratic equations, we must make a very important observation about *zero* as the result of multiplication.

The Zero Product Rule

If we are told that two quantities are multiplied together and the product is zero, what can we conclude? If the product of two factors is zero, then one or the other or both factors *must* be zero. We state this **zero product rule** in general as follows.

Zero Product Rule. If $A \cdot B = 0$, then $A = 0$ or $B = 0$ or both.

Example 1. Solve $(x + 3)(x - 5) = 0$ and check.

Solution. We observe that this product has two factors and we know that either one of them could be zero.

$$x + 3 = 0 \quad \textbf{or} \quad x - 5 = 0$$

Solving each of these first-degree equations, we have

$$x + 3 - 3 = 0 - 3 \quad \text{or} \quad x - 5 + 5 = 0 + 5$$

$$x = -3 \quad \text{or} \quad x = 5$$

Now we check each solution in the original equation.

CHECK: $x = -3$

$$(x + 3)(x - 5) = 0$$
$$(-3 + 3)(-3 - 5) \stackrel{?}{=} 0$$
$$(0)(-8) \stackrel{?}{=} 0$$
$$0 = 0$$

CHECK: $x = 5$

$$(x + 3)(x - 5) = 0$$
$$(5 + 3)(5 - 5) \stackrel{?}{=} 0$$
$$(8)(0) \stackrel{?}{=} 0$$
$$0 = 0$$

Each solution satisfies the original equation, so our solutions are $x = -3$ or $x = 5$.

Example 2. Solve $(x - 7)(x - 7) = 0$ and check.

Solution

$$(x - 7)(x - 7) = 0$$

$$x - 7 = 0 \quad \text{or} \quad x - 7 = 0$$

$$x = 7 \quad \text{or} \quad x = 7$$

Both solutions are the same, so we say that the solution is $x = 7$.

CHECK: $x = 7$

$$(x - 7)(x - 7) = 0$$

$$(7 - 7)(7 - 7) \stackrel{?}{=} 0$$

$$(0)(0) \stackrel{?}{=} 0$$

$$0 = 0$$

You check the solutions for Example 3.

Example 3. Solve $x(5 + x) = 0$ and check.

Solution

$$x(5 + x) = 0$$

$$x = 0 \quad \text{or} \quad 5 + x = 0$$

$$x = 0 \quad \text{or} \quad x = -5$$

CHECK: $x = 0$ CHECK: $x = -5$

Check your work on page 366. ▶

⫸ Trial Run

Solve and check.

_____ 1. $(x - 11)(x - 2) = 0$

_____ 2. $(y - 3)(y + 5) = 0$

_____ 3. $(a + 12)(a + 5) = 0$

_____ 4. $x(x - 1) = 0$

_____ 5. $(1 - x)(2 + x) = 0$

_____ 6. $(9 - x)(9 - x) = 0$

Answers are on page 367.

Solving Quadratic Equations by Factoring

We have just learned to solve an equation in which a product of factors is zero. Now we shall see how the zero product rule can help us solve a quadratic equation written in the **standard form** $ax^2 + bx + c = 0$. Let's solve

$$x^2 + 3x + 2 = 0$$

If we could rewrite the polynomial on the left-hand side as a *product*, we could use the zero product rule to solve the equation.

Example 4. Solve $x^2 + 3x + 2 = 0$ and check.

Solution

$$x^2 + 3x + 2 = 0$$

$$(x + 2)(x + 1) = 0 \qquad \text{Factor the polynomial.}$$

$$x + 2 = 0 \qquad \text{or} \qquad x + 1 = 0 \qquad \text{Use the zero product rule.}$$

$$x = -2 \qquad \text{or} \qquad x = -1 \qquad \text{Solve for } x.$$

CHECK: $x = -2$

$$x^2 + 3x + 2 = 0$$

$$(-2)^2 + 3(-2) + 2 \stackrel{?}{=} 0$$

$$4 - 6 + 2 \stackrel{?}{=} 0$$

$$0 = 0$$

CHECK: $x = -1$

$$x^2 + 3x + 2 = 0$$

$$(-1)^2 + 3(-1) + 2 \stackrel{?}{=} 0$$

$$1 - 3 + 2 \stackrel{?}{=} 0$$

$$0 = 0$$

The solutions are $x = -2$ or $x = -1$.

Example 5. Solve $3y^2 - 12 = 0$ and check.

Solution

$$3y^2 - 12 = 0$$

$$3(y^2 - 4) = 0$$

Note that we removed the *common factor* 3.

$$3(y + 2)(y - 2) = 0$$

Since the constant factor 3 can *never* be zero, we know

$$y + 2 = 0 \qquad \text{or} \qquad y - 2 = 0$$

$$y = -2 \qquad \text{or} \qquad y = 2$$

Now you check these solutions.

Now try Example 6.

CHECK: $y = -2$

$$3y^2 - 12 = 0$$

$$3(\underline{\qquad})^2 - 12 \stackrel{?}{=} 0$$

$$3(\underline{\qquad}) - 12 \stackrel{?}{=} 0$$

$$\underline{\qquad} - 12 \stackrel{?}{=} 0$$

$$\underline{\qquad} = 0$$

CHECK: $y = 2$

$$3y^2 - 12 = 0$$

$$3(\underline{\qquad})^2 - 12 \stackrel{?}{=} 0$$

$$3(\underline{\qquad}) - 12 \stackrel{?}{=} 0$$

$$\underline{\qquad} - 12 \stackrel{?}{=} 0$$

$$\underline{\qquad} = 0$$

Check your work on page 366. ▶

Example 6. Solve $5x^2 + 10x = 0$.

Solution

$$5x^2 + 10x = 0$$

$$5x(\underline{\hspace{2cm}}) = 0$$

$5x = 0$ or $\underline{\hspace{2cm}} = 0$

$x = 0$ or $x = \underline{\hspace{1cm}}$

Check your work on page 367. ▶

Try Example 7.

Example 7. Solve $0 = x^2 - 12x + 36$.

Solution

The solution is $x = \underline{\hspace{1cm}}$.

Check your work on page 367. ▶

⮕ **Trial Run**

Solve each equation.

$\underline{\hspace{2cm}}$ **1.** $x^2 - 4x - 5 = 0$

$\underline{\hspace{2cm}}$ **2.** $y^2 - 36 = 0$

$\underline{\hspace{2cm}}$ **3.** $x^2 - 7x = 0$

$\underline{\hspace{2cm}}$ **4.** $a^2 + 10a + 25 = 0$

$\underline{\hspace{2cm}}$ **5.** $3x^2 - 15x + 18 = 0$

$\underline{\hspace{2cm}}$ **6.** $27 + 6m - m^2 = 0$

Answers are on page 367.

More Quadratic Equations

Remember that solving quadratic equations depends very much on the zero product rule. To solve a quadratic equation, *we must write it in standard form* with the polynomial on one side and zero on the other side ($ax^2 + bx + c = 0$). Then we factor the polynomial, set each factor equal to zero, and solve for the variable.

Example 8. Solve $x^2 - x = 20$ and check.

Solution

$x^2 - x = 20$	Equation is not in standard form.
$x^2 - x - 20 = 20 - 20$	Subtract 20 from both sides.
$x^2 - x - 20 = 0$	Simplify.
$(x - 5)(x + 4) = 0$	Factor the polynomial.
$x - 5 = 0$ or $x + 4 = 0$	Use the zero product rule.
$x = 5$ or $x = -4$	Solve for x.

Now we check each solution in the *original* equation.

CHECK: $x = 5$

$$x^2 - x = 20$$
$$(5)^2 - 5 \stackrel{?}{=} 20$$
$$25 - 5 \stackrel{?}{=} 20$$
$$20 = 20$$

CHECK: $x = -4$

$$x^2 - x = 20$$
$$(-4)^2 - (-4) \stackrel{?}{=} 20$$
$$16 + 4 \stackrel{?}{=} 20$$
$$20 = 20$$

The solutions are $x = 5$ or $x = -4$.

Complete the missing steps in Example 9.

Example 9. Solve $x^2 - 10 = 3x$.

Solution

$x^2 - 10 = 3x$	Equation is not in standard form.
$x^2 - 10 - 3x = 3x - 3x$	Subtract $3x$ from both sides.
$x^2 - 10 - 3x = 0$	Simplify.
$x^2 - 3x - 10 = 0$	Rearrange terms.
$(\underline{\qquad})(\underline{\qquad}) = 0$	Factor.
$\underline{\qquad} = 0$ or $\underline{\qquad} = 0$	Use the zero product rule.
$x = \underline{\qquad}$ or $x = \underline{\qquad}$	Solve for x.

Check your work on page 367. ▶

You finish Example 10.

Example 10. Solve $y^2 = 16$.

Solution

$$y^2 = 16$$
$$y^2 - 16 = 0$$

Check your work on page 367. ▶

Example 11. Solve $8x^2 + 2x = 3$.

Solution

$$8x^2 + 2x = 3$$
$$8x^2 + 2x - 3 = 0$$
$$(4x + 3)(2x - 1) = 0$$
$$4x + 3 = 0 \quad \text{or} \quad 2x - 1 = 0$$
$$4x = -3 \qquad\qquad 2x = 1$$
$$x = -\frac{3}{4} \quad \text{or} \quad x = \frac{1}{2}$$

⫸ Trial Run

Solve each equation.

_____ **1.** $x^2 + 2x = 35$ _____ **2.** $y^2 = 64$

_____ **3.** $x^2 = 3x$ _____ **4.** $y^2 = 7y - 12$

_____ **5.** $3 = 10x^2 - 13x$ _____ **6.** $15 - 2a = a^2$

Answers are on page 367.

The method we have just learned will also work to solve equations containing parentheses or like terms that must be combined. Remember, to solve a quadratic equation, our aim is to have zero on one side of the equation and all other terms on the other side.

Example 12. Solve $3(x^2 - 1) - 9 = x(x + 10)$ and check.

    Solution

$3(x^2 - 1) - 9 = x(x + 10)$	
$3x^2 - 3 - 9 = x^2 + 10x$	Remove parentheses.
$3x^2 - 12 = x^2 + 10x$	Combine like terms.
$3x^2 - 12 - x^2 = x^2 + 10x - x^2$	Subtract x^2 from both sides.
$2x^2 - 12 = 10x$	Combine like terms.
$2x^2 - 12 - 10x = 10x - 10x$	Subtract $10x$ from both sides.
$2x^2 - 10x - 12 = 0$	Combine and rearrange terms.
$2(x^2 - 5x - 6) = 0$	Factor out common factor.
$2(x - 6)(x + 1) = 0$	Factor the polynomial.
$x - 6 = 0$ or $x + 1 = 0$	Use zero product rule.
$x = 6$ or $x = -1$	Solve for x.

CHECK: $x = 6$

$$3(x^2 - 1) - 9 = x(x + 10)$$

$$3[(6)^2 - 1] - 9 \stackrel{?}{=} 6(6 + 10)$$

$$3[36 - 1] - 9 \stackrel{?}{=} 6(16)$$

$$3[35] - 9 \stackrel{?}{=} 96$$

$$105 - 9 \stackrel{?}{=} 96$$

$$96 = 96$$

CHECK: $x = -1$

$$3(x^2 - 1) - 9 = x(x + 10)$$

$$3[(-1)^2 - 1] - 9 \stackrel{?}{=} -1(-1 + 10)$$

$$3[1 - 1] - 9 \stackrel{?}{=} -1(9)$$

$$3[0] - 9 \stackrel{?}{=} -9$$

$$0 - 9 \stackrel{?}{=} -9$$

$$-9 = -9$$

We can summarize the steps used to solve quadratic equations as follows.

Solving Quadratic Equations

1. Work with parentheses if necessary.
2. Rewrite the equation in standard form, with zero on one side and all other terms on the other side (combining like terms when possible).
3. Rewrite the equation with the polynomial in factored form.
4. Set each factor containing the variable equal to zero.
5. Solve for the variable and check each solution in the original equation.

Now you complete Examples 13 and 14.

Example 13. Solve $4x^2 + x = 3(x^2 - 4x)$.

Solution

$$4x^2 + x = 3(x^2 - 4x)$$

$$4x^2 + x = 3x^2 - 12x$$

$$x^2 + x = -12x$$

$$x^2 + 13x = 0$$

Example 14. Solve
$5x(x - 2) - 6 = 4x^2 - 2(5x + 1)$

Solution

$$5x(x - 2) - 6 = 4x^2 - 2(5x + 1)$$

$$5x^2 - 10x - 6 = 4x^2 - 10x - 2$$

$$x^2 - 10x - 6 = -10x - 2$$

$$x^2 - 6 = -2$$

Check your work on page 367. ▶

Check your work on page 367. ▶

IIII➡ Trial Run _____

Solve each equation.

_____ **1.** $4(x^2 - 2x) = 3x^2 - 15$ ⠀⠀⠀ _____ **2.** $7x^2 - 6x = 3x(2x + 3)$

_____ **3.** $3y(y - 6) - 21 = 2y^2 - 6(3y + 2)$

_____ **4.** $3(x^2 - 2) + 16 = x(x + 12)$ ⠀⠀⠀ _____ **5.** $3y(y - 5) = 2y(y - 3)$

_____ **6.** $5a(a - 1) + 36 = 3(2a^2 - 15) - 5a$

Answers are on page 367.

_____

▶ Examples You Completed _____

Example 3. Solve $x(5 + x) = 0$ and check.

CHECK: $x = 0$

$$x(5 + x) = 0$$
$$0(5 + 0) \overset{?}{=} 0$$
$$0(5) \overset{?}{=} 0$$
$$0 = 0$$

CHECK: $x = -5$

$$x(5 + x) = 0$$
$$-5[5 + (-5)] \overset{?}{=} 0$$
$$-5(0) \overset{?}{=} 0$$
$$0 = 0$$

Example 5. Solve $3y^2 - 12 = 0$ and check.

CHECK: $y = -2$

$$3y^2 - 12 = 0$$
$$3(-2)^2 - 12 \overset{?}{=} 0$$
$$3(4) - 12 \overset{?}{=} 0$$
$$12 - 12 \overset{?}{=} 0$$
$$0 = 0$$

CHECK: $y = 2$

$$3y^2 - 12 = 0$$
$$3(2)^2 - 12 \overset{?}{=} 0$$
$$3(4) - 12 \overset{?}{=} 0$$
$$12 - 12 \overset{?}{=} 0$$
$$0 = 0$$

Example 6. Solve $5x^2 + 10x = 0$.

Solution

$$5x^2 + 10x = 0$$
$$5x(x + 2) = 0$$
$$5x = 0 \quad \text{or} \quad x + 2 = 0$$
$$x = 0 \quad \text{or} \quad x = -2$$

Example 7. Solve $0 = x^2 - 12x + 36$.

Solution

$$0 = x^2 - 12x + 36$$
$$0 = (x - 6)(x - 6)$$
$$0 = x - 6 \quad \text{or} \quad 0 = x - 6$$
$$6 = x \quad \text{or} \quad 6 = x$$

The solution is $x = 6$.

Example 9. Solve $x^2 - 10 = 3x$.

Solution

$$x^2 - 10 = 3x$$
$$x^2 - 10 - 3x = 3x - 3x$$
$$x^2 - 10 - 3x = 0$$
$$x^2 - 3x - 10 = 0$$
$$(x - 5)(x + 2) = 0$$
$$x - 5 = 0 \quad \text{or} \quad x + 2 = 0$$
$$x = 5 \quad \text{or} \quad x = -2$$

Example 10. Solve $y^2 = 16$.

Solution

$$y^2 = 16$$
$$y^2 - 16 = 0$$
$$(y + 4)(y - 4) = 0$$
$$y + 4 = 0 \quad \text{or} \quad y - 4 = 0$$
$$y = -4 \quad \text{or} \quad y = 4$$

Example 13. Solve $4x^2 + x = 3(x^2 - 4x)$.

Solution

$$4x^2 + x = 3(x^2 - 4x)$$
$$4x^2 + x = 3x^2 - 12x$$
$$x^2 + x = -12x$$
$$x^2 + 13x = 0$$
$$x(x + 13) = 0$$
$$x = 0 \quad \text{or} \quad x + 13 = 0$$
$$x = 0 \quad \text{or} \quad x = -13$$

Example 14. Solve

$$5x(x - 2) - 6 = 4x^2 - 2(5x + 1)$$

Solution

$$5x(x - 2) - 6 = 4x^2 - 2(5x + 1)$$
$$5x^2 - 10x - 6 = 4x^2 - 10x - 2$$
$$x^2 - 10x - 6 = -10x - 2$$
$$x^2 - 6 = -2$$
$$x^2 - 4 = 0$$
$$(x + 2)(x - 2) = 0$$
$$x + 2 = 0 \quad \text{or} \quad x - 2 = 0$$
$$x = -2 \quad \text{or} \quad x = 2$$

Answers to Trial Runs

page 360   **1.** $x = 11$ or $x = 2$ **2.** $y = 3$ or $y = -5$ **3.** $a = -12$ or $a = -5$ **4.** $x = 0$ or $x = 1$
5. $x = 1$ or $x = -2$ **6.** $x = 9$

page 362   **1.** $x = 5$ or $x = -1$ **2.** $y = 6$ or $y = -6$ **3.** $x = 0$ or $x = 7$ **4.** $a = -5$
5. $x = 3$ or $x = 2$ **6.** $m = 9$ or $m = -3$

page 364   **1.** $x = -7$ or $x = 5$ **2.** $y = 8$ or $y = -8$ **3.** $x = 0$ or $x = 3$ **4.** $y = 4$ or $y = 3$
5. $x = \dfrac{3}{2}$ or $x = -\dfrac{1}{5}$ **6.** $a = -5$ or $a = 3$

page 366   **1.** $x = 3$ or $x = 5$ **2.** $x = 0$ or $x = 15$ **3.** $y = 3$ or $y = -3$ **4.** $x = 5$ or $x = 1$
5. $y = 0$ or $y = 9$ **6.** $a = 9$ or $a = -9$

EXERCISE SET 7.5

Solve each equation.

_____ 1. $(x - 3)(x - 2) = 0$

_____ 2. $(x - 7)(x - 5) = 0$

_____ 3. $(x + 0.2)(x - 0.6) = 0$

_____ 4. $(x + 0.5)(x - 1.1) = 0$

_____ 5. $(a + 13)(a + 5) = 0$

_____ 6. $(a + 1)(a + 9) = 0$

_____ 7. $x\left(x - \dfrac{7}{8}\right) = 0$

_____ 8. $x\left(x + \dfrac{9}{10}\right) = 0$

_____ 9. $(7 - x)(9 + x) = 0$

_____ 10. $(2 - x)(8 + x) = 0$

_____ 11. $2(y - 3)(y + 1) = 0$

_____ 12. $-3(y + 6)(y - 1) = 0$

_____ 13. $x^2 - 6x + 8 = 0$

_____ 14. $x^2 - 9x + 8 = 0$

_____ 15. $x^2 - \dfrac{25}{16} = 0$

_____ 16. $x^2 - \dfrac{25}{49} = 0$

_____ 17. $y^2 + 9y + 18 = 0$

_____ 18. $y^2 + 8y + 15 = 0$

_____ 19. $m^2 + 1.5m = 0$

_____ 20. $m^2 - 0.9m = 0$

_____ 21. $z^2 + 3z - 54 = 0$

_____ 22. $z^2 + z - 42 = 0$

_____ 23. $2x^2 - 10x = 0$

_____ 24. $3x^2 - 9x = 0$

_____ 25. $z^2 - 30z + 225 = 0$

_____ 26. $z^2 - 26z + 169 = 0$

_____ 27. $5y^2 - 500 = 0$

_____ 28. $3y^2 - 192 = 0$

_____ 29. $18n - 9n^2 = 0$

_____ 30. $20n - 5n^2 = 0$

_____ 31. $33 - 8y - y^2 = 0$

_____ 32. $80 - 2y - y^2 = 0$

_____ 33. $5x^2 + 13x - 6 = 0$

_____ 34. $6x^2 + 19x - 7 = 0$

_____ 35. $x^2 + 0.3x = 0.18$

_____ 36. $x^2 - 0.7x = 0.18$

_____ 37. $y^2 = 81$

_____ 38. $y^2 = 121$

_____ 39. $2x^2 = 5x$

_____ 40. $5x^2 = 3x$

_____ 41. $72 = x^2 - x$

_____ 42. $54 = x^2 - 3x$

_____ 43. $10 - 3a = a^2$

_____ 44. $33 - 8a = a^2$

_____ 45. $x^2 - 16x = -64$

_____ 46. $x^2 - 20x = -100$

_____ 47. $10x^2 + 4 = 22x$

_____ 49. $144 = a^2$

_____ 51. $y^2 = -17y$

_____ 53. $100 = y^2 - 15y$

_____ 55. $x^2 - 1.4x = -0.45$

_____ 57. $x^2 + 6 = -5x$

_____ 59. $18y^2 = 8$

_____ 61. $2x^2 - 12x = -18$

_____ 63. $5y^2 = 5y$

_____ 65. $54 + 12x = 2x^2$

_____ 67. $10y^2 = 360$

_____ 69. $5(x^2 - 1) = 4(x^2 + 5)$

_____ 71. $7a(3a - 1) = 2(3a - 1)$

_____ 73. $2x(x - 4) - 28 = x^2 - 2(3x - 10)$

_____ 75. $4(x^2 - 2x) + 25 = x(3x + 2)$

_____ 77. $5x(x - 1) + 8 = 3(x^2 + 2) - 9x$

_____ 78. $7x(x - 1) - 60 = 2(2x^2 + 15) - 4x$

_____ 79. $5(y^2 + 2) = 2(2y + 5)$

_____ 48. $12x^2 + 15 = 27x$

_____ 50. $225 = a^2$

_____ 52. $y^2 = -15y$

_____ 54. $34 = y^2 + 15y$

_____ 56. $x^2 - 1.1x = -0.28$

_____ 58. $x^2 + 24 = -11x$

_____ 60. $20y^2 = 5$

_____ 62. $3x^2 - 24x = -48$

_____ 64. $11y^2 = 22y$

_____ 66. $45 - 40x = 5x^2$

_____ 68. $9y^2 = 9$

_____ 70. $3(x^2 - 2) = 2(x^2 + 15)$

_____ 72. $5a(4a - 3) = 2a - 3$

_____ 74. $3x(2x - 1) + 8 = 5x^2 - 3(3x - 5)$

_____ 76. $5(x^2 - 4x) + 9 = 2x(2x - 7)$

_____ 80. $4(y^2 - 3) = 3(3y - 4)$

☆ Stretching the Topics

Solve each equation.

_____ 1. $0.05m^2 + 0.13m = -0.06$

_____ 2. $(2x + 1)(x - 2) + x^2 = -2(5x + 2)$

_____ 3. $(x - 1)^2 + (x + 1)^2 + (x - 1)(x + 1) - 28 = 0$

Check your answers in the back of your book.

If you can solve the equations in **Checkup 7.5**, you are ready to go on to Section 7.6.

CHECKUP 7.5

Solve each equation.

_____ 1. $2x(x - 13) = 0$ _____ 2. $a^2 - 2a - 63 = 0$

_____ 3. $y^2 - \dfrac{16}{25} = 0$ _____ 4. $3x^2 + 30x + 63 = 0$

_____ 5. $7x^2 - 2.8x = 0$ _____ 6. $x^2 - 3x = 40$

_____ 7. $3x^2 + 18 = 15x$ _____ 8. $a^2 = 6a - 9$

_____ 9. $5(x^2 - 4) = 4(x^2 + 11)$ _____ 10. $4y(y + 1) + 7 = 3(y^2 - 1) - 3y$

Check your answers in the back of your book.

If You Missed Problems:	You Should Review Examples:
1	1–3
3–5	4–7
6–8	8–11
9, 10	12–14

We choose the *positive* answer to represent the distance from the foot of the tree.

$$x = \underline{\qquad} \text{ feet}$$

The height of the tree-part left standing is $x - 14$ feet.

$$x - 14 = \underline{\qquad} - 14$$
$$= \underline{\qquad} \text{ feet}$$

The height of the tree before the storm is $26 + (x - 14)$ feet.

$$26 + (x - 14)$$
$$= 26 + \underline{\qquad} \text{ feet}$$
$$= \underline{\qquad} \text{ feet}$$

Before the storm, the tree's height was _____ feet.

Check your work on page 375. ▶

Let's work together on Example 4.

Example 4. If the product of two consecutive integers is 90, *find the integers*.

Solution

We Think	**We Write**
We need an expression for each of the two consecutive integers.	Let x = first integer
	$x + 1$ = next consecutive integer
The *product* of the consecutive integers is found by *multiplying*.	$x(x + 1)$ = product of the consecutive integers
We know the product is 90.	$x(x + 1) = 90$

Now we have an equation to be solved by the usual methods.

$$x(x + 1) = 90$$
$$x^2 + x = 90 \qquad \text{Remove parentheses.}$$
$$x^2 + x - 90 = 0 \qquad \text{Write equation in standard form.}$$
$$(x + 10)(x - 9) = 0 \qquad \text{Factor the trinomial.}$$
$$x + 10 = 0 \quad \text{or} \quad x - 9 = 0 \qquad \text{Use the zero product rule.}$$
$$x = -10 \quad \text{or} \quad x = 9 \qquad \text{Solve for } x.$$

Our equation has two solutions, and each solution gives a pair of consecutive integers.

If first integer: $x = -10$

then next integer: $x + 1 = -10 + 1 = -9$

CHECK: $(-10)(-9) = 90$

If first integer: $x = 9$

then next integer: $x + 1 = 9 + 1 = 10$

CHECK: $(9)(10) = 90$

The two pairs of consecutive integers are -10 and -9 **or** 9 and 10.

Example 5. The sum of the squares of two consecutive *odd* integers is 2 more than 5 times their sum. *Find the two odd integers.*

Solution

We Think	***We Write***
We need an expression for each of the two consecutive *odd* integers. Consecutive odd integers are *two* units apart.	Let x = first odd integer $x + 2$ = next odd integer
We need an expression for the *sum* of the *squares* of these consecutive odd integers.	$x^2 + (x + 2)^2$ = the sum of the squares
We need an expression for the *sum* of the consecutive odd integers.	$x + (x + 2) = 2x + 2$ = sum of integers
We need an expression for 5 *times* the sum of the consecutive odd integers.	$5(2x + 2)$ = five times sum of integers
We know that the sum of the squares of these integers is 2 *more than* five times their sum.	$x^2 + (x + 2)^2 = 5(2x + 2) + 2$

Now we have an equation to be solved by the usual methods.

$$x^2 + (x + 2)^2 = 5(2x + 2) + 2$$

$$x^2 + x^2 + 4x + 4 = 10x + 10 + 2 \qquad \text{Remove parentheses.}$$

$$2x^2 + 4x + 4 = 10x + 12 \qquad \text{Combine like terms.}$$

$$2x^2 - 6x - 8 = 0 \qquad \text{Write equation in standard form.}$$

$$2(x^2 - 3x - 4) = 0 \qquad \text{Remove common factor.}$$

$$2(x - 4)(x + 1) = 0 \qquad \text{Factor the trinomial.}$$

$$x - 4 = 0 \quad \text{or} \quad x + 1 = 0 \qquad \text{Use zero product rule.}$$

$$x = 4 \quad \text{or} \quad x = -1 \qquad \text{Solve for } x.$$

We reject the solution $x = 4$ because 4 is even rather than odd. Our only solution is

$$x = -1 \qquad \text{First odd integer.}$$

$$x + 2 = -1 + 2 \qquad \text{Next consecutive odd integer.}$$

$$= 1$$

The consecutive odd integers are -1 and 1. Let's check this solution using the words of the problem.

$$\text{CHECK:} \quad \text{sum of squares} \stackrel{?}{=} 2 + 5 \text{ times sum}$$

$$(-1)^2 + 1^2 \stackrel{?}{=} 2 + 5(-1 + 1)$$

$$1 + 1 \stackrel{?}{=} 2 + 5(0)$$

$$2 \stackrel{?}{=} 2 + 0$$

$$2 = 2$$

▶ Examples You Completed ─────────────────────

Example 2. Find the length of one side of a square checkerboard with area 64 square inches.

Solution

$$\text{Let } s = \text{length of one side}$$

$$\text{Then } s \cdot s = \text{area}$$

$$s \cdot s = 64$$

$$s^2 = 64$$

$$s^2 - 64 = 0$$

$$(s + 8)(s - 8) = 0$$

$$s + 8 = 0 \quad \text{or} \quad s - 8 = 0$$

$$s = -8 \quad \text{or} \quad s = 8$$

The length of one side is 8 inches.

Example 3. Finish solving

$$2x^2 - 28x - 480 = 0$$

Solution

$$2x^2 - 28x - 480 = 0$$

$$2(x^2 - 14x - 240) = 0$$

$$2(x - 24)(x + 10) = 0$$

$$x - 24 = 0 \quad \text{or} \quad x + 10 = 0$$

$$x = 24 \quad \text{or} \quad x = -10$$

Distance from foot of tree:

$$x = 24 \text{ feet}$$

Height of standing part:

$$x - 14 = 10 \text{ feet}$$

Height of tree before storm:

$$26 + (x - 14)$$

$$= 26 + 10$$

$$= 36 \text{ feet}$$

EXERCISE SET 7.6

Use an equation to solve each problem.

_____ 1. One number is 5 more than another. The square of the smaller added to 3 times the larger is 43. Find the numbers.

_____ 2. One number is 3 more than another. The square of the smaller added to 5 times the larger is 65. Find the numbers.

_____ 3. The width of a rectangle is 1 foot less than the length. The area is 72 square feet. Find the length and the width.

_____ 4. The width of a rectangle is 2 feet less than the length. The area is 63 square feet. Find the length and the width.

_____ 5. Three times the square of an integer added to 6 times the integer is 105. Find the integer.

_____ 6. If 4 times an integer is subtracted from twice the square of the integer, the result is 48. Find the integer.

_____ 7. If the area of a square is 81 square centimeters, find the length of a side of the square.

_____ 8. The area of a square is 36 square centimeters. Find the length of a side of the square.

_____ 9. One leg of a right triangle is 7 centimeters less than the other leg and the hypotenuse is 17 centimeters. Find the lengths of the two legs.

_____ 10. One leg of a right triangle is 4 centimeters more than the other leg and the hypotenuse is 20 centimeters. Find the lengths of the two legs.

_____ 11. The length of a rectangle is twice its width. The area is 128 square feet. Find the dimensions of the rectangle.

_____ 12. The length of a rectangle is 3 times its width. The area is 75 square feet. Find the dimensions of the rectangle.

_____ 13. The Haskins plan to buy a rectangular lot that is 50 feet longer than it is wide. If the area of the lot is 30,000 square feet, find its dimensions.

_____ 14. The Perrys plan to fence a dog pen that is to be 8 feet longer than it is wide. They have enough fence to enclose an area of 240 square feet. Find the dimensions of the pen.

_____ 15. A sail is to be manufactured in the shape of a right triangle. If the hypotenuse must be 13 feet long and one leg is 7 feet more than the other leg, what must be the dimensions of the sail?

_____ 16. A flower bed is in the shape of a right triangle. One side is 2 feet shorter than the other side and the hypotenuse is 10 feet. Find how many feet of picket fence will be needed to border the flower bed.

_____ **17.** Find two consecutive integers whose product is 56.

_____ **18.** Find two consecutive integers whose product is 72.

_____ **19.** The sum of the squares of two consecutive integers is 61. Find the integers.

_____ **20.** The sum of the squares of two consecutive integers is 113. Find the integers.

_____ **21.** The product of two consecutive even integers is 4 more than five times the smaller. Find the integers.

_____ **22.** The product of two consecutive odd integers is 7 more than their sum. Find the integers.

☆ Stretching the Topics

Use an equation to solve each problem.

_____ **1.** A 2-inch square is to be cut out of each corner of a square piece of tin. The sides are then turned up to form an open box with a volume of 288 cubic inches. What must be the length of a side of the original square of tin?

_____ **2.** The total area of a circular cylinder is represented by $S = 2\pi r^2 + 2\pi rh$. Factor this expression and use the factored form to find the total area of a cylinder whose radius is 6 inches and whose height is 16 inches. Use $\pi \doteq 3.14$.

_____ **3.** Find two consecutive even integers such that the square of the larger subtracted from 12 times the smaller is equal to their product.

Check your answers in the back of your book.

If you can do the problems in **Checkup 7.6**, you are ready to do the **Review Exercises** for Chapter 7.

✔ CHECKUP 7.6

Use an equation to solve each problem.

_____ **1.** The length of a rectangle is 2 centimeters more than the width and the area is 35 centimeters. Find the dimensions.

_____ **2.** The length of the Abells' living room is the same as the width. If it takes 225 square feet of carpet to cover the floor, find the length of a side of the room.

_____ **3.** The support beams of a carport form a right triangle. The hypotenuse is 25 feet and one leg is 5 feet less than the other. Find the length of the piece of lumber that would be needed to cut all three beams.

_____ **4.** One number is four more than another. The square of the smaller added to twice the larger is 32. Find the numbers.

_____ **5.** The sum of the squares of two consecutive integers is 13. Find the integers.

Check your answers in the back of your book.

If You Missed Problems:	You Should Review Example:
1	1
2	2
3	3
4, 5	4

Summary

In this chapter we learned to rewrite a polynomial as a product of factors. Different types of **factoring** were used in different situations, and we agreed to look for certain patterns in our search for factors.

Type of Factoring	How to Recognize	Examples
Common factor	Look for a factor common to all terms. $$AB + AC = A(B + C)$$	$$3x^2 - 6x^2 = 3x^2(x - 2)$$ $$x(a + b) - (a + b) = (a + b)(x - 1)$$
By grouping	Look for four terms that can be rewritten as two pairs of terms, each pair containing a common factor.	$$x^3 + x^2 + 2x + 2$$ $$= x^2(x + 1) + 2(x + 1)$$ $$= (x + 1)(x^2 + 2)$$
Difference of two squares	Look for two square terms with a minus sign between them. $$A^2 - B^2 = (A + B)(A - B)$$	$$4x^2 - 9 = (2x + 3)(2x - 3)$$ $$\frac{1}{100}x^2 - \frac{1}{9}y^2 = \left(\frac{1}{10}x + \frac{1}{3}y\right)\left(\frac{1}{10}x - \frac{1}{3}y\right)$$
Perfect square trinomial	Look for trinomial with first term A^2, third term B^2, and middle term $2AB$. $$A^2 + 2AB + B^2 = (A + B)^2$$ $$A^2 - 2AB + B^2 = (A - B)^2$$	$$x^2 - 16x + 64 = (x - 8)^2$$ $$\frac{1}{121}y^2 + \frac{2}{11}y + 1 = \left(\frac{1}{11}y + 1\right)^2$$
General trinomial	Look for two binomials, by trial and error, that multiply to give the trinomial.	$$x^2 - 4x - 21 = (x - 7)(x + 3)$$ $$6y^2 + 11xy - 10x^2 = (2y + 5x)(3y - 2x)$$

Remember that the order in which factors are written is not important, but the factors themselves are *unique*. A polynomial can be factored completely in only one way, and you can always check your factors by *multiplying*. If a polynomial cannot be written as a product of factors, we say that it is **prime**.

In this chapter we also learned to solve **quadratic** or second-degree equations using the process of factoring and the **zero product rule**. Solving quadratic equations required the following procedure.

1. Work with parentheses if necessary.
2. Rewrite the equation in standard form, with zero on one side and all other terms on the other side, combining like terms if possible.
3. Factor the polynomial.
4. Set each variable factor equal to zero.
5. Solve for the variable and check each solution in the original equation.

❑ Speaking the Language of Algebra _____

Complete each statement with the appropriate word or phrase.

1. The process of rewriting an expression as a product is called _____ .

2. In factoring, the first thing to look for is a _____ _____ .

3. A polynomial of the form $A^2 - B^2$ is called a _____ _____ _____ _____ .

4. A polynomial of the form $A^2 + 2AB + B^2$ or $A^2 - 2AB + B^2$ is called a _____ _____ _____ .

5. If a polynomial cannot be written as a product of factors, we say that it is _____ .

6. An equation in which the variable appears with an exponent of 2 is called a _____ equation or a _____ – _____ equation.

7. The rule which states that if a product of factors is zero, then at least one of the factors must be zero is called the _____ _____ rule.

Check your answers in the back of your book.

△ Writing About Mathematics _____

Write your response to each question in complete sentences.

1. Explain your aim in factoring an expression and tell how you can check to see if you have factored an expression correctly.

2. Explain, in words, how to recognize a polynomial that is:
   (a) a difference of two squares.
   (b) a perfect square trinomial.

3. Explain the zero product rule in your own words. Then describe the steps you would use to solve the equation: $x^2 - 3 = 2x$.

REVIEW EXERCISES for Chapter 7

Rewrite each expression as a product of factors.

_____ **1.** $10x - 15$

_____ **2.** $2y^3 - \dfrac{3}{5}y^2$

_____ **3.** $10x^2y + 4xy^2$

_____ **4.** $-3a^2b + 2a^3b^2 - 4a^4b^3$

_____ **5.** $4x + 4y + xz + yz$

_____ **6.** $25 - y^2$

_____ **7.** $3m^2 - 48n^2$

_____ **8.** $18x^4 - 8x^2y^2$

_____ **9.** $a^2 - 16ab + 64b^2$

_____ **10.** $x^2 - 8x + 15$

_____ **11.** $x^2y^2 - 3xy - 10$

_____ **12.** $4a^2 - 20ab + 2b^2$

_____ **13.** $6x^2 - 11x - 7$

_____ **14.** $-2x^2 + 4xy + 6y^2$

_____ **15.** $10m^2 + 15m - 220$

_____ **16.** $8x^2 - 11x + 3$

_____ **17.** $3y^4 + 63y^2 - 300$

_____ **18.** $x^3 - 4x^2 - x + 4$

_____ **19.** $5x^5 - 5x$

Solve each equation.

_____ **20.** $(x - 7)(x + 8) = 0$

_____ **21.** $x^2 - 169 = 0$

_____ **22.** $x^2 + 14x + 49 = 0$

_____ **23.** $m^2 + \dfrac{2}{3}m = 0$

_____ **24.** $162 - 2a^2 = 0$

_____ **25.** $z^2 + 17z + 52 = 0$

_____ **26.** $y^2 + 11y = -24$

_____ **27.** $y^2 = -2.3y$

_____ **28.** $x^2 + 63 = 24x$

_____ **29.** $5x^2 = 405$

_____ **30.** $5x(x - 2) - 18 = 4x^2 - 3(x - 4)$

_____ **31.** The length of a rectangular bathroom is 4 feet more than the width. If 60 square feet of linoleum will cover the floor, what are the dimensions of the bathroom?

_____ **32.** If the area of a square is 225 square centimeters, find the length of a side of the square.

_____ **33.** One integer is 4 less than another and the product of the two is 117. Find the integers.

_____ **34.** Hannah wants to border her antique shawl with lace. The shawl is in the shape of a right triangle with one side 1 foot longer than the other side, and the hypotenuse is 5 feet. How many feet of lace should she buy?

———— **35.** The product of two consecutive odd integers is 7 more than 4 times the larger. Find the integers.

Check your answers in the back of your book.

If You Missed Exercises:	You Should Review Examples:	
1–4	SECTION 7.1	1–9
5		10–17
6–8	SECTION 7.2	1–8
9	SECTION 7.3	13–16
10, 11		1–8
12–16		9–12
17		19–22
18, 19	SECTION 7.4	1–3
20	SECTION 7.5	1–3
21–25		4–7
26–29		8–11
30		12–14
31–35	SECTION 7.6	1–5

If you have completed the **Review Exercises** and corrected your errors, you are ready to take the **Practice Test** for Chapter 7.

PRACTICE TEST for Chapter 7

Rewrite each expression as a product of factors.

		SECTION	EXAMPLES

Rewrite each expression as a product of factors.   SECTION   EXAMPLES

_____ **1.** $-10x + 25$ 7.1   1

_____ **2.** $33x^2 - 77x$ 7.1   4

_____ **3.** $6y^5 + 3y^4 - 15y^2$ 7.1   6

_____ **4.** $-8a^3b + 14a^2b^2 - 8ab^3$ 7.1   8

_____ **5.** $ay - 3y + 5a - 15$ 7.1   14

_____ **6.** $16x^2 - 25a^2$ 7.2   2

_____ **7.** $0.36 - a^2b^2$ 7.2   1

_____ **8.** $7x^2 - 7$ 7.2   3

_____ **9.** $a^2 + 9ab + 20b^2$ 7.3   1

_____ **10.** $x^2 - 13x + 22$ 7.3   2

_____ **11.** $20 + 8x - x^2$ 7.3   4

_____ **12.** $20x^2 - 28x - 3$ 7.3   10

_____ **13.** $-3a^2b + 24ab^2 - 48b^3$ 7.3   11, 12

_____ **14.** $x^4 - 10x^2 + 9$ 7.3   19

Solve each equation.

_____ **15.** $8x(11 + x) = 0$ 7.5   3

_____ **16.** $x^2 + 1.4x = 0$ 7.5   6

_____ **17.** $3n^2 - 75 = 0$ 7.5   5

_____ **18.** $a^2 + 17a + 70 = 0$ 7.5   7

_____ **19.** $y^2 = \dfrac{25}{121}$ 7.5   10

_____ **20.** $x^2 = 8x - 16$ 7.5   7, 9

_____ **21.** $5y^2 - 10y = 40$ 7.5   8

_____ **22.** $5x(x - 2) - 12 = 4x^2 - 2(x - 4)$ 7.5   15

_____ **23.** The width of a rectangle is 6 meters less than the length. The area of   7.6   1
the rectangle is 40 square meters. Find the length and the width.

		SECTION	EXAMPLE

_____ **24.** The hypotenuse of a right triangle is 3 yards longer than twice one of the legs. If the remaining leg is 12 yards long, find the length of the hypotenuse.

SECTION 7.6   EXAMPLE 3

_____ **25.** One integer is 7 more than another integer and their product is 30. Find the integers.

SECTION 7.6   EXAMPLE 4

Check your answers in the back of your book.

SHARPENING YOUR SKILLS after Chapters 1–7

Perform the indicated operations and simplify.

SECTION

_____ 1. $\left(-\dfrac{4}{5}\right)^3$

2.4

_____ 2. $(21.3)(0.72)$

1.3

_____ 3. $\dfrac{5}{9} - \left(\dfrac{2}{5} - \dfrac{11}{15}\right)$

2.4

_____ 4. $-|3(-7)| + |-15|$

2.1

_____ 5. $(-6)^2(0.4)\left(-\dfrac{1}{2}k\right)$

3.3

_____ 6. $3[3x - (x - 4)] - [2(x + 2) - 5x]$

3.3

_____ 7. $(-2x^2)^3 \cdot (-xy^2)^2$

5.1

_____ 8. $(y^2 - 2yz + 2z^2) - 2(-y^2 + 3yz - z^2)$

6.1

_____ 9. $(5x^2y)(-3xy)(3x^4y^5)$

6.2

_____ 10. $-a^2b(5 - 2ab + 7a^2b^2)$

6.2

_____ 11. $(2a - 3b)(4a + 5b)$

6.3

_____ 12. $-y^2(y - 3)(y + 5)$

6.3

_____ 13. $\dfrac{15x^3y - 12x^2y^2 + 4xy^3}{-3xy}$

6.4

_____ 14. $(4x^2 - 17x - 14) \div (x - 5)$

6.4

Solve each equation.

_____ 15. $\dfrac{2}{3}x - 15 = 9$

4.2

_____ 16. $11 = 5(x + 3) + 6$

4.2

_____ 17. $-0.5x + 3 = 0.1x - 21$

4.2

_____ 18. Evaluate $-4a^2b^3$ when $a = 2$ and $b = -3$.

5.1

_____ 19. If the width of a rectangle is 10 less than 3 times the length, write an expression for the area. Find the area if the length is 15 meters.

3.1

_____ 20. Solve $A = p + prt$ for t.

4.2

Check your answers in the back of your book.

CHAPTER

Working with Rational Expressions

At the beginning of the semester the number of males in Harry's math class was 3 more than twice the number of females. After 5 males and 5 females had dropped the class, the ratio of females to males was 1 to 4. How many males and females were in Harry's math class at the beginning of the semester?

Before you complete this chapter, you will be able to solve this problem using a fractional equation. In Chapters 1 and 2 we learned to work with numerical fractions. Now that we have discussed the algebra of polynomials, we turn our attention to the algebra of *fractions* containing polynomials. In this chapter we learn to

1. Simplify rational algebraic expressions.
2. Multiply and divide rational algebraic expressions.
3. Add and subtract rational algebraic expressions.
4. Solve equations containing rational algebraic expressions.
5. Switch from word statements to equations containing rational algebraic expressions.

8.1 Simplifying Rational Algebraic Expressions

Recognizing Rational Algebraic Expressions

An algebraic expression that can be written as a fraction with a polynomial numerator and a polynomial denominator is called a **rational algebraic expression**. For instance,

$$\frac{5}{2x} \qquad \frac{x^2}{7x} \qquad \frac{y+3}{y+5} \qquad \frac{a^2+5}{2a} \qquad \frac{x^2+7x+6}{x^2-1}$$

are all examples of rational algebraic expressions.

We can *evaluate* rational expressions by substituting some numerical value for the variable.

Example 1. Evaluate $\dfrac{y+3}{y-5}$ when $y = 0$.

Solution

$$\frac{y+3}{y-5} = \frac{0+3}{0-5} \qquad \begin{array}{l}\text{Substitute}\\ \text{0 for } y.\end{array}$$

$$= \frac{3}{-5} \qquad \text{Simplify.}$$

$$= -\frac{3}{5}$$

You complete Example 2.

Example 2. Evaluate $\dfrac{x+3}{5x}$ when $x = 2$.

Solution

$$\frac{x+3}{5x} = \frac{2+3}{5 \cdot 2}$$

$$= \frac{}{10}$$

$$= \underline{\hspace{2em}}$$

Did you remember to reduce your fraction? Check your work on page 395. ▶

Now you try to evaluate the fraction in Example 3.

Example 3. Evaluate $\dfrac{y+3}{y-5}$ when $y = -3$.

Solution

$$\frac{y+3}{y-5} = \frac{-3+3}{-3-5}$$

$$= \frac{}{-8}$$

$$= \underline{\hspace{2em}}$$

Check your work on page 395. ▶

Example 4. Evaluate $\dfrac{y+3}{y-5}$ when $y = 5$.

Solution

$$\frac{y+3}{y-5} = \frac{5+3}{5-5}$$

$$= \frac{8}{0} \qquad \text{is undefined}$$

Notice that our answer in Example 4 was undefined, because we learned in Chapters 1 and 2 that **division by zero is always undefined**. In working with rational algebraic expressions, we must always avoid substituting for the variable any number that will make the denominator equal zero. The process of finding the value of the variable that will make the denominator zero is called **finding the restrictions on the variable**.

Let's find the restrictions on the variable for the fraction $\dfrac{5x}{x-7}$. Since the denominator is zero when

$$x - 7 = 0$$

$$x = 7$$

the restriction is $x \neq 7$.

Example 5. Find the restrictions on the variable for $\dfrac{5x}{3x - 2}$.

Solution. The denominator is zero when

$$3x - 2 = 0$$
$$3x = 2$$
$$x = \frac{2}{3}$$

so the restriction is

$$x \neq \frac{2}{3}$$

You complete Example 6.

Example 6. Find the restrictions on the variable for $\dfrac{x + 3}{4x}$.

 Solution. The denominator is zero when

$$4x = 0$$
$$x = \underline{\hspace{1cm}}$$

so the restriction is

$$x \neq \underline{\hspace{1cm}}$$

Check your work on page 395. ▶

Notice that some rational expressions require no restrictions. For instance, in the fraction $\dfrac{x + 3}{4}$, the denominator will never be zero. There are *no* restrictions on x.

||▶ **Trial Run**

Evaluate each rational expression and reduce if possible.

 _____ **1.** $\dfrac{x + 5}{6x}$ when $x = 3$

_____ **2.** $\dfrac{y - 5}{y + 2}$ when $y = 0$

_____ **3.** $\dfrac{a - 4}{a^2 - 4}$ when $a = 4$

Find the restrictions on the variable.

_____ **4.** $\dfrac{x - 2}{x + 3}$ _____ **5.** $\dfrac{y^2 + 1}{3y}$ _____ **6.** $\dfrac{3a}{2a - 9}$

Answers are on page 396.

Reducing Rational Algebraic Expressions Containing Monomials

In Chapter 1 you learned to reduce a rational number (or fraction) to lowest terms by dividing its numerator and denominator by common factors. For example,

$$\frac{15}{21} = \frac{\overset{1}{\cancel{3}} \cdot 5}{\underset{1}{\cancel{3}} \cdot 7} = \frac{5}{7}$$

To reduce a rational algebraic expression with a monomial numerator and a monomial denominator, we may again look for common factors *or* we may use the Fourth Law of Exponents.

Original Fraction	Reducing by Common Factors	Reducing by Fourth Law of Exponents
$\dfrac{x^3}{x}$	$\dfrac{x^3}{x} = \dfrac{\overset{1}{\cancel{x}} \cdot x \cdot x}{\underset{1}{\cancel{x}}} = x^2$	$\dfrac{x^3}{x} = x^{3-1} = x^2$
$\dfrac{y^2}{y^5}$	$\dfrac{y^2}{y^5} = \dfrac{\overset{1}{\cancel{y}} \cdot \overset{1}{\cancel{y}}}{\underset{1}{\cancel{y}} \cdot \underset{1}{\cancel{y}} \cdot y \cdot y \cdot y} = \dfrac{1}{y^3}$	$\dfrac{y^2}{y^5} = \dfrac{1}{y^{5-2}} = \dfrac{1}{y^3}$
$\dfrac{6a^2x}{3ax}$	$\dfrac{6a^2x}{3ax} = \dfrac{\overset{1}{\cancel{3}} \cdot 2 \cdot \overset{1}{\cancel{a}} \cdot a \cdot \overset{1}{\cancel{x}}}{\underset{1}{\cancel{3}} \cdot \underset{1}{\cancel{a}} \cdot \underset{1}{\cancel{x}}} = 2a$	$\dfrac{6a^2x}{3ax} = \dfrac{6}{3} \cdot \dfrac{a^2}{a} \cdot \dfrac{x}{x}$ $= 2a^{2-1}x^{1-1} = 2a$

Either approach to reducing monomial fractions yields the correct result. In this chapter we shall most often use the method of reducing by dividing the numerator and denominator by common factors.

Reducing Fractions

$$\frac{AC}{BC} = \frac{A}{B} \qquad (B \neq 0, C \neq 0)$$

Example 7. Reduce $\dfrac{-3x^3}{12x^7}$.

Solution

$$\frac{-3x^3}{12x^7} = \frac{\overset{-1}{\cancel{-3}}\overset{1}{\cancel{x^3}}}{\underset{4}{\cancel{12}}\underset{x^4}{\cancel{x^7}}} \qquad \begin{array}{l}\text{Divide numerator and denominator by 3.} \\ \text{Divide numerator and denominator by } x^3.\end{array}$$

$$= \frac{-1}{4x^4} \qquad \text{Simplify numerator and denominator.}$$

$$= -\frac{1}{4x^4}$$

Example 8. Reduce $\dfrac{22a^2bc^3}{55abc^5}$.

Solution

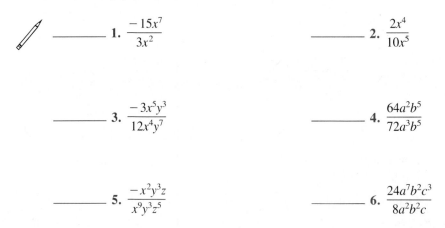

$$\dfrac{22a^2bc^3}{55abc^5} = \dfrac{\overset{2}{\cancel{22}}\overset{a}{\cancel{a^2}}\overset{1}{\cancel{b}}\overset{1}{\cancel{c^3}}}{\underset{5}{\cancel{55}}\underset{1}{\cancel{a}}\underset{1}{\cancel{b}}\underset{c^2}{\cancel{c^5}}}$$

Divide numerator and denominator by common factors 11, a, b, and c^3.

$$= \dfrac{2a}{5c^2}$$

Simplify numerator and denominator.

⇒ Trial Run

Reduce the fractions.

———— **1.** $\dfrac{-15x^7}{3x^2}$ ⠀⠀⠀⠀⠀⠀ ———— **2.** $\dfrac{2x^4}{10x^5}$

———— **3.** $\dfrac{-3x^5y^3}{12x^4y^7}$ ⠀⠀⠀⠀⠀ ———— **4.** $\dfrac{64a^2b^5}{72a^3b^5}$

———— **5.** $\dfrac{-x^2y^3z}{x^9y^3z^5}$ ⠀⠀⠀⠀⠀ ———— **6.** $\dfrac{24a^7b^2c^3}{8a^2b^2c}$

Answers are on page 396.

Reducing Rational Algebraic Expressions Containing Polynomials

Factoring will continue to provide the key to reducing fractions as we consider fractions that contain polynomials in the numerator and denominator. Remember, we reduce fractions by dividing the numerator and denominator by *common factors*.

> **Reducing Rational Expressions**
>
> **1.** Factor the numerator and denominator completely.
> **2.** Divide the numerator and denominator by all common factors.
> **3.** Write the final fraction with the remaining factors in the numerator over the remaining factors in the denominator.
>
> The final fraction is equivalent to the original fraction.

Let's try some examples, remembering to *factor the numerator and denominator completely before looking for common factors.*

Example 9. Reduce $\dfrac{2x + 2y}{2a - 4b}$.

Solution

$$\frac{2x + 2y}{2a - 4b} = \frac{\overset{1}{\cancel{2}}(x + y)}{\underset{1}{\cancel{2}}(a - 2b)}$$

$$= \frac{x + y}{a - 2b}$$

You try Example 10.

Example 11. Reduce $\dfrac{z^2 - 9}{2z + 6}$.

Solution

$$\frac{z^2 - 9}{2z + 6} = \frac{\overset{1}{\cancel{(z + 3)}}(z - 3)}{\underset{1}{2\cancel{(z + 3)}}}$$

$$= \frac{z - 3}{2}$$

Now complete Example 12.

Example 13. Reduce $\dfrac{3x^2 - 3xy - 6y^2}{x^3 - xy^2}$.

Solution

$$\frac{3x^2 - 3xy - 6y^2}{x^3 - xy^2}$$

$$= \frac{3(x^2 - xy - 2y^2)}{x(x^2 - y^2)}$$

$$= \frac{3(x - 2y)\overset{1}{\cancel{(x + y)}}}{x(x - y)\underset{1}{\cancel{(x + y)}}}$$

$$= \frac{3(x - 2y)}{x(x - y)}$$

You complete Example 14.

Example 10. Reduce $\dfrac{-2x - 2y}{3x + 3y}$.

Solution

$$\frac{-2x - 2y}{3x + 3y} = \frac{-2(x + y)}{3(\underline{\quad} + \underline{\quad})}$$

$$= \underline{\quad\quad}$$

Check your work on page 395. ▶

Example 12. Reduce $\dfrac{3x}{9x^2 - 15xy}$.

Solution

Check your work on page 395. ▶

Example 14. Reduce $\dfrac{x^2 - 5x - 14}{x^2 - 14x + 49}$.

Solution

Check your work on page 395. ▶

Notice that we generally write our final fraction with the numerator and denominator in factored form to be sure that we have not overlooked a common factor.

Let's take a close look at the steps we could use to reduce the fraction $\dfrac{5-x}{x-5}$.

$$\dfrac{5-x}{x-5} = \dfrac{-x+5}{x-5}$$ Rewrite numerator and denominator with terms in the same order.

$$= \dfrac{-1(x-5)}{x-5}$$ Factor -1 from the numerator.

$$= \dfrac{-1\cancel{(x-5)}^{1}}{\cancel{x-5}_{1}}$$ Divide numerator and denominator by the common factor $(x-5)$.

$$= \dfrac{-1}{1}$$

$$= -1$$ Simplify.

Similar reasoning can be used to reduce any fraction of the form $\dfrac{A-B}{B-A}$.

$$\dfrac{A-B}{B-A}$$ Original fraction.

$$= \dfrac{A-B}{-A+B}$$ Rewrite terms in same order.

$$= \dfrac{A-B}{-1(A-B)}$$ Factor -1 from denominator.

$$= \dfrac{\cancel{A-B}^{1}}{-1\cancel{(A-B)}_{1}}$$ Divide numerator and denominator by common factor $(A-B)$.

$$= -1$$ Simplify.

From now on, whenever you see opposite factors such as $A-B$ and $B-A$ in the numerator and denominator of a fraction, you will know that their quotient is -1, without going through the steps shown here.

$$\boxed{\dfrac{A-B}{B-A} = -1 \quad \text{provided that } B \neq A}$$

Try to complete Example 15.

Example 15. Reduce $\dfrac{9 - 3x}{x - 3}$.

Solution

$$\dfrac{9 - 3x}{x - 3} = \dfrac{3(3 - x)}{x - 3}$$

Check your work on page 396. ▶

Example 16. Reduce $\dfrac{36 - x^2}{x^2 - 3x - 18}$.

Solution

$$\dfrac{36 - x^2}{x^2 - 3x - 18} = \dfrac{(6 + x)(6 - x)}{(x - 6)(x + 3)}$$

$$= \dfrac{(6 + x)\overset{-1}{\cancel{(6 - x)}}}{\underset{1}{\cancel{(x - 6)}}(x + 3)}$$

$$= \dfrac{-1(6 + x)}{x + 3}$$

or

$$\dfrac{-6 - x}{x + 3}$$

⟱ Trial Run

Reduce each fraction.

_____ 1. $\dfrac{4x + 4y}{8}$

_____ 2. $\dfrac{3x + 3y}{3x - 3y}$

_____ 3. $\dfrac{4x + 4y}{8x + 8y}$

_____ 4. $\dfrac{x^2 + 3x}{2x + 6}$

_____ 5. $\dfrac{x^2 - 5x + 6}{x^2 - 4}$

_____ 6. $\dfrac{a^2 - 16}{12 - 3a}$

Answers are on page 396.

   By now you should agree that factoring is an important tool in our work with fractions. Once the numerator and denominator are in factored form, we may reduce a fraction with ease. It is important that you recognize the difference between factors and terms.

   Remember that *terms* are quantities that are being added together, but *factors* are quantities that are being multiplied together. *We reduce fractions only by dividing numerators and denominators by common factors.*

Cannot Be Reduced	Can Be Reduced
$\dfrac{x+5}{5}$	$\dfrac{5x}{5}=x$
$\dfrac{4+xy}{x}$	$\dfrac{4xy}{x}=4y$
$\dfrac{x+2}{x+3}$	$\dfrac{2(x+1)}{3(x+1)}=\dfrac{2}{3}$

▶ Examples You Completed

Example 2. Evaluate $\dfrac{x+3}{5x}$ when $x=2$.

Solution

$$\frac{x+3}{5x}=\frac{2+3}{5(2)}$$
$$=\frac{5}{10}$$
$$=\frac{1}{2}$$

Example 3. Evaluate $\dfrac{y+3}{y-5}$ when $y=-3$.

Solution

$$\frac{y+3}{y-5}=\frac{-3+3}{-3-5}$$
$$=\frac{0}{-8}$$
$$=0$$

Example 6. Find the restrictions on the variable for $\dfrac{x+3}{4x}$.

Solution. The denominator is zero when
$$4x=0$$
$$x=0$$
so the restriction is
$$x\neq 0$$

Example 10. Reduce $\dfrac{-2x-2y}{3x+3y}$.

Solution

$$\frac{-2x-2y}{3x+3y}=\frac{\overset{1}{-2\cancel{(x+y)}}}{\underset{1}{3\cancel{(x+y)}}}$$
$$=-\frac{2}{3}$$

Example 12. Reduce $\dfrac{3x}{9x^2-15xy}$.

Solution

$$\frac{3x}{9x^2-15xy}=\frac{\overset{1}{\cancel{3x}}}{\underset{1}{\cancel{3x}(3x-5y)}}$$
$$=\frac{1}{3x-5y}$$

Example 14. Reduce $\dfrac{x^2-5x-14}{x^2-14x+49}$.

Solution

$$\frac{x^2-5x-14}{x^2-14x+49}$$
$$=\frac{\overset{1}{\cancel{(x-7)}}(x+2)}{\underset{1}{\cancel{(x-7)}}(x-7)}$$
$$=\frac{x+2}{x-7}$$

Example 15. Reduce $\dfrac{9 - 3x}{x - 3}$.

Solution

$$\frac{9 - 3x}{x - 3} = \frac{\overset{-1}{\cancel{3(3 - x)}}}{\underset{1}{\cancel{x - 3}}}$$

$$= 3(-1)$$

$$= -3$$

Answers to Trial Runs

page 389   **1.** $\dfrac{4}{9}$ **2.** $-\dfrac{5}{2}$ **3.** 0 **4.** $x \neq -3$ **5.** $y \neq 0$ **6.** $a \neq \dfrac{9}{2}$

page 391   **1.** $-5x^5$ **2.** $\dfrac{1}{5x}$ **3.** $-\dfrac{x}{4y^4}$ **4.** $\dfrac{8}{9a}$ **5.** $-\dfrac{1}{x^7z^4}$ **6.** $3a^5c^2$

page 394   **1.** $\dfrac{x + y}{2}$ **2.** $\dfrac{x + y}{x - y}$ **3.** $\dfrac{1}{2}$ **4.** $\dfrac{x}{2}$ **5.** $\dfrac{x - 3}{x + 2}$ **6.** $\dfrac{-(a + 4)}{3}$

EXERCISE SET 8.1

Find the restrictions on the variable.

_____ 1. $\dfrac{6}{x^2}$

_____ 2. $\dfrac{-4}{x^2}$

_____ 3. $\dfrac{x^2}{x+1}$

_____ 4. $\dfrac{x^2}{x+2}$

_____ 5. $\dfrac{2(x+2)}{3x}$

_____ 6. $\dfrac{5(x-1)}{4x}$

_____ 7. $\dfrac{y-1}{8-y}$

_____ 8. $\dfrac{y-3}{5-y}$

_____ 9. $\dfrac{4x}{2x-1}$

_____ 10. $\dfrac{3x}{3x-2}$

Reduce the fractions.

_____ 11. $\dfrac{8x^5}{4x^2}$

_____ 12. $\dfrac{9x^7}{3x^3}$

_____ 13. $\dfrac{-5a^6}{20a^5}$

_____ 14. $\dfrac{-4a^7}{12a^6}$

_____ 15. $\dfrac{6x^5y^2}{2x^3y^4}$

_____ 16. $\dfrac{21x^3y}{7x^2y^6}$

_____ 17. $\dfrac{-15ab^4}{20a^3b^2}$

_____ 18. $\dfrac{-18a^2b}{27a^8b^2}$

_____ 19. $\dfrac{25x^2yz}{20x^4yz^2}$

_____ 20. $\dfrac{28x^3yz^2}{35x^5y^4z^2}$

_____ 21. $\dfrac{2x+2y}{6}$

_____ 22. $\dfrac{4x-4y}{12}$

_____ 23. $\dfrac{3a+15b}{3}$

_____ 24. $\dfrac{7a+21b}{7}$

_____ 25. $\dfrac{18xy-14x}{2x}$

_____ 26. $\dfrac{12xy-6x}{3x}$

_____ 27. $\dfrac{8x}{64x+4x^2}$

_____ 28. $\dfrac{15x}{27x+3x^2}$

_____ 29. $\dfrac{36a^2+36b^2}{30}$

_____ 30. $\dfrac{21a^2+21b^2}{49}$

_____ 31. $\dfrac{-25x^2 - 25y^2}{-5}$

_____ 32. $\dfrac{-24x^2 - 24y^2}{-8}$

_____ 33. $\dfrac{49x^3}{7x^2 + 7x}$

_____ 34. $\dfrac{81x^4}{9x^2 + 9x}$

_____ 35. $\dfrac{-12a^2b - 15abc}{18ab}$

_____ 36. $\dfrac{-10a^2b - 18abc}{22ab}$

_____ 37. $\dfrac{8x^2 + 12x + 4}{40}$

_____ 38. $\dfrac{15x^2 + 12x + 9}{39}$

_____ 39. $\dfrac{4a + 4b}{4x + 4y}$

_____ 40. $\dfrac{3a + 3b}{3x + 3y}$

_____ 41. $\dfrac{3x - 3y}{6x + 6y}$

_____ 42. $\dfrac{4x - 4y}{12x + 12y}$

_____ 43. $\dfrac{9x + 9y}{18x + 18y}$

_____ 44. $\dfrac{5x + 5y}{15x + 15y}$

_____ 45. $\dfrac{-4a - 4b}{12a + 12b}$

_____ 46. $\dfrac{-3a - 3b}{15a + 15b}$

_____ 47. $\dfrac{x^2 - y^2}{3x + 3y}$

_____ 48. $\dfrac{x^2 - y^2}{8x - 8y}$

_____ 49. $\dfrac{x^2 - 49}{4x + 28}$

_____ 50. $\dfrac{x^2 - 25}{4x + 20}$

_____ 51. $\dfrac{15x - 15y}{5x^2 - 5y^2}$

_____ 52. $\dfrac{12x - 12y}{3x^2 - 3y^2}$

_____ 53. $\dfrac{-7a - 7b}{3a^2 - 3b^2}$

_____ 54. $\dfrac{-5a - 5b}{2a^2 - 2b^2}$

_____ 55. $\dfrac{x^2 - 4y^2}{3x + 6y}$

_____ 56. $\dfrac{x^2 - 9y^2}{2x + 6y}$

_____ 57. $\dfrac{x^2 + 9x + 20}{x^2 - 16}$

_____ 58. $\dfrac{x^2 + 7x + 12}{x^2 - 9}$

_____ 59. $\dfrac{x^2 + 6x + 8}{x^2 + 7x + 10}$

_____ 60. $\dfrac{x^2 + 8x + 15}{x^2 + 5x + 6}$

_____ 61. $\dfrac{25x^2 - 225}{5x^2 - 50x + 105}$

_____ 62. $\dfrac{16x^2 - 144}{4x^2 - 20x + 24}$

_____ 63. $\dfrac{3x^2 + 14xy - 5y^2}{3x^2 + 2xy - y^2}$

_____ 64. $\dfrac{6x^2 + xy - y^2}{2x^2 - xy - y^2}$

_____ 65. $\dfrac{4a^2 - 8ab - 12b^2}{2a^2 - 12ab + 18b^2}$

_____ 66. $\dfrac{5a^2 - 4ab - b^2}{3a^2 - 6ab + 3b^2}$

_____ 67. $\dfrac{x^2 - 25}{x^2 + 25}$

_____ 68. $\dfrac{4x^2 - 9}{4x^2 + 9}$

_____ 69. $\dfrac{2b - 1}{4b^2 - 4b + 1}$

_____ 70. $\dfrac{3b + 1}{9b^2 + 6b + 1}$

_____ 71. $\dfrac{x^4 - 16y^4}{x^2 + 2xy - 8y^2}$

_____ 72. $\dfrac{x^4 - 81y^4}{x^2 + 6xy - 27y^2}$

_____ 73. $\dfrac{a - b}{9b - 9a}$

_____ 74. $\dfrac{2a - b}{5b - 10a}$

_____ 75. $\dfrac{x^2 - 25}{5 - x}$

_____ 76. $\dfrac{x^2 - 49}{7 - x}$

☆ Stretching the Topics

_____ 1. Find the restrictions on the variable for the expression $\dfrac{2m^2 - 5m - 12}{2m^3 + 11m^2 + 12m}$.

_____ 2. Reduce $\dfrac{6a^2b - 12abc + 6bc^2}{9c^2 - 9a^2}$.

_____ 3. Reduce $\dfrac{6x^n(b + c)^n}{3x^{n-1}(b + c)}$.

Check your answers in the back of your book.

If you can do **Checkup 8.1,** you are ready to go on to Section 8.2.

CHECKUP 8.1

Find the restrictions on the variable.

_____ **1.** $\dfrac{x + 5}{3x}$ _____ **2.** $\dfrac{x + 4}{2x - 1}$

Reduce the fractions.

_____ **3.** $\dfrac{x^{12}y^5}{x^9y^7}$ _____ **4.** $\dfrac{-5x^4}{20x^5}$

_____ **5.** $\dfrac{16xy - 20x}{4x}$ _____ **6.** $\dfrac{6xy}{30x^2y - 24xy^2}$

_____ **7.** $\dfrac{3a + 3b}{12a - 12b}$ _____ **8.** $\dfrac{5y - 5x}{15x^2 - 15y^2}$

_____ **9.** $\dfrac{m^2 - 4n^2}{2m - 4n}$ _____ **10.** $\dfrac{x^2 + 6x + 8}{x^2 + 7x + 10}$

Check your answers in the back of your book.

If You Missed Problems:	You Should Review Examples:
1, 2	5, 6
3, 4	7, 8
5, 6	12
7	9, 10
8	15, 16
9, 10	14

8.2 Multiplying and Dividing Rational Algebraic Expressions

Multiplying and Dividing Monomial Fractions

Recall that the product of fractions is found by multiplying their numerators and multiplying their denominators and then reducing. We continue to use this method to multiply monomial fractions and then reduce, first reducing the numerical part and then reducing the variable parts.

$$\boxed{\begin{array}{c} \textbf{Multiplying Fractions} \\[2mm] \dfrac{A}{B} \cdot \dfrac{C}{D} = \dfrac{A \cdot C}{B \cdot D} \qquad (B,\, D \neq 0) \end{array}}$$

Example 1. Find $\dfrac{-6x^2y}{5z} \cdot \dfrac{15xz^2}{2y^4}$.

Solution

$$\frac{-6x^2y}{5z} \cdot \frac{15xz^2}{2y^4} = \frac{-6 \cdot 15 \cdot x^2 \cdot x \cdot y \cdot z^2}{5 \cdot 2 \cdot y^4 \cdot z}$$

Multiply numerators.
Multiply denominators.

$$= \frac{\overset{-3}{\cancel{-6}} \cdot \overset{3}{\cancel{15}} \cdot x^3 \cdot \overset{1}{\cancel{y}} \cdot \overset{z}{\cancel{z^2}}}{\underset{1}{\cancel{5}} \cdot \underset{1}{\cancel{2}} \cdot \underset{y^3}{\cancel{y^4}} \cdot \underset{1}{\cancel{z}}}$$

Divide numerator and denominator by common factors.

$$= \frac{-9x^3z}{y^3}$$

As was true with numerical fractions, we may reduce *before* multiplying rational algebraic expressions. This is done by dividing any one numerator and any one denominator by the same factor. Then we multiply the factors remaining in the numerators and denominators.

Example 2. Find $\dfrac{7a^2x}{3b} \cdot \dfrac{9b^3}{14ax}$.

Solution

$$\frac{7a^2x}{3b} \cdot \frac{9b^3}{14ax} = \frac{\overset{1}{\cancel{7}}\overset{a}{\cancel{a^2}}\overset{1}{\cancel{x}}}{\underset{1}{\cancel{3}}\underset{1}{\cancel{b}}} \cdot \frac{\overset{3}{\cancel{9}}\overset{b^2}{\cancel{b^3}}}{\underset{2}{\cancel{14}}\underset{1}{\cancel{a}}\underset{1}{\cancel{x}}}$$

Divide numerator and denominator by common factors.

$$= \frac{3ab^2}{2}$$

Multiply remaining factors.

Since we know that division of fractions requires multiplication by the reciprocal of the divisor, we may use our multiplication skills to divide monomial fractions. Just remember *to invert the divisor and multiply.*

$$\boxed{\begin{array}{c} \textbf{Dividing Fractions} \\[2mm] \dfrac{A}{B} \div \dfrac{C}{D} = \dfrac{A}{B} \cdot \dfrac{D}{C} \qquad (B,\, C,\, D \neq 0) \end{array}}$$

Example 3. Find $\dfrac{6x^4}{5y} \div \dfrac{9x^3}{10y^4}$.

Solution

$$\dfrac{6x^4}{5y} \div \dfrac{9x^3}{10y^4}$$

$$= \dfrac{6x^4}{5y} \cdot \dfrac{10y^4}{9x^3} \quad \text{Invert divisor and multiply.}$$

$$= \dfrac{\overset{2}{\cancel{6}}\overset{x}{\cancel{x^4}}}{\underset{11}{\cancel{5y}}} \cdot \dfrac{\overset{2}{\cancel{10}}\overset{y^3}{\cancel{y^4}}}{\underset{3\;1}{\cancel{9x^3}}} \quad \text{Reduce.}$$

$$= \dfrac{4xy^3}{3} \quad \text{Multiply remaining factors.}$$

Now you try Example 4.

Example 4. Find $\dfrac{-7x^2y^4}{8z} \div \dfrac{5x^5y}{12z}$.

Solution

Check your work on page 405. ▶

⇒ Trial Run

Perform the indicated operation. Write answers in reduced form.

_____ **1.** $\dfrac{-8x^3y}{3z} \cdot \dfrac{9xz^3}{4y^4}$

_____ **2.** $\dfrac{a^2bc^3}{7} \cdot \dfrac{14}{a^4c^5}$

_____ **3.** $\dfrac{9x^2y}{5xz^3} \cdot \dfrac{25xz^2}{3xy} \cdot \dfrac{xy^2}{15}$

_____ **4.** $\dfrac{4x}{y} \div \dfrac{3y}{-x}$

_____ **5.** $\dfrac{12a^3x^5}{5y^2} \div \dfrac{16x^5}{10y^2}$

Answers are on page 405.

Multiplying and Dividing Polynomial Fractions

The methods we have already learned are all that we need to multiply fractions with polynomial numerators and denominators.

Multiplying Fractions

1. Factor all numerators and denominators.
2. Reduce, by dividing out all factors common to a numerator and a denominator.
3. Multiply remaining numerator factors and multiply remaining denominator factors.

Example 5. Find $\dfrac{x^2 - 4}{5x + 5} \cdot \dfrac{x + 1}{x + 2}$.

Solution

$$\frac{x^2 - 4}{5x + 5} \cdot \frac{x + 1}{x + 2}$$

$$= \frac{\overset{1}{\cancel{(x + 2)}}(x - 2)}{5\cancel{(x + 1)}} \cdot \frac{\overset{1}{\cancel{x + 1}}}{\cancel{x + 2}}$$

$$= \frac{x - 2}{5}$$

Now you try to complete Example 6.

Example 7. Find $\dfrac{5x + 5y}{6a - 3x} \cdot \dfrac{6x - 12a}{10x + 10y}$.

Solution

$$\frac{5x + 5y}{6a - 3x} \cdot \frac{6x - 12a}{10x + 10y}$$

$$= \frac{\overset{1}{\cancel{5}}\overset{1}{\cancel{(x + y)}}}{\cancel{3}\cancel{(2a - x)}} \cdot \frac{\overset{2}{\cancel{6}}\overset{-1}{\cancel{(x - 2a)}}}{\cancel{10}\cancel{(x + y)}}$$

$$= \frac{-2}{2}$$

$$= -1$$

Now try Example 8.

Example 6. Find $\dfrac{x^2 + 7x + 10}{x^2 + 2x} \cdot \dfrac{x^3}{4x + 20}$.

Solution

$$\frac{x^2 + 7x + 10}{x^2 + 2x} \cdot \frac{x^3}{4x + 20}$$

$$= \frac{(x + 5)(x + 2)}{x(x + 2)} \cdot \frac{x^3}{4(x + 5)}$$

Check your work on page 405. ▶

Example 8. Find

$$\frac{x^2 - 2x - 3}{x^2 - 1} \cdot \frac{2x^2 + 5x - 3}{x^2 - 9}$$

Solution

Check your work on page 405. ▶

Once again, we declare that if we can multiply polynomial fractions, we can surely divide them. We simply must remember to *invert the divisor and multiply*.

Dividing Fractions

1. Invert the divisor and change the operation to multiplication.
2. Factor numerators and denominators.
3. Reduce by dividing out common factors.
4. Multiply remaining factors in numerators and denominators.

Example 9. Find $\dfrac{x^2 - 25}{5} \div (x - 5)$.

Solution

$$\dfrac{x^2 - 25}{5} \div (x - 5).$$

$$= \dfrac{x^2 - 25}{5} \div \dfrac{x - 5}{1}$$

$$= \dfrac{x^2 - 25}{5} \cdot \dfrac{1}{x - 5}$$

$$= \dfrac{(x + 5)(x - 5)}{5} \cdot \dfrac{1}{x - 5}$$

$$= \dfrac{x + 5}{5}$$

Now complete Example 10.

Example 10. Find
$$\dfrac{x^2 + 12x + 36}{5x^3 + 30x^2} \div \dfrac{7x + 42}{10x^4}.$$

Solution

$$\dfrac{x^2 + 12x + 36}{5x^3 + 30x^2} \div \dfrac{7x + 42}{10x^4}$$

$$= \dfrac{x^2 + 12x + 36}{5x^3 + 30x^2} \cdot \dfrac{10x^4}{7x + 42}$$

$$= \dfrac{(x + 6)(\underline{\qquad})}{5x^2(\underline{\qquad})} \cdot \dfrac{10x^4}{7(\underline{\qquad})}$$

Check your work on page 405. ▶

⇒ Trial Run

Perform the indicated operation. Write answers in reduced form.

_____ 1. $\dfrac{2x + 2}{x + 3} \cdot \dfrac{1}{x + 1}$

_____ 2. $\dfrac{x^2 + 5x + 6}{x^2 + 3x} \cdot \dfrac{x^3}{3x + 6}$

_____ 3. $\dfrac{3x + 3y}{4a - 8b} \cdot \dfrac{8a - 16b}{12x + 12y}$

_____ 4. $\dfrac{x^2 - 3x + 2}{x^2 - 1} \cdot \dfrac{2x^2 + x - 1}{x^2 - 4x + 4}$

_____ 5. $\dfrac{16 - x^2}{3} \div \dfrac{x^2 - 8x + 16}{3x - 12}$

_____ 6. $\dfrac{x^2 - 2x - 35}{3x^2 + 15x} \div \dfrac{4x - 28}{9x^3}$

Answers are on page 405.

▶ Examples You Completed

Example 4. Find $\dfrac{-7x^2y^4}{8z} \div \dfrac{5x^5y}{12z}$.

Solution

$$\frac{-7x^2y^4}{8z} \div \frac{5x^5y}{12z}$$

$$= \frac{-7x^2y^4}{8z} \cdot \frac{12z}{5x^5y}$$

$$= \frac{-7x^2y^4}{8z} \cdot \frac{12z}{5x^5y}$$

$$= \frac{-21y^3}{10x^3}$$

Example 6. Find $\dfrac{x^2 + 7x + 10}{x^3 + 2x} \cdot \dfrac{x^3}{4x + 20}$.

Solution

$$\frac{x^2 + 7x + 10}{x^2 + 2x} \cdot \frac{x^3}{4x + 20}$$

$$= \frac{(x+5)(x+2)}{x(x+2)} \cdot \frac{x^2}{4(x+5)}$$

$$= \frac{x^2}{4}$$

Example 8. Find

$$\frac{x^2 - 2x - 3}{x^2 - 1} \cdot \frac{2x^2 + 5x - 3}{x^2 - 9}$$

Solution

$$\frac{x^2 - 2x - 3}{x^2 - 1} \cdot \frac{2x^2 + 5x - 3}{x^2 - 9}$$

$$= \frac{(x-3)(x+1)}{(x+1)(x-1)} \cdot \frac{(2x-1)(x+3)}{(x-3)(x+3)}$$

$$= \frac{2x - 1}{x - 1}$$

Example 10. Find

$$\frac{x^2 + 12x + 36}{5x^3 + 30x^2} \div \frac{7x + 42}{10x^4}$$

Solution

$$\frac{x^2 + 12x + 36}{5x^3 + 30x^2} \div \frac{7x + 42}{10x^4}$$

$$= \frac{x^2 + 12x + 36}{5x^3 + 30x^2} \cdot \frac{10x^4}{7x + 42}$$

$$= \frac{(x+6)(x+6)}{5x^2(x+6)} \cdot \frac{10x^4}{7(x+6)}$$

$$= \frac{2x^2}{7}$$

Answers to Trial Runs

page 402   **1.** $\dfrac{-6x^4z^2}{y^3}$ **2.** $\dfrac{2b}{a^2c^2}$ **3.** $\dfrac{x^2y^2}{z}$ **4.** $\dfrac{-4x^2}{3y^2}$ **5.** $\dfrac{3a^3}{2}$

page 404   **1.** $\dfrac{2}{x + 3}$ **2.** $\dfrac{x^2}{3}$ **3.** $\dfrac{1}{2}$ **4.** $\dfrac{2x - 1}{x - 2}$ **5.** $-(4 + x)$ or $-4 - x$ **6.** $\dfrac{3x^2}{4}$

EXERCISE SET 8.2

Perform the indicated operation. Write answers in reduced form.

_____ 1. $\dfrac{x}{y} \cdot \dfrac{1}{y}$

_____ 2. $\dfrac{-x}{y} \cdot \dfrac{1}{3}$

_____ 3. $\dfrac{3xy}{4z} \cdot \dfrac{16z}{12y}$

_____ 4. $\dfrac{5xy}{9z} \cdot \dfrac{27z}{15y}$

_____ 5. $\dfrac{3x}{2y} \cdot \dfrac{10y^2}{x^2} \cdot \dfrac{7}{x}$

_____ 6. $\dfrac{5x}{3y} \cdot \dfrac{21y^2}{x^2} \cdot \dfrac{2}{x}$

_____ 7. $\dfrac{1}{a} \div \dfrac{1}{ab}$

_____ 8. $\dfrac{1}{a} \div \dfrac{1}{a^2}$

_____ 9. $\dfrac{x}{3} \div (-3)$

_____ 10. $\dfrac{x}{5} \div (-15)$

_____ 11. $\dfrac{x^2y}{2z} \div \dfrac{xy}{8z^2}$

_____ 12. $\dfrac{x^3y}{3z^2} \div \dfrac{x^2y}{18z}$

_____ 13 $\dfrac{4ab^3}{c} \div \dfrac{4a^3}{5b^2c}$

_____ 14. $\dfrac{9a^3b^2}{c} \div \dfrac{9a}{7b^3c}$

_____ 15. $\dfrac{4y^3}{5x} \div \dfrac{2xy}{45}$

_____ 16. $\dfrac{9y^3}{7x} \div \dfrac{3xy}{28}$

_____ 17. $\dfrac{34b^3}{81a^3} \div \dfrac{17a^2}{18a^4b}$

_____ 18. $\dfrac{38b^3}{25a^4} \div \dfrac{19a}{10a^3b}$

_____ 19. $\dfrac{15x^2}{16y^3} \div \dfrac{9y^8}{8x^4}$

_____ 20. $\dfrac{12x^3}{49y^3} \div \dfrac{16y^9}{21x^3}$

_____ 21. $\dfrac{2x + 2}{x + 3} \cdot \dfrac{1}{x + 1}$

_____ 22. $\dfrac{3x + 3}{x + 4} \cdot \dfrac{1}{x + 1}$

_____ 23. $\dfrac{3x + 6}{x^2} \cdot \dfrac{x}{8x + 16}$

_____ 24. $\dfrac{5x + 10}{x^3} \cdot \dfrac{x^4}{15x + 30}$

_____ 25. $\dfrac{2a + 6b}{a^2b} \cdot \dfrac{ab^2}{a + 3b}$

_____ 26. $\dfrac{7a + 14b}{a^3b^2} \cdot \dfrac{a^2b^3}{a + 2b}$

_____ 27. $\dfrac{m^2 + n^2}{6} \cdot \dfrac{42}{3m^2 + 3n^2}$

_____ 28. $\dfrac{m^2 + n^2}{19} \cdot \dfrac{38}{5m^2 + 5n^2}$

_____ 29. $\dfrac{x^2 - 9y^2}{x^2 + 9y^2} \cdot \dfrac{x^2 + 9y^2}{x + 3y}$

_____ 30. $\dfrac{x^2 - 25y^2}{x^2 + 25y^2} \cdot \dfrac{x^2 + 25y^2}{x + 5y}$

_____ 31. $\dfrac{x^2 - 4}{3x - 9} \cdot \dfrac{6 - 2x}{x^2 + 4x + 4}$

_____ 32. $\dfrac{x^2 - 9}{2x - 4} \cdot \dfrac{6 - 3x}{x^2 - 6x + 9}$

_____ 33. $\dfrac{(x + 3)^2}{64yz} \cdot \dfrac{40y^2z}{x^2 - 9}$

_____ 34. $\dfrac{(x + 5)^2}{81yz^3} \cdot \dfrac{72y^5z}{x^2 - 25}$

_____ 35. $\dfrac{2x^2 + 3x + 1}{2x^2 + 5x + 3} \cdot \dfrac{2x^2 + 11x + 12}{2x^2 + 13x + 6}$

_____ 36. $\dfrac{20x^2 - 7x - 3}{4x^2 - 7x - 2} \cdot \dfrac{3x^2 - 5x - 2}{15x^2 - 4x - 3}$

_____ 37. $\dfrac{6x^2 + xy - y^2}{2x^2 - 9xy - 5y^2} \cdot \dfrac{x^2 - 25y^2}{15x - 5y}$

_____ 38. $\dfrac{8x^2 + 18xy - 5y^2}{16x^2 - 8xy + y^2} \cdot \dfrac{4x^2 + 3xy - y^2}{12x + 30y}$

_____ 39. $\dfrac{x^2 - 49}{y} \div \dfrac{x + 7}{y^2}$

_____ 40. $\dfrac{x^2 - 64}{y^4} \div \dfrac{x + 8}{y^3}$

_____ 41. $\dfrac{x^2 + 7x + 10}{63x} \div \dfrac{7x + 14}{9x}$

_____ 42. $\dfrac{x^2 - 12x + 32}{60x} \div \dfrac{5x - 20}{12x}$

_____ 43. $\dfrac{x + 2}{4x - 8} \div \dfrac{x^2 - 7x - 18}{2x - 4}$

_____ 44. $\dfrac{x + 7}{8x - 56} \div \dfrac{x^2 + 4x - 21}{4x - 28}$

_____ 45. $\dfrac{2 - x}{3} \div \dfrac{(x - 2)^2}{9}$

_____ 46. $\dfrac{5 - x}{4} \div \dfrac{(x - 5)^2}{16}$

_____ 47. $\dfrac{x^2 - 8x + 15}{x^2 - 25} \div \dfrac{x^2 - 9}{x + 5}$

_____ 48. $\dfrac{x^2 - 7x - 18}{x^2 - 81} \div \dfrac{x^2 + 4x + 4}{x + 9}$

_____ 49. $\dfrac{a^2 - 2ab + b^2}{10a^5b^5} \div \dfrac{a - b}{50a^3b^3}$

_____ 50. $\dfrac{a^2 - 6ab + 9b^2}{14a^4b^7} \div \dfrac{a - 3b}{70a^3b^2}$

_____ 51. $\dfrac{xy - y^2}{x^2 - 2xy + y^2} \div \dfrac{5xy - 3y^2}{5x^2 - 8xy + 3y^2}$

_____ 52. $\dfrac{x^2 - 7xy}{x^2 - 14xy + 49y^2} \div \dfrac{3x^2 + 2xy}{3x^2 - 19xy - 14y^2}$

☆ Stretching the Topics ────────────

Perform the indicated operations. Write answers in reduced form.

_____ 1. $\dfrac{-(3x^3y)^2}{10x} \cdot \dfrac{(-3x^2)^3}{(2x^2y)^2}$

_____ 2. $\dfrac{a^2 - b^2}{a^2} \cdot \dfrac{a^2b - b^3 + a^2c - cb^2}{ab + ac}$

_____ 3. $\dfrac{5x^2 + 10xy}{3xy - 3y^2} \cdot \dfrac{2x - 3y}{x^2 + 4xy + 4y^2} \div \dfrac{4x^2 - 9y^2}{4x^2 - 4y^2}$

Check your answers in the back of your book.

If you can simplify the products and quotients in **Checkup 8.2**, you are ready to go on to Section 8.3.

CHECKUP 8.2

Perform the indicated operation. Write answers in reduced form.

_____ **1.** $\dfrac{-8x^3}{3y} \cdot \dfrac{21xy^2}{4yz}$ _____ **2.** $\dfrac{15a^3b}{4c} \div \dfrac{21b^2}{8c^3}$

_____ **3.** $\dfrac{-9x^3y}{16z} \div \dfrac{7x^5y^2}{12z^3}$ _____ **4.** $\dfrac{2x + 8y}{xy^2} \cdot \dfrac{x^2y}{x^2 + 4xy}$

_____ **5.** $\dfrac{x^2 - 9}{3x - 12} \cdot \dfrac{2x - 8}{x^2 + 6x + 9}$.   _____ **6.** $\dfrac{3}{y - x} \cdot \dfrac{4x - 4y}{12}$

_____ **7.** $\dfrac{x^2 - 64}{y^2} \div \dfrac{x + 8}{y}$ _____ **8.** $\dfrac{6x + 18}{54x} \div \dfrac{x^2 + 9x + 18}{3x}$

_____ **9.** $\dfrac{x^2 - 121}{x^2 - 5x - 24} \cdot \dfrac{x^2 - 16x + 64}{x^2 + 14x + 33}$ _____ **10.** $\dfrac{n^2 + 3n - 18}{n^2 - 36} \div \dfrac{2n^2 - 5n - 3}{4n^2 - 1}$

Check your answers in the back of your book.

If You Missed Problems:	You Should Review Examples:
1	1, 2
2, 3	3, 4
4–6	5–7
7, 8	9, 10
9, 10	8, 10

8.3 Building Rational Algebraic Expressions

We spent time in Section 8.1 reducing algebraic fractions to simplest form by dividing numerators and denominators by common factors. For the work that lies ahead, we must review the methods for **building fractions**.

Building Monomial Fractions

We learned to build numerical fractions in Chapter 1. The same method works when we wish to build fractions containing monomials to new fractions having certain denominators.

$$\frac{5x}{3y} = \frac{?}{12y^4}$$ We must find a numerator for the fraction on the right.

$$\frac{5x}{3y} = \frac{?}{3y(4y^3)}$$ By what factor was the original denominator $(3y)$ multiplied to arrive at $12y^4$? The factor must be $4y^3$.

$$\frac{5x}{3y} = \frac{5x(4y^3)}{3y(4y^3)}$$ If we multiplied $3y$ by $4y^3$, we must *also* multiply $5x$ by $4y^3$ (because $\frac{4y^3}{4y^3} = 1$).

$$\frac{5x}{3y} = \frac{20xy^3}{12y^4}$$ We have found our new fraction.

To build a fraction to a new fraction with a certain denominator, we see that we must use these steps.

Building Fractions

1. Decide by what factor the original denominator was multiplied to arrive at the new denominator.
2. Then multiply the original numerator by the *same* factor to arrive at the new numerator.

Example 1. Find the new fraction.

$$\frac{5}{xy} = \frac{?}{3x^2y^3}$$

Solution

$$\frac{5}{xy} = \frac{?}{3x^2y^3}$$

$$\frac{5}{xy} = \frac{?}{(xy)(3xy^2)}$$

$$\frac{5}{xy} = \frac{5(3xy^2)}{(xy)(3xy^2)}$$

$$\frac{5}{xy} = \frac{15xy^2}{3x^2y^3}$$

Now complete Example 2.

Example 2. Find the new fraction.

$$\frac{-7a}{2bc} = \frac{?}{6b^2c^2}$$

Solution

$$\frac{-7a}{2bc} = \frac{?}{6b^2c^2}$$

$$\frac{-7a}{2bc} = \frac{?}{2bc(3bc)}$$

$$\frac{-7a}{2bc} = \frac{-7a(\underline{\quad})}{2bc(3bc)}$$

$$= \frac{}{6b^2c^2}$$

Check your work on page 412. ▶

⫸ Trial Run

Write the new fractions.

_____ **1.** $\dfrac{-5}{7x} = \dfrac{?}{14x^2}$ _____ **2.** $\dfrac{8x}{5y} = \dfrac{?}{15y^4}$

_____ **3.** $\dfrac{2a}{9bc} = \dfrac{?}{27abc}$ _____ **4.** $\dfrac{ax}{by} = \dfrac{?}{a^2b^2y}$

Answers are on page 412.

Building Polynomial Fractions

Building fractions containing polynomials requires a bit more care because the building factor may not be obvious. If the original and new denominators are written in factored form, however, the procedure used is exactly the same as we have already discussed.

$$\frac{3}{x + 2} = \frac{?}{x(x + 2)}$$

By what factor was the original denominator $(x + 2)$ multiplied to arrive at $x(x + 2)$? The factor must be x.

$$\frac{3}{x + 2} = \frac{x(3)}{x(x + 2)}$$

If we multiplied the original denominator by x, we must *also* multiply the original numerator by x.

$$\frac{3}{x + 2} = \frac{3x}{x(x + 2)}$$

We have found our new fraction. Notice that we leave the denominator in factored form because that form will be more useful in adding fractions.

Example 3. Find the new fraction.

$$\frac{2x - 5}{x - 1} = \frac{?}{(x - 1)(x + 1)}$$

Solution

$$\frac{2x - 5}{x - 1} = \frac{?}{(x - 1)(x + 1)}$$

$$\frac{2x - 5}{x - 1} = \frac{(2x - 5)(x + 1)}{(x - 1)(x + 1)}$$

$$\frac{2x - 5}{x - 1} = \frac{2x^2 - 3x + 5}{(x - 1)(x + 1)}$$

Example 4. Find the new fraction.

$$\frac{x + 2}{x + 3} = \frac{?}{x^2(x + 3)}$$

Solution

$$\frac{x + 2}{x + 3} = \frac{?}{x^2(x + 3)}$$

$$\frac{x + 2}{x + 3} = \frac{x^2(x + 2)}{x^2(x + 3)}$$

$$\frac{x + 2}{x + 3} = \frac{x^3 + 2x^2}{x^2(x + 3)}$$

If the new denominator is *not* in factored form, we must factor it before we can decide what the building factors must be.

$$\frac{7x}{x - 3} = \frac{?}{x^2 - 9}$$

We must find our new numerator.

$$\frac{7x}{x - 3} = \frac{?}{(x - 3)(x + 3)}$$

First we factor the new denominator. By what factor was the original denominator $(x - 3)$ multiplied to arrive at the new denominator $(x - 3)(x + 3)$? The factor must be $(x + 3)$.

$$\frac{7x}{x - 3} = \frac{7x(x + 3)}{(x - 3)(x + 3)}$$

We must multiply the original numerator by $x + 3$

$$\frac{7x}{x - 3} = \frac{7x^2 + 21x}{(x - 3)(x + 3)}$$

We have found our new fraction.

You try to complete Example 5.

Example 5. Find the new fraction.

$$\frac{x + 6}{2x + 3} = \frac{?}{2x^2 - 7x - 15}$$

Solution

$$\frac{x + 6}{2x + 3} = \frac{?}{2x^2 - 7x - 15}$$

$$\frac{x + 6}{2x + 3} = \frac{?}{(2x + 3)(x - 5)}$$

$$\frac{x + 6}{2x + 3} = \frac{(x + 6)(\underline{\hspace{1cm}})}{(2x + 3)(x - 5)}$$

$$\frac{x + 6}{2x + 3} = \frac{}{(2x + 3)(x - 5)}$$

Check your work on page 412. ▶

Example 6. Find the new fraction.

$$\frac{a - 8}{a + 2} = \frac{?}{3a^3 - 12a}$$

Solution

$$\frac{a - 8}{a + 2} = \frac{?}{3a^3 - 12a}$$

$$\frac{a - 8}{a + 2} = \frac{?}{3a(a^2 - 4)}$$

$$\frac{a - 8}{a + 2} = \frac{?}{3a(a - 2)(a + 2)}$$

$$\frac{a - 8}{a + 2} = \frac{3a(a - 2)(a - 8)}{3a(a - 2)(a + 2)}$$

$$= \frac{3a(a^2 - 10a + 16)}{3a(a - 2)(a + 2)}$$

$$\frac{a - 8}{a + 2} = \frac{3a^3 - 30a^2 + 48a}{3a(a - 2)(a + 2)}$$

⏸▶ Trial Run

Find the new fractions.

_____ **1.** $\dfrac{1}{3} = \dfrac{?}{3(a - 2)}$

_____ **2.** $\dfrac{-4}{x + 5} = \dfrac{?}{x(x + 5)}$

_____ **3.** $\dfrac{x}{2x - y} = \dfrac{?}{(2x - y)(2x + y)}$

_____ **4.** $\dfrac{5x}{x - 4} = \dfrac{?}{x^2 - 16}$

_____ **5.** $\dfrac{x + 1}{x + 4} = \dfrac{?}{x^3 + 4x^2}$

_____ **6.** $\dfrac{x - 3}{2x - 1} = \dfrac{?}{2x^2 + 9x - 5}$

Answers are on page 412.

▶ **Examples You Completed** _____

Example 2. Find the new fraction.

$$\frac{-7a}{2bc} = \frac{?}{6b^2c^2}$$

Solution

$$\frac{-7a}{2bc} = \frac{?}{6b^2c^2}$$

$$\frac{-7a}{2bc} = \frac{?}{2bc(3bc)}$$

$$\frac{-7a}{2bc} = \frac{-7a(3bc)}{2bc(3bc)}$$

$$= \frac{-21abc}{6b^2c^2}$$

Example 5. Find the new fraction.

$$\frac{x+6}{2x+3} = \frac{?}{2x^2 - 7x - 15}$$

Solution

$$\frac{x+6}{2x+3} = \frac{?}{2x^2 - 7x - 15}$$

$$\frac{x+6}{2x+3} = \frac{?}{(2x+3)(x-5)}$$

$$\frac{x+6}{2x+3} = \frac{(x+6)(x-5)}{(2x+3)(x-5)}$$

$$\frac{x+6}{2x+3} = \frac{x^2 + x - 30}{(2x+3)(x-5)}$$

Answers to Trial Runs _____

page 410 **1.** $\dfrac{-10x}{14x^2}$ **2.** $\dfrac{24xy^3}{15y^4}$ **3.** $\dfrac{6a^2}{27abc}$ **4.** $\dfrac{a^3bx}{a^2b^2y}$

page 411 **1.** $\dfrac{a-2}{3(a-2)}$ **2.** $\dfrac{-4x}{x(x+5)}$ **3.** $\dfrac{2x^2 + xy}{(2x+y)(2x-y)}$ **4.** $\dfrac{5x^2 + 20x}{(x+4)(x-4)}$ **5.** $\dfrac{x^3 + x^2}{x^2(x+4)}$

6. $\dfrac{x^2 + 2x - 15}{(2x-1)(x+5)}$

EXERCISE SET 8.3

Find the new fractions.

_____ 1. $\dfrac{2}{3} = \dfrac{?}{3x}$

_____ 2. $\dfrac{4}{5} = \dfrac{?}{5x}$

_____ 3. $\dfrac{-1}{5x} = \dfrac{?}{15x^2}$

_____ 4. $\dfrac{-1}{6x} = \dfrac{?}{24x^2}$

_____ 5. $\dfrac{4}{xy} = \dfrac{?}{2x^2y^2}$

_____ 6. $\dfrac{5}{xy} = \dfrac{?}{4x^2y^2}$

_____ 7. $\dfrac{3x}{7y} = \dfrac{?}{21y^3}$

_____ 8. $\dfrac{4x}{9y} = \dfrac{?}{36y^3}$

_____ 9. $\dfrac{2a}{-bc} = \dfrac{?}{15abc}$

_____ 10. $\dfrac{5a}{-xy} = \dfrac{?}{10axy}$

_____ 11. $x = \dfrac{?}{3y}$

_____ 12. $a = \dfrac{?}{2y}$

_____ 13. $\dfrac{-x^2}{y^2} = \dfrac{?}{3y^3}$

_____ 14. $\dfrac{-a^2}{b^2} = \dfrac{?}{12b^4}$

_____ 15. $-2b^2 = \dfrac{?}{4a^2b^2}$

_____ 16. $-3x^2 = \dfrac{?}{9x^2y^2}$

_____ 17. $\dfrac{2}{3} = \dfrac{?}{12(x-y)}$

_____ 18. $\dfrac{3}{4} = \dfrac{?}{20(a-b)}$

_____ 19. $\dfrac{5}{2x} = \dfrac{?}{10x^3}$

_____ 20. $\dfrac{3}{5x} = \dfrac{?}{15x^4}$

_____ 21. $\dfrac{-4}{a} = \dfrac{?}{a(a+b)}$

_____ 22. $\dfrac{-7}{x} = \dfrac{?}{x(x+y)}$

_____ 23. $\dfrac{3x}{x+y} = \dfrac{?}{x^2+xy}$

_____ 24. $\dfrac{2y}{x-y} = \dfrac{?}{xy-y^2}$

_____ 25. $\dfrac{2a}{b+1} = \dfrac{?}{b^2-1}$

_____ 26. $\dfrac{3a}{b-2} = \dfrac{?}{b^2-4}$

_____ 27. $\dfrac{6x^2}{x-5} = \dfrac{?}{x^2-3x-10}$

_____ 28. $\dfrac{5x}{x-7} = \dfrac{?}{x^2-9x+14}$

_____ 29. $\dfrac{x+5}{x+3} = \dfrac{?}{x^3+3x^2}$

_____ 30. $\dfrac{a+3}{a+2} = \dfrac{?}{a^4+2a^3}$

_____ 31. $\dfrac{a-1}{a-3} = \dfrac{?}{a^2-7a+12}$

_____ 32. $\dfrac{x-2}{x-5} = \dfrac{?}{x^2-8x+15}$

413

_____ 33. $\dfrac{-3y}{x^2 - x} = \dfrac{?}{x^3 - 2x^2 + x}$ _____ 34. $\dfrac{-2x}{y^3 - 2y^2} = \dfrac{?}{y^4 - 4y^2}$

_____ 35. $\dfrac{x - y}{2x + 2y} = \dfrac{?}{2x^2 - 4xy - 6y^2}$ _____ 36. $\dfrac{x + 2y}{3x - 3y} = \dfrac{?}{3x^2 + 9xy - 12y^2}$

☆ Stretching the Topics _____

Find the new fractions.

_____ 1. $\dfrac{x - 2}{2x^2 - 3x} = \dfrac{?}{6x^4 + x^3 - 15x^2}$

_____ 2. $\dfrac{5}{a + b} = \dfrac{?}{ax - ay + bx - by}$

_____ 3. $\dfrac{-3}{x - 2} = \dfrac{?}{x^4 - 3x^2 - 4}$

Check your answers in the back of your book.

If you can complete **Checkup 8.3**, you are ready to go to Section 8.4.

CHECKUP 8.3

Find the new fractions.

_____ **1.** $\dfrac{6}{ab} = \dfrac{?}{9a^3b^2}$

_____ **2.** $\dfrac{-5x}{2yz} = \dfrac{?}{10y^2z^3}$

_____ **3.** $\dfrac{-x}{6y^2} = \dfrac{?}{18y^4}$

_____ **4.** $a^2 = \dfrac{?}{16a^2b^2}$

_____ **5.** $\dfrac{x-5}{x+2} = \dfrac{?}{(x+2)(x+2)}$

_____ **6.** $\dfrac{2a-b}{a-3b} = \dfrac{?}{a^3(a-3b)}$

_____ **7.** $\dfrac{2a}{a+b} = \dfrac{?}{a^2b+ab^2}$

_____ **8.** $\dfrac{5x}{x+6} = \dfrac{?}{x^2-36}$

_____ **9.** $\dfrac{x-2}{x-1} = \dfrac{?}{x^2+3x-4}$

_____ **10.** $\dfrac{a-b}{5a+5b} = \dfrac{?}{5a^2+10ab+5b^2}$

Check your answers in the back of your book.

If You Missed Problems:	You Should Review Examples:
1–4	1, 2
5, 6	3, 4
7–10	5, 6

4

8.4 Adding and Subtracting Rational Algebraic Expressions

Adding and Subtracting Fractions with the Same Denominator

The procedure for combining algebraic fractions having the same denominator is exactly the same as the procedure for combining numerical fractions.

> **Combining Fractions with the Same Denominator**
>
> 1. Combine the numerators.
> 2. Keep the same denominator.
> 3. Factor the numerator and denominator and reduce the fraction if possible.

Example 1. Find $\dfrac{a}{5} + \dfrac{2a}{5}$.

Solution

$$\frac{a}{5} + \frac{2a}{5}$$

$$= \frac{a + 2a}{5}$$

$$= \frac{3a}{5}$$

You complete Example 2.

Example 2. Find $\dfrac{5x}{4y} - \dfrac{x}{4y} + \dfrac{3}{4y} + \dfrac{5}{4y}$.

Solution

$$\frac{5x}{4y} - \frac{x}{4y} + \frac{3}{4y} + \frac{5}{4y}$$

$$= \frac{5x - x + 3 + 5}{4y}$$

$$= \frac{4x + 8}{4y}$$

$$= \frac{4(\underline{\hspace{2cm}})}{4y}$$

$$= \underline{\hspace{2cm}}$$

Check your work on page 422. ▶

Example 3. Find $\dfrac{2a - b}{4a} - \dfrac{2a - 3b}{4a}$.

Solution. Be careful with the minus sign before the second fraction. It shows that the *entire* second numerator is being subtracted. Parentheses will help.

$$\frac{2a - b}{4a} - \frac{2a - 3b}{4a}$$

$$= \frac{2a - b - (2a - 3b)}{4a} \qquad \text{Write the difference of the numerators and keep the same denominator.}$$

$$= \frac{2a - b - 2a + 3b}{4a} \qquad \text{Remove parentheses.}$$

$$= \frac{2b}{4a} \qquad \text{Combine like terms.}$$

$$= \frac{b}{2a} \qquad \text{Reduce the fraction.}$$

Let's try adding fractions with more than one term in their denominators. As long as the denominators all match, our method will be the same Try Example 4.

Example 4. Find

$$\frac{x^2}{x+1} - \frac{2x}{x+1} - \frac{3}{x+1}$$

Solution

$$\frac{x^2}{x+1} - \frac{2x}{x+1} - \frac{3}{x+1}$$

$$= \frac{x^2 - 2x - 3}{x+1}$$

$$= \frac{(x-3)(\underline{\hspace{1cm}})}{x+1}$$

$$= \underline{\hspace{1cm}}$$

Check your work on page 422. ▶
Complete Example 5.

Example 5. Find $\dfrac{2x-3}{x^2-1} - \dfrac{x-4}{x^2-1}$.

Solution

$$\frac{2x-3}{x^2-1} - \frac{x-4}{x^2-1}$$

$$= \frac{2x - 3 - (x-4)}{x^2 - 1}$$

Check your work on page 422. ▶

⫸ **Trial Run**

Perform the indicated operations and reduce if possible.

_____ 1. $\dfrac{4x}{5y} + \dfrac{6x}{5y}$

_____ 2. $\dfrac{7x}{3y} + \dfrac{2}{3y} - \dfrac{x}{3y} + \dfrac{1}{3y}$

_____ 3. $\dfrac{7a-3b}{10b} - \dfrac{a-3b}{10b}$

_____ 4. $\dfrac{3}{x+7} + \dfrac{4}{x+7}$

_____ 5. $\dfrac{4x}{x-2y} - \dfrac{3x}{x-2y} + \dfrac{2y}{x-2y}$

_____ 6. $\dfrac{x^2}{x+2} - \dfrac{13x}{x+2} - \dfrac{30}{x+2}$

_____ 7. $\dfrac{2a^2}{a^2-8a} - \dfrac{17a}{a^2-8a} + \dfrac{8}{a^2-8a}$

_____ 8. $\dfrac{5x-3}{x^2-4} - \dfrac{4x-1}{x^2-4}$

Answers are on page 423.

Adding and Subtracting Fractions with Different Denominators

The method for adding and subtracting numerical fractions with different denominators works for algebraic fractions also.

Adding and Subtracting Fractions

1. Factor the denominators of the original fractions.
2. Choose the LCD. (It is the product of all the different factors appearing in the original denominators, with each factor used the most times that it appears in any single denominator.)
3. Build new fractions from the original fractions, with each new fraction having the LCD as its denominator.
4. Simplify each numerator.
5. Combine the numerators to form one fraction with the LCD as its denominator.
6. Simplify the numerator and reduce the fraction if possible.

You try to complete Example 6.

Example 6. Find $\dfrac{x}{3} + \dfrac{2x}{5} - \dfrac{x}{9}$.

Solution

$$\dfrac{x}{3} + \dfrac{2x}{5} - \dfrac{x}{9}$$

$$= \dfrac{x}{3} + \dfrac{2x}{5} - \dfrac{x}{3 \cdot 3} \qquad \text{LCD: } 3 \cdot 3 \cdot 5$$

$$= \dfrac{x \cdot 3 \cdot 5}{3 \cdot 3 \cdot 5} + \dfrac{2x \cdot 3 \cdot 3}{3 \cdot 3 \cdot 5} - \dfrac{x \cdot 5}{3 \cdot 3 \cdot 5}$$

$$= \dfrac{}{45} + \dfrac{}{45} - \dfrac{}{45}$$

$$= \dfrac{+ \quad -}{45}$$

$$= \dfrac{}{45}$$

Check your work on page 422. ▶

Example 7. Find $\dfrac{3}{y} + \dfrac{7}{x^2} + \dfrac{5}{3x}$.

Solution

$$\dfrac{3}{y} + \dfrac{7}{x^2} + \dfrac{5}{3x} \qquad \text{LCD: } 3x^2y$$

$$= \dfrac{3 \cdot 3 \cdot x^2}{3x^2y} + \dfrac{7 \cdot 3 \cdot y}{3x^2y} + \dfrac{5 \cdot x \cdot y}{3x^2y}$$

$$= \dfrac{9x^2}{3x^2y} + \dfrac{21y}{3x^2y} + \dfrac{5xy}{3x^2y}$$

$$= \dfrac{9x^2 + 21y + 5xy}{3x^2y}$$

Example 8. Find $\dfrac{1}{x+2} + \dfrac{1}{5x}$.

Solution

$$\dfrac{1}{x+2} + \dfrac{1}{5x} \qquad \text{LCD: } 5x(x+2)$$

$$= \dfrac{1 \cdot 5x}{5x(x+2)} + \dfrac{1(x+2)}{5x(x+2)}$$

$$= \dfrac{5x}{5x(x+2)} + \dfrac{x+2}{5x(x+2)}$$

$$= \dfrac{5x + x + 2}{5x(x+2)}$$

$$= \dfrac{6x + 2}{5x(x+2)}$$

$$= \dfrac{2(3x+1)}{5x(x+2)}$$

You complete Example 9.

Example 9. Find $\dfrac{4}{2x-1} - \dfrac{2}{2x^2-x}$.

Solution

$$\dfrac{4}{2x-1} - \dfrac{2}{2x^2-x}$$

$$= \dfrac{4}{2x-1} - \dfrac{2}{x(2x-1)}$$

$$\text{LCD: } x(2x-1)$$

$$= \dfrac{4 \cdot x}{x(2x-1)} - \dfrac{2}{x(2x-1)}$$

Check your work on page 423. ▶

Notice that we factored the numerator in the final step of Example 8 to see whether a factor common to the numerator and denominator might appear.

Example 10. Find $\dfrac{x}{(x+3)^2} + \dfrac{2x}{x^2-9}$.

Solution

$$\dfrac{x}{(x+3)^2} + \dfrac{2x}{x^2-9}$$

$$= \dfrac{x}{(x+3)^2} + \dfrac{2x}{(x+3)(x-3)}$$

Notice that $(x+3)$ appears *twice* in the first denominator. LCD: $(x+3)^2(x-3)$.

$$= \dfrac{x(x-3)}{(x+3)^2(x-3)} + \dfrac{2x(x+3)}{(x+3)^2(x-3)}$$

$$= \dfrac{x^2-3x}{(x+3)^2(x-3)} + \dfrac{2x^2+6x}{(x+3)^2(x-3)}$$

$$= \dfrac{x^2 - 3x + 2x^2 + 6x}{(x+3)^2(x-3)}$$

$$= \dfrac{3x^2 + 3x}{(x+3)^2(x-3)}$$

$$= \dfrac{3x(x+1)}{(x+3)^2(x-3)}$$

Example 11. Find $\dfrac{7x}{x^2 - x - 2} - \dfrac{2}{3x - 6}$.

Solution

$$\dfrac{7x}{x^2 - x - 2} - \dfrac{2}{3x - 6}$$

$$= \dfrac{7x}{(x + 1)(x - 2)} - \dfrac{2}{3(x - 2)} \qquad \text{LCD: } 3(x + 1)(x - 2)$$

$$= \dfrac{7x \cdot 3}{3(x + 1)(x - 2)} - \dfrac{2(x + 1)}{3(x + 1)(x - 2)}$$

$$= \dfrac{21x}{3(x + 1)(x - 2)} - \dfrac{2x + 2}{3(x + 1)(x - 2)}$$

$$= \dfrac{21x - (2x + 2)}{3(x + 1)(x - 2)}$$

$$= \dfrac{19x - 2}{3(x + 1)(x - 2)}$$

Adding fractions is not a quick procedure, but each step we have used is necessary for obtaining the correct result.

⫸ Trial Run ▬▬▬▬▬▬▬▬▬▬▬▬▬▬▬▬▬▬▬▬▬▬▬▬▬▬▬▬▬▬▬▬▬▬▬▬

Perform the indicated operations and reduce if possible.

———— 1. $\dfrac{x}{4} + \dfrac{3x}{5}$ ———— 2. $\dfrac{5x}{9} - \dfrac{x}{6} + \dfrac{1}{4}$

———— 3. $\dfrac{3}{x} + \dfrac{2}{x^2} - \dfrac{1}{3x}$ ———— 4. $\dfrac{5}{2x} - \dfrac{4}{x + 3}$

———— 5. $\dfrac{1}{x - 1} + \dfrac{3}{x + 2}$ ———— 6. $\dfrac{x}{x^2 - 4} - \dfrac{1}{x^2 - 2x}$

Answers are on page 423.

▶ **Examples You Completed** _____

Example 2. Find $\dfrac{5x}{4y} - \dfrac{x}{4y} + \dfrac{3}{4y} + \dfrac{5}{4y}$.

Solution

$$\dfrac{5x}{4y} - \dfrac{x}{4y} + \dfrac{3}{4y} + \dfrac{5}{4y}$$

$$= \dfrac{5x - x + 3 + 5}{4y}$$

$$= \dfrac{4x + 8}{4y}$$

$$= \dfrac{\overset{1}{\cancel{4}}(x + 2)}{\underset{1}{\cancel{4}y}}$$

$$= \dfrac{x + 2}{y}$$

Example 4. Find $\dfrac{x^2}{x + 1} - \dfrac{2x}{x + 1} - \dfrac{3}{x + 1}$.

Solution

$$\dfrac{x^2}{x + 1} - \dfrac{2x}{x + 1} - \dfrac{3}{x + 1}$$

$$= \dfrac{x^2 - 2x - 3}{x + 1}$$

$$= \dfrac{(x - 3)\overset{1}{\cancel{(x + 1)}}}{\underset{1}{\cancel{x + 1}}}$$

$$= x - 3$$

Example 5. Find $\dfrac{2x - 3}{x^2 - 1} - \dfrac{x - 4}{x^2 - 1}$.

Solution

$$\dfrac{2x - 3}{x^2 - 1} - \dfrac{x - 4}{x^2 - 1}$$

$$= \dfrac{2x - 3 - (x - 4)}{x^2 - 1}$$

$$= \dfrac{2x - 3 - x + 4}{x^2 - 1}$$

$$= \dfrac{x + 1}{x^2 - 1}$$

$$= \dfrac{\overset{1}{\cancel{x + 1}}}{\underset{1}{\cancel{(x + 1)}}(x - 1)}$$

$$= \dfrac{1}{x - 1}$$

Example 6. Simplify $\dfrac{x}{3} + \dfrac{2x}{5} - \dfrac{x}{9}$.

Solution

$$\dfrac{x}{3} + \dfrac{2x}{5} - \dfrac{x}{9}$$

$$= \dfrac{x}{3} + \dfrac{2x}{5} - \dfrac{x}{3 \cdot 3} \qquad \text{LCD: } 3 \cdot 3 \cdot 5$$

$$= \dfrac{x \cdot 3 \cdot 5}{3 \cdot 3 \cdot 5} + \dfrac{2x \cdot 3 \cdot 3}{3 \cdot 3 \cdot 5} - \dfrac{x \cdot 5}{3 \cdot 3 \cdot 5}$$

$$= \dfrac{15x}{45} + \dfrac{18x}{45} - \dfrac{5x}{45}$$

$$= \dfrac{15x + 18x - 5x}{45}$$

$$= \dfrac{28x}{45}$$

Example 9. Find $\dfrac{4}{2x-1} - \dfrac{2}{2x^2-x}$.

Solution

$$\dfrac{4}{2x-1} - \dfrac{2}{2x^2-x}$$

$$= \dfrac{4}{2x-1} - \dfrac{2}{x(2x-1)} \qquad \text{LCD: } x(2x-1)$$

$$= \dfrac{4 \cdot x}{x(2x-1)} - \dfrac{2}{x(2x-1)}$$

$$= \dfrac{4x-2}{x(2x-1)}$$

$$= \dfrac{2(\cancel{2x-1})^{1}}{x(\cancel{2x-1})_{1}}$$

$$= \dfrac{2}{x}$$

Answers to Trial Runs

page 418 **1.** $\dfrac{2x}{y}$ **2.** $\dfrac{2x+1}{y}$ **3.** $\dfrac{3a}{5b}$ **4.** $\dfrac{7}{x+7}$ **5.** $\dfrac{x+2y}{x-2y}$ **6.** $x-15$ **7.** $\dfrac{2a-1}{a}$

8. $\dfrac{1}{x+2}$

page 421 **1.** $\dfrac{17x}{20}$ **2.** $\dfrac{14x+9}{36}$ **3.** $\dfrac{2(4x+3)}{3x^2}$ **4.** $\dfrac{-3(x-5)}{2x(x+3)}$ **5.** $\dfrac{4x-1}{(x-1)(x+2)}$ **6.** $\dfrac{x+1}{x(x+2)}$

EXERCISE SET 8.4

Perform the indicated operations and reduce if possible.

_____ 1. $\dfrac{5x}{3} + \dfrac{4x}{3}$

_____ 2. $\dfrac{9x}{7} + \dfrac{5x}{7}$

_____ 3. $\dfrac{2a}{7} + \dfrac{3a}{7} - \dfrac{6}{7}$

_____ 4. $\dfrac{5a}{9} + \dfrac{2a}{9} - \dfrac{4}{9}$

_____ 5. $\dfrac{7}{5x} - \dfrac{2}{5x}$

_____ 6. $\dfrac{15}{11x} - \dfrac{4}{11x}$

_____ 7. $\dfrac{5x}{2y} - \dfrac{3}{2y} - \dfrac{x}{2y} + \dfrac{1}{2y}$

_____ 8. $\dfrac{8x}{5y} - \dfrac{3}{5y} + \dfrac{2x}{5y} - \dfrac{2}{5y}$

_____ 9. $\dfrac{2a - 3}{3b} - \dfrac{2a + 3}{3b}$

_____ 10. $\dfrac{5a - 2}{7b} - \dfrac{a - 2}{7b}$

_____ 11. $\dfrac{4y}{x + 3y} + \dfrac{5y}{x + 3y}$

_____ 12. $\dfrac{6x}{x + 2y} - \dfrac{4x}{x + 2y}$

_____ 13. $\dfrac{3x}{x - 1} - \dfrac{3}{x - 1}$

_____ 14. $\dfrac{5x}{x - 2} - \dfrac{10}{x - 2}$

_____ 15. $\dfrac{3x}{2x + 5} - \dfrac{x}{2x + 5}$

_____ 16. $\dfrac{7x}{3x - 1} - \dfrac{4x}{3x - 1}$

_____ 17. $\dfrac{x^2}{x^2 - 4} + \dfrac{2x}{x^2 - 4}$

_____ 18. $\dfrac{x^2}{x^2 - 9} + \dfrac{3x}{x^2 - 9}$

_____ 19. $\dfrac{3x^2}{x + 5} - \dfrac{75}{x + 5}$

_____ 20. $\dfrac{2x^2}{x + 4} - \dfrac{32}{x + 4}$

_____ 21. $\dfrac{2a^2}{a^2 + 7a} + \dfrac{11a}{a^2 + 7a} - \dfrac{21}{a^2 + 7a}$

_____ 22. $\dfrac{2a^2}{a^2 + 3a} + \dfrac{a}{a^2 + 3a} - \dfrac{15}{a^2 + 3a}$

_____ 23. $\dfrac{7x - y}{3x^2 - 6xy} - \dfrac{4x - y}{3x^2 - 6xy}$

_____ 24. $\dfrac{6x - y}{4x^2 - 8xy} - \dfrac{2x - y}{4x^2 - 8xy}$

_____ 25. $\dfrac{x}{6} - \dfrac{y}{10}$

_____ 26. $\dfrac{x}{9} - \dfrac{y}{6}$

_____ 27. $\dfrac{3}{x} + \dfrac{5}{x^2} - \dfrac{2}{5x^2}$

_____ 28. $\dfrac{5}{x} + \dfrac{6}{x} - \dfrac{7}{3x^2}$

_____ 29. $7 - \dfrac{1}{y}$

_____ 30. $9 - \dfrac{1}{y}$

_____ 31. $\dfrac{4}{x} - 2x$

_____ 32. $\dfrac{6}{x} - 3x$

_____ 33. $\dfrac{5}{x} + \dfrac{2}{y}$

_____ 34. $\dfrac{7}{x} - \dfrac{11}{y}$

_____ 35. $\dfrac{2}{x} - \dfrac{3}{y} + \dfrac{2}{5x} - \dfrac{1}{3y}$

_____ 36. $\dfrac{3}{x} - \dfrac{2}{y} + \dfrac{1}{4x} - \dfrac{4}{3y}$

_____ 37. $\dfrac{4}{x^2 y} - \dfrac{5}{xy^2}$

_____ 38. $\dfrac{3}{x^2 y} - \dfrac{7}{xy^2}$

_____ 39. $\dfrac{x - 2y}{3x^2 y} - \dfrac{x + 2y}{4xy^2} + \dfrac{1}{6xy}$

_____ 40. $\dfrac{2x - y}{2x^2 y} - \dfrac{2x + y}{5xy^2} - \dfrac{8}{5xy}$

_____ 41. $\dfrac{7}{x + 2} - \dfrac{5}{(x + 2)^2}$

_____ 42. $\dfrac{11}{x + 3} - \dfrac{4}{(x + 3)^2}$

_____ 43. $\dfrac{3}{x + 4} - \dfrac{5}{x - 4}$

_____ 44. $\dfrac{4}{x + 5} - \dfrac{3}{x - 5}$

_____ 45. $\dfrac{x}{3} - \dfrac{2}{x + 1}$

_____ 46. $\dfrac{x}{4} - \dfrac{3}{x - 1}$

_____ 47. $\dfrac{4}{3x + 3} - \dfrac{5}{x + 1}$

_____ 48. $\dfrac{7}{2x + 2} - \dfrac{4}{x + 1}$

_____ 49. $\dfrac{7}{x - 2} - \dfrac{5}{x + 2}$

_____ 50. $\dfrac{4}{x - 3} - \dfrac{5}{x + 3}$

_____ 51. $\dfrac{x}{x^2 - 25} - \dfrac{1}{x^2 - 5x}$

_____ 52. $\dfrac{x}{x^2 - 49} - \dfrac{3}{x^2 - 7x}$

_____ 53. $\dfrac{1}{x^2 - 2x} - \dfrac{2}{x^3 - 2x^2}$

_____ 54. $\dfrac{1}{x^2 - 3x} - \dfrac{3}{x^3 - 3x^2}$

_____ 55. $\dfrac{4}{x^2 - 49} - \dfrac{1}{x^2 + 14x + 49}$

_____ 56. $\dfrac{5}{x^2 - 36} - \dfrac{1}{x^2 + 12x + 36}$

_____ 57. $\dfrac{x + 6}{x^2 + 2x - 8} + \dfrac{3}{x + 4}$

_____ 58. $\dfrac{x - 8}{x^2 - 11x + 18} - \dfrac{4}{x - 9}$

_____ 59. $\dfrac{2x - 6}{2x^2 - 7x + 3} - \dfrac{5}{2x^2 - x}$

_____ 60. $\dfrac{3x - 21}{3x^2 - 22x + 7} - \dfrac{4}{3x^2 - x}$

_____ 61. $\dfrac{3}{x^2 - 81} - \dfrac{1}{x^2 - 18x + 81}$

_____ 62. $\dfrac{4}{x^2 - 64} - \dfrac{2}{x^2 - 16x + 64}$

_____ 63. $\dfrac{x}{2x^2 + x - 6} - \dfrac{1}{2x^2 - x - 3}$

_____ 64. $\dfrac{x}{2x^2 - 11x + 15} + \dfrac{2}{2x^2 - 9x + 10}$

☆ Stretching the Topics _____

Perform the indicated operations and reduce if possible.

_____ **1.** $\dfrac{5}{n^2 - 8n + 15} - \dfrac{3}{n - 5} + \dfrac{2}{3 - n}$

_____ **2.** $\dfrac{x}{x^2 + 4x - 21} + \dfrac{4}{x^2 + 7x} + \dfrac{2x}{x^2 - 9}$

_____ **3.** $\dfrac{3a + 3b}{xa - 3ya + xb - 3yb} + \dfrac{5}{3y - x}$

Check your answers in the back of your book.

If you can complete **Checkup 8.4,** you are ready to go on to Section 8.5.

✓ CHECKUP 8.4

Perform the indicated operations and reduce if possible.

_____ 1. $\dfrac{7x}{4} - \dfrac{9x}{4}$

_____ 2. $\dfrac{6a}{5b} - \dfrac{4}{5b} + \dfrac{4a}{5b} - \dfrac{1}{5b}$

_____ 3. $\dfrac{x^2}{x-1} - \dfrac{9x}{x-1} + \dfrac{8}{x-1}$

_____ 4. $\dfrac{7x-10}{2x^2-6x} - \dfrac{2x+5}{2x^2-6x}$

_____ 5. $\dfrac{3}{xy} + \dfrac{2}{y^2}$

_____ 6. $\dfrac{4}{x-3} + \dfrac{7}{x}$

_____ 7. $\dfrac{7}{2x+2} + \dfrac{4}{x+1}$

_____ 8. $\dfrac{4}{x^2-25} - \dfrac{1}{x^2+10x+25}$

_____ 9. $\dfrac{3}{3x^2-2x} - \dfrac{2}{3x^3-2x^2}$

_____ 10. $\dfrac{x+8}{x^2-2x-63} - \dfrac{1}{x+7}$

Check your answers in the back of your book.

If You Missed Problems:	You Should Review Examples:
1, 2	1, 2
3, 4	4, 5
5, 6	7, 8
7	9
8, 9	10
10	11

8.5 Working with Complex Fractions

A **complex fraction** contains fractions in its numerator and/or its denominator. Simplifying complex fractions requires that you be able to add, subtract, multiply, and divide fractions.

Example 1. Simplify the complex fraction $\dfrac{\dfrac{8a^3}{3b^2}}{\dfrac{16a^5}{15b^8}}$.

Solution

$$\dfrac{\dfrac{8a^3}{3b^2}}{\dfrac{16a^5}{15b^8}} = \frac{8a^3}{3b^2} \div \frac{16a^5}{15b^8}$$ Rewrite the complex fraction as a *division problem.*

$$= \frac{8a^3}{3b^2} \cdot \frac{15b^8}{16a^5}$$ Invert the divisor and change the operation to multiplication.

$$= \frac{\overset{1}{\cancel{8}}\overset{1}{\cancel{a^3}}}{\cancel{3b^2}} \cdot \frac{\overset{5}{\cancel{15}}\overset{b^6}{\cancel{b^8}}}{\underset{2\ a^2}{\cancel{16a^5}}}$$ Reduce before multiplying.

$$= \frac{5b^6}{2a^2}$$ Multiply.

Now you try Example 2.

Example 2. Simplify $\dfrac{\dfrac{x^2 - 4}{x^3}}{\dfrac{5x + 10}{x^2 + x}}$.

Solution

$$\dfrac{\dfrac{x^2 - 4}{x^3}}{\dfrac{5x + 10}{x^2 + x}} = \frac{x^2 - 4}{x^3} \div \frac{5x + 10}{x^2 + x}$$

$$= \frac{x^2 - 4}{x^3} \cdot \frac{x^2 + x}{5x + 10}$$

$$= \frac{(x + 2)(\underline{\qquad})}{x^3} \cdot \frac{x(x + 1)}{5(\underline{\qquad})}$$

$$= \frac{(\underline{\qquad})(\underline{\qquad})}{5x^2}$$

Check your work on page 433. ▶

Trial Run

Simplify each complex fraction.

_____ **1.** $\dfrac{\dfrac{5a^2}{2b^3}}{\dfrac{25a^3}{12b^7}}$

_____ **2.** $\dfrac{\dfrac{1}{x+6}}{\dfrac{1}{2x+12}}$

_____ **3.** $\dfrac{\dfrac{2x-1}{6x-3}}{5}$

_____ **4.** $\dfrac{\dfrac{2x+2}{x^2}}{\dfrac{x^2-1}{3x^3}}$

Answers are on page 433.

Now we must develop a procedure for simplifying a complex fraction in which the numerator and/or denominator is a sum or difference containing fractional expressions. Let's try to simplify such a complex fraction.

$$\frac{y-\dfrac{1}{y}}{y-1} = \left[\frac{y}{1}-\frac{1}{y}\right] \div [y-1]$$ Rewrite the complex fraction as a division problem.

$$= \left[\frac{y^2}{y}-\frac{1}{y}\right] \div \left[\frac{y-1}{1}\right]$$ Rewrite each bracketed quantity as a *single fraction*, using methods for adding fractions.

$$= \left[\frac{y^2-1}{y}\right] \div \left[\frac{y-1}{1}\right]$$

$$= \left[\frac{y^2-1}{y}\right] \cdot \left[\frac{1}{y-1}\right]$$ Invert the divisor and change the operation to multiplication.

$$= \frac{(y+1)\overset{1}{\cancel{(y-1)}}}{y} \cdot \frac{1}{\underset{1}{\cancel{y-1}}}$$ Factor and reduce.

$$= \frac{y+1}{y}$$ Multiply remaining factors.

The steps for simplifying complex fractions can be summarized as follows.

Simplifying Complex Fractions (Method 1)

1. Rewrite the complex fraction as a division problem.
2. Express each part of the division problem as a single fraction.
3. Proceed with the method for dividing fractions.

Example 3. Simplify $\dfrac{3x - \dfrac{14x + 5}{x}}{\dfrac{x}{5} - \dfrac{5}{x}}$.

Solution

$$\frac{3x - \dfrac{14x + 5}{x}}{\dfrac{x}{5} - \dfrac{5}{x}} = \left[\frac{3x}{1} - \frac{14x + 5}{x}\right] \div \left[\frac{x}{5} - \frac{5}{x}\right]$$

$$= \left[\frac{3x \cdot x}{x} - \frac{14x + 5}{x}\right] \div \left[\frac{x \cdot x}{5x} - \frac{5 \cdot 5}{5x}\right]$$

$$= \left[\frac{3x^2}{x} - \frac{14x + 5}{x}\right] \div \left[\frac{x^2}{5x} - \frac{25\cdot}{5x}\right]$$

$$= \left[\frac{3x^2 - 14x - 5}{x}\right] \div \left[\frac{x^2 - 25}{5x}\right]$$

$$= \frac{3x^2 - 14x - 5}{x} \cdot \frac{5x}{x^2 - 25}$$

$$= \frac{(3x + 1)(\overset{1}{\cancel{x - 5}})}{\underset{1}{\cancel{x}}} \cdot \frac{\overset{1}{5\cancel{x}}}{(x + 5)(\underset{1}{\cancel{x - 5}})}$$

$$= \frac{5(3x + 1)}{x + 5}$$

There is a second approach that some prefer for simplifying complex fractions. In that approach, both the numerator and denominator are multiplied by the quantity that is the LCD for *both* the numerator and denominator. Let's rework Example 3 using this method.

$$\frac{3x - \dfrac{14x + 5}{x}}{\dfrac{x}{5} - \dfrac{5}{x}}$$

Choose LCD for both numerator and denominator. LCD: $5x$.

$$= \frac{5x\left[3x - \dfrac{14x + 5}{x}\right]}{5x\left[\dfrac{x}{5} - \dfrac{5}{x}\right]}$$

Multiply both numerator and denominator by LCD: $5x$.

$$= \frac{5x(3x) - \dfrac{5x}{1}\left(\dfrac{14x + 5}{x}\right)}{\dfrac{5x}{1}\left(\dfrac{x}{5}\right) - \dfrac{5x}{1}\left(\dfrac{5}{x}\right)}$$

Use the distributive property.

$$= \frac{15x^2 - 5(14x + 5)}{x^2 - 25}$$

Simplify each term.

$$= \frac{15x^2 - 70x - 25}{x^2 - 25}$$

Remove parentheses.

$$= \frac{5(3x^2 - 14x - 5)}{(x + 5)(x - 5)}$$

Factor numerator and denominator.

$$= \frac{5(3x + 1)\cancel{(x - 5)}^{1}}{(x + 5)\cancel{(x - 5)}_{1}}$$ Factor numerator completely.

$$= \frac{5(3x + 1)}{x + 5}$$ Reduce the fraction.

As you can see, the result is the same as that obtained in Example 3.

Simplifying Complex Fractions (Method 2)

1. Choose the LCD for *both* the numerator and denominator.
2. Multiply both numerator and denominator by that LCD.
3. Simplify the numerator and denominator, and reduce if possible.

Example 4. Use method 2 to simplify $\dfrac{x + \dfrac{2}{x}}{x - \dfrac{4}{x^3}}$.

Solution

$$\frac{x + \dfrac{2}{x}}{x - \dfrac{4}{x^3}}$$ Choose the LCD for both numerator and denominator. LCD: x^3.

$$= \frac{x^3\left(x + \dfrac{2}{x}\right)}{x^3\left(x - \dfrac{4}{x^3}\right)}$$ Multiply numerator and denominator by LCD: x^3.

$$= \frac{x^3(x) + \dfrac{x^3}{1}\left(\dfrac{2}{x}\right)}{x^3(x) - \dfrac{x^3}{1}\left(\dfrac{4}{x^3}\right)}$$ Use the distributive property.

$$= \frac{x^4 + 2x^2}{x^4 - 4}$$ Simplify each term.

$$= \frac{x^2\cancel{(x^2 + 2)}^{1}}{\cancel{(x^2 + 2)}_{1}(x^2 - 2)}$$ Factor numerator and denominator.

$$= \frac{x^2}{x^2 - 2}$$ Reduce.

Since either of the two methods for simplifying complex fractions is acceptable, you should decide which method you prefer and learn to use it successfully.

⫸ Trial Run

Simplify each complex fraction.

_____ 1. $\dfrac{x - \dfrac{9}{x}}{x + 3}$

_____ 2. $\dfrac{1 + \dfrac{1}{x}}{1 - \dfrac{1}{x^2}}$

_____ 3. $\dfrac{2x + \dfrac{5x - 3}{x}}{\dfrac{x}{3} - \dfrac{3}{x}}$

_____ 4. $\dfrac{\dfrac{1}{x} + \dfrac{1}{y}}{\dfrac{1}{x} - \dfrac{1}{y}}$

Answers are below.

▶ Examples You Completed

Example 2. Simplify $\dfrac{\dfrac{x^2 - 4}{x^3}}{\dfrac{5x + 10}{x^2 + x}}$.

Solution

$$\dfrac{\dfrac{x^2 - 4}{x^3}}{\dfrac{5x + 10}{x^2 + x}} = \frac{x^2 - 4}{x^3} \div \frac{5x + 10}{x^2 + x}$$

$$= \frac{x^2 - 4}{x^3} \cdot \frac{x^2 + x}{5x + 10}$$

$$= \frac{\overset{1}{\cancel{(x + 2)}}(x - 2)}{\underset{x^2}{\cancel{x^3}}} \cdot \frac{\overset{1}{\cancel{x}}(x + 1)}{5\underset{1}{\cancel{(x + 2)}}}$$

$$= \frac{(x - 2)(x + 1)}{5x^2}$$

Answers to Trial Runs

page 430   1. $\dfrac{6b^4}{5a}$ 2. 2 3. $\dfrac{5}{3}$ 4. $\dfrac{6x}{x - 1}$

page 433   1. $\dfrac{x - 3}{x}$ 2. $\dfrac{x}{x - 1}$ 3. $\dfrac{3(2x - 1)}{x - 3}$ 4. $\dfrac{y + x}{y - x}$

Name _____   Date _____

EXERCISE SET 8.5

Simplify the complex fractions. Write answers in reduced form.

_____ 1. $\dfrac{\frac{5}{12}}{\frac{35}{48}}$

_____ 2. $\dfrac{\frac{8}{15}}{\frac{21}{56}}$

_____ 3. $\dfrac{\frac{x^9}{y^3}}{\frac{4x^4}{y^2}}$

_____ 4. $\dfrac{\frac{x^{12}}{2y^5}}{\frac{3x^7}{10y^3}}$

_____ 5. $\dfrac{\frac{x-3y}{4x^2}}{\frac{x^2-9y^2}{12x^3}}$

_____ 6. $\dfrac{\frac{x+5y}{3y^5}}{\frac{x^2-25y^2}{9y^6}}$

_____ 7. $\dfrac{\frac{1}{x^2-16}}{\frac{4}{5x+20}}$

_____ 8. $\dfrac{\frac{1}{x^2-36}}{\frac{9}{2x-12}}$

_____ 9. $\dfrac{\frac{x^2+7x}{x^2-49}}{x}$

_____ 10. $\dfrac{\frac{x^2+9x}{x^2-81}}{x}$

_____ 11. $\dfrac{\frac{9x^2-9x}{3}}{x^4-x^3}$

_____ 12. $\dfrac{\frac{4x^2+4x}{2}}{x^3+x^2}$

_____ 13. $\dfrac{\frac{25x^2-4}{9x^3}}{\frac{5x+2}{18x}}$

_____ 14. $\dfrac{\frac{9x^2-16}{12x^4}}{\frac{3x-4}{9x}}$

_____ 15. $\dfrac{3+\frac{3}{4}}{1-\frac{3}{8}}$

_____ 16. $\dfrac{5-\frac{1}{3}}{2-\frac{5}{6}}$

_____ 17. $\dfrac{x-\frac{49}{x}}{x+7}$

_____ 18. $\dfrac{x-\frac{25}{x}}{x-5}$

_____ 19. $\dfrac{\frac{1}{x}}{1-\frac{1}{x}}$

_____ 20. $\dfrac{\frac{2}{x^2}}{1+\frac{2}{x^2}}$

434

_____ 21. $\dfrac{5 - \dfrac{1}{x}}{25 - \dfrac{1}{x^2}}$

_____ 22. $\dfrac{9 - \dfrac{1}{x}}{81 - \dfrac{1}{x^2}}$

_____ 23. $\dfrac{\dfrac{1}{2} - \dfrac{1}{5}}{\dfrac{2}{3} - \dfrac{4}{15}}$

_____ 24. $\dfrac{\dfrac{1}{3} - \dfrac{1}{4}}{\dfrac{1}{2} + \dfrac{1}{8}}$

_____ 25. $\dfrac{\dfrac{1}{x^2} - \dfrac{1}{9}}{\dfrac{1}{x} - \dfrac{1}{3}}$

_____ 26. $\dfrac{\dfrac{1}{x^2} - \dfrac{1}{16}}{\dfrac{1}{x} - \dfrac{1}{4}}$

_____ 27. $\dfrac{6 - \dfrac{5x + 25}{x^2}}{\dfrac{3}{5} + \dfrac{1}{x}}$

_____ 28. $\dfrac{10 - \dfrac{x + 3}{x^2}}{\dfrac{x}{3} - \dfrac{1}{5}}$

_____ 29. $\dfrac{1 - \dfrac{4}{x} - \dfrac{45}{x^2}}{\dfrac{x}{9} - \dfrac{9}{x}}$

_____ 30. $\dfrac{1 + \dfrac{1}{x} - \dfrac{72}{x^2}}{\dfrac{x}{8} - \dfrac{8}{x}}$

☆ Stretching the Topics _____

Simplify the complex fractions. Write answers in reduced form.

_____ 1. $\dfrac{\dfrac{4x}{y} - 5 + \dfrac{x}{y}}{\dfrac{x}{y} - \dfrac{8y}{x} - 2}$

_____ 2. $\dfrac{\dfrac{3y - 1}{y - 3} + 5}{\dfrac{y + 3}{y - 2}}$

_____ 3. $\dfrac{\dfrac{2}{x^2 - 4x - 21} - \dfrac{1}{x - 7}}{\dfrac{-1}{x - 7} + \dfrac{3}{x + 3}}$

Check your answers in the back of your book.

If you can complete **Checkup 8.5**, you are ready to go on to Section 8.6.

✓ **CHECKUP 8.5**

Simplify the complex fractions. Write answers in reduced form.

_____ 1. $\dfrac{\dfrac{x^9}{y^3}}{\dfrac{x^{12}}{3y^2}}$

_____ 2. $\dfrac{\dfrac{15x^8}{4}}{\dfrac{5x^5}{6y^4}}$

_____ 3. $\dfrac{\dfrac{3x-5}{5x}}{\dfrac{9x^2-25}{20x^3}}$

_____ 4. $\dfrac{\dfrac{x^2-9}{x}}{\dfrac{3x-9}{x^2+x}}$

_____ 5. $\dfrac{\dfrac{x^2+8x}{x^2-64}}{3}$

_____ 6. $\dfrac{x-\dfrac{16}{x}}{x-4}$

_____ 7. $\dfrac{25-\dfrac{16}{a^2}}{5-\dfrac{4}{a}}$

_____ 8. $\dfrac{\dfrac{1}{x^2}-\dfrac{1}{81}}{\dfrac{1}{x}-\dfrac{1}{9}}$

_____ 9. $\dfrac{5-\dfrac{9x-4}{x^2}}{\dfrac{5}{4}-\dfrac{1}{x}}$

_____ 10. $\dfrac{1-\dfrac{2}{x}-\dfrac{24}{x^2}}{\dfrac{x}{4}-\dfrac{4}{x}}$

Check your answers in the back of your book.

If You Missed Problems:	You Should Review Examples:
1, 2	1
3–5	2
6–8	4
9	3
10	3, 4

8.6 Solving Fractional Equations

Fractional equations are equations that contain rational expressions. From our earlier work with equations we recall that we can

1. Add (or subtract) the same quantity to (or from) both sides of an equation.
2. Multiply (or divide) both sides of an equation by the same nonzero quantity.

To solve a fractional equation, our first goal is to *get rid of all the denominators* in the equation. We can accomplish this by multiplying both sides of the equation by the LCD for *all* the fractions in the equation.

$$\text{Solve} \qquad \frac{x}{3} + \frac{5}{6} = \frac{9x}{2} \qquad\qquad \text{LCD: } 6 \left(\text{or } \frac{6}{1}\right).$$

$$\frac{6}{1}\left(\frac{x}{3} + \frac{5}{6}\right) = \frac{6}{1} \cdot \frac{9x}{2} \qquad \text{Multiply both sides by LCD: } 6.$$

$$\frac{6}{1}\left(\frac{x}{3}\right) + \frac{6}{1}\left(\frac{5}{6}\right) = \frac{6}{1} \cdot \frac{9x}{2} \qquad \begin{array}{l}\text{Use the distributive property on} \\ \text{the left side.}\end{array}$$

$$\frac{\overset{2}{\cancel{6}}}{1} \cdot \frac{x}{\underset{1}{\cancel{3}}} + \frac{\overset{1}{\cancel{6}}}{1} \cdot \frac{5}{\underset{1}{\cancel{6}}} = \frac{\overset{3}{\cancel{6}}}{1} \cdot \frac{9x}{\underset{1}{\cancel{2}}} \qquad \begin{array}{l}\text{Multiply fractions in each term} \\ \text{by the usual method.}\end{array}$$

$$2x + 5 = 27x \qquad \text{Simplify. Now this is a first-degree equation.}$$

$$5 = 25x \qquad \text{Subtract } 2x \text{ from both sides.}$$

$$\frac{5}{25} = \frac{25x}{25} \qquad \text{Divide both sides by 25.}$$

$$\frac{1}{5} = x \qquad \text{Simplify.}$$

Example 1. Solve $\dfrac{3a}{4} = 10 + \dfrac{a}{3}$ and check.

Solution. The LCD is 12.

$$\frac{\overset{3}{\cancel{12}}}{1}\left(\frac{3a}{\cancel{4}}\right) = 12(10) + \frac{\overset{4}{\cancel{12}}}{1}\left(\frac{a}{\cancel{3}}\right)$$

$$9a = 120 + 4a$$

$$5a = 120$$

$$\frac{5a}{5} = \frac{120}{5}$$

$$a = 24$$

CHECK:

$$\frac{3a}{4} = 10 + \frac{a}{3}$$

$$\frac{3(24)}{4} \overset{?}{=} 10 + \frac{24}{3}$$

$$3(6) \overset{?}{=} 10 + 8$$

$$18 = 18$$

Example 2. Solve $\dfrac{7}{x^2} - \dfrac{3}{x} = \dfrac{5}{2x}$.

Solution. The LCD is $2x^2$, but remember that we agreed to multiply both sides of equations by *nonzero* quantities only. We must find the *restrictions on the variable* so that the LCD is never zero. Here we must find the value of x for which

$$2x^2 = 0$$
$$x^2 = 0$$
$$x = 0$$

Therefore, we agree that $x \neq 0$ and multiply both sides of the equations by $2x^2$.

$$\frac{\overset{2}{\cancel{2x^2}}}{1}\left(\frac{7}{\cancel{x^2}}\right) - \frac{\overset{2x}{\cancel{2x^2}}}{1}\left(\frac{3}{\cancel{x}}\right) = \frac{\overset{x}{\cancel{2x^2}}}{1}\left(\frac{5}{\cancel{2x}}\right)$$

$$14 - 6x = 5x$$
$$14 = 11x$$
$$\frac{14}{11} = x$$

From now on we shall find the *restrictions on the variable* as soon as we have chosen the LCD for a fractional equation.

▶ Trial Run

Solve.

_____ 1. $\dfrac{x}{5} + \dfrac{7}{10} = \dfrac{3x}{2}$

_____ 2. $\dfrac{2a}{7} = 5 + \dfrac{a}{4}$

_____ 3. $\dfrac{4}{x^2} - \dfrac{2}{x} = \dfrac{3}{2x}$

_____ 4. $\dfrac{y}{5} = 4 - y$

_____ 5. $\dfrac{3}{x} - \dfrac{5}{x} = \dfrac{7}{2}$

_____ 6. $\dfrac{3x-2}{x^2} - \dfrac{4}{3x} = 0$

Answers are on page 441.

We summarize the steps we have developed for solving fractional equations.

Solving Fractional Equations

1. Choose the LCD for the entire equation.
2. Find the restrictions on the variable to avoid any value that makes the LCD zero.
3. Multiply each term on both sides of the equation by the LCD.
4. Remove parentheses and combine like terms.
5. Solve the equation and check your solutions against the restrictions on the variable.

Try to complete Example 3.

Example 3. Solve $\dfrac{5}{2x} = \dfrac{2}{x + 7}$.

Solution. The LCD is $2x(x + 7)$.
Restrictions are $x \neq 0$, $x \neq$ _____ .

$$\dfrac{2x(x + 7)}{1}\left(\dfrac{5}{2x}\right) = \dfrac{2x(x + 7)}{1}\left(\dfrac{2}{x + 7}\right)$$

$$5(x + 7) = 2x \cdot 2$$

Check your work on page 441. ▶

Example 4. Solve $\dfrac{x}{x - 3} + \dfrac{x - 6}{x - 3} = 1$.

Solution. The LCD is $x - 3$.
The restriction is $x \neq 3$.

$$\dfrac{x - 3}{1}\left(\dfrac{x}{x - 3}\right) + \dfrac{x - 3}{1}\left(\dfrac{x - 6}{x - 3}\right)$$

$$= (x - 3)(1)$$
$$x + x - 6 = x - 3$$
$$2x - 6 = x - 3$$
$$x - 6 = -3$$
$$x = 3$$

Our restriction was $x \neq 3$. Therefore, this equation has *no solution*.

In Example 4 the solution $x = 3$ was rejected because $x \neq 3$ in the original equation. A solution that satisfies a new version of an equation but does not satisfy the original equation is called an **extraneous solution**. If we were to substitute 3 in the original equation, we would find ourselves dividing by zero, and we know that divison by zero is not possible.

We must always *check solutions to fractional equations against the restrictions* in order to avoid extraneous solutions.

Example 5. Solve $\dfrac{5x}{x + 2} = \dfrac{4x + 3}{x^2 - 4}$.

Solution. First we factor the denominators so that we can identify the LCD.

$$\dfrac{5x}{x + 2} = \dfrac{4x + 3}{(x + 2)(x - 2)}$$

LCD: $(x + 2)(x - 2)$
Restrictions: $x \neq -2$, $x \neq 2$.

$$\dfrac{(x + 2)(x - 2)}{1}\left[\dfrac{5x}{x + 2}\right] = \dfrac{(x + 2)(x - 2)}{1}\left[\dfrac{4x + 3}{(x + 2)(x - 2)}\right]$$

Multiply each fraction by the LCD.

$$5x(x - 2) = 4x + 3$$

Reduce the fractions.

$$5x^2 - 10x = 4x + 3$$ Remove parentheses. This is a second-degree equation.

$$5x^2 - 14x - 3 = 0$$ Set the quadratic expression equal to zero.

$$(5x + 1)(x - 3) = 0$$ Factor the trinomial.

$$5x + 1 = 0 \quad \text{or} \quad x - 3 = 0$$ Use the zero product rule.

$$5x = -1$$

$$x = -\frac{1}{5} \quad \text{or} \quad x = 3$$ Solve for x. Neither solution is a restricted x-value.

Notice that a fractional equation may become a quadratic (second-degree) equation after the denominators are eliminated. Both solutions must be checked against the restrictions before they are accepted.

Example 6. Solve $\dfrac{4}{x^2 + x - 2} - \dfrac{2}{x + 2} = \dfrac{1}{3x - 3}$.

Solution. First we factor the denominators.

$$\frac{4}{(x + 2)(x - 1)} - \frac{2}{x + 2} = \frac{1}{3(x - 1)}$$ LCD: $3(x + 2)(x - 1)$
Restrictions: $x \neq -2,\ x \neq 1$.

$$\frac{3(x+2)(x-1)}{1}\left[\frac{4}{(x+2)(x-1)}\right] - \frac{3(x+2)(x-1)}{1}\left(\frac{2}{x+2}\right) = \frac{3(x+2)(x-1)}{1}\left[\frac{1}{3(x-1)}\right]$$

$$12 - 6(x - 1) = x + 2$$

$$12 - 6x + 6 = x + 2$$

$$18 - 6x = x + 2$$

$$18 = 7x + 2$$

$$16 = 7x$$

$$\frac{16}{7} = x$$

⏹⏹⏹➡ Trial Run

Solve each equation.

_____ **1.** $\dfrac{x}{x - 6} + \dfrac{3}{2} = \dfrac{4}{x - 6}$ _____ **2.** $\dfrac{4}{x - 5} - \dfrac{5}{x + 1} = \dfrac{3}{2x + 2}$

_____ **3.** $\dfrac{10}{3x} = \dfrac{2}{x + 9}$ _____ **4.** $\dfrac{x}{x - 2} + \dfrac{7}{3} = \dfrac{2}{x - 2}$

_____ 5. $\dfrac{x}{x-4} - \dfrac{8}{x-4} = 2$ _____ 6. $\dfrac{5}{x+3} = \dfrac{7}{x^2-9}$

Answers are below.

▶ Example You Completed

Example 3. Solve $\dfrac{5}{2x} = \dfrac{2}{x+7}$.

Solution. The LCD is $2x(x+7)$. Restrictions are $x \neq 0$, $x \neq -7$.

$$\frac{\overset{1}{\cancel{2x}(x+7)}}{1}\left(\frac{5}{\cancel{2x}}\right) = \frac{2x\overset{1}{\cancel{(x+7)}}}{1}\left(\frac{2}{\cancel{x+7}}\right)$$

$$5(x+7) = 2x \cdot 2$$

$$5x + 35 = 4x$$

$$x + 35 = 0$$

$$x = -35$$

Answers to Trial Runs

page 438   **1.** $x = \dfrac{7}{13}$ **2.** $a = 140$ **3.** $x = \dfrac{8}{7}$ **4.** $y = \dfrac{10}{3}$ **5.** $x = -\dfrac{4}{7}$ **6.** $x = \dfrac{6}{5}$

page 440   **1.** $x = \dfrac{26}{5}$ **2.** $x = \dfrac{73}{5}$ **3.** $x = -\dfrac{45}{2}$ **4.** No solution   **5.** $x = 0$ **6.** $x = \dfrac{22}{5}$

EXERCISE SET 8.6

Solve each fractional equation.

_____ 1. $\dfrac{x}{5} = \dfrac{7}{10}$

_____ 2. $\dfrac{x}{8} = \dfrac{9}{20}$

_____ 3. $\dfrac{x}{4} = \dfrac{x - 2}{2}$

_____ 4. $\dfrac{y}{10} = \dfrac{y - 10}{5}$

_____ 5. $\dfrac{2x}{7} = \dfrac{x - 2}{6}$

_____ 6. $\dfrac{3x}{5} = \dfrac{x - 1}{4}$

_____ 7. $\dfrac{y + 5}{5} = \dfrac{y - 5}{10}$

_____ 8. $\dfrac{y + 6}{6} = \dfrac{y - 6}{18}$

_____ 9. $\dfrac{x}{9} + \dfrac{x}{2} = 5$

_____ 10. $\dfrac{x}{3} - \dfrac{x}{9} = 2$

_____ 11. $\dfrac{x + 1}{3} - \dfrac{x}{9} = 4$

_____ 12. $\dfrac{x + 3}{5} - \dfrac{x}{25} = 1$

_____ 13. $\dfrac{3}{y} - \dfrac{7}{y} = 4$

_____ 14. $\dfrac{5}{y} - \dfrac{11}{y} = 6$

_____ 15. $\dfrac{7}{x} - \dfrac{3}{2} = \dfrac{5}{2x}$

_____ 16. $\dfrac{11}{x} - \dfrac{4}{3} = \dfrac{1}{3x}$

_____ 17. $\dfrac{3}{x - 4} = \dfrac{2}{x - 1}$

_____ 18. $\dfrac{-1}{x + 2} = \dfrac{2}{x + 5}$

_____ 19. $\dfrac{z - 1}{z + 1} = 2$

_____ 20. $\dfrac{z + 4}{z - 4} = 3$

_____ 21. $\dfrac{2x - 1}{x^2} - \dfrac{5}{3x} = 0$

_____ 22. $\dfrac{3x + 1}{x^2} - \dfrac{7}{5x} = 0$

_____ 23. $\dfrac{x}{x - 7} + \dfrac{5}{4} = \dfrac{1}{x - 7}$

_____ 24. $\dfrac{x}{x - 9} + \dfrac{1}{2} = \dfrac{5}{x - 9}$

_____ 25. $\dfrac{8}{x - 3} - \dfrac{2}{x + 4} = \dfrac{5}{2x - 6}$

_____ 26. $\dfrac{7}{x - 5} - \dfrac{3}{x + 1} = \dfrac{2}{3x + 3}$

_____ 27. $\dfrac{11}{4x} = \dfrac{5}{x + 4}$

_____ 28. $\dfrac{3}{7x} = \dfrac{2}{x + 7}$

_____ 29. $\dfrac{a}{a - 4} + \dfrac{a - 8}{a - 4} = 3$

_____ 30. $\dfrac{2a}{a - 3} + \dfrac{a - 9}{a - 3} = 1$

_____ 31. $\dfrac{4}{x - 7} = \dfrac{5}{x^2 - 49}$

_____ 32. $\dfrac{3}{x - 5} = \dfrac{2}{x^2 - 25}$

_____ 33. $\dfrac{4}{2x-1}+\dfrac{1}{2x+1}=\dfrac{2}{4x^2-1}$

_____ 34. $\dfrac{5}{3x+1}-\dfrac{2}{3x-1}=\dfrac{1}{9x^2-1}$

_____ 35. $\dfrac{7}{a-2}=\dfrac{3}{a^2-7a+10}$

_____ 36. $\dfrac{3}{a-9}=\dfrac{2}{a^2-6a-27}$

_____ 37. $\dfrac{3b}{4b^2-1}=\dfrac{-1}{2b-1}$

_____ 38. $\dfrac{4b}{9b^2-4}=\dfrac{-2}{3b+2}$

_____ 39. $\dfrac{3}{x^2-x-12}-\dfrac{2}{x-4}=\dfrac{1}{2x+6}$

_____ 40. $\dfrac{1}{x^2-6x-7}-\dfrac{3}{x-7}=\dfrac{2}{3x+3}$

_____ 41. $\dfrac{5}{x-1}=\dfrac{3x-2}{x^2+15x-16}$

_____ 42. $\dfrac{3}{x-2}=\dfrac{7x-4}{x^2-8x+12}$

_____ 43. $\dfrac{5x+3}{x-7}=2$

_____ 44. $\dfrac{6x+5}{x-9}=3$

_____ 45. $\dfrac{a-2}{a+5}=\dfrac{a-1}{a+1}$

_____ 46. $\dfrac{a-7}{a-2}=\dfrac{a+3}{a+2}$

_____ 47. $\dfrac{x}{x-5}+\dfrac{2}{5}=\dfrac{5}{x-5}$

_____ 48. $\dfrac{x}{x-4}+\dfrac{3}{2}=\dfrac{4}{x-4}$

_____ 49. $\dfrac{2x}{2x-3}-\dfrac{6}{2x-3}=2$

_____ 50. $\dfrac{3x}{3x-1}-\dfrac{4}{3x-1}=4$

_____ 51. $x+\dfrac{40}{x}=13$

_____ 52. $x-\dfrac{28}{x}=3$

_____ 53. $2-\dfrac{15}{x^2}=\dfrac{7}{x}$

_____ 54. $3+\dfrac{5}{x}=\dfrac{2}{x^2}$

_____ 55. $\dfrac{y}{y+12}=\dfrac{1}{y}$

_____ 56. $\dfrac{y}{y+6}=\dfrac{1}{y}$

_____ 57. $\dfrac{3x}{2x-1}=\dfrac{2}{x-1}$

_____ 58. $\dfrac{2x}{x+6}=\dfrac{1}{x-5}$

_____ 59. $\dfrac{x}{x+1}=\dfrac{x+3}{x^2-1}$

_____ 60. $\dfrac{2x}{x-3}=\dfrac{x-2}{x^2-9}$

☆ Stretching the Topics ─────────────────────

Solve each fractional equation.

_____ **1.** $\dfrac{6}{2x^2 + 5x - 12} - \dfrac{2}{4x^2 - 9} = \dfrac{4}{2x^2 + 11x + 12}$

_____ **2.** $\dfrac{x}{a} - \dfrac{2x}{b} + \dfrac{3x}{c} = k$ for x

_____ **3.** $I = \dfrac{E}{\dfrac{Rn + r}{n}}$ for n

Check your answers in the back of your book.

If you can complete **Checkup 8.6,** you are ready to go on to Section 8.7.

✓ **CHECKUP 8.6**

Solve each fractional equation.

_____ 1. $\dfrac{14}{y} = \dfrac{7}{9}$

_____ 2. $\dfrac{x}{2} = \dfrac{x-3}{5}$

_____ 3. $\dfrac{x-4}{5} = x + 8$

_____ 4. $\dfrac{y+5}{2} = \dfrac{y-4}{3}$

_____ 5. $\dfrac{6}{a} - \dfrac{4}{9} = \dfrac{8}{9a}$

_____ 6. $\dfrac{5}{11x} = \dfrac{2}{x+11}$

_____ 7. $\dfrac{6x}{x-3} - \dfrac{x+15}{x-3} = 6$

_____ 8. $\dfrac{2x-1}{x+2} - \dfrac{x-3}{x-2} = 1$

_____ 9. $\dfrac{x-1}{2x+4} - \dfrac{x+3}{2x-4} = \dfrac{2}{x^2-4}$

_____ 10. $\dfrac{2x-1}{x+4} - \dfrac{x}{x+3} = \dfrac{-1}{x^2+7x+12}$

Check your answers in the back of your book.

If You Missed Problems:	You Should Review Examples:
1–5	1, 2
6, 7	3, 4
8–10	5, 6

8.7 Switching from Word Statements to Fractional Equations

Suppose that we again practice switching from word statements to equations. Remember that in each problem we must identify the variable and translate the word phrases into variable expressions before coming up with an equation to solve.

Example 1. Maria and Tony earn the same monthly salary. Maria saves $\frac{1}{5}$ of her salary and Tony saves $\frac{1}{6}$ of his salary. Together they save \$550 each month. What is the monthly salary?

Solution

We Think	*We Write*
Maria and Tony earn the same salary per month.	Let x = monthly salary (dollars)
Maria saves $\dfrac{1}{5}$ of her salary.	$\dfrac{1}{5}x$ = Maria's savings (dollars)
Tony saves $\dfrac{1}{6}$ of his salary.	$\dfrac{1}{6}x$ = Tony's savings (dollars)
Their total savings is the *sum* of the amounts saved.	$\dfrac{1}{5}x + \dfrac{1}{6}x$ = total savings (dollars)
We know they save \$550 each month.	$\dfrac{1}{5}x + \dfrac{1}{6}x = 550$

Now we have an equation to be solved by the usual methods.

$$\frac{1}{5}x + \frac{1}{6}x = 550$$

$$\frac{x}{5} + \frac{x}{6} = \frac{550}{1} \qquad \text{LCD: } 5 \cdot 6 = 30$$

$$\frac{30}{1}\left(\frac{x}{5}\right) + \frac{30}{1}\left(\frac{x}{6}\right) = 30(550)$$

$$6x + 5x = 16{,}500$$

$$11x = 16{,}500$$

$$\frac{11x}{11} = \frac{16{,}500}{11}$$

$$x = 1500$$

The monthly salary is \$1500.

Example 2. On long trips, Mary drives her car 5 miles per hour faster than her sister Jo. If Mary can drive 312 miles in the same amount of time that Jo drives 282 miles, how fast does each sister drive?

Solution. The formulas relating distance to rate and time are

distance = rate × time	rate = $\dfrac{distance}{time}$	time = $\dfrac{distance}{rate}$

We know each sister's distance, and we know that they travel those distances in the same amount of time, but we do not know their rates. The rates are unknown quantities.

For Jo	*For Mary*
r = rate	$r + 5$ = rate
282 = distance	312 = distance
$\dfrac{282}{r}$ = time	$\dfrac{312}{r+5}$ = time

Since the sisters' driving times are the *same*, we know

$$\text{Jo's time} = \text{Mary's time}$$
$$\frac{282}{r} = \frac{312}{r+5} \qquad \text{LCD: } r(r+5)$$
$$\frac{r(r+5)}{1}\left(\frac{282}{r}\right) = \frac{r(r+5)}{1}\left(\frac{312}{r+5}\right)$$
$$282(r+5) = 312r$$
$$282r + 1410 = 312r$$
$$1410 = 30r$$
$$\frac{1410}{30} = \frac{30r}{30}$$
$$47 = r$$

Jo's rate	*Mary's rate*
r = 47 mph	$r + 5$ = 52 mph

Jo drives at a rate of 47 mph and Mary drives at a rate of 52 mph.

Solving Ratio and Proportion Problems

Ratios are handy tools to use in solving everyday mathematical problems.

A **ratio** is a way of comparing two quantities by means of a fraction.

First Quantity	Second Quantity	Ratio	Interpretation
23 males	11 females	$\dfrac{\text{males}}{\text{females}} = \dfrac{23}{11}$	For every 23 males, there are 11 females.
$250 saved	$750 spent	$\dfrac{\text{spending}}{\text{saving}} = \dfrac{750}{250} = \dfrac{3}{1}$	For every $3 spent, $1 was saved.
7 games won	5 games lost	$\dfrac{\text{wins}}{\text{losses}} = \dfrac{7}{5}$	For every 7 games won, there were 5 games lost.

You try to set up the ratio in Example 3.

Example 3. Martin earns $1500 per month and Clarence earns $1200 per month. Find the ratio of Martin's salary to Clarence's salary. Interpret your answer.

Solution

$$\frac{\text{Martin's salary}}{\text{Clarence's salary}} = \frac{1500}{}$$

$$= \frac{5}{}$$

For every ___ dollars that Martin earns, Clarence earns ___ dollars.

Check your work on page 453. ▶

Example 4. If Martin's salary is raised to $1800, what must be Clarence's new salary so that the ratio will continue to be $\dfrac{5}{4}$?

Solution. Let C be Clarence's new salary. Then

$$\frac{1800}{C} = \frac{5}{4} \qquad \text{LCD: } 4C$$

$$\frac{4C}{1}\left(\frac{1800}{C}\right) = \frac{4C}{1}\left(\frac{5}{4}\right)$$

$$4(1800) = 5C$$

$$7200 = 5C$$

$$\frac{7200}{5} = \frac{5C}{5}$$

$$1440 = C$$

Clarence's new salary must be $1440.

In Example 4, we wrote an equation that said that two ratios were equal. Such an equation is called a **proportion**.

A **proportion** is an equation that states that two ratios are equal.

It is easier to solve fractional equations that are proportions by using the definition of equal fractions.

Equality of Fractions

If $\dfrac{A}{B} = \dfrac{C}{D}$ then $AD = BC$

(provided that $B \neq 0$, $D \neq 0$).

A numerical example should convince you that this definition makes sense.

$$\frac{9}{12} = \frac{3}{4}$$

$$9(4) = 12(3)$$

$$36 = 36$$

We shall use the definition of equality of fractions (sometimes called the method of **cross products**) to solve all our proportion problems. You must realize, however, that we can use this method only when solving a fractional equation that states that *one* fraction is equal to *one* other fraction. We cannot use this method if either side of a fractional equation contains more than one fraction. In those situations, we *must* multiply both sides by the LCD.

Example 5. In an orchard of 95 trees, the ratio of pear trees to apple trees is 2 to 3. How many pear trees and how many apple trees are in the orchard?

Solution

We Think	**We Write**
The number of each kind of tree is unknown.	Let x = number of pear trees
The number of apple trees can be found by *subtracting* the number of pear trees (x) from the total number of trees (95).	$95 - x$ = number of apple trees
We would like to express the ratio of pear trees to apple trees using our variable expressions.	$\dfrac{x}{95 - x}$ = ratio of pears to apples
We know this ratio equals $\dfrac{2}{3}$, so we write a proportion.	$\dfrac{x}{95 - x} = \dfrac{2}{3}$

Now we can solve this proportion by the usual methods.

$$\frac{x}{95 - x} = \frac{2}{3}$$

$$3x = 2(95 - x) \qquad \text{Use equality of fractions.}$$

$$3x = 190 - 2x \qquad \text{Remove parentheses.}$$

$$5x = 190 \qquad \text{Isolate the variable term.}$$

$$\frac{5x}{5} = \frac{190}{5} \qquad \text{Isolate the variable.}$$

$$x = 38 \qquad \text{Simplify.}$$

There are 38 pear trees in the orchard. There are $95 - 38 = 57$ apple trees.

Now we are ready to solve the problem stated at the beginning of the chapter.

Example 6. At the beginning of the semester the number of males in Harry's math class was 3 more than twice the number of females. After 5 males and 5 females had dropped the class, the ratio of females to males was 1 to 4. How many males and females were in Harry's math class at the beginning of the semester?

Solution. Since the original number of males is described in terms of the number of females, we let

$$x = \text{original number of females}$$

	Beginning of Semester	End of Semester
Number of females	x	$x - 5$
Number of males	$2x + 3$	$(2x + 3) - 5$ $= 2x - 2$

Since we know the ratio of females to males is $\dfrac{1}{4}$ at the *end* of the semester,

$$\frac{x - 5}{2x - 2} = \frac{1}{4}$$

$4(x - 5) = 2x - 2$	Use equality of fractions.
$4x - 20 = 2x - 2$	Remove parentheses.
$2x = \underline{\hspace{1cm}}$	Isolate the variable term.
$x = \underline{\hspace{1cm}}$	Isolate the variable.

	Beginning of Semester	End of Semester
Number of females	$x = \underline{\hspace{1cm}}$	$x - 5 = \underline{\hspace{1cm}} - 5$ $= \underline{\hspace{1cm}}$
Number of males	$2x + 3 = 2(\underline{\hspace{1cm}}) + 3$ $= \underline{\hspace{1cm}}$	$2x - 2 = 2(\underline{\hspace{1cm}}) - 2$ $= \underline{\hspace{1cm}}$

There were _____ females and _____ males in Harry's math class at the beginning of the semester. Check your work on page 453. ▶

Solving Work Problems

If two or more people or machines work together to complete a job, we expect that less time will be needed than if one person or machine were to complete the same job alone. The amount of work done by each person or machine is found by multiplying rate of work by time worked. The *rate* of work for a person or machine is the portion of the job that is done per one unit of time.

> Amount of work = (rate of work) · (time worked)

Let's see how fractional equations can be used to solve **work problems**.

Example 7. It takes Linda 4 hours to rake the lawn but it takes her older sister Nancy only 2 hours to rake the same lawn. If they rake *together*, how long will it take them?

Solution

We Think	We Write
What portion of the lawn can Linda rake per hour? This is her rate.	$\frac{1}{4}$ lawn per hour is Linda's rate
What portion of the lawn can Nancy rake per hour? This is her rate.	$\frac{1}{2}$ lawn per hour is Nancy's rate
We know Nancy and Linda work together for the *same* number of hours.	Let t = hours worked
Work done by each sister is found by *multiplying* rate of work by time worked.	$\frac{1}{4} \cdot t$ = Linda's work $\frac{1}{2} \cdot t$ = Nancy's work
The total work done is the *sum* of the sisters' work. The total work must equal 1 raked lawn.	$\frac{1}{4}t + \frac{1}{2}t = 1$

Now we have a fractional equation to solve by the usual methods.

$$\frac{1}{4}t + \frac{1}{2}t = 1$$

$$\frac{t}{4} + \frac{t}{2} = 1 \qquad \text{LCD: } 4$$

$$\frac{4}{1} \cdot \frac{t}{4} + \frac{4}{1} \cdot \frac{t}{2} = 4 \cdot 1$$

$$t + 2t = 4$$

$$3t = 4$$

$$t = \frac{4}{3}$$

Together the sisters can rake the lawn in $\frac{4}{3}$ or $1\frac{1}{3}$ hours (or 1 hour 20 minutes). Notice that the amount of time needed is *less* than either sister's time alone. This will always be true.

Example 8. A tub can be filled with hot water in 15 minutes and with cold water in 10 minutes. How long will it take to fill the tub if both faucets are allowed to run at the given rate?

Solution. We let t = time for both faucets (in minutes).

	Rate	Time	Work
Hot water	$\dfrac{1}{15}$	t	$\dfrac{1}{15} \cdot t$
Cold water	$\dfrac{1}{10}$	t	$\dfrac{1}{10} \cdot t$

$$\frac{1}{15}t + \frac{1}{10}t = 1 \text{ tub}$$

$$\frac{t}{15} + \frac{t}{10} = 1 \qquad\qquad \text{LCD: } 30$$

$$\frac{30}{1} \cdot \frac{t}{15} + \frac{30}{1} \cdot \frac{t}{10} = 30(1)$$

$$2t + 3t = 30$$

$$5t = 30$$

$$t = 6$$

The tub will fill in 6 minutes.

▶ **Examples You Completed** _____

Example 3 (*Solution*)

$$\frac{\text{Martin's salary}}{\text{Clarence's salary}} = \frac{1500}{1200}$$

$$= \frac{5}{4}$$

For every 5 dollars that Martin earns, Clarence earns 4 dollars.

Example 6 (*Solution*)

$$\frac{x - 5}{2x - 2} = \frac{1}{4}$$

$$4(x - 5) = 2x - 2$$

$$4x - 20 = 2x - 2$$

$$2x = 18$$

$$x = 9$$

	Beginning of Semester	End of Semester
Number of females	$x = 9$	$x - 5 = 9 - 5$ $= 4$
Number of males	$2x + 3 = 2(9) + 3$ $= 21$	$2x - 2 = 2(9) - 2$ $= 16$

There were 9 females and 21 males in Harry's math class at the beginning of the semester.

EXERCISE SET 8.7

Use an equation to solve each problem.

_____ **1.** Eric and Kate each own interest in the Specialty Shop. Each month Eric receives $\frac{1}{10}$ of the profits and Kate receives $\frac{1}{8}$ of the profits. Together one month they received $630. How much profit did the Specialty shop make that month?

_____ **2.** Each month Marco saves $\frac{1}{10}$ of his paycheck in the credit union and $\frac{1}{12}$ of his paycheck in his savings account. If he saves a total of $154 each month, what is Marco's monthly paycheck?

_____ **3.** Louise's average driving speed is 6 miles per hour faster than her husband's average speed. If Louise can drive 280 miles in the same time that her husband drives 250 miles, how fast does each drive?

_____ **4.** Driving 3 miles per hour faster than his normal rate, Henry can travel 638 miles in the same time it takes him to travel 605 miles at his normal rate. What is Henry's normal rate of speed?

_____ **5.** José and Rick assembled 36 parts of a carburetor in an hour. The ratio of the number Rick did to the number José did is 4 to 5. Find how many parts each assembled during the hour.

_____ **6.** The ratio of seniors to graduate students in an upper division course is 7 to 3. If there are 16 more seniors than graduate students, how many graduate students are in the class?

_____ **7.** On a road map, 2 inches represents 30 miles. If the distance between Birmingham and Chattanooga is 11 inches on the map, how far apart are the two cities?

_____ **8.** The ratio of students to teachers at Banton High School is 3 teachers for every 70 students. If the school presently has 24 teachers, what is the enrollment of the student body?

_____ **9.** At State University the ratio of in-state students to out-of-state students is 5 to 2. If there are 3750 more in-state students than out-of-state students, how many out-of-state students are enrolled?

_____ **10.** In the freshman class at P.J.C., there are 50 more computer science majors than there are math majors. If the ratio of computer science majors to math majors is 7 to 2, find how many freshmen are computer science majors.

_____ **11.** The width of a rectangle is $\frac{2}{3}$ of its length. If the perimeter is 60 feet, find the dimensions. (*Remember*: $P = 2l + 2w$.)

_____ **12.** The width of a rectangular lot is $\frac{2}{5}$ of its length. If the perimeter of the rectangle is 140 feet, find the dimensions of the lot.

_____ **13.** A number is added to the numerator, and twice the same number is subtracted from the denominator of the fraction $\frac{4}{15}$. The resulting fraction is equal to $\frac{7}{9}$. Find the number.

_____ 14. If the denominator of $\dfrac{x}{12}$ is decreased by x, the new fraction is equal to $\dfrac{1}{2}$. Find the value of x.

_____ 15. It takes 5 hours to fill a watering tank for the elephants at the zoo using a pump. The tank can be filled in 10 hours using a hose. How long will it take to fill the tank using both the pump and the hose at the same time?

_____ 16. Sam can paint Sherry's house in 5 days. Sherry can paint it by herself in 7 days. How long will it take them to paint the house if they work together?

_____ 17. Louis can clean the dog pens at the Humane Shelter in 8 hours and Margaret can do it in 6 hours. How long will it take them to clean the pens if they work together?

_____ 18. Charlie can clean the theater after a play in 6 hours and Ricky can clean the theater in 4 hours. If they work together, how long will it take them to clean the theater?

_____ 19. If an $8000 investment earns $480 in 6 months, how much will $15,000 earn invested at the same rate for the same period of time?

_____ 20. The interest owed on a loan of $5000 borrowed for 6 months is $300. If $12,000 is borrowed for the same period of time at the same rate of interest, how much interest will be owed?

☆ Stretching the Topics _____

Use an equation to solve each problem.

_____ 1. In a fraction the denominator is 4 greater than the numerator. If 24 is added to the numerator, the fraction will be equal to the reciprocal of the original fraction. What is the original numerator of the fraction?

_____ 2. Two pipes are pouring water into an irrigation reservoir. The first pipe can fill it in 20 days but if both pipes are used the reservoir can be filled in 12 days. How many days would it take the second pipe to fill the reservoir if the first pipe were not open?

_____ 3. Carlita's scout troop hiked 10 miles to Horseshoe Cave for a picnic. They walked 3 miles per hour faster going than they did returning. The total time spent hiking was 7 hours. How fast did the troop hike on the return trip?

Check your answers in the back of your book.

If you can solve the problems in **Checkup 8.7**, you are ready to go on to the **Review Exercises** for Chapter 8.

Name _____   **Date** _____

✓ CHECKUP 8.7

Use an equation to solve each problem.

_____ 1. Sidney owns $\frac{3}{8}$ interest in a farm and his brother, Wallis, owns $\frac{1}{4}$ interest. When the farm was sold, their combined share was $150,000. What was the total selling price of the farm?

_____ 2. In a certain math course the ratio of success to failure is 20 to 3. If there are 115 students enrolled in the course, how many can be expected to succeed?

_____ 3. Faye's average jogging speed is 2 miles per hour faster than her daughter's average speed. If Faye can jog 9 miles in the same length of time that her daughter can jog 6 miles, how fast does each jog?

_____ 4. If the denominator of the fraction $\frac{x}{7}$ is decreased by x, the new fraction is equal to $\frac{2}{5}$. Find the value of x.

_____ 5. It takes Daphne 4 hours to mow the playground with her riding lawn mower, but it takes Sara 6 hours with her push mower. If they both begin mowing together, how long will it take them to finish?

Check your answers in the back of your book.

If You Missed Problem:	You Should Review Example:
1	1
2	5
3	2
4	6
5	7

456

Summary

In this chapter we learned to work with **rational algebraic expressions**, which are fractions with polynomial numerators and polynomial denominators.

In Order to	We Must	Examples
Restrict the variable in a rational expression	Exclude any value of the variable that makes the denominator zero.	(1) For $\dfrac{3x}{y^2}$, $\quad y \neq 0$ (2) For $\dfrac{3}{x^2 + 2x} = \dfrac{3}{x(x + 2)}$, $x \neq 0$, $x \neq -2$
Reduce a rational expression	Factor numerator and denominator. Divide numerator and denominator by any common factors.	(1) $\dfrac{x^9 y}{x^2 y^3} = \dfrac{x^7}{y^2}$ (2) $\dfrac{3x + 6}{x^2 - 4} = \dfrac{3(x + 2)}{(x + 2)(x - 2)}$ $= \dfrac{3}{x - 2}$
Multiply rational expression	Factor each numerator and denominator. Reduce if possible. Then multiply remaining numerators and denominators.	(1) $\dfrac{x^2 y}{3z} \cdot \dfrac{2z^2}{x} = \dfrac{2xyz}{3}$ (2) $\dfrac{x^2 + x}{10} \cdot \dfrac{5x - 5}{x^2 - 1}$ $= \dfrac{x(x + 1)}{10} \cdot \dfrac{5(x - 1)}{(x + 1)(x - 1)}$ $= \dfrac{x}{2}$
Divide rational expressions	Invert the divisor and multiply.	(1) $\dfrac{2x}{3} \div \dfrac{x^2}{6} = \dfrac{2x}{3} \cdot \dfrac{6}{x^2}$ $= \dfrac{4}{x}$ (2) $\dfrac{a - 4}{a^2} \div \dfrac{a^2 - 3a - 4}{7a}$ $= \dfrac{a - 4}{a^2} \cdot \dfrac{7a}{(a - 4)(a + 1)}$ $= \dfrac{7}{a(a + 1)}$

In Order to	We Must	Examples
Build a rational expression	Multiply numerator and denominator by the same factor.	$\dfrac{3}{x+3} = \dfrac{?}{x^2+5x+6}$ $= \dfrac{?}{(x+3)(x+2)}$ $= \dfrac{3(x+2)}{(x+3)(x+2)}$ $= \dfrac{3x+6}{x^2+5x+6}$
Add (or subtract) rational expressions	Choose the LCD and build each fraction to a new fraction having the LCD. Combine numerators over LCD. Simplify and reduce if possible.	$\dfrac{x}{x+1} - \dfrac{3}{x+2}$ LCD: $(x+1)(x+2)$ $= \dfrac{x(x+2)}{(x+1)(x+2)} - \dfrac{3(x+1)}{(x+1)(x+2)}$ $= \dfrac{x^2+2x-3x-3}{(x+1)(x+2)}$ $= \dfrac{x^2-x-3}{(x+1)(x+2)}$
Simplify a complex fraction (Method 1)	Rewrite the complex fraction as a division problem. Write the dividend and the divisor as single fractions. Invert the divisor and multiply.	$\dfrac{\frac{2}{x}+\frac{1}{3}}{\frac{2x+12}{x^2}}$ $= \left[\dfrac{2}{x}+\dfrac{1}{3}\right] \div \left[\dfrac{2x+12}{x^2}\right]$ $= \left[\dfrac{6}{3x}+\dfrac{x}{3x}\right] \div \left[\dfrac{2x+12}{x^2}\right]$ $= \dfrac{6+x}{3x} \cdot \dfrac{x^2}{2x+12}$ $= \dfrac{\overset{1}{\cancel{6+x}}}{\cancel{3x}} \cdot \dfrac{\overset{x}{\cancel{x^2}}}{2\cancel{(x+6)}} = \dfrac{x}{6}$
Simplify a complex fraction (Method 2)	Multiply numerator and denominator by LCD for *both* numerator and denominator.	$\dfrac{\frac{2}{x}+\frac{1}{3}}{\frac{2x+12}{x^2}}$ LCD: $3x^2$ $= \dfrac{3x^2\left[\frac{2}{x}+\frac{1}{3}\right]}{3x^2\left[\frac{2x+12}{x^2}\right]}$ $= \dfrac{6x+x^2}{3(2x+12)} = \dfrac{x\overset{1}{\cancel{(6+x)}}}{6\cancel{(x+6)}} = \dfrac{x}{6}$

In Order to	We Must	Examples
Solve a fractional equation	Multiply both sides of the equation by the LCD. Solve and check solutions against restrictions on the variable.	$\dfrac{3}{x} + \dfrac{5}{x+1} = \dfrac{1}{2x}$ LCD: $2x(x+1)$ $x \neq 0, x \neq -1$ $$\dfrac{2x(x+1)}{1}\left[\dfrac{3}{x}\right] + \dfrac{2x(x+1)}{1}\left[\dfrac{5}{x+1}\right]$$ $$= \dfrac{2x(x+1)}{1}\left[\dfrac{1}{2x}\right]$$ $$6(x+1) + 10x = x+1$$ $$6x + 6 + 10x = x+1$$ $$15x = -5$$ $$x = \dfrac{-5}{15}$$ $$x = -\dfrac{1}{3}$$
Solve a proportion	Use the definition of equality of fractions (cross products). Solve and check solutions against restrictions.	$\dfrac{x+3}{x} = \dfrac{6}{5}$ $x \neq 0$ $$5(x+3) = 6x$$ $$5x + 15 = 6x$$ $$15 = x$$

Having mastered the skills for working with rational algebraic expressions, we practiced switching from word statements to fractional equations.

☐ Speaking the Language of Algebra _____

Complete each statement with the appropriate word or phrase.

1. A fraction with a polynomial numerator and a polynomial denominator is called a _____ _____ _____ .

2. We must sometimes find restrictions on the variable in a rational expression because division by _____ is _____ .

3. To multiply rational algebraic expressions, we multiply _____ and multiply _____ .

4. In addition and subtraction of rational algebraic expressions, we must be sure the fractions have the same _____ .

5. After a complex fraction has been rewritten as a division problem, the dividend and divisor must each be expressed as a _____ _____ .

6. To solve a fractional equation, we multiply both sides of the equation by the _____ _____ _____ for all the fractions in the equation.

7. A solution for a new version of an equation that does *not* satisfy the original equation is called an _____ solution.

8. An equation that states that two ratios are equal is called a _____ .

Check your answers in the back of your book.

△ **Writing About Mathematics** _____

Write your response to each question in complete sentences.

1. Explain why it is necessary to state restrictions on a variable appearing in the denominator of a rational algebraic expression.

2. Explain the steps you would use to reduce a rational algebraic expression.

3. Explain the difference between the method you would use to add fractions and the method you would use to solve a fractional equation.

REVIEW EXERCISES for Chapter 8

Evaluate each rational expression and reduce if possible.

_____ 1. $\dfrac{a-3}{a+6}$ when $a = 12$

_____ 2. $\dfrac{x-5}{x^2-25}$ when $x = -10$

Find the restrictions on the variable in each expression.

_____ 3. $\dfrac{x^2}{x+4}$

_____ 4. $\dfrac{5x}{2x-1}$

Reduce the fractions.

_____ 5. $\dfrac{12x^7}{6x^2}$

_____ 6. $\dfrac{45x^4y^3}{36x^2y}$

_____ 7. $\dfrac{5a+15b}{5}$

_____ 8. $\dfrac{64x^5}{8x^3+8x}$

_____ 9. $\dfrac{-7a-7b}{35a+35b}$

_____ 10. $\dfrac{x^2-16}{5x^2+20x}$

_____ 11. $\dfrac{5a+6b}{25a^2-36b^2}$

_____ 12. $\dfrac{9x-9y}{3y^2-3x^2}$

_____ 13. $\dfrac{x^2-10x+25}{2x^2-13x+15}$

_____ 14. $\dfrac{12x^2+8xy-4y^2}{12x-4y}$

Perform the indicated operation. Write answers in reduced form.

_____ 15. $\dfrac{3x}{7y} \cdot \dfrac{49y^2}{x^3} \cdot \dfrac{x^2y}{18}$

_____ 16. $\dfrac{a^3b^2}{3c} \div \dfrac{a^2b}{27c^2}$

_____ 17. $\dfrac{10a+20b}{a^3b} \cdot \dfrac{a^4b^5}{a^2+2ab}$

_____ 18. $\dfrac{(x+4)^2}{64yz^2} \div \dfrac{3x^2-48}{24y^3z}$

_____ 19. $\dfrac{10x^2-11x-6}{2x^2+5x-12} \cdot \dfrac{3x^2+11x-4}{15x^2-11x+2}$

_____ 20. $\dfrac{4x^2-8xy}{x^2-4y^2} \div \dfrac{16x^2-24xy}{2x^2+xy-6y^2}$

Find the new fraction.

_____ 21. $\dfrac{3x}{7y} = \dfrac{?}{35y^2}$

_____ 22. $\dfrac{3y}{y-2} = \dfrac{?}{y^2-4}$

Perform the indicated operations and reduce if possible.

_____ 23. $\dfrac{7x}{2y} + \dfrac{5}{2y} - \dfrac{3x}{2y} + \dfrac{1}{2y}$

_____ 24. $\dfrac{2x^2-8x}{6x^2+7x-3} + \dfrac{x-15}{6x^2+7x-3}$

_____ 25. $\dfrac{8}{x} + \dfrac{1}{x^2} - \dfrac{3}{4x^2}$

_____ 26. $\dfrac{x+6}{3x} - \dfrac{3x-4}{2x}$

_____ 27. $\dfrac{x+y}{4x^2 y} - \dfrac{x-2y}{3xy^2} - \dfrac{21}{12xy}$

_____ 28. $\dfrac{5x}{6x-18} - \dfrac{3x}{2x-6}$

_____ 29. $\dfrac{2a}{a^2-4} - \dfrac{3}{a^2-a-2}$

_____ 30. $\dfrac{x+5}{x^2+3x-18} - \dfrac{4}{x+6}$

_____ 31. $\dfrac{\dfrac{a^3 b^2}{6c^3}}{\dfrac{a^4 b^7}{15c}}$

_____ 32. $\dfrac{\dfrac{1}{x^2} - \dfrac{1}{49}}{\dfrac{1}{x} - \dfrac{1}{7}}$

Solve each fractional equation.

_____ 33. $\dfrac{x-3}{5} - \dfrac{x}{10} = 1$

_____ 34. $\dfrac{2x-5}{x^2} - \dfrac{7}{3x} = 0$

_____ 35. $\dfrac{y}{y-3} + \dfrac{7}{3} = \dfrac{3}{y-3}$

_____ 36. $\dfrac{a+3}{a+1} = \dfrac{a-3}{a-2}$

_____ 37. $\dfrac{a}{a-4} + \dfrac{2}{a} = \dfrac{16}{a^2-4a}$

_____ 38. $\dfrac{x+3}{x+5} + \dfrac{2}{x-9} = \dfrac{-20}{x^2-4x-45}$

Use an equation to solve each problem.

_____ 39. If 4 gallons of paint will cover 600 square feet, how many gallons will be needed to paint 1800 square feet?

_____ 40. The number of games the Braves lost last year was 5 less than the number they won. If the ratio of the games won to the games lost is 3 to 2, find how many games the Braves played last year.

_____ 41. It takes Megan 3 hours to prepare the medicine for the patients in the north wing of the hospital, but Fred can do it in 2 hours. If they work together, how long will it take them to prepare the medicine?

_____ 42. On a vacation it took Maurice the same time to drive 275 miles as it took Denise to drive 325 miles. Maurice's rate of speed was 10 miles per hour less than Denise's. How fast was Denise driving?

Check your answers in the back of your book.

If You Missed Exercises:	You Should Review Examples:	
1, 2	SECTION 8.1	1–4
3, 4		5, 6
5, 6		7, 8
7–9		9–12
10–14		13–16
15, 16	SECTION 8.2	1–5
17–20		6–11
21	SECTION 8.3	1, 2
22		3–6
23, 24	SECTION 8.4	2–5
25–27		6–8
28–30		9–11
31	SECTION 8.5	1, 2
32		3, 4
33, 34	SECTION 8.6	1, 2
35–38		3–6
39–42	SECTION 8.7	1–8

If you have completed the **Review Exercises** and corrected your errors, you are ready to take the **Practice Test** for Chapter 8.

PRACTICE TEST for Chapter 8

		SECTION	EXAMPLES
_____	1. Evaluate $\dfrac{3x - 2}{x + 5}$ when $x = 3$.	8.1	1, 2
_____	2. Find the restrictions on the variable for the expression $\dfrac{y + 2}{3y + 1}$.	8.1	5

Reduce each fraction.

_____	3. $\dfrac{-11x^3y^4}{44x^6y^3}$	8.1	7, 8
_____	4. $\dfrac{36x^4}{6x^3 + 12x^2}$	8.1	12
_____	5. $\dfrac{2x^2 - 10x}{x^2 - 25}$	8.1	13, 14
_____	6. $\dfrac{3x^2 - 3y^2}{-9x - 9y}$	8.1	15, 16

Perform the indicated operations. Write answers in reduced form.

_____	7. $\dfrac{2a^2}{3b} \cdot \dfrac{6b^3}{a^5} \cdot \dfrac{ab}{10}$	8.2	2
_____	8. $\dfrac{27y}{20x^4} \div \dfrac{15y^5}{8x^2}$	8.2	4
_____	9. $\dfrac{x^2 - 5x - 6}{3x + 3} \div \dfrac{5x^2 - 30x}{6x^2}$	8.2	10
_____	10. $\dfrac{6x^2 - 7x - 5}{2x^2 - 9x - 5} \cdot \dfrac{3x^2 - 14x - 5}{9x^2 - 12x - 5}$	8.2	8
_____	11. $\dfrac{\dfrac{4x^2 - 1}{x^5}}{\dfrac{6x - 3}{x^2 + x}}$	8.5	2
_____	12. $\dfrac{4x}{5a} + \dfrac{3}{5a} + \dfrac{x}{5a} - \dfrac{8}{5a}$	8.4	2
_____	13. $\dfrac{2x^2 - x}{3x^2 + 5x - 2} + \dfrac{x^2 - 3x + 1}{3x^2 + 5x - 2}$	8.4	5
_____	14. $\dfrac{4}{a^2b} - \dfrac{3}{5ab} + \dfrac{1}{3ab^2}$	8.4	7

		Section	Examples

_____ 15. $\dfrac{6}{y + 5} - \dfrac{6}{y - 5}$ — 8.4 — 8, 9

_____ 16. $\dfrac{3}{a^2 + a - 12} - \dfrac{2}{a^2 - 3a}$ — 8.4 — 11

Solve each fractional equation.

_____ 17. $\dfrac{x + 2}{9} - \dfrac{x}{27} = 1$ — 8.6 — 1

_____ 18. $\dfrac{5}{3x} - \dfrac{4}{x + 7} = 0$ — 8.6 — 3

_____ 19. $\dfrac{x - 5}{5x^2 - 3x - 2} = \dfrac{2}{x - 1}$ — 8.6 — 5, 6

Use an equation to solve each problem.

_____ 20. On a recent plane flight, the ratio of first class passengers to tourist class passengers was 2 to 11. If the number of tourist passengers was 6 more than 5 times the number of first class passengers, how many first class passengers were on the flight? — 8.6 — 6

_____ 21. Amy allows herself to consume a certain number of calories each day. Today she has consumed $\frac{1}{6}$ of her allotment at breakfast, $\frac{1}{4}$ of her allotment at lunch and $\frac{1}{2}$ of her allotment at supper. If she has consumed 990 calories so far, how many calories are left for a midnight snack? — 8.6 — 1

_____ 22. Mel can clean the oven at the pizza parlor in 5 hours, but Luigi can do the same job in 4 hours. How long will it take them if they work together? — 8.6 — 7

Check your answers in the back of your book.

SHARPENING YOUR SKILLS after Chapters 1–8

Perform the indicated operations.

SECTION

_____ **1.** $\dfrac{5(-2)^2 - 3 \cdot 4}{(-6)(4) + 3 \cdot 2^3}$

2.3

_____ **2.** $\left(\dfrac{2}{3}\right)^2 \div \left(\dfrac{3}{5} \cdot \dfrac{10}{27}\right)$

2.4

_____ **3.** $-5[3(2x - y) - 7(x + 2y)]$

3.3

_____ **4.** $(-2x^2y)^3(xy^3)^2$

5.1

_____ **5.** $\left(\dfrac{-27a^2b^5}{28a^7}\right)^2$

5.2

_____ **6.** $(-4a^3b)(3ab^2)^2$

6.2

_____ **7.** $(2a - 3b)^2$

6.3

_____ **8.** $(x - 0.5y)(x + 0.5y)$

6.3

_____ **9.** $-2x(x - 5y)(x + 2y)$

6.3

Rewrite each expression as a product of factors.

_____ **10.** $15x^3y - 65x^2y^2 + 20xy^3$

7.1

_____ **11.** $2x^2 - 98$

7.2

_____ **12.** $x^2 - 10xy + 25y^2$

7.3

_____ **13.** $-15x^2 - 80x + 60$

7.3

Solve each equation.

_____ **14.** $(4x - 3) - (2x + 6) = -9$

4.2

_____ **15.** $2(5y - 1) - 5 = 3(y + 2)$

4.2

_____ **16.** $4y(3y - 4) = 5y(y + 1)$

7.5

Use an equation to solve each problem.

_____ **17.** One leg of a right triangle is 7 inches less than the other leg. The hypotenuse is 13 inches. Find the two legs.

7.6

_____ **18.** In a basketball game, the winning team's score was 17 less than twice the score of the losing team. If the winner's score was 59, what was the loser's score?

4.3

Check your answers in the back of your book.

9

Working with Inequalities

To qualify for federal funding, a special program must enroll between 150 and 210 students. Moreover, the number of rural students must be twice the number of nonrural students. How many of each type of student must be enrolled in the special program?

In Chapter 1 we learned that numbers can be compared easily by describing their locations on the number line. Statements that compare two quantities using phrases such as

"is less than"
"is greater than"
"is between"

are called **inequalities**.

In this chapter we learn how to

1. Write and graph numerical inequalities and variable inequalities.
2. Simplify and solve first-degree inequalities.
3. Combine inequalities containing variables.
4. Switch from word statements to variable inequalities.

9.1 Writing and Graphing Inequalities

Mathematicians have invented some handy symbols to be used in writing inequalities. If we are given any two real numbers, a and b, it is always possible to compare them.

Symbols	Words	Number Line
$a < b$	a is less than b	a lies to the left of b
$a > b$	a is greater than b	a lies to the right of b
$a = b$	a is equal to b	a and b are at the same point

Let's practice comparing some numbers using symbols.

Example 1. Fill in the blanks with $<$, $>$, or $=$.

$$2 \underline{\quad} 10$$
$$-5 \underline{\quad} 0$$
$$\frac{1}{2} \underline{\quad} \frac{1}{3}$$

Solution

$$2 < 10$$
$$-5 < 0$$
$$\frac{1}{2} > \frac{1}{3}$$

You complete Example 2.

Example 2. Fill in the blanks with $<$, $>$, or $=$.

$$1 \underline{\quad} 0$$
$$-3 \underline{\quad} -7$$
$$-\frac{5}{2} \underline{\quad} -\frac{10}{4}$$

Solution

$$1 \underline{\quad} 0$$
$$-3 \underline{\quad} -7$$
$$-\frac{5}{2} \underline{\quad} -\frac{10}{4}$$

Check your work on page 472. ▶

When one number lies *between* two other numbers, we may use a **three-part inequality** to express that fact. For example, to say that 5 is between 2 and 9, we may write

$$2 < 5 < 9$$

This inequality says that 5 is greater than 2 *and* less than 9.

Example 3. Write an inequality to say that 0 is between -1 and 3.

Solution

$$-1 < 0 < 3$$

Now you try Example 4.

Example 4. Write an inequality to say that $-\frac{1}{2}$ is between $-\frac{3}{4}$ and $-\frac{1}{4}$.

Solution

$$-\frac{3}{4} < \underline{\quad} < \underline{\quad}$$

Check your work on page 472. ▶

Suppose that we are asked to find a number that is less than 7. One student might respond with the answer 6. Another might suggest 0. Still others might offer $\frac{5}{2}$ or -3 or -659. Which answer is correct? All of them are correct, of course. In fact, there is an *infinite* list of numbers that are less than 7.

In the problem described, we were asked to find all values of x that satisfy the inequality

$$x < 7$$

We discovered that there is no practical way to *list* all the solutions to such a **variable inequality**. One way to picture the numbers that are in this infinite list is to use the number line.

On the number line, all the numbers that satisfy the inequality $x < 7$ are the numbers that lie to the *left* of 7. We illustrate those numbers by

Such a picture is called a **graph** of the inequality. Notice the **open dot** at the point 7 which shows that 7 itself is *not* a solution to the inequality but that any number up to 7 (such as 6.9 or 6.999 and so on) *does* satisfy the inequality. Notice also that the arrow pointing to the left on the graph shows that it extends forever in that direction.

Let's practice graphing solutions to inequalities.

Example 5. Graph the solutions for

$$a > -2$$

   Solution. Our solutions are all numbers to the _____ of -2.

Check your work on page 473. ▶

Example 6. Graph the solutions for

$$-1 < x < 3$$

   Solution. This inequality says that the middle quantity, x, lies *between* the other quantities.

Suppose that we are asked to find all numbers that are less than 5 *or* equal to 5. In symbols, this is written $x \le 5$. On the number line, we need all the numbers that lie to the *left* of 5 or *at* the point 5.

Note the **solid dot** at 5 to show that 5 itself is a solution to the inequality.

Symbols	Words	Number Line
$x \le a$	x is less than *or* equal to a	*(graph: open arrow left, solid dot at a)*
$x \ge a$	x is greater than *or* equal to a	*(graph: open arrow right, solid dot at a)*

Example 7. Graph the solutions for $y \geq 0$.

Solution. We need numbers to the *right* of 0 or *at* 0.

-3 -2 -1 0 1 2 3

You graph Example 8.

Example 9. Graph the solutions for

$$-\frac{1}{2} \leq x < 4$$

Solution

-2 -1 $-\frac{1}{2}$ 0 1 2 3 4 5

Now try Example 10.

Example 8. Graph the solutions for
$$x \leq -3$$

Solution. We need numbers to the _____ of -3 or at -3.

-6   -5   -4   -3   -2   -1   0

Check your work on page 473. ▶

Example 10. Graph the solutions for

$$0 < x \leq \frac{3}{2}$$

Solution

-1   0   1   2   3   4

Check your work on page 473. ▶

⫸ Trial Run

Graph the solutions on the number lines.

1. $a > 5$

2. $x \leq -2$

3. $0 \leq x \leq \frac{5}{2}$

4. $-2 < x \leq 1$

Answers are on page 473.

▶ Examples You Completed

Example 2 (*Solution*)

$$1 > 0$$

$$-3 > -7$$

$$-\frac{5}{2} = -\frac{10}{4}$$

Example 4. Write an inequality to say that $-\frac{1}{2}$ is between $-\frac{3}{4}$ and $-\frac{1}{4}$.

Solution

$$-\frac{3}{4} < -\frac{1}{2} < -\frac{1}{4}$$

Example 5. Graph the solutions for $a > -2$.

    Solution. Our solutions are all numbers to the *right* of -2.

Example 8. Graph the solutions for $x \le -3$.

    Solution. We need numbers to the *left* of -3 or *at* -3.

Example 10. Graph the solutions for

$$0 < x \le \frac{3}{2}$$

Solution

Answers to Trial Run

page 472   **1.**

2.

3.

4.

EXERCISE SET 9.1

Fill each blank with the correct symbol ($<$, $>$, or $=$).

1. -11 _____ 0

2. -9 _____ -1

3. 3 _____ -19

4. 5 _____ -23

5. $\dfrac{1}{3}$ _____ $\dfrac{4}{12}$

6. $\dfrac{10}{15}$ _____ $\dfrac{2}{3}$

7. -23 _____ -2

8. -45 _____ 0

9. $\dfrac{-800}{50}$ _____ -15

10. $\dfrac{-600}{30}$ _____ -19

11. -2 _____ -10

12. 5 _____ -35

13. -1.5 _____ -5.9

14. -1.7 _____ -3.8

Write an inequality for each statement.

_____ **15.** -5 is between -11 and 0.

_____ **16.** -6 is between -9 and -1.

_____ **17.** 3 is greater than -2 and less than 7.

_____ **18.** 4 is greater than 0 and less than 10.

_____ **19.** $-\dfrac{2}{5}$ is between 0 and $-\dfrac{4}{5}$.

_____ **20.** $-\dfrac{3}{8}$ is between $-\dfrac{7}{8}$ and $-\dfrac{1}{8}$.

Graph the solutions for each inequality on a number line.

21. $x > 1$

22. $x > 3$

23. $a \leq 3$

24. $a \leq 0$

25. $y \geq -2$

26. $y \geq -3$

27. $b < 0$

28. $b < -6$

29. $0 < x < 6$

30. $-2 < x < 3$

31. $-1 \leq x < \dfrac{7}{2}$

32. $0 \leq x < \dfrac{5}{2}$

33. $-6 \leq a \leq -2$

34. $-5 \leq a \leq -3$

35. $-\dfrac{3}{2} < y \leq 0$

36. $-1 < y \leq \dfrac{1}{2}$

37. $3 < x < 4$

38. $-1 < x < 0$

39. $-5 \leq x \leq 1$

40. $-4 \leq x \leq 2$

474

☆ Stretching the Topics

Graph the solutions for each inequality.

1. $-\dfrac{1}{4} > x$

2. $x \neq -\dfrac{1}{2}$

3. $\dfrac{1}{2}x < -3.2$

Check your answers in the back of your book.

If you complete **Checkup 9.1**, you are ready to go on to Section 9.2.

CHECKUP 9.1

Fill each blank with the correct symbol (<, >, or =).

1. -7 ____ 1

2. $\dfrac{6}{10}$ ____ $\dfrac{15}{25}$

3. -18 ____ -22

4. 0 ____ $-\dfrac{4}{5}$

Write an inequality for each statement.

_____ 5. -2 is between -5 and 0.

_____ 6. 3 is between $\dfrac{3}{2}$ and $\dfrac{7}{2}$.

Graph the solutions for each inequality on a number line.

7. $a > -1$ ⟷

8. $x \leq 0$ ⟷

9. $-3 \leq y \leq \dfrac{9}{2}$ ⟷

10. $-2 < x \leq 2$ ⟷

Check your answers in the back of your book.

If You Missed Problems:	You Should Review Examples:
1–4	1, 2
5, 6	3, 4
7, 8	5, 7, 8
9, 10	6, 9, 10

9.2 Solving Inequalities

We have already learned to use the operations of addition, subtraction, multiplication, and division on both sides of an *equation*. Is an *inequality* still true after we perform those same operations on both sides? We must investigate before we can decide.

Operating on Numerical Inequalities

Let's operate on both sides of the inequality

$$6 < 10$$

first with some positive numbers

Add 3	Subtract 15	Multiply by 3	Divide by 2
$6 + 3 \overset{?}{<} 10 + 3$	$6 - 15 \overset{?}{<} 10 - 15$	$6 \cdot 3 \overset{?}{<} 10 \cdot 3$	$\dfrac{6}{2} \overset{?}{<} \dfrac{10}{2}$
$9 \overset{?}{<} 13$	$-9 \overset{?}{<} -5$	$18 \overset{?}{<} 30$	$3 \overset{?}{<} 5$
True	True	True	True
$9 < 13$	$-9 < -5$	$18 < 30$	$3 < 5$

and then with some negative numbers.

Add -7	Subtract -4	Multiply by -5	Divide by -2
$6 + (-7) \overset{?}{<} 10 + (-7)$	$6 - (-4) \overset{?}{<} 10 - (-4)$	$6(-5) \overset{?}{<} 10(-5)$	$\dfrac{6}{-2} \overset{?}{<} \dfrac{10}{-2}$
$-1 \overset{?}{<} 3$	$10 \overset{?}{<} 14$	$-30 \overset{?}{<} -50$	$-3 \overset{?}{<} -5$
True	True	False	False
$-1 < 3$	$10 < 14$	$-30 > -50$	$-3 > -5$

These examples may help you see that

> If we add (or subtract) the same positive or negative quantity to (or from) both sides of an inequality, the *direction* of the resulting inequality will remain the *same*.
>
> If we multiply (or divide) both sides of an inequality by the same *positive* number, the *direction* of the resulting inequality will remain the *same*.
>
> If we multiply (or divide) both sides of an inequality by the same *negative* number, the *direction* of the resulting inequality must be *reversed*.

Example 1. Perform the indicated operation on both sides of the given inequality.

Inequality	Operation	Result
$-1 < 5$	add 3	$-1 + 3 < 5 + 3$ ___ < ___
$12 > 7$	subtract 5	$12 - 5 > 7 - 5$ ___ > ___
$-2 < -1$	multiply by 5	$5(-2) < 5(-1)$ ___ < ___
$12 < 21$	divide by 3	$\dfrac{12}{3} < \dfrac{21}{3}$ ___ < ___
$18 > 9$	divide by -3	$\dfrac{18}{-3} < \dfrac{9}{-3}$ ___ < ___
$16 > -7$	multiply by -1	$-1(16) < -1(-7)$ ___ < ___

Check your work on page 482. ▶

Solving Variable Inequalities

Now that we know how to operate on both sides of an inequality, we may solve variable inequalities such as

$$3x - 4 \geq 17$$

This is a **first-degree inequality** because the variable term has an exponent of 1. As with a first-degree equation, our goal is to *isolate the variable*.

$3x - 4 \geq 17$	The original inequality.
$3x - 4 + 4 \geq 17 + 4$	Add 4 to both sides.
$3x \geq 21$	Simplify the sums.
$\dfrac{3x}{3} \geq \dfrac{21}{3}$	Divide both sides by 3.
$x \geq 7$	Simplify the quotients.

We may graph the solutions on a number line.

Solving a First-Degree Inequality

1. Perform any necessary additions and/or subtractions on both sides of the inequality to isolate the variable term.

2. Perform any necessary multiplications and/or divisions on both sides of the inequality, remembering to *reverse* the direction of the inequality whenever multiplying or dividing by a *negative* number.

Example 2. Solve and graph the solutions for

$$1 - y < -3$$

Solution

$$1 - y < -3$$

$$-y < -4 \qquad \text{Subtract 1.}$$

$$\frac{-1y}{-1} > \frac{-4}{-1} \qquad \text{Divide by } -1.$$

$$y > 4 \qquad \text{Simplify quotients.}$$

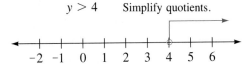

Notice that we reversed the direction of the inequality when dividing both sides by -1.

You try Example 3.

Example 4. Solve $\dfrac{x + 3}{2} > -\dfrac{1}{3}$.

Solution

$$\frac{x + 3}{2} > -\frac{1}{3} \qquad \text{LCD: } 2 \cdot 3 \text{ or } 6$$

$$\frac{6}{1}\left(\frac{x + 3}{2}\right) > \frac{6}{1}\left(-\frac{1}{3}\right) \qquad \begin{array}{l}\text{Multiply by}\\ \text{LCD.}\end{array}$$

$$3(x + 3) > 2(-1) \qquad \text{Simplify.}$$

$$3x + 9 > -2 \qquad \begin{array}{l}\text{Remove}\\ \text{parentheses.}\end{array}$$

$$3x > -11 \qquad \text{Subtract 9.}$$

$$x > -\frac{11}{3} \qquad \text{Divide by 3.}$$

Now you complete Example 5.

Example 3. Solve and graph the solutions for

$$\frac{x}{2} - 4 \le 1$$

Solution

$$\frac{x}{2} - 4 \le 1$$

$$\frac{x}{2} \le 5 \qquad \text{Add 4.}$$

$$\frac{2}{1}\left(\frac{x}{2}\right) \le 2(\underline{}) \qquad \text{Multiply by 2.}$$

$$x \le \underline{} \qquad \text{Simplify products.}$$

Check your work on page 482. ▶

Example 5. Solve $-\dfrac{3x}{2} + 5 \ge 2$.

Solution

$$-\frac{3x}{2} + 5 \ge 2$$

$$-\frac{3x}{2} \ge -3 \qquad \text{Subtract 5.}$$

$$-\frac{2}{3}\left(-\frac{3x}{2}\right) \le -\frac{2}{3}\left(\frac{}{1}\right) \qquad \begin{array}{l}\text{Multiply by re-}\\ \text{ciprocal of } -\dfrac{3}{2}.\end{array}$$

$$x \le \underline{} \qquad \text{Simplify.}$$

Check your work on page 483. ▶

To solve inequalities containing parentheses or variables on both sides, we use the same methods we used for equations. We must remove parentheses and then get all variable terms on one side and constant terms on the other side.

Example 6. Solve $5(x + 1) - 3 \leq 12$.

Solution

$5(x + 1) - 3 \leq 12$

$5x + 5 - 3 \leq 12$ Remove parentheses.

$5x + 2 \leq 12$ Combine like terms.

$5x \leq 10$ Subtract 2.

$\dfrac{5x}{5} \leq \dfrac{10}{5}$ Divide by 5.

$x \leq 2$ Simplify.

Now you complete Example 7.

Example 7. Solve

$$4(x - 1) - 10 \geq 7x + 1$$

Solution

$4(x - 1) - 10 \geq 7x + 1$

$4x - 4 - 10 \geq 7x + 1$

$4x - 14 \geq 7x + 1$

Check your work on page 483. ▶

◉▶ **Trial Run**

Solve each inequality.

———— **1.** $-3x > 15$

———— **2.** $a + 2.1 < -7.1$

———— **3.** $\dfrac{2a}{5} - 3 \geq 1$

———— **4.** $9 - 6y \leq -6$

———— **5.** $2y - 3(y + 1) > 5$

———— **6.** $5y - 7 \leq 2y + 5$

Answers are on page 483.

Solving Three-Part Inequalities

Consider this three-part inequality in which x is not yet isolated in the middle.

$$-1 < 3x + 5 < 8$$

Remember that this inequality says that

$$3x + 5 > -1 \quad and \quad 3x + 5 < 8$$

Let's solve both inequalities.

$$3x > -6 \qquad \text{Subtract 5} \qquad 3x < 3$$
$$x > -2 \qquad \text{Divide by 3} \qquad x < 1$$

We conclude that x must be greater than -2 *and* less than 1, and we write the solution as

$$-2 < x < 1$$

A shorter way of solving the original inequality allows us to operate on all three parts at the same time. We decide what operations must be performed to isolate the variable in the middle, and then perform those operations on all three parts of the inequality.

$-1 < \quad 3x + 5 \quad < 8$	The original inequality.
$-1 - 5 < 3x + 5 - 5 < 8 - 5$	Subtract 5.
$-6 < \quad 3x \quad < 3$	Simplify the sums.
$\dfrac{-6}{3} < \dfrac{3x}{3} < \dfrac{3}{3}$	Divide by 3.
$-2 < \quad x \quad < 1$	Simplify the quotients.

Notice that this solution matches the earlier solution.

Example 8. Solve $-3 \le -5x - 3 < 2$.

Solution

$$-3 \le \quad -5x - 3 \quad < 2$$
$$-3 + 3 \le -5x - 3 + 3 < 2 + 3$$
$$0 \le \quad -5x \quad < 5$$
$$\frac{0}{-5} \ge \frac{-5x}{-5} \qquad > \frac{5}{-5}$$
$$0 \ge \quad x \quad > -1$$

It is customary to write three-part inequalities with the smaller number on the left. Here we write

$$-1 < x \le 0$$

Now you try Example 9.

Example 9. Solve $-5 < \dfrac{1}{2}x + 3 \le 0$.

Solution

$$-5 < \frac{1}{2}x + 3 \le 0$$

Check your work on page 483. ▶

⫸ Trial Run

Solve each inequality.

_____ **1.** $1 \le 2x + 1 \le 7$ _____ **2.** $-3 < \dfrac{1}{2}x - 4 < 1$

_____ **3.** $1 \le 2 - y < 4$ _____ **4.** $-1 \le 2 - \dfrac{3}{4}x \le 8$

Answers are on page 483.

▶ Examples You Completed

Example 1 (*Results*)

$$-1 + 3 < 5 + 3$$
$$2 < 8$$

$$12 - 5 > 7 - 5$$
$$7 > 2$$

$$5(-2) < 5(-1)$$
$$-10 < -5$$

$$\frac{12}{3} < \frac{21}{3}$$
$$4 < 7$$

$$\frac{18}{-3} < \frac{9}{-3}$$
$$-6 < -3$$

$$-1(16) < -1(-7)$$
$$-16 < 7$$

Example 3. Solve and graph the solutions for

$$\frac{x}{2} - 4 \le 1$$

Solution

$$\frac{x}{2} - 4 \le 1$$

$$\frac{x}{2} \le 5$$

$$2\left(\frac{x}{2}\right) \le 2(5)$$

$$x \le 10$$

Example 5. Solve $-\dfrac{3x}{2} + 5 \geq 2$.

Solution

$$-\frac{3x}{2} + 5 \geq 2$$

$$-\frac{3x}{2} \geq -3$$

$$-\frac{2}{3}\left(-\frac{3x}{2}\right) \leq -\frac{2}{3}\left(\frac{-3}{1}\right)$$

$$x \leq 2$$

Example 7. Solve $4(x - 1) - 10 \geq 7x + 1$.

Solution

$$4(x - 1) - 10 \geq 7x + 1$$

$$4x - 4 - 10 \geq 7x + 1$$

$$4x - 14 \geq 7x + 1$$

$$-3x - 14 \geq 1$$

$$-3x \geq 15$$

$$\frac{-3x}{-3} \leq \frac{15}{-3}$$

$$x \leq -5$$

Example 9. Solve $-5 < \dfrac{1}{2}x + 3 \leq 0$.

Solution

$$-5 < \frac{1}{2}x + 3 \quad \leq 0$$

$$-5 - 3 < \frac{1}{2}x + 3 - 3 \leq 0 - 3$$

$$-8 < \frac{1}{2}x \quad \leq -3$$

$$2(-8) < \frac{2}{1}\left(\frac{1}{2}x\right) \quad \leq 2(-3)$$

$$-16 < \quad x \quad \leq -6$$

Answers to Trial Runs

page 480   **1.** $x < -5$ **2.** $a < -9.2$ **3.** $a \geq 10$ **4.** $y \geq \dfrac{5}{2}$ **5.** $y < -8$ **6.** $y \leq 4$

page 482   **1.** $0 \leq x \leq 3$ **2.** $2 < x < 10$ **3.** $-2 < y \leq 1$ **4.** $-8 \leq x \leq 4$

EXERCISE SET 9.2

Perform the indicated operation on both sides of the given inequality.

_____ 1. $-3 < 5$, add 4

_____ 2. $2 < 7$, add -2

_____ 3. $13 > 2$, add -7

_____ 4. $3 > -2$, add 6

_____ 5. $2 > -3$, multiply by 4

_____ 6. $4 > -5$, multiply by 3

_____ 7. $-18 < -6$, divide by 3

_____ 8. $-21 < -14$, divide by 7

_____ 9. $13 > 6$, subtract 20

_____ 10. $12 > 3$, subtract 15

_____ 11. $10 > -5$, divide by -5

_____ 12. $12 > -16$, divide by -4

_____ 13. $-13 < -1$, multiply by -1

_____ 14. $-19 < -3$, divide by -1

_____ 15. $-8 < 16$, divide by -2

_____ 16. $16 > 10$, multiply by -3

Solve each inequality and graph the solutions.

_____ 17. $x + 3 \leq 7$

_____ 18. $x + 5 \leq 7$

_____ 19. $3x > -15$

_____ 20. $4x > -16$

_____ 21. $-2a > 7$

_____ 22. $-3a > 9$

_____ 23. $3a - 3 \geq 12$

_____ 24. $4a - 5 \geq 15$

_____ 25. $\dfrac{x}{3} - 1 < -2$

_____ 26. $\dfrac{x}{5} - 3 < -4$

_____ 27. $-5z + 12 \leq 17$

_____ 28. $-6z + 14 \leq 26$

_____ 29. $3 - y > 1$

_____ 30. $4 - y > 1$

_____ 31. $\dfrac{5x}{2} + 1 \geq 6$

_____ 32. $\dfrac{4x}{3} + 5 \geq 9$

Solve each inequality.

_____ 33. $4(x - 5) \geq -2$

_____ 34. $3(x - 6) \geq -4$

_____ 35. $4y - 2(y - 5) < 10$

_____ 36. $7y - 3(y - 4) < 12$

_____ 37. $\dfrac{2x - 9}{3} > 5$

_____ 38. $\dfrac{5x + 1}{4} > 9$

_____ 39. $2a - 4 > 5a - 1$

_____ 40. $5a - 8 > 7a - 12$

_____ 41. $5x - (x - 4) \leq 2x + 5$

_____ 42. $3x - (x - 5) \leq x + 7$

_____ 43. $\dfrac{x-4}{3} > \dfrac{2x-1}{2}$ _____ 44. $\dfrac{3x-1}{5} > \dfrac{4x-3}{2}$

_____ 45. $2(z-2) - 3 < 5(z+5) - 23$ _____ 46. $3(z+1) + 5 < 9(2z+3) + 1$

_____ 47. $\dfrac{-x}{4} + 2 \geq \dfrac{x}{3} - 5$ _____ 48. $\dfrac{-x}{3} + 2 \geq \dfrac{x}{5} - 6$

_____ 49. $\dfrac{x}{4} - \dfrac{2x}{3} \geq \dfrac{1}{12}$ _____ 50. $\dfrac{x}{5} - \dfrac{5x}{3} \geq \dfrac{4}{15}$

_____ 51. $4 < x + 6 < 9$ _____ 52. $-5 < x - 3 < 1$

_____ 53. $-1 \leq -2x \leq 8$ _____ 54. $-6 \leq -4x \leq 4$

_____ 55. $-5 \leq 3x + 4 \leq 16$ _____ 56. $-7 \leq 5x + 3 \leq 28$

_____ 57. $7 \leq \dfrac{1}{2}x + 9 < 17$ _____ 58. $-7 \leq \dfrac{1}{3}x - 6 < -5$

_____ 59. $4 \leq 6 - 3x \leq 9$ _____ 60. $-5 \leq 7 - 2x \leq 5$

☆ Stretching the Topics _____

Solve each inequality.

_____ 1. $7x - 3[4 - (5 - x)] \leq 7x - (5 - 2x)$

_____ 2. $\dfrac{4(y-6)}{-3} > \dfrac{6-2y}{5}$

_____ 3. $-1 \leq -3x - 5[2(x-1) - (3x+4)] < 7$

Check your answers in the back of your book.

If you can solve the inequalities in **Checkup 9.2**, you are ready to go on to Section 9.3.

CHECKUP 9.2

Perform the indicated operation on both sides of the given inequality.

_____ 1. $-2 > -9$, add 5

_____ 2. $4 > -6$, divide by -2

Solve each inequality and graph the solutions.

_____ 3. $x - 5 \geq -7$

_____ 4. $-5a \leq 15$

_____ 5. $-\dfrac{4x}{3} + 7 < 3$

_____ 6. $-6 \leq 5y - 1 \leq 9$

Solve each inequality.

_____ 7. $9(a - 2) \leq -3$

_____ 8. $\dfrac{4x - 1}{2} > -3$

_____ 9. $5y - 2 < 9y$

_____ 10. $-2 \leq \dfrac{3x}{2} - 5 \leq 4$

Check your answers in the back of your book.

If You Missed Problems:	You Should Review Examples:
1, 2	1
3–5	2–5
6	8
7, 8	6
9	7
10	9

9.3 Solving Pairs of Variable Inequalities (Optional)

Sometimes it is necessary to find the values of a variable that satisfy more than one inequality at the same time. The graphs of the solutions for each inequality will help us decide on the correct solutions for more than one inequality.

Satisfying One Inequality and Another Inequality

To find solutions satisfying one inequality *and* another, let's graph both inequalities on the *same* number line and see where the graphs *overlap*.

Example 1. Solve $x < 3$ and $x > -1$.

Solution

The graphs overlap *between* -1 and 3. The inequality $-1 < x < 3$ describes all numbers satisfying *both* of the original inequalities.

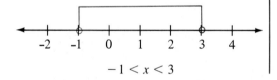

$$-1 < x < 3$$

Example 2. Solve $x > 2$ and $x > 5$.

Solution

Both graphs appear at the same time to the *right* of 5. The inequality $x > 5$ describes all numbers satisfying *both* of the original inequalities.

$$x > 5$$

Example 3. Solve $x < 7$ and $x \le -1$.

Solution

Both graphs appear at the same time to the *left* of -1 or *at* -1. The inequality $x \le -1$ describes all numbers satisfying *both* of the original inequalities.

$$x \le -1$$

Example 4. Solve $x > 2$ and $x \le 0$.

Solution

The graphs *never* overlap here. There are *no* numbers that satisfy *both* inequalities at the same time. There is *no* solution.

Satisfying One Inequality *and* Another Inequality by Graphing

1. Isolate the variable in each inequality.
2. Graph the solutions for the inequalities on the same number line.
3. Find the numbers that are included in *both* graphs. Look for the overlap.
4. Write a single inequality that describes the numbers satisfying both of the original inequalities.

Example 5. Solve $2 + 3x \geq 11$ and $5x - 1 > 2$.

Solution

$$2 + 3x \geq 11 \quad \text{and} \quad 5x - 1 > 2$$

$$3x \geq 9 \qquad\qquad 5x > 3 \qquad \text{Isolate variable terms.}$$

$$\frac{3x}{3} \geq \frac{9}{3} \qquad\qquad \frac{5x}{5} > \frac{3}{5} \qquad \text{Isolate the variables.}$$

$$x \geq 3 \quad \text{and} \quad x > \frac{3}{5} \qquad \text{Simplify.}$$

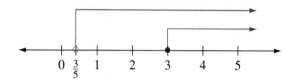

The inequality $x \geq 3$ describes the numbers satisfying both inequalities.

Try to complete Example 6.

Example 6. Solve $1 - x < 8$ and $\frac{2}{3}x \leq 0$.

Solution

$$1 - x < 8 \qquad \text{and} \qquad \frac{2}{3}x \leq 0$$

$$-x < 7 \qquad\qquad\qquad\qquad\qquad \text{Isolate variable terms.}$$

$$\frac{-1x}{-1} > \frac{7}{-1} \qquad \frac{3}{2}\left(\frac{2}{3}x\right) \leq \frac{3}{2}(0) \qquad \text{Isolate the variables.}$$

$$x > -7 \quad \text{and} \quad x \leq 0 \qquad \text{Simplify.}$$

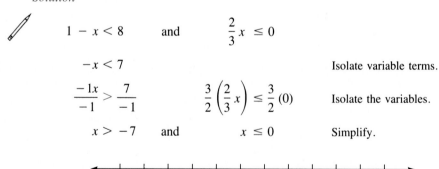

The inequality _____ describes the numbers satisfying both inequalities. This inequality states that x is _____ than -7 and _____ than or equal to 0.
Check your work on page 492. ▶

⫸ Trial Run

Solve each pair of inequalities.

_____ 1. $x < -3$ and $x \leq 2$ _____ 2. $x \leq 7$ and $x > -1$

_____ 3. $5x + 6 > -4$ and $4 - x \leq 8$ _____ 4. $2x - 1 < 0$ and $2x - 3 \geq 1$

Answers are on page 492.

Satisfying One Inequality or Another Inequality

Even in everyday English, the word *and* is much more demanding than the word *or*. If we are told that we must meet one condition *and* another, we shall be successful only when we have met *both* of those conditions. On the other hand, if we are told that we must meet one condition *or* another, we shall be successful if we meet *either* the first condition *or* the second condition or perhaps both.

The same is true for satisfying one inequality *or* another; any numbers that satisfy *either* of the inequalities (or perhaps both) will be acceptable solutions. To find those solutions, we can graph both inequalities on the same number line and see where *either* one or the other (or both) of the inequalities is satisfied.

Example 7. Solve $x > 1$ or $x \geq 5$.

Solution

All the numbers that satisfy *either* (or both) of the inequalities lie to the *right* of 1. We describe those numbers with the inequality $x > 1$.

$x > 1$

Example 8. Solve $x \leq 0$ or $x \leq 2$.

Solution

All the numbers that satisfy *either* (or both) of the inequalities lie to the *left* of 2 or *at* 2. We describe those numbers with the inequality $x \leq 2$.

$x \leq 2$

Example 9. Solve $x < -1$ or $x > \frac{3}{2}$.

Solution

All the numbers that satisfy *either* of the inequalities lie to the *left* of -1 *or* to the *right* of $\frac{3}{2}$. Our solution cannot be written as a single inequality. We describe the solutions by

$$x < -1 \qquad \text{or} \qquad x > \frac{3}{2}$$

Example 10. Solve $x < 5$ or $x > 3$.

Solution

All the numbers on the number line satisfy either (or both) of the inequalities. Our solutions are *all real numbers*.

Satisfying One Inequality *or* Another Inequality by Graphing

1. Isolate the variable in each inequality.
2. Graph the solutions for the inequalities on the same number line.
3. Find the numbers that are included in *either* graph.
4. Describe the numbers satisfying either of the original inequalities by writing a single inequality or a pair of inequalities (using the word *or*).

Example 11. Solve $-3 + 7x > 4$ or $-\frac{5}{2}x \le 0$.

Solution

$$-3 + 7x > 4 \qquad \text{or} \qquad -\frac{5}{2}x \le 0$$

$$7x > 7 \qquad\qquad -\frac{2}{5}\left(-\frac{5}{2}x\right) \ge -\frac{2}{5}(0)$$

$$\frac{7x}{7} > \frac{7}{7}$$

$$x > 1 \qquad \text{or} \qquad x \ge 0$$

Our solution is $x \ge 0$.

Try to complete Example 12.

Example 12. Solve $5x - 1 > 14$ or $-2x + 5 \geq 0$.

Solution

The solution is _____ .

Check your work on page 492. ▶

▶ Trial Run

Solve each pair of inequalities.

_____ **1.** $6x - 5 \geq 7$ or $3 - x < 8$ _____ **2.** $3x - 1 < 2$ or $x - 9 \leq -6$

_____ **3.** $2x + 13 < 9$ or $7x + 9 > 16$ _____ **4.** $\dfrac{x}{5} + 6 \leq 7$ or $\dfrac{2}{3}x < 0$

Answers are on page 492.

Remember, to satisfy one inequality *and* another inequality, we must find those numbers that satisfy *both* inequalities at the same time. To satisfy one inequality *or* another inequality, we must find those numbers that satisfy either the first inequality *or* the second inequality (or perhaps both).

▶ Examples You Completed

Example 6 (*Solution*)

$$x > -7 \quad \text{and} \quad x \le 0$$

The inequality $-7 < x \le 0$ describes the numbers satisfying both inequalities. This inequality states that x is greater than -7 and less than or equal to 0.

Example 12. Solve $5x - 1 > 14$ or $-2x + 5 \ge 0$.

Solution

$$
\begin{array}{lll}
5x - 1 > 14 & \text{or} & -2x + 5 \ge 0 \\
5x > 15 & & -2x \ge -5 \\
\dfrac{5x}{5} > \dfrac{15}{5} & & \dfrac{-2x}{-2} \le \dfrac{-5}{-2} \\
x > 3 & \text{or} & x \le \dfrac{5}{2}
\end{array}
$$

The solution is $x \le \dfrac{5}{2}$ or $x > 3$.

Answers to Trial Runs

page 489   **1.** $x < -3$ **2.** $-1 < x \le 7$ **3.** $x > -2$ **4.** No solution

page 491   **1.** $x > -5$ **2.** $x \le 3$ **3.** $x < -2$ or $x > 1$ **4.** $x \le 5$

EXERCISE SET 9.3

Find the solution for each pair of inequalities.

_____ 1. $x \geq -2$ and $x \leq 3$ _____ 2. $x \geq -6$ and $x \leq 2$

_____ 3. $x > -6$ and $x > 0$ _____ 4. $x \geq -5$ and $x > -1$

_____ 5. $x < 5$ or $x < -2$ _____ 6. $x < -1$ or $x < 3$

_____ 7. $x > 1$ or $x < -2$ _____ 8. $x > 0$ or $x < -3$

_____ 9. $x > 5$ and $x < -1$ _____ 10. $x > -1$ and $x < -3$

_____ 11. $x > \frac{1}{2}$ and $x < 4$ _____ 12. $x > -2$ and $x < 3$

_____ 13. $x > 0$ or $x > 4$ _____ 14. $x > -3$ or $x > 2$

_____ 15. $x + 6 < 8$ and $x + 2 > 0$ _____ 16. $x + 5 < 6$ and $x + 5 > 1$

_____ 17. $3x + 15 \geq 21$ and $5x - 2 < 13$ _____ 18. $2x + 9 \geq 11$ and $3x - 27 < -15$

_____ 19. $7x + 10 > 10$ or $9 - 2x \leq 19$ _____ 20. $11x - 3 > 8$ or $5x - 7 \geq 3$

_____ 21. $7 - 3x > 10$ or $\frac{2}{3}x > 0$ _____ 22. $12 - 5x > 22$ or $\frac{4}{5}x > 0$

_____ 23. $\frac{x}{2} + 3 > 5$ and $8 - \frac{x}{3} \leq 7$ _____ 24. $\frac{x}{3} - 1 \geq 0$ and $5 - \frac{x}{2} < -6$

_____ 25. $5x - 3 < 12$ or $2x - 3 < -5$ _____ 26. $2x - 13 < -7$ or $3x + 11 < 5$

_____ 27. $3x + 1 < 7$ or $2x - 3 \leq 3$ _____ 28. $7x + 2 \leq 9$ or $3x - 1 < -4$

_____ 29. $-x - 6 < 3$ and $\frac{2x}{3} - 1 > 1$ _____ 30. $-x + 9 < 5$ and $\frac{5x}{2} - 4 > 1$

_____ 31. $3x - 5 < 2x + 6$ and $4x - 1 \geq 2x + 9$

_____ 32. $2x + 5 > -4x - 7$ and $2x - 11 \leq -6x + 13$

_____ 33. $2(x - 5) > 3(x - 2)$ or $3(x - 1) > 2(x + 5)$

_____ 34. $4(x - 3) > 5(x - 1)$ or $7(x - 5) > 2(3x - 5)$

_____ 35. $\frac{5}{6}x - \frac{2}{3}x \geq \frac{1}{2}$ and $\frac{1}{5}x - \frac{3}{10}x < \frac{1}{2}$ _____ 36. $\frac{2}{7}x - \frac{1}{2}x > \frac{3}{14}$ and $\frac{1}{5}x - \frac{2}{3}x \leq \frac{14}{15}$

_____ 37. $\frac{-x}{2} + \frac{x}{8} \leq 3$ or $\frac{2x}{3} - \frac{x}{4} \geq 5$ _____ 38. $\frac{-x}{5} + \frac{x}{10} \geq 3$ or $\frac{3x}{4} - \frac{x}{2} \geq 2$

☆ Stretching the Topics

Solve each pair of inequalities.

_____ **1.** $14 - 3(2x - 1) < 5[x - (2x + 4)]$ and $-[8 - 2(x - 5)] < 6 - 2(x + 5)$

_____ **2.** $\dfrac{3x - 2}{5} + \dfrac{x}{3} \le 2x - 1$ or $\dfrac{x - 5}{2} - \dfrac{2x}{7} \ge x - 3$

_____ **3.** $2x - (4 - x) \ge 11$ or $9 < 2x + 1 < 15$

Check your answers in the back of your book.

If you can solve the inequalities in **Checkup 9.3**, you are ready to go on to Section 9.4.

Name _____ Date _____

✓ **CHECKUP 9.3**

Find the solution for each pair of inequalities.

_____ 1. $x \geq -1$ and $x > 3$

_____ 2. $x > -2$ or $x \geq 3$

_____ 3. $a \leq -2$ or $a \geq 4$

_____ 4. $a > -2$ and $a < -5$

_____ 5. $x < 0$ or $x \geq -6$

_____ 6. $x < 4$ and $x > 0$

_____ 7. $4 - 3x > -8$ and $2x - 5 > -13$

_____ 8. $2x + 2 < 5$ and $5x - 7 > -12$

_____ 9. $\dfrac{x}{5} - 2 > -1$ or $7 - \dfrac{x}{3} < 1$

_____ 10. $4x - 3 > 13$ or $-2x + 3 > 0$

Check your answers in the back of your book.

If You Missed Problems:	You Should Review Examples:
1	2
2	7
3	9
4	4
5	10
6	1
7, 8	5, 6
9, 10	11, 12

495

9.4 Switching from Word Statements to Inequalities

Many everyday math problems involve inequalities rather than equations.

Example 1. Write the inequality that describes each word statement.

Words	Inequality
To qualify as a sophomore, a student must have completed at least 30 hours of coursework.	h ____ 30
The age of a Miss America contestant must be at least 18 but not more than 25 years.	18 ____ a ____ 25

Check your answers on page 499. ▶

Suppose that we try solving some word problems using inequalities. We must continue with our usual methods: identify the variable, write any expression containing the variable, write an inequality from the information in the problem, and solve.

Example 2. Mr. Angelo, a retiree, will lose some retirement benefits if he earns over $8000 per year as a night watchman. If he has already earned $7200, for how many more hours may he work at $5 per hour without losing any retirement benefits?

Solution

We Think	*We Write*
We don't know the number of hours to be worked.	Let h = number of hours to be worked
The amount Mr. Angelo will earn is found by *multiplying* his hourly rate times h hours.	$5h$ = dollars to be earned
His total earnings are the *sum* of the amount he has already earned and the amount he will earn.	$7200 + 5h$ = total dollars earned
Total earnings must be $8000 *or less*.	$7200 + 5h \le 8000$

Now we have an inequality to solve by the usual methods.

$$7200 + 5h \le 8000$$
$$5h \le 800$$
$$\frac{5h}{5} \le \frac{800}{5}$$
$$h \le 160$$

Mr. Angelo may work 160 or fewer hours.

Example 3. In order to receive a B in her math course, Henrietta must have an average of at least 80 on six tests. Her grades on the first five tests were 73, 85, 80, 79, and 72. What grade must she receive on the sixth test in order to earn a B in the course?

Solution

We Think	*We Write*
We don't know her grade on the sixth test.	Let G = grade on sixth test
The average grade for six tests is found by *adding* the grades and *dividing* the sum by 6.	$\dfrac{73 + 85 + 80 + 79 + 72 + G}{6}$ = average
Henrietta wants an average grade of *at least* 80.	$\dfrac{73 + 85 + 80 + 79 + 72 + G}{6} \geq 80$

Now we must solve the inequality.

$$\frac{73 + 85 + 80 + 79 + 72 + G}{6} \geq 80$$

$$\frac{389 + G}{6} \geq 80$$

$$\frac{6}{1}\left(\frac{389 + G}{6}\right) \geq 6(80)$$

$$389 + G \geq 480$$

$$G \geq 480 - 389$$

$$G \geq 91$$

Henrietta must receive a 91 or better on her last test.

Let's solve the problem stated at the beginning of this chapter.

Example 4. To qualify for federal funding, a special program must enroll between 150 and 210 students. Moreover, the number of rural students enrolled must be twice the number of nonrural students. How many of each type of student must be enrolled in the special program?

Solution

Let $\qquad x$ = number of nonrural students

$2x$ = number of rural students

$x + 2x$ = total number of students

Since we are working with a "between" statement, we can use a three-part inequality.

$$150 \leq x + 2x \leq 210$$

$$150 \leq \quad 3x \quad \leq 210$$

$$\frac{150}{3} \leq \quad \frac{3x}{3} \quad \leq \frac{210}{3}$$

$$50 \leq \quad x \quad \leq 70$$

Between 50 and 70 nonrural students must be enrolled. To find the number of rural students, we must look at $2x$. Since

$$50 \leq x \leq 70$$

we can multiply all parts of the inequality by 2 to obtain $2x$ in the middle

$$2(50) \leq 2 \cdot x \leq 2(70)$$

$$100 \leq 2x \leq 140$$

Between 100 and 140 rural students must be enrolled in the program.

Example 5. A total of 200 spaces is available for a guided tour of Cape Canaveral. If an adult's ticket sells for $5 and a child's ticket for $3, use an inequality to state the possible number of adults' tickets that can be sold in order for the money from ticket sales to total at least $900.

Solution. We shall let a represent the number of adults' tickets and use a chart to organize our information.

	Price per Ticket (Dollars)	Number of Tickets	Dollars from Tickets
Adults	5	a	$5 \cdot a$
Children	3	$200 - a$	$3(200 - a)$

The total dollars from ticket sales must be at least $900.

$$5a + 3(200 - a) \geq 900$$

$$5a + 600 - 3a \geq 900$$

$$2a + 600 \geq 900$$

$$2a \geq 300$$

$$a \geq 150$$

Since there are only 200 spaces, the possible number of adults' tickets is described by

$$150 \leq a \leq 200$$

At least 150 adults' tickets must be sold, but what about children's tickets?

$$200 - a = 200 - 150$$
$$= 50$$

Therefore, 50 *or fewer* children's tickets must be sold. Remember that the total number of tickets available is just 200.

▶ Example You Completed

Example 1. Write the inequality that describes each word statement.

Words	Inequality
To qualify as a sophomore, a student must have completed at least 30 hours of coursework.	$h \geq 30$
The age of a Miss America contestant must be at least 18 but not more than 25 years.	$18 \leq a \leq 25$

EXERCISE SET 9.4

Use an inequality to solve each problem.

_____ 1. In order for the Barminskis to have good television reception, the top of their antenna must be at least 24 feet above the ground. If the height of their house is 16 feet, what antenna lengths would assure good reception?

_____ 2. The total passenger weight of the persons riding the elevator at Holt's Department Store must not exceed 2175 pounds. If the average weight of an adult is 145 pounds, how many adults are allowed in the elevator?

_____ 3. In addition to her yearly salary of $12,380, Ms. Greenwell sells vacuum cleaners to supplement her income. She receives a commission of $48.75 on each vacuum cleaner she sells. If she wants to keep her total yearly income below $20,960 for tax purposes, how many vacuum cleaners should she sell each year?

_____ 4. Ella's diet requires her to have no more than 1100 calories per day. For lunch she can have 100 more calories than she has for breakfast, and for dinner she can have twice the number of calories she has for breakfast. Find the possible number of calories that she can have for each meal.

_____ 5. The perimeter of a triangle can be no more than 36 feet. One side is 7 feet and the other two sides are the same. Find the possible lengths for the two equal sides.

_____ 6. If 5 times a number added to 9 is less than or equal to 4 times the number added to 12, find the values the number can have.

_____ 7. The sum of 7 times a number and 13 is greater than 27 and less than 48. Find the possible values for the number.

_____ 8. The Ovell Memorial Theater has only 400 seats. If the theater company charges admissions of $4.25 for adults and $3 for children, use an inequality to state the possible number of adults' tickets they can sell and still cover their production expenses of $1600.

_____ 9. The Quarterback Club is selling candy bars to raise money. Large bars sell for $2 and small bars for $1. They decide that they can sell 650 candy bars. Use an inequality to state the possible number of $1 bars they can sell so that their total sales will be greater than or equal to $1000.

_____ 10. If the product of 8 and a number is added to 15, the result is greater than or equal to 11 times the number. Find the possible values for the number.

_____ 11. The length of a cattle-loading pen is to be 6 feet longer than the width. The perimeter must be less than or equal to 108 feet, which is the amount of fence available. Find the possible widths of the pen.

_____ 12. The perimeter of a triangle must be less than 75 centimeters. One side is 22 centimeters and the second side is to be 3 centimeters more than the third side. Find the possible lengths of the third side.

_____ 13. In order for the Associated Student Government to break even on a concert, they must take in at least $34,800. The price of a ticket sold in advance is $6 and the price is $8 at the door. If ASG sells 5000 tickets altogether, use an inequality to state the possible number of tickets that must be sold at the door so that they do not lose money.

_____ 14. The length of a rectangle is to be 2 meters less than 5 times the width. If the difference between the two dimensions is not more than 4 meters, find the possible values for the width.

_____ 15. Mr. Geiger wishes to use no more than 90 feet of fencing for a dog pen. If he wishes the length to be 7 more feet than the width, what are the possible dimensions of the pen?

_____ 16. In order to receive an A in her math course, Irene must have an average of at least 90 on five tests. If her grades on the first four tests were 96, 82, 98, and 80, what grades can she receive on the fifth test in order to earn an A in the course?

_____ 17. The length of a rectangle is to be 7 meters more than twice the width. If the perimeter must be no more than 48 meters, find the possible values for the width.

_____ 18. If Lamar works for an hourly wage of $4.25, for what number of hours can he work so that his gross pay will be at least $153?

_____ 19. The constitution of Associated Student Government states that the entertainment committee may have no fewer than 12 members and no more than 30. It also states that the number of senior members shall be twice the number of junior members. Find how many juniors must be appointed to the committee.

_____ 20. The Press Club is having a dinner for its 150 members. It will cost them $4.25 to serve a chicken plate and $5.50 to serve a ham plate. If they can spend no more than $765, use an inequality to state the number of ham plates that could be served.

☆ Stretching the Topics _____

Use an inequality to solve each problem.

_____ 1. Juan earns a salary of $15,000 a year plus a 12 percent commission on all sales he makes. If Juan's income last year was more than $16,200 but at most $18,000, what was the total amount of his sales?

_____ 2. Frank is enclosing a rectangular garden by using his garage for one length and fence for the other three sides. The length is to be 20 feet more than $\frac{2}{3}$ the width. Find the possible widths if the maximum amount of fencing available is 164 feet.

Check your answers in the back of your book.

If you can solve the problems in **Checkup 9.4**, you are ready to do the **Review Exercises** for Chapter 9.

CHECKUP 9.4

Use an inequality to solve each problem.

_____ 1. A rental agency charges a flat fee of $15, plus $2.75 an hour for use of a garden tiller. If Irma rents the tiller, for how many hours can she use it and spend no more than $26?

_____ 2. Weight Worrier Bob has lost 3, 2, 4, and 2 pounds during the first 4 weeks of his diet. How many pounds will he have to lose during the fifth week so that his average weight loss will be at least 3 pounds per week?

_____ 3. The length of a rectangular playground is to be twice its width. The dimensions must be such that the playground can be enclosed with no more than 327 feet of available fence. Find the possible dimensions the playground can have.

_____ 4. The Shanklin Theater has a seating capacity of 1200. If the drama group charges admissions of $6 for adults and $4 for children, use an inequality to state the possible number of adults' tickets they can sell and at least cover their production costs of $6700.

_____ 5. If George earns $37.50, $58.25, $61.15, and $42.35 during the first 4 days of his yard sale, how much will he have to sell during the fifth day so that his sales average is at least $50 per day?

Check your answers in the back of your book.

If You Missed Problem:	You Should Review Example:
1	2
2	3
3	2
4	5
5	3

Summary

In this chapter we learned to solve and graph **first-degree inequalities**.

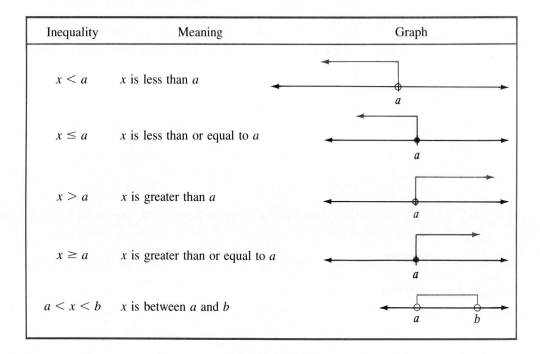

Inequality	Meaning	Graph
$x < a$	x is less than a	
$x \leq a$	x is less than or equal to a	
$x > a$	x is greater than a	
$x \geq a$	x is greater than or equal to a	
$a < x < b$	x is between a and b	

We discovered that first-degree inequalities can be solved by methods similar to those used for solving first-degree equations.

Operation	Statement	Examples
Addition (or subtraction)	We may add (or subtract) the same quantity to (or from) both sides of an inequality. The direction of the inequality remains the same.	(1) $x - 2 < 8$ $x < 10$ (2) $x + 3 \geq -1$ $x \geq -4$
Multiplication (or division) by a *positive* quantity	We may multiply (or divide) both sides of an inequality by a *positive* quantity. The direction of the inequality remains the same.	(1) $\dfrac{x}{2} \leq 5$ $x \leq 10$ (2) $3x > -1$ $x > -\dfrac{1}{3}$
Multiplication (or division) by a *negative* quantity	We may multiply (or divide) both sides of an inequality by a *negative* quantity. The direction of the inequality is *reversed*.	(1) $-4x > 8$ $x < -2$ (2) $\dfrac{-x}{3} \leq -1$ $-x \leq -3$ $x \geq 3$

In an optional section we discussed solutions to pairs of inequalities. We learned that the numbers satisfying one inequality *and* another inequality must satisfy *both* inequalities at the same time. Numbers satisfying one inequality *or* another inequality may satisfy either one inequality *or* the other (or both). The number-line graphs of the inequalities helped us find the correct solutions.

Finally, we used our knowledge of inequalities to solve word problems involving phrases such as at least, no more than, fewer than, more than, and less than.

☐ Speaking the Language of Algebra ―――――――――――――――

Complete each statement with the appropriate word or phrase.

1. Statements such as $x < 2$ and $y \geq 3$ are called ―――――― .

2. To write an inequality stating that the variable x is between 2 and 5, we write ―――― .

3. On the number line, all numbers satisfying the inequality $x < -7$ lie to the ―――――― of -7.

4. We must remember to reverse the direction of an inequality whenever we ―――――― or ―――――― both sides by a ―――――― quantity.

5. To write an inequality stating that 3 times some number, n, is at least 18.7, we write ―――――― .

Check your answers in the back of your book.

△ Writing About Mathematics ―――――――――――――――――――

Write your response to each question in complete sentences.

1. Describe the similarities and differences in the steps used to solve a first-degree equation and the steps used to solve a first-degree inequality.

2. Write a word statement that would lead you to write each of the following inequalities:

   (a) $x \leq 500$

   (b) $x > 21$

   (c) $\$5000 < x \leq \$7500$

3. Explain how you decide whether to use an open dot or a closed dot in graphing an inequality on the number line.

REVIEW EXERCISES for Chapter 9

Fill each blank with the correct symbols (<, >, or =).

1. -27 _____ -3 **2.** 0.3 _____ -1.1

Write an inequality for each of the following statements.

_____ **3.** -7 is between -10 and 3.   _____ **4.** $\dfrac{5}{3}$ is between 0 and 4.

Graph the solutions for each inequality on a number line.

_____ **5.** $y \geq -5$ _____ **6.** $a > 0$

_____ **7.** $-3 < x \leq 7$ _____ **8.** $-2 \leq y \leq \dfrac{5}{2}$

Perform the indicated operation on both sides of the given inequality.

_____ **9.** $-5 < 4$, add 9 _____ **10.** $7 < 10$, multiply by -3

_____ **11.** $10 > -4$, divide by -2 _____ **12.** $-3 > -9$, subtract 6

Solve each inequality and graph the solutions.

_____ **13.** $3x + 9 \leq 21$ _____ **14.** $8 - y > 3$

_____ **15.** $\dfrac{a}{2} + 2 \geq -2$ _____ **16.** $\dfrac{-3x}{2} + 6 < -3$

Solve each inequality.

_____ **17.** $2x - 5(x + 5) < -10$ _____ **18.** $8x - 9 \geq 3x + 1$

_____ **19.** $\dfrac{x - 7}{4} < \dfrac{3x + 1}{2}$ _____ **20.** $\dfrac{-x}{3} + 2 \geq \dfrac{x}{2} - 1$

_____ **21.** $3(y - 1) - y \geq 4(y + 2) + 1$ _____ **22.** $-3 \leq 3y - 1 \leq 5$

_____ **23.** $-8 \leq 4 - 2x < 0$ _____ **24.** $-3 \leq \dfrac{5x}{3} + 2 \leq 7$

Find the solutions for each pair of inequalities.

_____ **25.** $x + 4 > 7$ or $x - 6 \geq -1$ _____ **26.** $\dfrac{-x}{3} \leq -2$ and $\dfrac{3x}{2} - 5 > 1$

_____ **27.** $4(x - 2) > 2(x - 1)$ or $5(3 - x) < -2(3x - 7)$

Use an inequality to solve each problem.

_____ 28. Carl's salary for a 40-hour week is $140, but he finds that he needs to earn at least $218 a week to pay his bills. If he can earn $5.20 per hour working overtime, for how many hours could he work overtime per week and still pay all his bills?

_____ 29. The perimeter of a rectangle can be no more than 72 feet and the length must be twice the width. Find the possible dimensions of the rectangle.

_____ 30. The Band Boosters Club is having a dinner to raise money for the purchase of uniforms. They decide that they can serve 500 people. If the tickets are $5 for adults and $3 for children, use an inequality to state the possible number of adults' tickets the club can sell so that total sales will be at least $2100.

Check your answers in the back of your book.

If You Missed Exercises:	You Should Review Examples:	
1, 2	SECTION 9.1	1, 2
3, 4		3, 4
5–8		5–10
9–12	SECTION 9.2	1
13–16		2–5
17–21		6, 7
22–24		8, 9
25, 27	SECTION 9.3	7–12
26		1–6
28–30	SECTION 9.4	1–5

If you have completed the **Review Exercises** and corrected your errors, you are ready to take the **Practice Test** for Chapter 9.

PRACTICE TEST for Chapter 9

		SECTION	EXAMPLE

_____ 1. Write an inequality that says that $-\frac{1}{2}$ is between -1 and $\frac{1}{4}$.   **9.1**   **3**

2. Graph the solutions for $x \le -1$ on the number line.   **9.1**   **8**

$\longleftarrow\!\!\!\!\!\!\!\longrightarrow$

3. Graph the solutions for $-2 < a \le \frac{5}{3}$ on the number line.   **9.1**   **9**

$\longleftarrow\!\!\!\!\!\!\!\longrightarrow$

Solve each inequality and graph the solutions.

_____ 4. $7 - x \ge 5$ **9.2**   **2**

$\longleftarrow\!\!\!\!\!\!\!\longrightarrow$

_____ 5. $\frac{2x}{5} - 1 > 1$ **9.2**   **5**

$\longleftarrow\!\!\!\!\!\!\!\longrightarrow$

Solve each inequality.

_____ 6. $5x - 3(2x + 4) < -3$ **9.2**   **6**

_____ 7. $8x + 7 > 5x + 7$ **9.2**   **7**

_____ 8. $\frac{3x + 1}{5} < \frac{2x + 1}{4}$ **9.2**   **4**

_____ 9. $6(a - 3) - 4a \ge 3(a - 2) + 5$ **9.2**   **7**

_____ 10. $\frac{3y}{2} + 5 \le \frac{-y}{9} + 1$ **9.2**   **4**

_____ 11. $-5 \le 4x + 3 \le -1$ **9.2**   **8**

_____ 12. $-3 < 5 - 2x < 3$ **9.2**   **8**

_____ 13. $-8 \le \frac{5}{4}x + 2 \le 2$ **9.2**   **9**

_____ 14. $2 - x \ge 1$ and $\frac{x}{5} + 3 < 2$ **9.3**   **5**

_____ 15. $4(x + 2) < 3(x - 1)$ or $7(2 - x) < 11 - 4(x + 3)$ **9.3**   **12**

	SECTION	EXAMPLE

_____ **16.** For income tax purposes, typewriter salesman Thomas wishes his yearly gross income to remain below $30,000. If he is paid a monthly salary of $2000 plus a bonus of $75 for each typewriter he sells, how many typewriters should he sell this year?   9.4   2

_____ **17.** The perimeter of a rectangular pool is to be between 80 and 98 feet. If the length is to be 10 feet more than the width, find the possible widths for the pool.   9.4   4

Check your answers in the back of your book.

SHARPENING YOUR SKILLS after Chapters 1–9

SECTION

_____ 1. Change $\frac{5}{8}$ to a percent.

1.3

_____ 2. Compare $\frac{4}{5}$ and $\frac{5}{7}$ using $<$ or $>$.

2.1

_____ 3. Find $12\frac{1}{2}\%$ of $3540.

1.3

_____ 4. Evaluate $(x - 2y)^2 - 4xy$ when $x = 1$ and $y = -2$.

5.1

_____ 5. Find the restrictions on the variable for $\frac{x - 5}{x^2 - 16}$.

8.1

Perform the indicated operations and simplify.

_____ 6. $\frac{2}{3}[2(6x - 2y) - (9x - y)]$

3.3

_____ 7. $(9xy^3)(-x^2y)(-2xy^2)$

6.2

_____ 8. $(x - 1)(x^2 + x + 1)$

6.3

_____ 9. $-4x(2 - x)(3 + x)$

6.3

_____ 10. $\dfrac{m^2 - 9n^2}{2m + 6n}$

8.1

_____ 11. $\dfrac{x^2 - 49}{4x - 8} \cdot \dfrac{3x - 6}{x^2 + 14x + 49}$

8.2

_____ 12. $\dfrac{5}{4x + 4} + \dfrac{2}{x + 1}$

8.4

_____ 13. $\dfrac{\dfrac{x - 2}{x^3y^4}}{\dfrac{x^2 - 4}{x^2y}}$

8.5

Rewrite each expression as a product of factors.

_____ 14. $16a^3b^3 - 24a^2b^2 + 8ab$

7.3

_____ 15. $6m^4 - 96$

7.2

_____ 16. $-x^3y - 8x^2y^2 - 15xy^3$

7.3

Solve each equation.

_____ **17.** $\dfrac{1}{2}[3x - (x + 2)] = 8$ 4.2

_____ **18.** $0.2(5y - 1) = 0.3(7 + y) + 0.5$ 4.2

_____ **19.** $x^2 + 10x - 39 = 0$ 7.5

_____ **20.** $3x^2 = 27x$ 7.5

_____ **21.** $\dfrac{x}{3} = \dfrac{x - 2}{6}$ 8.6

_____ **22.** $\dfrac{4}{a - 3} = \dfrac{-2a}{a^2 + 5a - 24}$ 8.6

Use an equation to solve each problem.

_____ **23.** Earl has 56 feet of border to use for enclosing his tulip bed. If he wishes the bed to be 3 times as long as it is wide, what should be the dimensions of the bed?   4.3

_____ **24.** The sum of the squares of two consecutive integers is 181. Find the integers.   7.6

_____ **25.** Driving her car Louise's average speed is 6 miles per hour faster than her husband's average speed. If Louise can drive 280 miles in the same length of time her husband drives 250 miles, how fast does each drive?   8.6

Check your answers in the back of your book.

10

Working with Equations in Two Variables

Doug and Bob plan to take turns driving on a 600-mile trip. If Doug drives at a rate of 50 miles per hour and Bob drives at 60 miles per hour, write three ordered pairs that describe the number of hours that each might drive.

So far in our study of algebra we have concentrated on expressions and equations containing *one* variable. We have learned to solve first-degree and second-degree (quadratic) equations by finding the values of the variable that make the equations true statements.

In this chapter we deal with first-degree equations that contain *two* variables, and we learn how to

1. Write ordered pairs.
2. Write equations containing two variables.
3. Find solutions of equations containing two variables.
4. Graph solutions of equations containing two variables.
5. Graph solutions of constant equations.

511

10.1 Switching from Word Statements to Equations Containing Two Variables

Suppose that a student wishes to earn $72 in a week, working at two part-time jobs. One job pays $3 per hour and one pays $4 per hour. Let's write an expression for the amount of money the student can earn in 1 week. Since we cannot assume that the number of hours spent at one job is the same as the number of hours spent at the other job, we must use **two variables**.

Let

$$x = \text{hours per week spent at first job}$$

$$y = \text{hours per week spent at second job}$$

$$3x = \text{dollars per week earned at first job}$$

$$4y = \text{dollars per week earned at second job}$$

$$3x + 4y = \text{total earnings in 1 week}$$

If the student wishes to earn $72 per week, we may write

$$3x + 4y = 72$$

There are many possible values of x and y that will satisfy this equation.

Example 1. If the student works 4 hours at the first job, how many hours must she spend at the second job?

Solution. Let $x = 4$ and find y.

$$3x + 4y = 72$$
$$3(4) + 4y = 72$$
$$12 + 4y = 72$$
$$4y = 60$$
$$y = 15$$

She can work 4 hours at the first job and 15 hours at the second job.

Now you complete Example 2.

Example 2. If she works 12 hours at the first job, how many hours must she spend at the second job?

Solution. Let $x =$ _____ and find y.

$$3x + 4y = 72$$
$$3(\underline{\hspace{1cm}}) + 4y = 72$$

She can work _____ hours at the first job and _____ hours at the second job.

Check your work on page 517. ▶

Finding Ordered Pairs

In each part of the student's work problem, we found an x-value *and* a y-value that would satisfy the equation $3x + 4y = 72$. Sometimes we choose to write these corresponding x-values and y-values in the form of **ordered pairs** enclosed in parentheses with a comma between.

Corresponding Values	Ordered Pairs
$x = 4,$ $y = 15$	$(4, 15)$
$x = 12,$ $y = 9$	$(12, 9)$
$x = 0,$ $y = 18$	$(0, 18)$

The first number in the ordered pair is always the *x*-value and the second number in the ordered pair is always the corresponding *y*-value. An ordered pair **satisfies** an equation if we can make the equation *true* by substituting the *x*-value and *y*-value from the ordered pair into the equation. In other words, *solutions for equations containing two variables are ordered pairs of numbers.*

Example 3. Does (3, 1) satisfy

$$x + 2y = 5?$$

Solution. From (3, 1) we know that $x = 3$ and $y = 1$.

$$x + 2y = 5$$
$$3 + 2(1) \stackrel{?}{=} 5$$
$$3 + 2 \stackrel{?}{=} 5$$
$$5 = 5$$

Yes, (3, 1) satisfies $x + 2y = 5$.

You complete Example 4.

Example 4. Does (0, −5) satisfy

$$9x - 2y = 6?$$

Solution. From (0, −5) we know that $x = \underline{\hspace{1cm}}$ and $y = \underline{\hspace{1cm}}$.

$$9x - 2y = 6$$
$$9(\underline{\hspace{0.8cm}}) - 2(\underline{\hspace{0.8cm}}) \stackrel{?}{=} 6$$

Check your work on page 517. ▶

Suppose that we wish to find several values of *x* and *y* that satisfy a particular equation. Let's find some ordered pairs that satisfy

$$y = 2x + 3$$

We may *choose any x-value* that we wish, substitute that value for *x* in the equation, and then find the corresponding *y*-value. We can use a table to organize our work.

x	$y = 2x + 3$	(x, y)
−2	$y = 2(-2) + 3 = -4 + 3 = -1$	$(-2, -1)$
0	$y = 2(0) + 3 = 0 + 3 = 3$	$(0, 3)$
$\frac{1}{2}$	$y = 2(\frac{1}{2}) + 3 = 1 + 3 = 4$	$(\frac{1}{2}, 4)$
0.4	$y = 2(0.4) + 3 = 0.8 + 3 = 3.8$	$(0.4, 3.8)$

Since we could continue in this way forever, we agree that there is an *infinite number* of ordered pairs that will satisfy the equation $y = 2x + 3$.

Finding ordered pairs to satisfy the equation $5x + y = 16$ looks a bit more difficult. First we can *isolate y*.

$$5x + y = 16 \qquad \text{Original equation.}$$
$$y = 16 - 5x \qquad \text{Subtract } 5x \text{ from both sides to isolate } y.$$

Now to find a *y*-value corresponding to some *x*-value, we must multiply the *x*-value by 5 and subtract that quantity from 16. For variety we often choose a negative *x*-value, a positive *x*-value, and zero.

Example 5. Complete the table of ordered pairs that satisfy $y = 16 - 5x$.

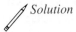 *Solution*

x	$y = 16 - 5x$	(x, y)
-2	$y = 16 - 5(-2) = 16 + \underline{\hspace{2em}} = \underline{\hspace{2em}}$	$(-2, \underline{\hspace{2em}})$
0	$y = 16 - 5(0) = 16 - \underline{\hspace{2em}} = \underline{\hspace{2em}}$	$(0, \underline{\hspace{2em}})$
$\frac{2}{5}$	$y = 16 - 5(\frac{2}{5}) = 16 - \underline{\hspace{2em}} = \underline{\hspace{2em}}$	$(\frac{2}{5}, \underline{\hspace{2em}})$
1	$y = 16 - 5(1) = 16 - \underline{\hspace{2em}} = \underline{\hspace{2em}}$	$(1, \underline{\hspace{2em}})$

Check your work on page 518. ▶

Example 6. Complete a table of three ordered pairs that satisfy $2x + 3y = 0$.

Solution. First we isolate y.

$$2x + 3y = 0 \qquad \text{Original equation.}$$

$$3y = -2x \qquad \text{Subtract } 2x \text{ to isolate } y\text{-term.}$$

$$y = \frac{-2x}{3} \qquad \text{Divide by 3 to isolate } y.$$

x	$y = \dfrac{-2x}{3}$	(x, y)
-3	$y = \dfrac{-2(-3)}{3} = \dfrac{6}{3} = 2$	$(-3, 2)$
0	$y = \dfrac{-2(0)}{3} = \dfrac{0}{3} = 0$	$(0, 0)$
1	$y = \dfrac{-2(1)}{3} = -\dfrac{2}{3}$	$\left(1, -\dfrac{2}{3}\right)$

You may have noticed that first-degree equations in two variables can always be written in the form "numerical coefficient times x, plus numerical coefficient times y, equals constant." Using A, B, and C to represent constants, we note:

General First-Degree Equation in Two Variables

$$Ax + By = C$$

where A and B are not both 0.

To find several ordered pairs that satisfy a first-degree equation in two variables, we have used the following procedure.

> **Finding Ordered Pairs**
>
> **1.** Isolate y in the original equation.
> **2.** Choose any x-values and find the corresponding y-values.
> **3.** Write each solution as an ordered pair (x, y).

▥▶ Trial Run

Decide if the given ordered pair satisfies the given equation.

_____ **1.** $5x - 3y = 2$; $(1, 1)$ _____ **2.** $x - 2y = 0$; $(4, -2)$

_____ **3.** $y = \dfrac{1}{3}x - 4$; $(-3, -5)$ _____ **4.** $-2x + 7y = -9$; $(-1, 1)$

Find three ordered pairs that satisfy each equation. There are lots of correct answers.

_____ **5.** $y = 2x + 5$ _____ **6.** $x - y = -2$

_____ **7.** $3x + y = 4$ _____ **8.** $5x - 2y = 1$

Answers are on page 518.

Writing Equations from Word Statements

The student's work hours problem pointed out the everyday usefulness of equations in two variables. The methods we have just learned will help us find ordered pairs that satisfy such equations. As always when dealing with word problems, we must be sure to *identify the variables in each situation*.

Example 7. Aaron Gray wishes to fence a rectangular portion of his yard for a dog run. If he has 60 feet of fencing to use, what are three possible pairs of dimensions for the run?

Solution. Let's use an illustration.

Length: y

Width: x

Fencing: $x + y + x + y = 60$

$$2x + 2y = 60$$

Now we isolate y to make it easier to find ordered pairs.

$$2x + 2y = 60$$

$$2y = 60 - 2x \qquad \text{Subtract } 2x \text{ to isolate } y\text{-term.}$$

$$\frac{2y}{2} = \frac{60 - 2x}{2} \qquad \text{Divide by 2 to isolate } y.$$

$$y = \frac{60}{2} - \frac{2x}{2} \qquad \text{Divide each polynomial term by 2.}$$

$$y = 30 - x \qquad \text{Reduce.}$$

In finding three pairs of dimensions, keep in mind that *zero* and *negative* x-values would make no sense in a problem dealing with length and width.

x	$y = 30 - x$	(x, y)	Width	Length
10	$y = 30 - 10 = 20$	$(10, 20)$	10 feet	20 feet
15	$y = 30 - 15 = 15$	$(15, 15)$	15 feet	15 feet
$8\frac{1}{2}$	$y = 30 - 8\frac{1}{2} = 21\frac{1}{2}$	$(8\frac{1}{2}, 21\frac{1}{2})$	$8\frac{1}{2}$ feet	$21\frac{1}{2}$ feet

Keep in mind that there are infinitely many other possibilities.

Now let's solve the problem stated at the beginning of the chapter.

Example 8. Doug and Bob plan to take turns driving on a 600-mile trip. If Doug drives at a rate of 50 miles per hour and Bob drives at 60 miles per hour, write three ordered pairs that describe the number of hours that each might drive.

Solution. Remember that distance = rate × time.

Let x = hours Doug drives   Then $50x$ = miles Doug drives

y = hours Bob drives $60y$ = miles Bob drives

$50x + 60y$ = total miles driven

The equation we must use is

$$50x + 60y = 600$$

To find ordered pairs, we isolate y.

$60y = 600 - 50x$ Subtract $50x$ to isolate y-term.

$y = \dfrac{600 - 50x}{60}$ Divide by 60 to isolate y.

$y = \dfrac{600}{60} - \dfrac{50x}{60}$ Divide each polynomial term by 60.

$y = 10 - \dfrac{5}{6}x$ Simplify each term.

x	$y = 10 - \dfrac{5}{6}x$	(x, y)	Doug's Time	Bob's Time
12	$y = 10 - \dfrac{5}{6}(12) = 10 - 10 = 0$	(12, 0)	12 hours	0 hours
6	$y = 10 - \dfrac{5}{6}(6) = 10 - 5 = 5$	(6, 5)	6 hours	5 hours
0	$y = 10 - \dfrac{5}{6}(0) = 10 - 0 = 10$	(0, 10)	0 hours	10 hours

▶ Examples You Completed

Example 2. If she works 12 hours at the first job, how many hours must she spend at the second job?

Solution. Let $x = 12$ and find y.

$3x + 4y = 72$

$3(12) + 4y = 72$

$36 + 4y = 72$

$4y = 36$

$y = 9$

She can work 12 hours at the first job and 9 hours at the second job.

Example 4. Does $(0, -5)$ satisfy $9x - 2y = 6$?

Solution. From $(0, -5)$, we know that $x = 0$ and $y = -5$.

$9x - 2y = 6$

$9(0) - 2(-5) \stackrel{?}{=} 6$

$0 + 10 \stackrel{?}{=} 6$

$10 \neq 6$

No, $(0, -5)$ does *not* satisfy the equation $9x - 2y = 6$.

Example 5. Complete the table of ordered pairs that satisfy $y = 16 - 5x$.

Solution

x	$y = 16 - 5x$	(x, y)
-2	$y = 16 - 5(-2) = 16 + 10 = 26$	$(-2, 26)$
0	$y = 16 - 5(0) = 16 - 0 = 16$	$(0, 16)$
$\frac{2}{5}$	$y = 16 - 5(\frac{2}{5}) = 16 - 2 = 14$	$(\frac{2}{5}, 14)$
1	$y = 16 - 5(1) = 16 - 5 = 11$	$(1, 11)$

Answers to Trial Run

page 515   **1.** Yes   **2.** No   **3.** Yes   **4.** No

EXERCISE SET 10.1

Decide if the given ordered pair satisfies the given equation.

_____ **1.** $6x - 5y = 2$; $(2, 2)$ _____ **2.** $4x - 3y = 1$; $(1, 1)$

_____ **3.** $2x - 4y = 0$; $(2, -1)$ _____ **4.** $3x - 9y = 0$; $(3, -1)$

_____ **5.** $y = \frac{1}{5}x - 6$; $(5, -5)$ _____ **6.** $y = \frac{1}{2}x - 5$; $(2, -4)$

_____ **7.** $-3x + 2y = 7$; $(1, 2)$ _____ **8.** $-4x + 5y = 11$; $(-4, 5)$

_____ **9.** $5x + 3y = 8$; $(0, 0)$ _____ **10.** $7x + 11y = 18$; $(0, 0)$

_____ **11.** $2x + 3y = -6$; $(3, -4)$ _____ **12.** $3x + 8y = -21$; $(1, -3)$

Find three ordered pairs that satisfy each equation.

_____ **13.** $x + y = 2$ _____ **14.** $x - y = 3$

_____ **15.** $2x + y = 4$ _____ **16.** $x + 2y = 6$

_____ **17.** $2x - 3y = 1$ _____ **18.** $4x - 5y = 1$

_____ **19.** $x - 2y = 0$ _____ **20.** $2x + y = 0$

_____ **21.** $x + 3y = 3$ _____ **22.** $4x + y = 5$

_____ **23.** $y = x + 6$ _____ **24.** $y = x - 5$

_____ **25.** $y = 3x - 5$ _____ **26.** $y = 4x - 7$

_____ **27.** $4x + 3y = -6$ _____ **28.** $3x + 2y = -5$

Write each word statement as an equation containing two variables.

_____ **29.** One number is 8 more than 4 times another.

_____ **30.** One number is 3 less than 6 times another.

_____ **31.** The sum of two numbers is 28.

_____ **32.** The sum of one number and 10 is twice another number.

_____ **33.** The length of a rectangle is 5 less than twice its width.

_____ **34.** The length of a rectangle is 2 more than three times the width.

_____ **35.** The Math Club treasurer has a stack of both $1 and $5 bills. Their total value is $100.

_____ **36.** Steve bought a stack of books at a yard sale. Some cost $2 each and some cost $3 each. The total cost was $38.

_____ **37.** The perimeter of a rectangle is 48 feet.

_____ **38.** The perimeter of a rectangle is 64 feet.

_____ **39.** Rick Yunkus has savings accounts in two different banks. The total amount of his savings is $12,500.

_____ **40.** The value of the nickels and dimes Peggy has in her change purse is $1.25.

_____ **41.** Henry earned $147.20 last week, receiving $3.20 per hour for regular time and $4.80 per hour for overtime.

_____ **42.** For the basketball game, some student and some adult tickets were sold. Student tickets sell for $1.50 and adult tickets for $3. The total gate receipts were $1500.

_____ **43.** In the Humane Society animal shelter, the number of cats is 2 less than 5 times the number of dogs.

_____ **44.** In the Chemistry Club, the number of boys is 7 less than 3 times the number of girls.

_____ **45.** Kelly walked 20 miles in the Walk-a-Thon to raise money for the Special Olympics. Part of the time she averaged 3 miles per hour and the remainder she averaged 2 miles per hour.

_____ **46.** When Duane went on vacation, he made part of the trip by bus and the other part by train. He averaged 50 miles per hour on the bus and 60 miles per hour by train. The total trip was 600 miles.

☆ Stretching the Topics _____

_____ **1.** Does the ordered pair $(\frac{1}{2}, -\frac{5}{6})$ satisfy the equation $\frac{1}{3}x - \frac{2}{5}y = \frac{1}{2}$?

_____ **2.** Complete the following table and write the equation showing the relationship that exists between x and y.

x	1	2	3	4	5	?
y	4	7	10	?	16	19

_____ **3.** Write an equation containing two variables that describes the following situation. Carlos has a certain amount of money to invest. He wishes to invest part of the money at $8\frac{1}{2}$ percent and part at 12 percent so that he can obtain an annual income of 10 percent on the entire amount.

Check your answers in the back of your book.

If you can complete **Checkup 10.1**, you are ready to go on to Section 10.2.

✓ CHECKUP 10.1

Decide if the given ordered pair satisfies the given equation.

_____ **1.** $2x - 6y = 4$; $(1, -1)$ _____ **2.** $y = \dfrac{1}{3}x - 2$; $(6, -1)$

_____ **3.** $-4x + y = 0$; $(0, 0)$ _____ **4.** $5x - y = -4$; $(-2, 6)$

Find three ordered pairs that satisfy each equation.

_____ **5.** $x - y = -2$

_____ **6.** $2x - y = 0$

_____ **7.** $y = 3x - 2$

Write each word statement as an equation containing two variables.

_____ **8.** The amount of money Mr. Brinkley has invested at Union Bank is $500 less than twice the amount invested at Farmers State Bank.

_____ **9.** Janyce bought a box of antique bottles. Some cost $2 each and some cost $3 each. The total cost was $17.

_____ **10.** The perimeter of a rectangular lot is 196 feet.

Check your answers in the back of your book.

If You Missed Problems:	You Should Review Examples:
1–4	3, 4
5–7	5, 6
8, 9	8
10	7

10.2 Graphing First-Degree Equations in Two Variables

Earlier we learned to use the **number line** to graph solutions for first-degree equations and inequalities containing one variable. We would like to graph solutions for first-degree equations in two variables. One number line will not work very well, however, since every solution is an ordered *pair* of numbers. Every solution pairs an *x*-value with a corresponding *y*-value.

Over three hundred years ago, a mathematician and philosopher named René Descartes thought of a way to graph ordered pairs using *two* number lines. He placed the two number lines at right angles (perpendicular) to each other and labeled the horizontal line the ***x*-axis** and the vertical line the **y-axis**. Naming the point at which the two lines intersect the **origin**, he marked off the units in the usual way. Descartes came up with the following diagram, called the **Cartesian coordinate plane** in his honor.

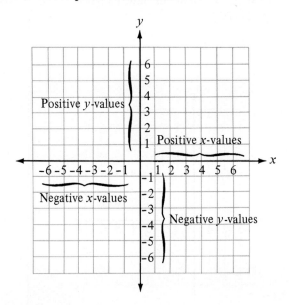

Graphing Ordered Pairs

Now we must discover how to use the Cartesian coordinate plane to graph ordered pairs. To graph an ordered pair such as (2, 3) we must keep in mind that

$$(2, 3) \quad \text{means} \quad \text{"when } x = 2, \text{ then } y = 3\text{"}$$

On the Cartesian plane, we may use a dotted vertical line to show where *x* is 2 and a dashed horizontal line to show where *y* is 3.

Since we wish *x* to be 2 and *y* to be 3 *at the same time*, the point where the dotted line crosses the dashed line must represent the graph of the ordered pair (2, 3).

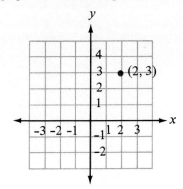

When plotting points in the Cartesian plane, we shall not sketch the dotted and dashed lines. You should do that mentally and then locate the point in the plane. All points should be labeled with an ordered pair. The point corresponding to an ordered pair contains two numbers called **coordinates**.

> **Coordinates of Ordered Pairs**
>
> *x*-coordinate: the *x*-value in an ordered pair
> *y*-coordinate: the *y*-value in an ordered pair

Example 1. Graph the ordered pairs (5, 1) and (1, 5).

Solution

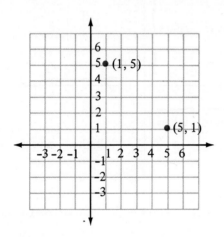

Notice that (5, 1) and (1, 5) do *not* represent the same point.

Now graph the ordered pairs in Example 2.

Example 2. Graph the ordered pairs (1, −2), (−3, 4), (−5, −5), and (2, $\frac{7}{2}$).

Solution

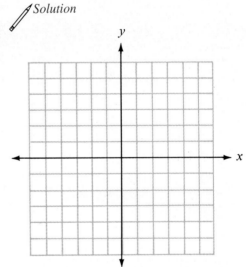

Check your work on page 530. ▶

Suppose that we plot some points for which *zero* is the *x*-coordinate and/or the *y*-coordinate.

Example 3. Graph the ordered pairs (0, 0), (3, 0), and (−5, 0).

Solution

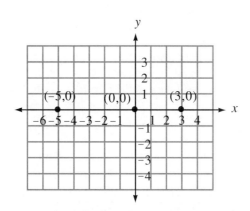

Example 4. Graph the ordered pairs (0, 0), (0, 2), and (0, −3).

Solution

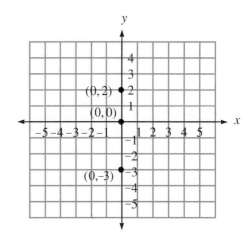

If a point has a *y*-coordinate of 0, that point lies on the *x*-axis.

If a point has an *x*-coordinate of 0, that point lies on the *y*-axis.

From our work in this section, we observe that each ordered pair of numbers corresponds to *one and only one* point in the Cartesian coordinate plane. Moreover, each point in the plane corresponds to *one and only one* ordered pair of numbers.

⫸ Trial Run

1. Graph the ordered pairs (0, −2), (−1, −3), (−1, 2), and ($\frac{3}{2}$, 2).

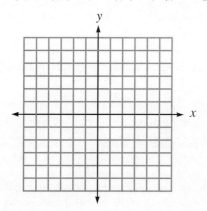

2. Give the coordinates of each point.

A ———   C ———

B ———   D ———

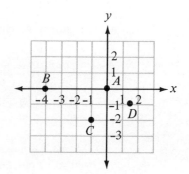

Answers are on page 530.

Graphing Equations by "Choose an x, Find a y" Method

In Section 10.1 we learned how to find ordered pairs of numbers that would satisfy equations in two variables. We considered first-degree equations of the form $Ax + By = C$. By choosing different x-values and finding corresponding y-values, we then came up with several ordered pairs that satisfied such equations.

For the equation $x + y = 7$, some solutions are

$$(7, 0) \quad (5, 2) \quad (10, -3) \quad (3, 4) \quad (0, 7) \quad (-2, 9) \quad (1, 6)$$

and so on. We agree that this equation is satisfied by an *infinite* number of ordered pairs, each found by choosing an x-value and finding a corresponding y-value.

Suppose that we graph these ordered pairs in the Cartesian coordinate plane, remembering to label each point.

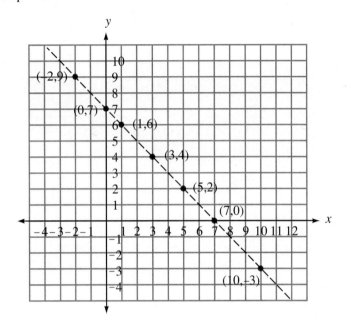

It is not just a nice coincidence that all these points seem to lie in a **straight line**. Indeed, *any* ordered pair that satisfies the equation

$$x + y = 7$$

will lie on the same straight line. That line is called the **graph of the equation**.

> **Graph of an Equation.** The graph of an equation is the set of all points corresponding to the ordered pairs that satisfy the equation.

Since it is impossible to list *all* points that satisfy an equation, we shall settle for a reasonable number of ordered pairs, say 3 or 4. Although it is true that two points are all that you need to construct a straight line, we shall use at least three points just to check our arithmetic. If three points that satisfy a first-degree equation do *not* lie in a straight line, that is a signal that you have made an error in substitution, and you should check each of your points again.

Example 5. Graph $2x - y = 6$.

Solution. In order to "choose an x and find a y," we discovered that it is helpful to isolate y right away.

$$2x - y = 6 \qquad \text{Original equation.}$$

$$-y = -2x + 6 \qquad \text{Subtract } 2x \text{ from both sides.}$$

$$y = 2x - 6 \qquad \text{Divide both sides by } -1.$$

Now we find ordered pairs, organizing our work in a table.

x	$y = 2x - 6$	(x, y)
-2	$y = 2(-2) - 6 = -4 - 6 = -10$	$(-2, -10)$
0	$y = 2(0) - 6 = 0 - 6 = -6$	$(0, -6)$
3	$y = 2(3) - 6 = 6 - 6 = 0$	$(3, 0)$
5	$y = 2(5) - 6 = 10 - 6 = 4$	$(5, 4)$

Referring to the last column, we plot the four points.

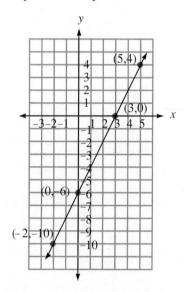

Notice that the ordered pairs satisfying the equation $2x - y = 6$ all lie in a straight line.

Graph of a First-Degree Equation. The graph of an equation of the type

$$Ax + By = C$$

is a straight line (provided A and B are not *both* 0).

You complete Example 6.

Example 6. Graph $y = 5x$.

Solution. Since y is isolated already, we shall "choose an x and find a y" to obtain points. Complete the last column of the table.

x	$y = 5x$	(x, y)
-1	$y = 5(-1) = -5$	$(-1, -5)$
0	$y = 5(0) = 0$	(_____ , _____)
$\frac{1}{2}$	$y = 5(\frac{1}{2}) = \frac{5}{2}$	(_____ , _____)
1	$y = 5(1) = 5$	(_____ , _____)

Now plot these points.

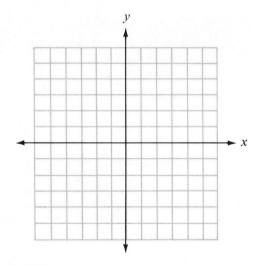

Check your work on page 530. ▶

The equation $y = 5x$ may not appear to be of the general first-degree form

$$Ax + By = C$$

but by subtracting $5x$ from both sides of the equation we may rewrite $y = 5x$ as

$$-5x + y = 0 \quad \text{or} \quad 5x - y = 0$$

which *is* general first-degree form. Note that in this equation C is 0.

Because their graphs are *lines*, we call such equations **linear equations**.

Linear Equation. An equation of the type

$$Ax + By = C$$

is called a linear equation (provided A and B are not both 0).

The method we have developed for graphing linear equations is sometimes called the **arbitrary point** method.

Graphing Linear Equations ($Ax + By = C$) Using Arbitrary Points

1. Isolate y in the equation.
2. Choose at least three x-values and find the corresponding y-values by substituting into the equation with y isolated.
3. Write each solution as an ordered pair (x, y).
4. In the Cartesian plane plot and label the point corresponding to each ordered pair.
5. Draw a straight line through the points.

▶ Trial Run

Graph each linear equation by the "choose an x, find a y" method.

1. $x + y = -3$

2. $y = 2x - 1$

3. $2x - y = 3$

4. $y = -16x$

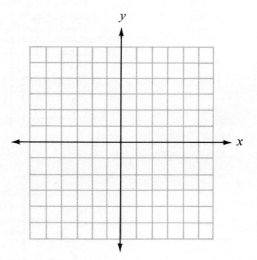

Answers are on page 530.

▶ Examples You Completed

Example 2. Graph the ordered pairs $(1, -2)$, $(-3, 4)$, $(-5, -5)$, and $(2, \frac{7}{2})$.

Solution

Example 6 (*Solution*)

x	$y = 5x$	(x, y)
-1	$y = 5(-1) = -5$	$(-1, -5)$
0	$y = 5(0) = 0$	$(0, 0)$
$\frac{1}{2}$	$y = 5(\frac{1}{2}) = (\frac{5}{2})$	$(\frac{1}{2}, \frac{5}{2})$
1	$y = 5(1) = 5$	$(1, 5)$

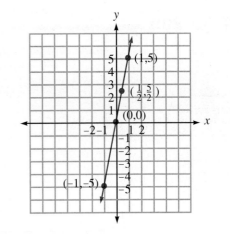

Answers to Trial Runs

page 525   **1.**

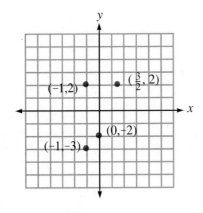

2. $A(0, 0)$; $B(-4, 0)$; $C(-1, -2)$; $D(\frac{3}{2}, -1)$

page 529   **1.**

2.

3.

4.

EXERCISE SET 10.2

Graph the following points in the Cartesian plane.

1. A: $(2, 0)$ 2. B: $(-3, 1)$

3. C: $(0, -5)$ 4. D: $(-3, -4)$

5. E: $(\frac{1}{2}, 4)$ 6. F: $(0, 0)$

7. G: $(5, 5)$ 8. H: $(3, -2)$

Give the coordinates of each point as an ordered pair.

_____ 9. A _____ 10. B

_____ 11. C _____ 12. D

_____ 13. E _____ 14. F

_____ 15. G _____ 16. H

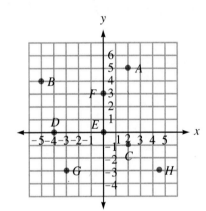

Graph each equation by the "choose an x, find a y" method.

17. $x + y = 2$ 18. $x + y = 5$ 19. $y = 25x$ 20. $3y = 15x$

21. $y = 4x - 2$ 22. $y = 3x - 1$ 23. $2x + y = 1$ 24. $3x + y = 2$

25. $x + 6y = 8$ 26. $x + 4y = 5$ 27. $2x - y = -5$ 28. $3x - y = -6$

☆ Stretching the Topics _____

Graph each equation.

1. $7 - (3 - 3y) = -5 + 2x$ 2. $0.04x + 0.2y = 0.7$ 3. $\frac{2}{3}x + \frac{1}{6}y = 4$

Check your answers in the back of your book.

If you can complete **Checkup 10.2**, you are ready to go on to Section 10.3.

✓ **CHECKUP 10.2**

Graph each of the following points in the Cartesian plane.

1. *A*: (3, 0)

2. *B*: (−2, 1)

3. *C*: (2, −3)

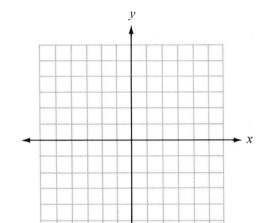

Give the coordinates of each point as an ordered pair.

_____ **4.** *A*

_____ **5.** *B*

_____ **6.** *C*

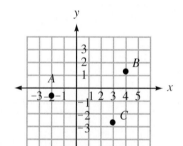

Graph each equation by the "choose an x, find a y" method.

7. $y = 2x$

8. $x + y = 1$

9. $x - 2y = 4$

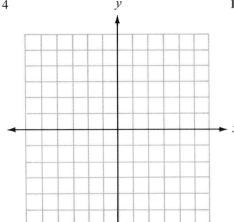

10. $3x - y = 0$

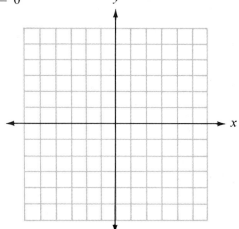

Check your answers in the back of your book.

If You Missed Problems:	You Should Review Examples:
1–6	1–4
7	6
8–10	5

10.3 Graphing Linear Equations by Other Methods

Sometimes when we isolate y in an equation of the form $Ax + By = C$, substituting into the final form becomes awkward. Perhaps we can find a less complicated method for graphing linear equations in which y is not already isolated.

Graphing Linear Equations by the Intercepts Method

If you glance back at the linear equations we have already graphed, you will notice that each straight line crosses the x-axis at some point and also crosses the y-axis at some point. These points are especially important; they determine the **intercepts** for a graph.

Intercepts

x-intercept: the x-coordinate of the point at which a graph crosses the x-axis

y-intercept: the y-coordinate of the point at which a graph crosses the y-axis

---

Earlier we observed that *each point on the x-axis has a y-coordinate of 0* and that *each point on the y-axis has an x-coordinate of 0*. Putting this observation together with the idea of intercepts, we may say that

> The x-intercept for a graph is the x-coordinate of the point that has a y-coordinate of 0.
>
> The y-intercept for a graph is the y-coordinate of the point that has an x-coordinate of 0.

To find the x-intercept for a graph, we must let $y = 0$ in the given equation and find the corresponding x-value. To find the y-intercept for a graph, we must let $x = 0$ in the given equation and find the corresponding y-value.

Example 1. Use intercepts to graph $3x + 2y = 6$.

Solution

To find the x-intercept, we substitute 0 for y.

$$3x + 2(0) = 6$$
$$3x + 0 = 6$$
$$3x = 6$$
$$x = 2$$

The graph crosses the x-axis at the point $(2, 0)$.

To find the y-intercept, we substitute 0 for x.

$$3(0) + 2y = 6$$
$$0 + 2y = 6$$
$$2y = 6$$
$$y = 3$$

The graph crosses the y-axis at the point $(0, 3)$.

Because there is less chance of error when we substitute 0 for a variable, we trust these two points to be accurate and use them to graph our line.

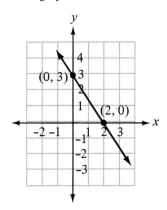

Now you help complete Example 2.

Example 2. Use intercepts to graph $5x - 2y = 15$.

✎ *Solution*

To find the *x*-intercept, we substitute 0 for _____ .

$$5x - 2(0) = 15$$

The graph crosses the *x*-axis at (_____ , 0).

To find the *y*-intercept, we substitute 0 for _____ .

$$5(0) - 2y = 15$$

The graph crosses the *y*-axis at (0, _____).

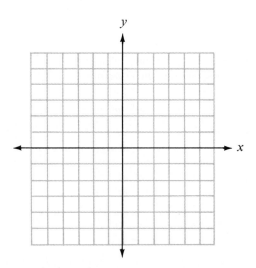

Check your work on page 543. ▶

Example 3. Graph $2x + 4y = 0$.

Solution

Find x-intercept: let $y = 0$.

$$2x + 4(0) = 0$$
$$2x = 0$$
$$x = 0$$

The graph crosses the x-axis at $(0, 0)$.

Find y-intercept: let $x = 0$.

$$2(0) + 4y = 0$$
$$4y = 0$$
$$y = 0$$

The graph crosses the y-axis at $(0, 0)$.

Here the x-intercept and y-intercept both give the point $(0, 0)$. Since one point will not give us a straight line, we must choose another x-value and find the corresponding y-value. Suppose that we let $x = 2$. If $x = 2$, then

$$2(2) + 4y = 0$$
$$4 + 4y = 0$$
$$4y = -4$$
$$y = -1$$

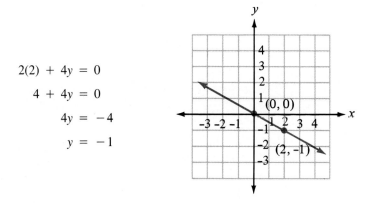

Our second point is $(2, -1)$.

From Example 3 we see that whenever the x-intercept and y-intercept both give the point $(0, 0)$ we must find a second point on the graph by choosing an x and finding the corresponding y.

Graphing Linear Equations Using Intercepts

1. Let $y = 0$ to find the x-intercept.
2. Let $x = 0$ to find the y-intercept.
3. Plot both intercepts and draw a straight line through both points.
4. If both intercepts give the point $(0, 0)$, find another ordered pair that satisfies the equation and draw a straight line through that point and the point $(0, 0)$.

⇒ Trial Run

Use intercepts to graph each linear equation.

1. $5x - 3y = -15$

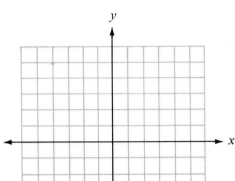

2. $-2x - 3y = 6$

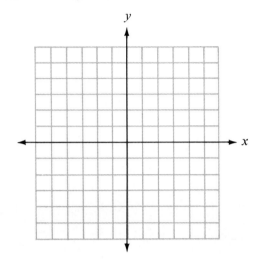

3. $3x + 8y = 12$

4. $4x - y = 0$

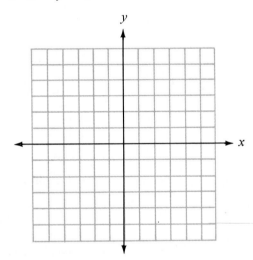

Answers are on page 544.

Graphing Constant Equations

In the general first-degree equation for a straight line,

$$Ax + By = C$$

we agree that A, B, and C represent constants. We have graphed several equations in which C was 0. Until now, however, we have not considered any equations in which A is zero or B is zero. In other words, we have not discussed equations in which the x-term or the y-term is missing.

We must investigate what happens when we use the Cartesian coordinate plane to graph equations such as

$$y = 4 \quad \text{or} \quad x = 1$$

Example 4. Graph $y = 4$.

Solution. The equation $y = 4$ states that y is always 4. What about x? This equation says that "no matter what x is, y is *always* 4." Let's look at some ordered pairs.

x	$y = 4$	(x, y)
-2	$y = 4$	$(-2, 4)$
0	$y = 4$	$(0, 4)$
$\frac{7}{8}$	$y = 4$	$(\frac{7}{8}, 4)$
3	$y = 4$	$(3, 4)$

Plotting these points, we have

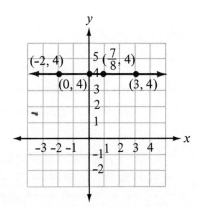

The graph of $y = 4$ in the Cartesian plane is a **horizontal line** (parallel to the x-axis) crossing the y-axis at $(0, 4)$.

Example 5. Graph $y + 3 = 0$.

Solution. First we isolate y.

$$y + 3 = 0$$

$$y = -3$$

No matter what x is, y is *always* -3. Our graph should be a horizontal line, crossing the y-axis at $(0, -3)$.

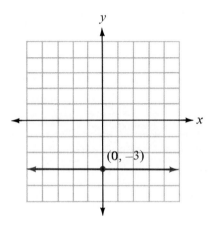

You complete Example 6.

Example 6. Graph $2y - 1 = 0$.

Solution. First we isolate y.

$$2y - 1 = 0$$

$$2y = 1$$

$$y = \frac{1}{2}$$

No matter what x is, y is *always* _____ . Our graph is a _____ line, crossing the y-axis at $(0, $ _____ $)$.

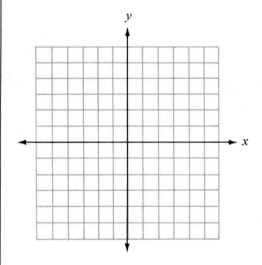

Check your graph on page 543. ▶

The graph of an equation of the form

$$y = k$$

where k is a constant, is a **horizontal line** (parallel to the x-axis) crossing the y-axis at the point $(0, k)$.

Now let us consider the graph of a linear equation in which the y-term is missing.

Example 7. Graph $x = -3$.

Solution. The equation $x = -3$ states that x is always -3. No matter what y is, x is *always* -3.

Some ordered pairs that satisfy the equation are

(x, y)
$(-3, -1)$
$(-3, 0)$
$(-3, \frac{1}{2})$
$(-3, 2)$

The graph of the equation is

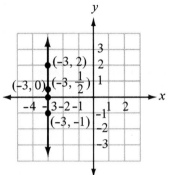

The graph of $x = -3$ is a **vertical line** (parallel to the y-axis) crossing the x-axis at $(-3, 0)$.

Example 8. Graph $3x + 4 = 0$.

Solution. First we isolate x.

$$3x + 4 = 0$$
$$3x = -4$$
$$x = -\frac{4}{3}$$

No matter what y is, x is *always* $-\frac{4}{3}$. Our graph should be a vertical line, crossing the x-axis at $(-\frac{4}{3}, 0)$.

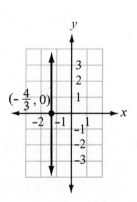

Now you try Example 9.

Example 9. Graph $x - 6 = 0$.

Solution. First we isolate x.

$$x - 6 = 0$$
$$x = 6$$

No matter what y is, x is *always* _____ . Our graph is a _____ line, crossing the x-axis at (_____ , 0).

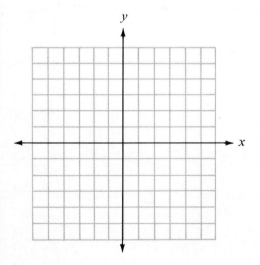

Check your work on page 543. ▶

> The graph of an equation of the form
>
> $$x = k$$
>
> where k is a constant, is a **vertical line** (parallel to the y-axis) crossing the x-axis at the point $(k, 0)$.

⟶ Trial Run

Graph each equation.

1. $y = 3$

2. $3y + 8 = 6$

3. $x + 5 = 0$

4. $4x - 8 = 0$

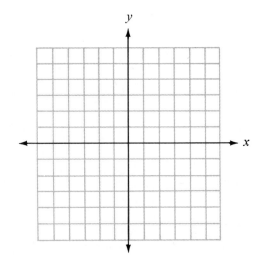

Answers are on page 544.

▶ Examples You Completed _____

Example 2. Use intercepts to graph $5x - 2y = 15$.

Solution

To find the x-intercept, we substitute 0 for y.

$$5x - 2(0) = 15$$
$$5x = 15$$
$$x = 3$$

The graph crosses the x-axis (3, 0).

To find the y-intercept, we substitute 0 for x.

$$5(0) - 2y = 15$$
$$-2y = 15$$
$$y = -\tfrac{15}{2}$$

The graph crosses the y-axis at $(0, -\tfrac{15}{2})$.

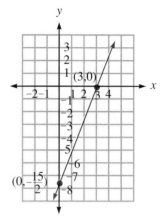

Example 6. Graph $2y - 1 = 0$.

Solution. First we isolate y.

$$2y - 1 = 0$$
$$2y = 1$$
$$y = \frac{1}{2}$$

No matter what x is, y is *always* $\tfrac{1}{2}$. Our graph is a horizontal line, crossing the y-axis at $(0, \tfrac{1}{2})$.

Example 9. Graph $x - 6 = 0$.

Solution. First we isolate x.

$$x - 6 = 0$$
$$x = 6$$

No matter what y is, x is *always* 6. Our graph is a vertical line, crossing the x-axis at (6, 0).

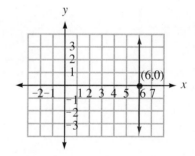

Answers to Trial Runs

page 538   **1.**

2.

3.

4.

page 542   **1.**

2.

3.

4.

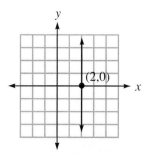

EXERCISE SET 10.3

Graph each equation by the intercepts method.

1. $x + y = 3$

2. $x + y = -2$

3. $x - y = -5$

4. $x - y = 4$

5. $3x + y = 6$

6. $5x + y = 5$

7. $x - 2y = 8$

8. $x - 3y = -6$

9. $5x + 3y = -15$

10. $3x + 4y = 12$

11. $x + 2y = 5$

12. $x + 4y = 6$

13. $3x - y = 7$

14. $2x - y = -3$

15. $3x - 4y = -14$

16. $2x - 3y = 10$

17. $-5x + 2y = 15$

18. $-7x + 3y = 14$

19. $-x - y = 3$

20. $-x - y = 1$

Graph each equation.

21. $y = 4$

22. $y = -3$

23. $x - 2 = 0$

24. $x - 5 = 0$

25. $y + 4 = 0$

26. $y + 1 = 0$

27. $3y - 9 = 0$

28. $5y - 10 = 0$

29. $2x = 7$

30. $3x = 5$

31. $2x + 5 = 0$

32. $4x + 18 = 0$

33. $6y - 16 = 0$

34. $8y - 28 = 0$

35. $9x - 27 = 0$

36. $7x + 35 = 0$

☆ Stretching the Topics _____

1. Graph $-1.2x + 3.6y = 2.4$ using the intercepts method.

2. Graph $-\frac{1}{5}x + \frac{2}{3}y = 2$ using the intercepts method.

3. Graph $\dfrac{x - 5}{2} = \dfrac{x - 3}{5}$.

Check your answers in the back of your book.

If you can graph each equation in **Checkup 10.3**, you are ready to go on to Section 10.4.

✓ CHECKUP 10.3

Graph each equation.

1. $2x - y = 4$

2. $x + 2y = 5$

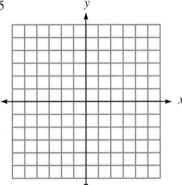

3. $-2x + 3y = -12$

4. $y - 4 = 0$

5. $3x = 8$

6. $3y + 6 = 0$

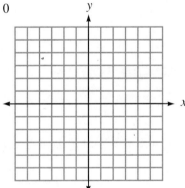

Check your answers in the back of your book.

If You Missed Problems:	You Should Review Examples:
1–3	1–3
4	5
5	8
6	6

10.4 Graphing Lines Using the Slope

Let's take a look at the graphs of several linear equations.

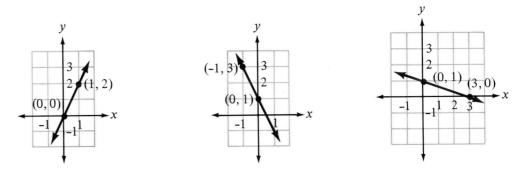

If we wish to describe the **steepness** of these lines, we might say that lines (1) and (2) seem to be equally steep, but in opposite directions. Line (3) seems to be less steep than the other two lines.

Defining the Slope of a Line

We can measure the steepness or **slope** of a line by locating any two points on that line and then comparing the units of change in the y-values (sometimes called the **rise**) to the corresponding units of change in the x-values (sometimes called the **run**).

$$\text{Slope} = \frac{\text{rise}}{\text{run}}$$

$$= \frac{\text{vertical change}}{\text{horizontal change}}$$

$$= \frac{\text{change in } y}{\text{change in } x}$$

To find the numerical value of the slope of a line through two points, we can let the first point be (x_1, y_1) and the second point be (x_2, y_2).

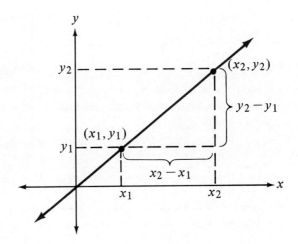

It is customary to use the letter **m** to stand for the slope of the line.

> **Definition of Slope.** If (x_1, y_1) and (x_2, y_2) are two points on a line, then the slope of that line is defined by
>
> $$m = \frac{y_2 - y_1}{x_2 - x_1}$$
>
> provided that $x_2 \neq x_1$.

Example 1. Find the slope of the line through $(1, 3)$ and $(4, 7)$.

Solution

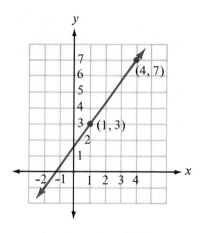

Let $(x_1, y_1) = (1, 3)$

and $(x_2, y_2) = (4, 7)$

Then $m = \dfrac{y_2 - y_1}{x_2 - x_1}$

$= \dfrac{7 - 3}{4 - 1}$

$m = \dfrac{4}{3}$

Notice that this line slopes *upward* from left to right. To move from the first point to the second point, we must go to the *right* 3 units and *up* 4 units.

You complete Example 2.

Example 2. Find the slope of the line through $(-4, 1)$ and $(3, -2)$.

Solution

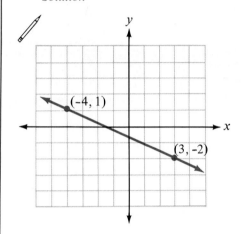

Let $(x_1, y_1) = (-4, 1)$

and $(x_2, y_2) = (3, -2)$

Then $m = \dfrac{y_2 - y_1}{x_2 - x_1}$

$= \dfrac{-2 - 1}{3 - (-4)}$

$m = $ _____

Notice that this line slopes _____ from left to right. To move from the first point to the second point, we must go to the _____ 7 units and _____ 3 units.

Check your work on page 557. ▶

From Examples 1 and 2 we make an important observation.

> If a line slopes *upward* from left to right, its slope is *positive*.
>
> If a line slopes *downward* from left to right, its slope is *negative*.

Example 3. Find the slope of the line through (1, 4) and (−2, 4).

Solution

Let $(x_1, y_1) = (1, 4)$

and $(x_2, y_2) = (-2, 4)$

Then $m = \dfrac{y_2 - y_1}{x_2 - x_1}$

$= \dfrac{4 - 4}{-2 - 1}$

$= \dfrac{0}{-3}$

$m = 0$

Example 4. Find the slope of the line through (2, 3) and (2, −1).

Solution

Let $(x_1, y_1) = (2, 3)$

and $(x_2, y_2) = (2, -1)$

Then $m = \dfrac{y_2 - y_1}{x_2 - x_1}$

$= \dfrac{-1 - 3}{2 - 2}$

$= \dfrac{-4}{0}$

m is undefined.

From Examples 3 and 4 we observe that

The slope of a *horizontal* line is 0.

The slope of a *vertical* line is undefined.

▶ **Trial Run**

Find the slope of the line through the given points.

_____ **1.** (2, 1) and (−3, 0)

_____ **2.** (−4, 3) and (−5, 8)

_____ **3.** (−3, 5) and (4, 5)

_____ **4.** (2, 2) and (−7, −7)

_____ **5.** (9, −3) and (3, 1)

_____ **6.** (7, −1) and (7, 5)

Answers are on page 558.

Using the Slope for Graphing

If we know *one point* on a line and the *slope* of that line, we can find another point on the same line using the fact that

$$\text{slope} = \frac{\text{change in } y}{\text{change in } x}$$

For instance, if we know that the point (2, 1) is on a line with slope $\frac{3}{4}$, we know we must start at (2, 1) and then go to the *right* 4 units and *up* 3 units to arrive at another point on the same line.

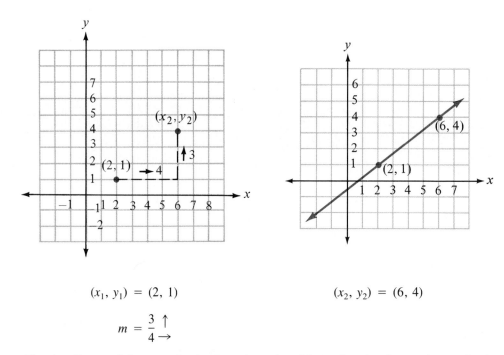

$$(x_1, y_1) = (2, 1) \qquad\qquad (x_2, y_2) = (6, 4)$$

$$m = \frac{3}{4} \begin{array}{c} \uparrow \\ \rightarrow \end{array}$$

The coordinates of the second point (x_2, y_2) are found by noting the changes in x and y that are described by the slope.

Old Coordinate	Change	New Coordinate
$x_1 = 2$	$+4$	$x_2 = 2 + 4 = 6$
$y_1 = 1$	$+3$	$y_2 = 1 + 3 = 4$

Example 5. Graph the line through the origin with slope 3.

Solution. Our first point is the origin $(0, 0)$ and we write our slope 3 as a fraction.

$$m = 3 = \frac{3}{1} \begin{array}{l} \uparrow \\ \rightarrow \end{array}$$

$$(x_1, y_1) = (0, 0)$$

Starting at the point $(0, 0)$ our slope says we must move to the *right* 1 unit and *up* 3 units.

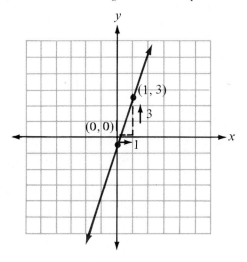

Now you try Example 6.

Example 6. Graph the line through $(3, -2)$ with slope $-\frac{1}{4}$.

Solution

$$(x_1, y_1) = (\underline{\hspace{1cm}}, \underline{\hspace{1cm}})$$

$$m = -\frac{1}{4} = \frac{-1}{4} \begin{array}{l} \downarrow \\ \rightarrow \end{array}$$

Starting at the point $(\underline{\hspace{1cm}}, \underline{\hspace{1cm}})$ our slope says we must move to the $\underline{\hspace{1.5cm}}$ 4 units and $\underline{\hspace{1.5cm}}$ 1 unit.

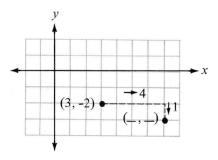

Check your work on page 557. ▶

Example 7. Graph the line through $(5, -2)$ with undefined slope.

Solution. Our first point is $(5, -2)$ and the slope is undefined. Recall that if the slope of a line is undefined, the line is *vertical*.

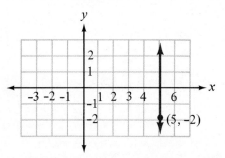

What is the equation for this line? The equation for a vertical line is always $x = k$, a constant. Here the equation must be

$$x = 5$$

Now you try Example 8.

Example 8. Graph the line through $(5, -2)$ with slope 0.

Solution. Our first point is $(\underline{\hspace{1cm}}, \underline{\hspace{1cm}})$ and the slope is $\underline{\hspace{1cm}}$. Recall that if the slope of a line is 0, the line is $\underline{\hspace{1.5cm}}$.

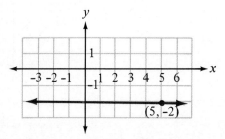

What is the equation for this line? The equation for a horizontal line is always $\underline{\hspace{1.5cm}}$. Here the equation must be

$$\underline{\hspace{1cm}} = \underline{\hspace{1cm}}$$

Check your work on page 557. ▶

⫸ Trial Run

Graph the line through the given point and with the given slope.

 1. $(0, 0);\ m = \dfrac{5}{2}$ **2.** $(4, 2);\ m = 0$

 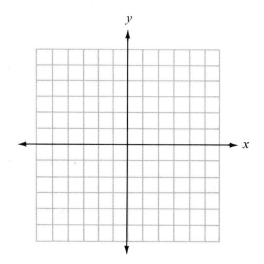

3. $(-2, -4);\ m = -1$ **4.** $(-3, 2);\ m$ is undefined.

 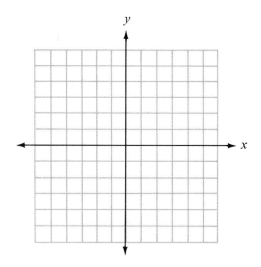

Answers are on page 558.

Graphing Lines by the Slope-Intercept Method

Let's take another look at the graphs at the beginning of Section 10.4. For each graph the
y-intercept has been found by the usual method. Let's find the slope of each line using the
points labeled on each graph.

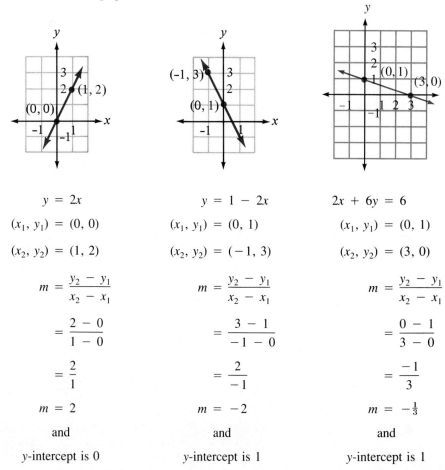

$$y = 2x$$

$$(x_1, y_1) = (0, 0)$$

$$(x_2, y_2) = (1, 2)$$

$$m = \frac{y_2 - y_1}{x_2 - x_1}$$

$$= \frac{2 - 0}{1 - 0}$$

$$= \frac{2}{1}$$

$$m = 2$$

and

y-intercept is 0

$$y = 1 - 2x$$

$$(x_1, y_1) = (0, 1)$$

$$(x_2, y_2) = (-1, 3)$$

$$m = \frac{y_2 - y_1}{x_2 - x_1}$$

$$= \frac{3 - 1}{-1 - 0}$$

$$= \frac{2}{-1}$$

$$m = -2$$

and

y-intercept is 1

$$2x + 6y = 6$$

$$(x_1, y_1) = (0, 1)$$

$$(x_2, y_2) = (3, 0)$$

$$m = \frac{y_2 - y_1}{x_2 - x_1}$$

$$= \frac{0 - 1}{3 - 0}$$

$$= \frac{-1}{3}$$

$$m = -\tfrac{1}{3}$$

and

y-intercept is 1

Let's rewrite each line's equation in the same form, with y isolated on the left-hand side
and the x-term and constant term on the right-hand side.

$$y = 2x \qquad\qquad y = 1 - 2x \qquad\qquad 2x + 6y = 6$$

becomes   becomes   becomes

$$y = 2x + 0 \qquad\qquad y = -2x + 1 \qquad\qquad 6y = -2x + 6$$

$$y = \frac{-2x}{6} + \frac{6}{6}$$

$$y = -\tfrac{1}{3}x + 1$$

Now we summarize the information about these equations and their slopes and y-intercepts.

Equation	Slope	y-Intercept
$y = 2x + 0$	2	0
$y = -2x + 1$	-2	1
$y = -\frac{1}{3}x + 1$	$-\frac{1}{3}$	1

It is not just a coincidence that when y is isolated in a linear equation, the slope of its graph matches the coefficient of the x-term and the y-intercept matches the constant term. In fact (unless the coefficient of y is 0) every linear equation can be put into this form.

Slope-Intercept Form. If a linear equation is written

$$y = mx + b$$

the slope of the line is m and the y-intercept is b.

Slope-intercept form is useful because it allows us to graph a line just by looking at the equation without substituting any values for either variable.

Graphing Linear Equations Using Slope-Intercept Method

1. Rewrite the equation in slope-intercept form

$$y = mx + b$$

2. Note the y-intercept, b, and plot the point $(0, b)$.
3. Use the slope, m, to plot and label a second point.
4. Draw a straight line through the two points.

Example 9. Graph $y = -3x + 2$ by the slope-intercept method.

Solution. $y = -3x + 2$ tells us that

$$m = -3$$

$$= \frac{-3}{1}$$

and $b = 2$. We start at the point $(0, 2)$ and move to the *right* 1 unit and *down* 3 units.

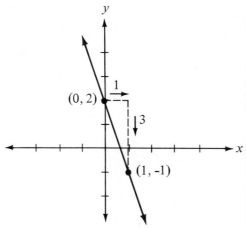

In this example, we chose to write the slope -3 as $\frac{-3}{1}$. We could also have used $\frac{3}{-1}$, in which case we'd have begun at $(0, 2)$ and moved to the *left* 1 unit and *up* 3 units. Our second point then would have been $(-1, 5)$, which is a different point on the *same line*. In this text, however, we shall always let the numerator carry the sign of the slope fraction.

Now you try to complete Example 10.

Example 11. Graph $2x - 3y = 9$ by the slope-intercept method.

Solution. First we must isolate y.

$$2x - 3y = 9$$

$$-3y = -2x + 9 \qquad \text{Subtract } 2x.$$

$$y = \frac{-2x + 9}{-3} \qquad \text{Divide by } -3.$$

$$y = \frac{-2x}{-3} + \frac{9}{-3} \qquad \begin{array}{l}\text{Rewrite the quotient}\\\text{on the right.}\end{array}$$

$$y = \frac{2}{3}x - 3 \qquad \text{Slope-intercept form.}$$

Now we know that

$$m = \frac{2}{3} \begin{array}{l}\uparrow\\\rightarrow\end{array}$$

$$b = -3$$

Example 10. Graph $x + y = 0$ by the slope-intercept method.

Solution. First we must isolate y in the equation.

$$x + y = 0$$

$$y = -x$$

$$y = -1x + 0$$

Now we know that

$$m = \underline{} = \frac{-1}{1}$$

$$b = \underline{}$$

We start at $(0, 0)$ and move to the *right* 1 unit and $\underline{}$ 1 unit.

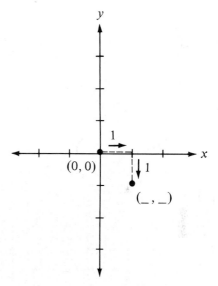

Check your work on page 557. ▶

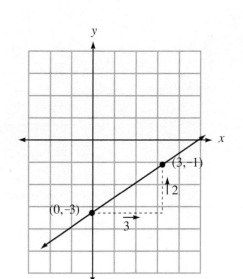

⟱ **Trial Run**

Graph each equation by the slope-intercept method.

1. $y = -3x + 6$

2. $x - y = 4$

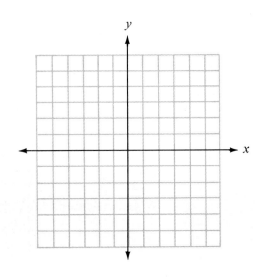

3. $x + 4y = -8$

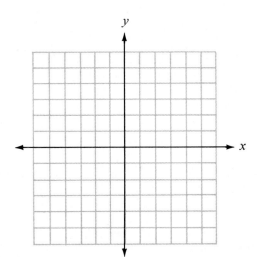

4. $3x - 4y = 12$

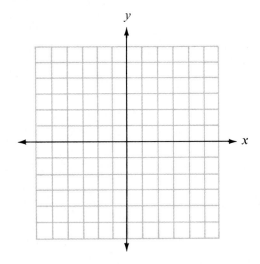

Answers are on page 558.

▶ Examples You Completed

Example 2. Find the slope of the line through $(-4, 1)$ and $(3, -2)$.

Solution

$$\text{Let } (x_1, y_1) = (-4, 1)$$

$$\text{and } (x_2, y_2) = (3, -2)$$

$$\text{Then } m = \frac{y_2 - y_1}{x_2 - x_1}$$

$$= \frac{-2 - 1}{3 - (-4)}$$

$$m = \frac{-3}{7}$$

Notice that this line slopes downward from left to right. To move from the first point to the second point, we must go to the *right* 7 units and *down* 3 units.

Example 8. Graph the line through $(5, -2)$ with slope zero.

Solution. Our first point is $(5, -2)$ and the slope is 0. Recall that if the slope of a line is 0, the line is *horizontal*.

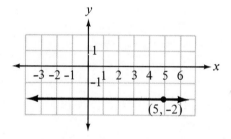

What is the equation for this line? The equation for a horizontal line is always $y = k$, a constant. Here the equation must be

$$y = -2$$

Example 6. Graph the line through $(3, -2)$ with slope $-\frac{1}{4}$.

Solution

$$(x_1, y_1) = (3, -2)$$

$$m = -\frac{1}{4} = \frac{-1}{4} \begin{matrix} \downarrow \\ \rightarrow \end{matrix}$$

Starting at the point $(3, -2)$, our slope says that we must move to the *right* 4 units and *down* 1 unit.

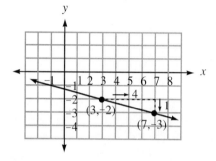

Example 10. Graph $x + y = 0$ by the slope-intercept method.

Solution. First we must isolate y in the equation.

$$x + y = 0$$

$$y = -x$$

$$y = -1x + 0$$

Now we know that

$$m = -1 = \frac{-1}{1}$$

$$b = 0$$

We start at $(0, 0)$ and move to the *right* 1 unit and *down* 1 unit.

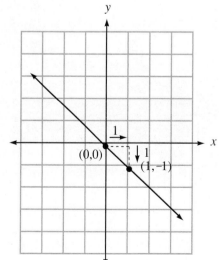

Answers to Trial Runs

page 549   **1.** $\frac{1}{5}$ **2.** -5 **3.** 0   **4.** 1   **5.** $-\frac{2}{3}$ **6.** Undefined

page 552

1.

2.

3.

4.

page 556

1.

2.

3.

4.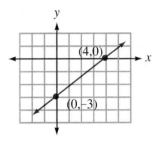

EXERCISE SET 10.4

Find the slope of the line through the given points.

_____ 1. (0, 5) and (−3, 0)

_____ 3. (5, −1) and (5, 4)

_____ 5. (6, −3) and (8, −5)

_____ 7. (3, 5) and (−2, 5)

_____ 9. (2, −3) and (5, 6)

_____ 2. (0, −7) and (4, 0)

_____ 4. (2, −7) and (2, 3)

_____ 6. (9, −7) and (5, −3)

_____ 8. (−4, −3) and (5, −3)

_____ 10. (1, −2) and (−3, 6)

Graph the line through the given point and with the given slope.

11. (1, 3); $m = 2$

13. (5, −1); $m = -3$

15. (3, 5); $m = 0$

17. (−2, 3); $m = -\frac{1}{2}$

19. (5, −4); m is undefined

12. (2, 5); $m = 3$

14. (4, −3); $m = -2$

16. (1, −4); $m = 0$

18. (−1, −1); $m = -\frac{1}{4}$

20. (3, −6); m is undefined

Use the slope-intercept method to graph each equation.

21. $y = -2x + 1$

23. $x - y = 9$

25. $x + 3y = 15$

27. $3x - 2y = 6$

29. $3x + 5y = 0$

22. $y = -3x + 4$

24. $x - y = 6$

26. $x + 6y = 12$

28. $5x - 2y = 10$

30. $4x + 3y = 0$

☆ Stretching the Topics

_____ 1. Find the slope of the line through $(\frac{1}{2}, -\frac{2}{3})$ and $(-\frac{3}{4}, \frac{1}{6})$.

_____ 2. If the coordinates of two points are represented by $(3k, -2c)$ and $(7k, 5c)$, find an expression for the slope of a line through the two points.

_____ 3. Find the missing coordinate if a line through $(-5, 3)$ and $(x, -2)$ has a slope of $-\frac{1}{2}$.

Check your answers in the back of your book.

If you can graph the equations in **Checkup 10.4**, you are ready to do the **Review Exercises** for Chapter 10.

✓ CHECKUP 10.4

Find the slope of the line through the given points.

_____ **1.** (16, −4) and (9, 3)   _____ **2.** (1, 5) and (−3, 5)

Graph the line through the given point and with the given slope.

3. (2, −7); $m = -\dfrac{3}{2}$

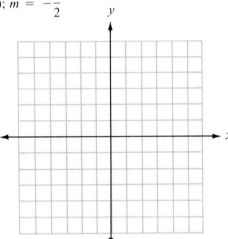

4. (5, −1); $m = 0$

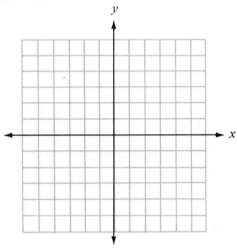

Use the slope-intercept method to graph each equation.

5. $2x - y = 3$

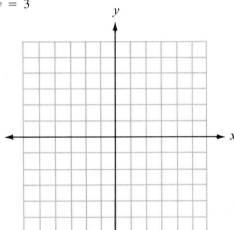

6. $3x - 4y = 0$

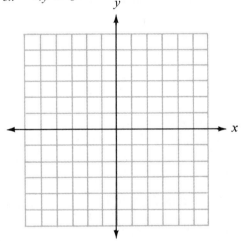

Check your answers in the back of your book.

If You Missed Problems:	You Should Review Examples:
1, 2	1–4
3	6
4	8
5	11
6	10

Summary

In this chapter we have considered equations containing two variables. In particular, we investigated first-degree equations of the form $Ax + By = C$, where A and B are not both zero.

We discovered that solutions to such equations are ordered pairs of numbers (x, y) that can be graphed as points in the Cartesian coordinate plane. Every first-degree equation will yield a graph that is a straight line. The **slope** m of a straight line through the points (x_1, y_1) and (x_2, y_2) can be found by $m = \dfrac{y_2 - y_1}{x_2 - x_1}$.

A straight line can be graphed by three different methods.

Method	Procedure
Arbitrary points	Find three points by choosing x-values and finding corresponding y-values. Plot the points and join with a straight line.
Intercepts	Find the x-intercept (by letting $y = 0$). Find the y-intercept (by letting $x = 0$). Plot both intercepts and join with a straight line.
Slope-intercept	Write equation in the form $y = mx + b$, where m is slope and b is y-intercept. Plot the y-intercept and use the slope to find second point. Join points with a straight line.

We also learned to recognize and graph constant equations.

Equation	Graph
$y = k$, a constant	Horizontal line, crossing the y-axis at $(0, k)$
$x = k$, a constant	Vertical line, crossing the x-axis at $(k, 0)$

❑ Speaking the Language of Algebra

Complete each statement with the appropriate word or phrase.

1. For the point $(3, -2)$, the x-coordinate is _____ and the y-coordinate is _____ .

2. The x-coordinate of the point at which a graph crosses the x-axis is called its _____. The y-coordinate of the point at which a graph crosses the y-axis is called its _____ .

3. The point $(0, 0)$ is called the _____ .

4. The graph of all points (x, y) that satisfy an equation $Ax + By = C$ is a _____ .

5. If (x_1, y_1) and (x_2, y_2) are two points on a line, the slope m of the line is found by $m =$ _____ .

6. The slope of a horizontal line is _____ . Every horizontal line corresponds to an equation of the form _____ .

7. The slope of a vertical line is _____ . Every vertical line corresponds to an equation of the form _____ .

8. If a linear equation is written in the form $y = mx + b$, then m is the _____ of the line and b is its _____ .

Check your answers in the back of your book.

△ Writing About Mathematics ──────────────

Write your response to each question in complete sentences.

1. Write three word statements that would lead you to write the following equation in two variables: $3x + 5y = 120$.

2. Explain how you would graph $3x + 5y = 120$
   (a) by the intercepts method.
   (b) by the slope-intercept method.

3. Explain why each of the following statements is true.
   (a) A line with a positive slope will slope upward from left to right.
   (b) A line with a negative slope will slope downward from left to right.
   (c) A line with slope 0 is horizontal.
   (d) A line with undefined slope is vertical.

REVIEW EXERCISES for Chapter 10

Decide if the given point satisfies the given equation.

_____ **1.** $y = \dfrac{1}{4}x - 3; (-8, -5)$ _____ **2.** $4x - y = 5; (0, -5)$

Find three ordered pairs that satisfy each equation.

_____ **3.** $2x - y = 6$ _____ **4.** $x - 3y = 0$

_____ **5.** $y = -2x + 5$ _____ **6.** $x - y = -2$

Write each word statement as an equation containing two variables. Then list three ordered pairs that satisfy the equation.

_____ **7.** The length of a rectangle is 5 less than 3 times the width.

_____ **8.** Regina sold some of her blouses at a yard sale. For some she got \$5 and for others \$3. The total amount she made from the sale of her blouses was \$45.

9. Graph the following points.

$$A: (-4, 2) \qquad B: (0, -5) \qquad C: (-2, -4)$$

10. Give the coordinates of each point.

____ A

____ B

____ C

____ D

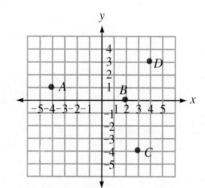

Graph each equation using arbitrary points. Label each point.

11. $x + y = 5$ **12.** $y = -3x$

13. $4x - y = 0$ **14.** $3x - y = 6$

Graph each equation using the intercepts method. Label the intercepts.

15. $x + y = 6$ **16.** $x - 3y = -6$

17. $2x - y = 7$ **18.** $5x - 3y = 15$

19. $x = 4$ **20.** $2y + 12 = 0$

Find the slope of the line through the given points.

_____ **21.** (3, 2) and (0, −1)   _____ **22.** (−7, 3) and (4, 3)

_____ **23.** (−1, −3) and (−5, 4)   _____ **24.** (−5, 3) and (−5, 10)

Graph the line through the given point and with the given slope.

25. (0, 3); $m = -4$ **26.** (−1, 2); $m = 0$

27. (3, −3); m is undefined   **28.** (4, −2); $m = \dfrac{3}{2}$

Graph each equation using the slope-intercept method. Label points.

29. $y = -5x + 4$ **30.** $x + y = 7$

31. $x + 4y = 20$ **32.** $2x - 7y = 14$

33. $3y = 2x$ **34.** $4x - 5y = 0$

Check your answers in the back of your book.

If You Missed Exercises:	You Should Review Examples:	
1, 2	SECTION 10.1	3, 4
3–6		5, 6
7, 8		7, 8
9, 10	SECTION 10.2	1–4
11–14		5, 6
15–18	SECTION 10.3	1–3
19, 20		4–7
21–24	SECTION 10.4	1–4
25–28		5–8
29–34		9–11

If you have completed the **Review Exercises** and corrected your errors, you are ready to take the **Practice Test** for Chapter 10.

PRACTICE TEST for Chapter 10

	SECTION	EXAMPLES

_____ **1.** Does the ordered pair $(-4, 3)$ satisfy the equation $y = \frac{1}{2}x + 1$?　　　10.1　　　3, 4

_____ **2.** Does the ordered pair $(2, -3)$ satisfy the equation $5x + y = 7$?　　　10.1　　　3, 4

Write each word statement as an equation containing two variables.

_____ **3.** Mary's weight is 15 pounds more than half her father's weight.　　　10.1　　　7–9

_____ **4.** Clay spent $15.35 at the post office on 25-cent stamps and 15-cent stamps.　　　10.1　　　7–9

Graph each equation using three arbitrary points. Label each point.

5. $x - y = 4$　　　　　　　　　　　　　　　　10.2　　　5, 7

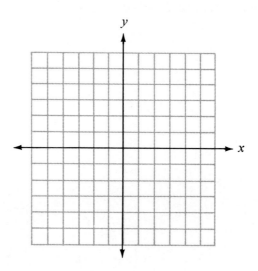

6. $y = 6 - 2x$　　　　　　　　　　　　　　　　10.2　　　6

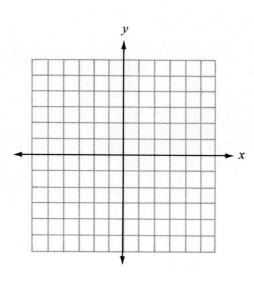

	SECTION	EXAMPLES
7. $2x + y = 0$	10.2	5, 7

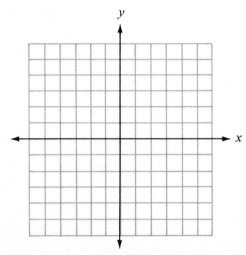

Graph each equation using the intercepts method. Label the intercepts.

	SECTION	EXAMPLES
8. $x - 5y = -10$	10.3	2

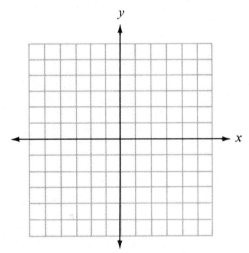

	SECTION	EXAMPLES
9. $4x + 3y = 8$	10.3	1

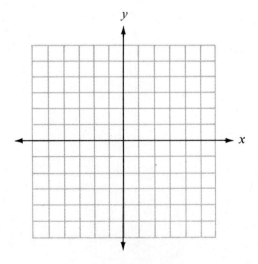

10. $-3x - 2y = 0$

10.3   3

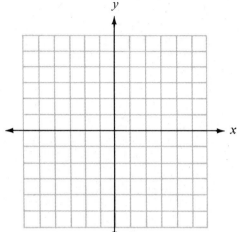

Graph each equation by inspection. Label the intercept.

11. $y - 4 = 0$

10.3   5

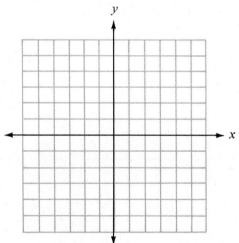

12. $3x + 2 = 0$

10.3   8

	SECTION	EXAMPLES

_____ **13.** Find the slope of the line through $(5, 2)$ and $(-3, 1)$.

SECTION 10.4   EXAMPLES 1, 2

14. Graph the line through the point $(2, -3)$ with slope $\frac{4}{3}$.

10.4   6

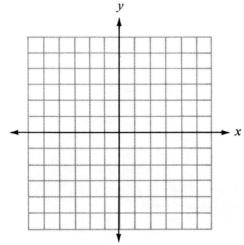

Graph each equation using the slope-intercept method. Label points.

15. $y = x + 4$

10.4   9

16. $6x + 4y = -12$

10.4   11

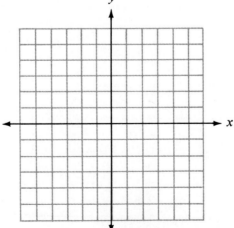

Check your answers in the back of your book.

SHARPENING YOUR SKILLS after Chapters 1–10

	SECTION

_____ **1.** Change 19.3 percent to a decimal number.   1.3

Perform the indicated operations and simplify.

_____ **2.** $0.1107 \div 1.23$ 1.3

_____ **3.** $\dfrac{2}{3} - \dfrac{7}{6} + \dfrac{4}{9}$ 2.4

_____ **4.** $0.3[-2.1 - (3 - 5.6)]$ 2.4

_____ **5.** $\dfrac{4}{5}(20x - 15y)$ 3.3

_____ **6.** $(-6)\left(-\dfrac{1}{2}\right)^2(y)$ 3.3

_____ **7.** $\dfrac{(-2a^2b^6)^4}{(a^3b^2)^5}$ 5.2

_____ **8.** $(-2x^{-1}y^2)^3(x^2y^3)^{-1}$ 5.3

_____ **9.** $5x^2(2x^3y)^3$ 6.2

_____ **10.** $-5ax(2 - 3y) + 10ax(1 - 2y)$ 6.2

_____ **11.** $(x^2 - 3)(x^2 + 3)$ 6.3

_____ **12.** $\dfrac{9x^2 - 45x - 36}{9x}$ 6.4

_____ **13.** $(x^3 - x + 5) \div (x + 2)$ 6.4

_____ **14.** $\dfrac{2x + 3y}{4x^2 + 12xy + 9y^2}$ 8.1

_____ **15.** $\dfrac{(3y - 1)^2}{14x^2y^2} \cdot \dfrac{21xy}{9y^2 - 1}$ 8.2

_____ **16.** $\dfrac{\dfrac{-36x^2y^3}{z^4}}{\dfrac{15x^4y}{z^7}}$ 8.5

_____ **17.** $\dfrac{3}{x + 4} - \dfrac{2}{5x}$ 8.4

Name _____   Date _____

569

Solve each equation or inequality.

———————— 18. $2(5y - 1) - 5 = 3(7 + y)$ 4.2

———————— 19. $63 + 12x - 3x^2 = 0$ 7.5

———————— 20. $3y(y + 2) + 10 = 2(y^2 - 5) - 3y$ 7.5

———————— 21. $\dfrac{a + 3}{a + 1} = \dfrac{a - 3}{a - 2}$ 8.6

———————— 22. $\dfrac{2}{3b - 1} + \dfrac{3}{b + 2} = 0$ 8.6

———————— 23. $\dfrac{5x - 2}{3} > -2$ 9.2

———————— 24. $-2 \leq 3y - 5 \leq 7$ 9.2

———————— 25. $5 - 2x > 1 \quad \text{or} \quad 3x - 6 < 3$ 9.3

———————— 26. Solve $A = P + Prt$ for P.   4.2

———————— 27. Solve and graph the inequality $-\dfrac{3}{2}x + 1 \leq 4$. ⟵——————————⟶   9.2

———————— 28. If Abe earns \$270 per week, for how many weeks must he work to earn at least \$1890?   9.4

———————— 29. The width of a rectangle is 4 centimeters less than the length and the area is 45 square centimeters. Find the dimensions.   7.6

———————— 30. If the denominator of the fraction $\dfrac{x}{7}$ is decreased by x, the new fraction is equal to $\dfrac{2}{5}$. Find the value of x.   8.6

Check your answers in the back of your book.

11

Solving Systems of Linear Equations

José works part-time mowing lawns and lifeguarding at the swimming pool. When he mows for 5 hours and guards for 6 hours, he earns $42. When he mows for 11 hours and guards for 4 hours, he earns $51. What is his hourly wage for each job?

In this problem there are two unknown quantities and two conditions that must be satisfied. To solve this problem, we would like to find out what ordered pair of values (if any) will satisfy two equations at the same time. This is called **solving a system of linear equations** and in this chapter we learn how to

1. Solve a system of linear equations by graphing.
2. Solve a system of linear equations by substitution.
3. Solve a system of linear equations by elimination (or addition).
4. Switch from words to systems of linear equations.

11.1 Solving a System of Equations by Graphing

One way in which we can try to locate common solutions for a system of two equations is by looking at the *graphs* of the two equations. We shall graph two equations in the same Cartesian plane and see what conclusions we can draw.

Consider the following system of two linear equations and graph each line by the intercepts method.

$$(1)\ x + y = 7 \qquad \text{and} \qquad (2)\ 2x - y = 5$$

$$\text{if } x = 0, y = 7 \qquad\qquad \text{if } x = 0, y = -5$$

$$(0, 7) \qquad\qquad\qquad (0, -5)$$

$$\text{if } y = 0, x = 7 \qquad\qquad \text{if } y = 0, x = \tfrac{5}{2}$$

$$(7, 0) \qquad\qquad\qquad (\tfrac{5}{2}, 0)$$

Now we graph both equations in the same plane, labeling the lines (1) and (2).

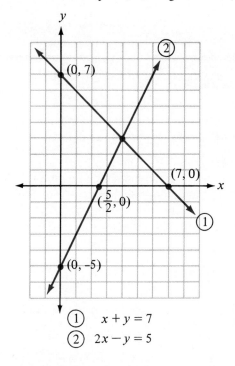

① $x + y = 7$
② $2x - y = 5$

These lines have just one point in common—the point at which the lines cross. That point appears to be (4, 3). We can see if that point is shared by both equations by checking the ordered pair (4, 3) in each one.

$$(1)\ x + y = 7 \qquad (2)\ 2x - y = 5$$
$$4 + 3 \overset{?}{=} 7 \qquad 2(4) - 3 \overset{?}{=} 5$$
$$7 = 7 \qquad\qquad 8 - 3 \overset{?}{=} 5$$
$$5 = 5$$

Indeed, the ordered pair (4, 3) does satisfy both equations. The ordered pair (4, 3) is the only solution for the system of equations.

Try Example 1.

Example 1. Use graphing to solve the system. Then check your solution.

$$(1)\ 2x + 3y = 12$$

$$(2)\ \ x - \ y = \ 1$$

Solution. Use intercepts.

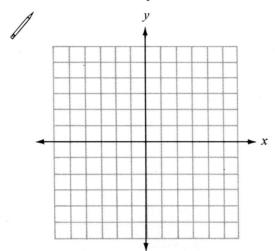

(1) $2x + 3y = 12$

   if $x = 0$, $y =$ ____

   $(0,$ ___$)$

   if $y = 0$, $x =$ ____

   $($___$, 0)$

(2) $x - y = 1$

   if $x = 0$, $y =$ ____

   $(0,$ ___$)$

   if $y = 0$, $x =$ ____

   $($___$, 0)$

The point $($___ , ___$)$ seems to be the solution for the system.

CHECK: We think the solution is $($___ , ___$)$.

(1) $2x + 3y = 12$

   $2($___$) + 3($___$) \overset{?}{=} 12$

   ___ $+$ ___ $\overset{?}{=} 12$

   ___ $= 12$

(2) $x - y = 1$

   ___ $-$ ___ $\overset{?}{=} 1$

   ___ $= 1$

Check your work on page 575. ▶

Example 2. Use graphing to solve the system.

$$(1)\ x + y = \ \ \ 3$$

$$(2)\ x + y = -1$$

Solution. Use intercepts.

(1) $x + y = 3$

   if $x = 0$, $y = 3$

   $(0, 3)$

   if $y = 0$, $x = 3$

   $(3, 0)$

(2) $x + y = -1$

   if $x = 0$, $y = -1$

   $(0, -1)$

   if $y = 0$, $x = -1$

   $(-1, 0)$

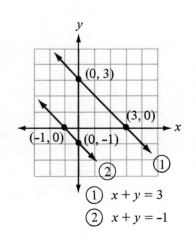

① $x + y = 3$

② $x + y = -1$

Our graphs do not cross; the lines are *parallel*. A system such as this has *no solution* and is called an **inconsistent system**.

Example 3. Use graphing to solve the system.

$$\text{(1) } y = 5x - 2$$

$$\text{(2) } y = 3$$

Solution. For equation (1) we shall use the slope-intercept method.

(1) $y = 5x - 2$

slope: $m = 5 = \dfrac{5}{1}$

y-intercept: $b = -2$

$(0, -2)$

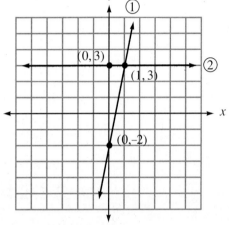

We recognize equation (2) as a constant equation.

(2) $y = 3$

The graph is a horizontal line crossing the y-axis at the point $(0, 3)$.

The point $(1, 3)$ appears to be the common solution. You should check it by substitution in each equation.

Example 4. Use graphing to solve the system.

$$\text{(1) } x + y = 3$$

$$\text{(2) } 2x + 2y = 6$$

Solution. Use intercepts.

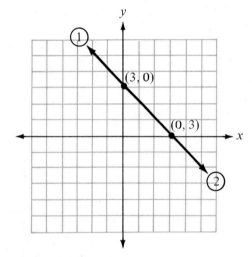

(1) $x + y = 3$

if $x = 0$, $y = 3$

$(0, 3)$

if $y = 0$, $x = 3$

$(3, 0)$

(2) $2x + 2y = 6$

if $x = 0$, $y = 3$

$(0, 3)$

if $y = 0$, $x = 3$

$(3, 0)$

The graph for both equations is the *same line*! Any solution for one of the equations will be a solution for the other equation. Such a system is said to be a **dependent system**. Here we say the solutions for the system are all ordered pairs that satisfy $x + y = 3$ (or $2x + 2y = 6$).

Notice that equation (2) is a constant multiple of equation (1) in Example 4. This will always be true if the equations in a *dependent* system are both written in standard form.

⫸ Trial Run

Use graphing to solve each system.

1. $x - y = 3$
$\quad\ y = 2x - 7$

2. $y = 2x$
$\quad\ x + 2y = 10$

3. $x - y = 5$
$\quad\ y = x + 2$

4. $2x + 4y = 2$
$\quad\ 3x + 6y = 3$

Answers are on page 576.

▶ Example You Completed

Example 1. Use graphing to solve the system. Then check your solution.

$$(1)\ \ 2x + 3y = 12$$
$$(2)\ \ x - y = 1$$

Solution. Use intercepts.

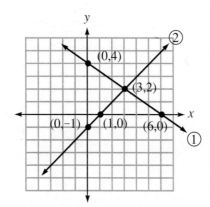

$$(1)\ \ 2x + 3y = 12$$
$$\text{if } x = 0, y = 4$$
$$(0, 4)$$
$$\text{if } y = 0, x = 6$$
$$(6, 0)$$
$$(2)\ \ x - y = 1$$
$$\text{if } x = 0, y = -1$$
$$(0, -1)$$
$$\text{if } y = 0, x = 1$$
$$(1, 0)$$

The point (3, 2) seems to be the solution for the system.

CHECK: We think the solution is (3, 2).

$$(1)\ \ 2x + 3y = 12 \qquad\qquad (2)\ x - y = 1$$
$$2(3) + 3(2) \overset{?}{=} 12 \qquad\qquad 3 - 2 \overset{?}{=} 1$$
$$6 + 6 \overset{?}{=} 12 \qquad\qquad\qquad 1 = 1$$
$$12 = 12$$

wers to Trial Run

page 575

1.

Solution: (4, 1)

2.

Solution: (2, 4)

3.

No solution.

4.

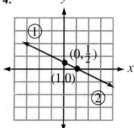

Solution: All ordered pairs that satisfy $2x + 4y = 2$ (or $3x + 6y = 3$).

EXERCISE SET 11.1

Solve each system by graphing.

1. $x - y = -2$
   $x + y = 4$

2. $x + y = 6$
   $x - y = -4$

3. $3x + 2y = 6$
   $2x + y = 2$

4. $5x - 4y = -60$
   $x + y = 6$

5. $3x - y = 6$
   $y = 6x$

6. $-2x + y = 8$
   $y = 4x$

7. $x + y = -5$
   $-2x - 2y = 8$

8. $x - y = 3$
   $3x - 3y = -6$

9. $4x - y = 8$
   $2x + y = 4$

10. $-3x + y = 9$
    $-2x - y = 6$

11. $2x - y = 2$
    $-6x + 3y = -6$

12. $-3x + 3y = 6$
    $2x - 2y = -4$

13. $3x - 2y = 0$
    $x - y = -1$

14. $4x + 3y = 0$
    $x + y = -1$

15. $2x - y = 10$
    $y = 4$

16. $3x + y = 9$
    $y = -3$

17. $2x + y = 9$
    $x - y = 3$

18. $6x + y = -3$
    $x + y = 7$

19. $4x - 2y = 16$
    $-2x + y = -4$

20. $-3x + 2y = -6$
    $6x - 4y = 8$

☆ Stretching the Topics _____

Solve each system by graphing.

1. $x + 2(y + 5) = 3(x + y + 1)$
   $0 = 5y - 3(x - 10)$

2. $\dfrac{2x - y}{4} = x - y$
   $2x - 3y = 0$

Check your answers in the back of your book.

If you can solve the systems in **Checkup 11.1**, you are ready to go on to Section 11.2.

✔ CHECKUP 11.1

Solve each system by graphing.

1. $x - y = 2$
   $x + y = 0$

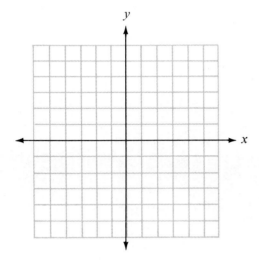

2. $-4x + 2y = 8$
   $2x - y = 2$

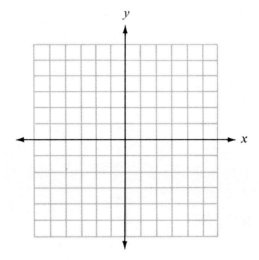

3. $2x + 3y = 6$
   $x = -3$

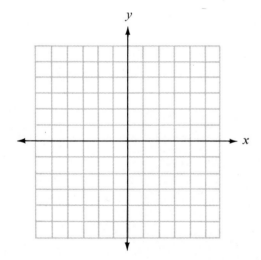

4. $-x + 2y = 4$
   $5x - 10y = -20$

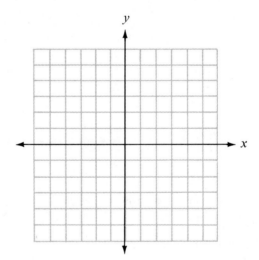

Check your answers in the back of your book.

If You Missed Problem:	You Should Review Example:
1	1
2	2
3	3
4	4

11.2 Solving a System of Equations by Substitution

We have seen that the solution for a system of equations can often be found by graphing each equation and finding the point common to both graphs. Sometimes, however, it is difficult (if not impossible) to find the exact coordinates of the point where two lines cross. We need a method to find solutions without graphing.

In finding the solution for a system of equations, remember that we are looking for a common x-value and a common y-value that satisfy *both* equations. The first method we shall use is called **solving a system by substitution**. The method of substitution works especially well when one variable is already isolated in one of the original equations or when one of the variables has a coefficient of 1.

Example 1. Use substitution to solve the system.

$$(1)\ \ y = 2x - 1$$

$$(2)\ \ 2x + 3y = 5$$

Solution. Equation (1) states that y and $2x - 1$ are the same. Wherever y is used, we may substitute $2x - 1$. In particular, we can substitute the expression $2x - 1$ for y in equation (2).

$2x + 3y = 5$	Original equation (2).
$2x + 3(2x - 1) = 5$	Substitute $2x - 1$ for y in equation (2).
$2x + 6x - 3 = 5$	Remove parentheses.
$8x - 3 = 5$	Combine like terms.
$8x = 8$	Add 3 to isolate x-term.
$x = 1$	Divide by 8 to isolate x.

Now we know the x-value for the common solution. To find the corresponding y-value, we can substitute 1 for x in *either* of the original equations.

(1) $y = 2x - 1$	or	(2) $2x + 3y = 5$
$y = 2(1) - 1$		$2(1) + 3y = 5$
$y = 2 - 1$		$2 + 3y = 5$
$y = 1$		$3y = 3$
		$y = 1$

The solution for the system is the ordered pair (1, 1). Graphically, this means that the two lines intersect at the point (1, 1).

Solving a System by Substitution

1. Look for a variable appearing in an equation with a coefficient of 1, and isolate that variable in that equation.
2. Substitute the expression obtained for the isolated variable into the other equation.
3. Solve for the value of the remaining variable.
4. Substitute that value into either of the original equations and solve for the other variable.
5. Write the solution as an ordered pair.

Example 2. Use substitution to solve the system.

$$(1)\ 5x + 2y = 9$$

$$(2)\ \ x + 3y = 4$$

Solution. We isolate x in equation (2).

$$(2)\ x + 3y = 4$$

$$x = 4 - 3y$$

Now substitute $4 - 3y$ for x in equation (1).

$$(1)\ 5x + 2y = 9$$

becomes

$$5(4 - 3y) + 2y = 9$$

$$20 - 15y + 2y = 9$$

$$20 - 13y = 9$$

$$-13y = -11$$

$$y = \frac{-11}{-13}$$

$$y = \frac{11}{13}$$

To find the corresponding x-value, return to equation (2) with x isolated and substitute $\frac{11}{13}$ for y.

$$x = 4 - 3y$$

$$x = 4 - 3\left(\frac{11}{13}\right)$$

$$= 4 - \frac{33}{13}$$

$$= \frac{52}{13} - \frac{33}{13}$$

$$x = \frac{19}{13}$$

The solution is $(\frac{19}{13}, \frac{11}{13})$.

Now try Example 3.

Example 3. Use substitution to solve the system.

$$(1)\ 2x + 3y = 4$$

$$(2)\ 3x + \ y = 7$$

Solution. We isolate y in equation (2).

$$(2)\ 3x + y = 7$$

$$y = 7 - 3x$$

Now we substitute _____ for y in equation (1).

$$(1)\ 2x + 3y = 4$$

becomes

$$2x + 3(\text{_____}) = 4$$

To find the corresponding y-value, return to equation (2) with y isolated and substitute _____ for x.

$$y = 7 - 3x$$

$$y = 7 - 3(\text{\_\_\_\_})$$

The solution is (_____ , _____).

Check your work on page 583. ▶

Can you imagine how difficult it would have been to locate the common points obtained in Examples 2 and 3 using the method of *graphing*? That is why we needed a method for solving a system *algebraically* without relying on a graph.

Example 4. Use substitution to solve the system.

$$(1) \quad y = 3x + 2$$

$$(2) \quad 3x - y = 3$$

Solution. Since y is isolated in equation (1), we can substitute $3x + 2$ for y in equation (2).

$$
\begin{aligned}
(2) \qquad 3x - y &= 3 \\
3x - (3x + 2) &= 3 \\
3x - 3x - 2 &= 3 \\
-2 &= 3
\end{aligned}
$$

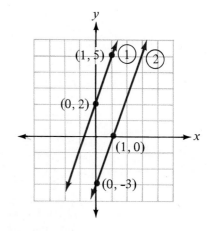

This result is disturbing, since we know that $-2 \neq 3$. To determine what this result means for our system, look at the graph.

Our two lines are *parallel*; the equations have no ordered pair in common. There is *no solution* for this system.

In general, if the method of substitution yields a *false* statement, we must conclude that there is *no* ordered pair that satisfies the system of equations. The system is *inconsistent* and the graphs of the equations are parallel lines.

Example 5. Use substitution to solve the system.

$$(1) \quad 4x + 2y = 10$$

$$(2) \quad y = 5 - 2x$$

Solution. Since y is isolated in equation (2), we can substitute $5 - 2x$ for y in equation (1).

$$
\begin{aligned}
(1) \qquad 4x + 2y &= 10 \\
4x + 2(5 - 2x) &= 10 \\
4x + 10 - 4x &= 10 \\
10 &= 10
\end{aligned}
$$

This is certainly a true statement, but it contains no variable, so we have not found either coordinate of a common point. To determine what this result means for our system, look at the graph.

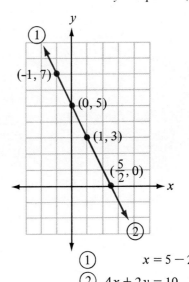

$$
\begin{aligned}
&\text{①} \qquad x = 5 - 2x \\
&\text{②} \quad 4x + 2y = 10
\end{aligned}
$$

Both equations graph into the *same line*; any solution for one equation is a solution for the other equation. Here any ordered pair that satisfies $y = 5 - 2x$ (or $4x + 2y = 10$) will satisfy both equations.

In general, whenever the method of substitution yields a statement that is *always true*, we conclude that the equations have *all* ordered pairs in common, and the system is *dependent*.

Let's note here the results that can occur when we solve a system by substitution.

If substitution yields *one* ordered pair, the graphs intersect in *one point* and the system is said to be **independent and consistent**.

If substitution yields a *false* statement, the system has *no solution*. The graphs are **parallel lines** and the system is **inconsistent**.

If substitution yields a statement that is *always true*, the system has an *infinite number* of solutions. Any solution for one equation is a solution for both equations. The graphs are the **same line** and the system is **dependent**.

⫸ Trial Run

Use substitution to solve each system.

_____ **1.** $y = 7x - 3$
$\quad\quad\quad y = 6 - 2x$

_____ **2.** $y = 3x$
$\quad\quad\quad 3x - 2y = 6$

_____ **3.** $x + 4y = 5$
$\quad\quad\quad 2y - x = 7$

_____ **4.** $y + 2 = 4x$
$\quad\quad\quad 3y + 2x = 1$

_____ **5.** $y = 3x + 7$ _____ **6.** $x = 4y + 1$
      $2y - 6x = 5$ $8y = 2x - 2$

Answers are below.

▶ **Example You Completed** _____

Example 3. Use substitution to solve the system.

$$(1)\ 2x + 3y = 4$$

$$(2)\ 3x +\ y = 7$$

Solution. We isolate y in equation (2).

$$(2)\ 3x + y = 7$$

$$y = 7 - 3x$$

Now we substitute $7 - 3x$ for y in equation (1).

$$(1)\ 2x + 3y = 4$$

becomes

$$2x + 3(7 - 3x) = 4$$

$$2x + 21 - 9x = 4$$

$$21 - 7x = 4$$

$$-7x = -17$$

$$x = \frac{-17}{-7}$$

$$x = \frac{17}{7}$$

To find the corresponding y-value, return to equation (2) with y isolated and substitute $\frac{17}{7}$ for x.

$$y = 7 - 3x$$

$$y = 7 - 3\left(\frac{17}{7}\right)$$

$$= 7 - \frac{51}{7}$$

$$= \frac{49}{7} - \frac{51}{7}$$

$$y = -\frac{2}{7}$$

The solution is $(\frac{17}{7}, -\frac{2}{7})$.

Answers to Trial Run _____

page 582   **1.** $(1, 4)$ **2.** $(-2, -6)$ **3.** $(-3, 2)$ **4.** $(\frac{1}{2}, 0)$ **5.** No solution
6. All ordered pairs that satisfy $x = 4y + 1$ (or $8y = 2x - 2$)

EXERCISE SET 11.2

Solve each system by the substitution method.

_____ **1.** $8x - 3y = 7$
$\qquad\qquad x = 2$

_____ **2.** $7x - 11y = -1$
$\qquad\qquad\quad x = 3$

_____ **3.** $3x - 8y = 13$
$\qquad\qquad\quad y = 2x$

_____ **4.** $5x - 3y = 28$
$\qquad\qquad\quad y = -3x$

_____ **5.** $y = 4x - 5$
$\qquad\quad 5x - y = 11$

_____ **6.** $y = 3x + 8$
$\qquad\quad 7x - y = -4$

_____ **7.** $2x - 5y = 9$
$\qquad\quad\; x - 3y = 2$

_____ **8.** $3x - 2y = 15$
$\qquad\quad\; x - 2y = 1$

_____ **9.** $10x - 5y = 20$
$\qquad\qquad\; y = 2x - 4$

_____ **10.** $3x - 15y = 15$
$\qquad\qquad\; x = 5y + 5$

_____ **11.** $4x - 3y = 7$
$\qquad\quad\; 2x - y = -1$

_____ **12.** $5x - 2y = 12$
$\qquad\quad\; 4x - y = 3$

_____ **13.** $\; x + 2y = -1$
$\qquad\quad 4x + 6y = 5$

_____ **14.** $\; x - 3y = 5$
$\qquad\quad 3x - 7y = 4$

_____ **15.** $3x - 2y = 12$
$\qquad\quad\; x + y = 6$

_____ **16.** $2x - 5y = 11$
$\qquad\quad\; x + y = 3$

_____ **17.** $x + 3y = 10$
$\qquad\quad x - 3 = 0$

_____ **18.** $x + 9y = 1$
$\qquad\quad x + 5 = 0$

_____ **19.** $14x - 2y = 10$
$\qquad\quad\; 7x - y = 2$

_____ **20.** $10x + 2y = 3$
$\qquad\quad\; 5x + y = -1$

_____ **21.** $\; x - y = -2$
$\qquad\quad 11x - 3y = 2$

_____ **22.** $\; x - y = -4$
$\qquad\quad 7x + 5y = -4$

_____ **23.** $\; x - 2y = -3$
$\qquad\quad 3x - 7y = 1$

_____ **24.** $\; x + 5y = 1$
$\qquad\quad 2x + 8y = -6$

_____ **25.** $3x + y = 2$
$\qquad\quad 4x - 2y = -3$

_____ **26.** $5x + y = -1$
$\qquad\quad 13x + 3y = 2$

☆ Stretching the Topics _____

Solve each system by the substitution method.

_____ **1.** $8x - 5y = 0.9$
$\qquad\quad 6x + y = 0.2$

_____ **2.** $3x + 9y = 21$
$\qquad\quad \frac{2}{3}x + y = \frac{1}{3}$

_____ **3.** Solve for x and y in terms of a.
$\qquad\quad ax + 3ay = 5a^2$
$\qquad\qquad x - y = a$

Check your answers in the back of your book.

If you can solve the systems in **Checkup 11.2,** you are ready to go on to Section 11.3.

✓ **CHECKUP 11.2**

Solve each system by the substitution method.

_____ **1.** $5x - 2y = 3$
$y = 2x$

_____ **2.** $x - y = -3$
$-4x + 4y = 1$

_____ **3.** $x - 6 = 0$
$3x - 4y = 8$

_____ **4.** $9x - 4y = 10$
$2x + y = 6$

_____ **5.** $4x + 7y = 16$
$x + 4y = 7$

Check your answers in the back of your book.

If You Missed Problems:	You Should Review Examples:
1	1
2	4
3	1
4, 5	2, 3

11.3 Solving a System of Equations by Elimination (or Addition)

Besides graphing and substitution, there is another method for solving systems of equations. This method uses two properties of equations with which we are already familiar.

---

1. We may multiply both sides of an equation by any constant (except 0).
2. We may add equal quantities to both sides of an equation without changing the solutions of the original equation.

Suppose that we wish to solve the system

$$(1)\ \ 3x + 2y = 9$$

$$(2)\ \ \ x - 2y = 7$$

Notice what happens when we add the left-hand side of equation (1) to the left-hand side of equation (2) *and* add the right-hand side of equation (1) to the right-hand side of equation (2). (We know that this is permissible because of the second property stated above.)

$$(1)\ \ 3x + 2y = \ \ 9$$
$$(2)\ \ \underline{\ \ x - 2y = \ \ 7}$$
$$4x + 0\ = 16$$

Notice that we have completely eliminated the variable y; our new equation contains the single variable x. Let's solve for x.

$$4x = 16$$

$$x = 4$$

This is the x-coordinate of the solution for the system. We may find the corresponding y-value by substituting 4 for x in either equation (1) or equation (2). Let's use equation (2) and then check our solution in equation (1).

$$(2)\ x - 2y = 7 \qquad\qquad \text{CHECK}$$
$$4 - 2y = 7 \qquad\qquad (1)\qquad 3x + 2y = 9$$
$$-2y = 3 \qquad\qquad\qquad 3(4) + 2\left(-\frac{3}{2}\right) \overset{?}{=} 9$$
$$y = -\frac{3}{2} \qquad\qquad\qquad\qquad 12 - 3 \overset{?}{=} 9$$
$$(x, y) = (4, -\tfrac{3}{2}) \qquad\qquad\qquad\qquad 9 = 9$$

The ordered pair $(4, -\tfrac{3}{2})$ satisfies both equations.

This method, called the **method of elimination** (or the **method of addition**), works nicely if one of the variables appears in the equations with *coefficients that are opposites*. This is an important observation.

Now you complete Example 1.

Example 1. Use elimination to solve the system.

$$(1) \quad x + 3y = 5$$
$$(2) \quad -x - 7y = 3$$

Solution. Since the coefficients of x are *opposites*, we shall eliminate the variable x and solve for y.

$$(1) \quad x + 3y = 5$$
$$(2) \quad \underline{-x - 7y = 3}$$
$$0 - 4y = 8$$
$$-4y = 8$$
$$y = \underline{\hspace{1cm}}$$

Now we can substitute _____ for y in equation (1) and check our solution in equation (2).

$$(1) \qquad x + 3y = 5$$
$$x + 3(\underline{\hspace{0.8cm}}) = 5$$
$$x - \underline{\hspace{0.8cm}} = 5$$
$$x = \underline{\hspace{0.8cm}}$$
$$(x, y) = (\underline{\hspace{0.8cm}}, \underline{\hspace{0.8cm}})$$

CHECK

$$(2) \qquad -x - 7y = 3$$
$$\underline{\hspace{0.8cm}} - 7(\underline{\hspace{0.8cm}}) \stackrel{?}{=} 3$$
$$\underline{\hspace{0.8cm}} + \underline{\hspace{0.8cm}} \stackrel{?}{=} 3$$
$$\underline{\hspace{0.8cm}} = 3$$

The ordered pair (_____ , _____) satisfies both equations.

Check your work on page 593. ▶

Elimination works well when one of the variables has opposite coefficients in the two equations. But can the elimination method be used when the original coefficients are not opposites? Yes, provided that we *make the coefficients opposites*.

Example 2. Use elimination to solve the system.

$$(1) \; 2x + 3y = 7$$
$$(2) \quad x + 5y = 4$$

Solution. Here neither variable seems ready for elimination. Suppose that we wish to eliminate the variable x. To do that, the x-term in equation (2) would have to be $-2x$. If we multiply equation (2) by -2, we will have what we want. (Remember that it is permissible to multiply both sides of an equation by any constant except 0.)

$$(2) \qquad x + 5y = 4$$

becomes

$$-2[x + 5y = 4]$$
$$-2x - 10y = -8$$

and our new system is

$$(1) \quad 2x + 3y = 7$$
$$(2) \quad \underline{-2x - 10y = -8}$$
$$-7y = -1$$
$$y = \frac{-1}{-7}$$
$$y = \frac{1}{7}$$

Substituting $\frac{1}{7}$ for y in the original equation (2), we have

$$(2) \qquad x + 5y = 4$$
$$x + 5\left(\frac{1}{7}\right) = 4$$
$$x + \frac{5}{7} = 4$$
$$x = 4 - \frac{5}{7}$$
$$x = \frac{28}{7} - \frac{5}{7}$$
$$x = \frac{23}{7}$$

The solution for the system is $\left(\frac{23}{7}, \frac{1}{7}\right)$.

Example 3. Use elimination to solve the system.

$$(1) \; 4x - 7y = 20$$
$$(2) \; 5x + 2y = 3$$

Solution. To eliminate either variable in these equations, we must multiply *both* equations by constants. We may choose to work with either variable. Since the y-coefficients are already opposite in sign, let's work to eliminate the variable y.

$$(1) \; 4x - 7y = 20$$
$$(2) \; 5x + 2y = 3$$

To eliminate y, the coefficients must be opposite multiples of 7 and 2. The smallest multiple of 7 and 2 is 14, so we must make the coefficients -14 and 14.

Multiply equation (1) by 2.   Multiply equation (2) by 7.

$$2[4x - 7y = 20]$$ $$7[5x + 2y = 3]$$
$$8x - 14y = 40$$ $$35x + 14y = 21$$

Our new system is

$$(1) \quad 8x - 14y = 40$$
$$(2) \quad \underline{35x + 14y = 21}$$
$$43x + 0 = 61$$
$$43x = 61$$
$$x = \frac{61}{43}$$

The thought of substituting this value for x into either equation is not very appealing. It might be easier to start all over again and eliminate the variable x, so that we are left with the variable y.

$$(1) \ 4x - 7y = 20$$

$$(2) \ 5x + 2y = \ 3$$

To eliminate x, the coefficients must be opposite multiples of 4 and 5. The smallest multiple of 4 and 5 is 20, so we must make the coefficients 20 and -20.

Multiply equation (1) by 5.   Multiply equation (2) by -4.

$$5[4x - \ 7y = 20] \qquad\qquad -4[5x + 2y = 3]$$

$$20x - 35y = 100 \qquad\qquad -20x - 8y = -12$$

$$
\begin{aligned}
(1) \quad & 20x - 35y = \ 100 \\
(2) \quad & \underline{-20x - \ \ 8y = -12} \\
& 0 \ - 43y = \ \ \ 88
\end{aligned}
$$

$$y = -\frac{88}{43}$$

The solution is the ordered pair $(\frac{61}{43}, -\frac{88}{43})$.

Example 4. Use elimination to solve the system.

$$(1) \ 7x + 6y = 8$$

$$(2) \ 5x + 4y = 6$$

Solution. Let's eliminate y. To do so, the coefficients of y must be opposite multiples of 6 and 4. The smallest multiple of 6 and 4 is 12, so we make the coefficients 12 and -12.

Multiply equation (1) by 2.   Multiply equation (2) by -3.

$$2[7x + \ 6y = 8] \qquad\qquad -3[5x + \ 4y = 6]$$

$$14x + 12y = 16 \qquad\qquad -15x - 12y = -18$$

$$
\begin{aligned}
(1) \quad & 14x + 12y = \ \ \ 16 \\
(2) \quad & \underline{-15x - 12y = -18} \\
& -x \qquad\quad = \ -2
\end{aligned}
$$

$$x = 2$$

Now we can substitute 2 for x in either of the original equations. Let's use equation (2).

$$
\begin{aligned}
(2) \quad 5x + 4y &= 6 \\
5(2) + 4y &= 6 \\
10 + 4y &= 6 \\
4y &= -4 \\
y &= -1
\end{aligned}
$$

The solution is the ordered pair $(2, -1)$.

In general, to solve a system of equations by the method of elimination, we should keep these steps in mind.

Solving a System by Elimination (Addition)

1. Write both equations in the form $Ax + By = C$.
2. Look to see if the coefficient of one of the variables in one equation is the opposite of the coefficient of the same variable in the other equation.
3. If the coefficients are not opposites, multiply one or both equations by constants that will make the coefficients opposites.
4. Add the equations together to eliminate one variable.
5. Solve for the value of the remaining variable.
6. Substitute that value into either of the original equations and solve for the other variable; *or* repeat the addition process, eliminating the other variable.
7. Write the solution as an ordered pair.

⫸ Trial Run

Use elimination to solve each system.

_____ **1.** $3x - y = 2$
$2x + y = 8$

_____ **2.** $8x - 5y = -4$
$-8x + 11y = 4$

_____ **3.** $-2x + y = 1$
$4x + 3y = 8$

_____ **4.** $7x - 3y = -9$
$2x + 5y = 4$

Answers are on page 593.

_____

Let's consider some more systems of equations.

Example 5. Use elimination to solve the system.

$$(1)\ 2x + 2y = 5$$

$$(2)\ 3x + 3y = 1$$

Solution. To eliminate the variable x, we can multiply equation (1) by 3 and multiply equation (2) by -2.

$$3[2x + 2y = 5]$$

$$-2[3x + 3y = 1]$$

$$(1)\quad 6x + 6y = 15$$

$$(2)\quad \underline{-6x - 6y = -2}$$

$$0 + 0 = 13$$

$$0 = 13$$

The elimination process has yielded a *false statement*. As before, we must conclude that there is *no ordered pair* common to both equations. Indeed, if we were to graph these equations, the two lines would be *parallel*. There is *no solution* for this *inconsistent* system.

Example 6. Use elimination to solve the system.

$$(1) \ y = 3x + 1$$

$$(2) \ 6x - 2y + 2 = 0$$

Solution. First we line up like variables beneath each other.

$$(1) \ -3x + \ y = \ \ \ 1$$

$$(2) \ \ \ \ 6x - 2y = -2$$

We can eliminate the variable y by multiplying equation (1) by 2.

$$2[-3x + \ y = 1]$$

$$-6x + 2y = 2$$

Our new system is

$$(1) \ -6x + 2y = \ \ \ 2$$

$$(2) \ \ \ \underline{\ \ 6x - 2y = -2}$$

$$0 \ + 0 \ = \ \ \ 0$$

$$0 \ = \ \ \ 0$$

We seem to have eliminated both variables! What does this mean? The statement $0 = 0$ is *always true*, but what is the solution for our system? If we were to graph these equations, they would both be the *same line*. Our conclusion is that these equations have *all* ordered pairs in common. All ordered pairs that satisfy $y = 3x + 1$ (or $\ 6x - 2y + 2 = 0$) will satisfy this *dependent* system.

Let's review the results that can occur when elimination is used to solve a system of equations.

If elimination yields *one* ordered pair, the graphs intersect in *one point*, and the system is **independent and consistent**.

If elimination yields a *false* statement, the system has *no solution*. The graphs are *parallel lines* and the system is **inconsistent**.

If elimination yields a statement that is *always true*, the system has an infinite number of solutions. Any solution for one equation is a solution for both equations. The graphs are the *same line* and the system is **dependent**.

▸ Trial Run

Use elimination to solve each system.

_____ **1.** $3x - 12y = 15$
$x - 4y = -7$

_____ **2.** $5x + 6y = 4$
$4x - 3y = -2$

_____ **3.** $5x - 10y = -15$
$-4x + 8y = 12$

_____ **4.** $-9x + 12y = -6$
$15x - 20y = 1$

Answers are below.

▶ Example You Completed

Example 1. Use elimination to solve the system.

$$(1) \quad x + 3y = 5$$
$$(2) \quad -x - 7y = 3$$

Solution. Since the coefficients of x are *opposites*, we shall eliminate the variable x and solve for y.

$$(1) \quad x + 3y = 5$$
$$(2) \quad \underline{-x - 7y = 3}$$
$$0 - 4y = 8$$
$$-4y = 8$$
$$y = -2$$

Now we can substitute -2 for y in equation (1) and check our solution in equation (2).

$(1) \quad x + 3y = 5$	CHECK
$x + 3(-2) = 5$	$(2) \quad -x - 7y = 3$
$x - 6 = 5$	$-11 - 7(-2) \overset{?}{=} 3$
$x = 11$	$-11 + 14 \overset{?}{=} 3$
$(x, y) = (11, -2)$	$3 = 3$

The ordered pair $(11, -2)$ satisfies both equations.

Answers to Trial Runs

page 591   **1.** $(2, 4)$ **2.** $(-\frac{1}{2}, 0)$ **3.** $(\frac{1}{2}, 2)$ **4.** $(-\frac{33}{41}, \frac{46}{41})$

page 593   **1.** No solution   **2.** $(0, \frac{2}{3})$ **3.** All ordered pairs that satisfy $5x - 10y = -15$
4. No solution

EXERCISE SET 11.3

Solve each system by the elimination method.

_____ 1. $3x - y = 4$
$2x + y = 1$

_____ 2. $4x - y = 5$
$3x + y = 9$

_____ 3. $-2x - y = 3$
$2x + 7y = -9$

_____ 4. $4x - 5y = 6$
$-4x - y = 6$

_____ 5. $2x - 4y = 1$
$4y = 7$

_____ 6. $4x - 9y = -9$
$9y = 1$

_____ 7. $3x + 4y = -2$
$2x - 4y = 12$

_____ 8. $5x + 3y = -1$
$x - 3y = 7$

_____ 9. $6x - 4y = 1$
$-6x - 4y = 7$

_____ 10. $-5x - 3y = 2$
$5x - 3y = 10$

_____ 11. $3x - 2y = 0$
$5x + 2y = 4$

_____ 12. $4x + 5y = 0$
$4x + y = 2$

_____ 13. $7x - 10y = 4$
$x + 5y = 7$

_____ 14. $8x - 4y = 6$
$3x + 2y = 4$

_____ 15. $3x + 4y = 5$
$-9x - 7y = 5$

_____ 16. $5x + 3y = 7$
$-10x - 13y = 7$

_____ 17. $2x - 8y = 9$
$-x + 4y = -6$

_____ 18. $-3x + 3y = -1$
$x - y = 8$

_____ 19. $5x - 2y = 5$
$4x - 6y = 15$

_____ 20. $10x - 9y = 12$
$2x - 3y = 4$

_____ 21. $3x + 4y = 9$
$9x + 4y = 5$

_____ 22. $7x + 6y = 10$
$7x + 12y = 9$

_____ 23. $5x - 4y = 1$
$2x + 3y = 5$

_____ 24. $7x - 6y = 20$
$3x + 5y = 1$

_____ 25. $14x + 7y = 7$
$-6x - 3y = -3$

_____ 26. $5x - 15y = 10$
$-6x + 18y = -12$

_____ 27. $11x - 9y = 14$
$7x + 3y = 6$

_____ 28. $3x - 5y = 15$
$11x + 15y = -5$

_____ 29. $7x - 5y = 2$
$2x - 3y = -1$

_____ 30. $2x - 3y = 5$
$5x - 2y = -4$

☆ Stretching the Topics _____

Solve each system by the elimination method.

_____ 1. $0.8b = 0.2a + 6.4$
$0.7a - 0.2b = 1$

_____ 2. $\frac{3}{5}x - \frac{2}{3}y = 3$
$\frac{2}{5}x + \frac{1}{3}y = -2$

_____ 3. Find a value of k for which the following system would be dependent.

$4x - 2y = 7$
$12x - ky = 21$

Check your answers in the back of your book.

If you can solve the systems in **Checkup 11.3**, you are ready to go on to Section 11.4.

✓ **CHECKUP 11.3**

Solve each system by the elimination method.

_____ 1. $3x + y = 6$
$5x - y = 10$

_____ 2. $3a + 5b = -1$
$2a - b = 8$

_____ 3. $2x - 3y = 5$
$-4x + 6y = -7$

_____ 4. $8x + 5y = 9$
$6x + 9y = 5$

_____ 5. $9x + 5y = 53$
$5x - 6y = -32$

Check your answers in the back of your book.

If You Missed Problem:	You Should Review Example:
1	1
2	2
3	5
4	4
5	3

11.4 Switching from Word Statements to Systems of Equations

Throughout our study of algebra, we have practiced switching word statements into variable equations. Sometimes we may be asked to solve a problem containing more than one unknown quantity; in such a situation we may wish to use *two* variables. In dealing with two variables, we must translate the given information into *two* equations and solve the resulting system.

Example 1. For a recent chili supper a total of 191 tickets were sold. Adults' tickets sold for $2 each and children's tickets sold for $1 each. If $304 worth of tickets were sold, how many adults' tickets and how many children's tickets were sold?

Solution

We Think	*We Write*
We must find *two* unknown quantities, so we use two variables.	Let x = number of adults' tickets y = number of children's tickets
We know the *total number* of tickets is 191.	(1) $x + y = 191$
Now we need expressions for the *money* from the two kinds of tickets.	$2 \cdot x$ = dollars from adults' tickets $1 \cdot y$ = dollars from children's tickets
We know the *total money* from all the tickets is 304 dollars.	(2) $2x + y = 304$

Now we have a system of two equations that we can solve by eliminating y.

$$(1) \quad x + y = 191$$
$$(2) \quad 2x + y = 304$$

(1) $-x - y = -191$	Multiply equation (1) by -1.
(2) $\underline{\;2x + y = 304\;}$	Equation (2).
$x + 0 = 113$	Add the equations.
$x = 113$	Solve for x.

We know that 113 adults' tickets were sold and we can substitute 113 for x in equation (1) to find y. Then we can check our solution in equation (2).

$$(1) \quad x + y = 191$$
$$113 + y = 191$$
$$y = 78$$

CHECK

$$(2) \quad 2x + y = 304$$
$$2(113) + 78 \stackrel{?}{=} 304$$
$$226 + 78 \stackrel{?}{=} 304$$
$$304 = 304$$

There were 113 adults' tickets and 78 children's tickets sold.

Let's return to the problem stated at the beginning of the chapter and use a system of equations to find José's hourly wages.

Example 2. José works part-time mowing lawns and lifeguarding at the swimming pool. When he mows for 5 hours and guards for 6 hours, he earns $42. When he mows for 11 hours and guards for 4 hours, he earns $51. What is his hourly wage for each job?

Solution. Our problem gives us information about José's earnings in two different situations.

We Think	*We Write*
We must find *two* unknown quantities, so we use two variables.	Let x = hourly mowing rate (dollars) y = hourly guarding rate (dollars)
In the first situation, José mows for 5 hours and guards for 6 hours. His *total* pay is 42 dollars.	(1) $5x + 6y = 42$
In the second situation, José mows for 11 hours and guards for 4 hours. His *total* pay is 51 dollars.	(2) $11x + 4y = 51$

Now we have a system of two equations that we can solve by elimination. To eliminate y, we note that the smallest multiple of 6 and 4 is 12.

$$(1)\quad 5x + 6y = 42$$
$$(2)\quad 11x + 4y = 51$$

(1) $2[\ 5x + 6y\ = 42]$	Multiply equation (1) by 2.
(2) $-3[11x + 4y\ = 51]$	Multiply equation (2) by -3.
(1) $10x + 12y =\quad 84$	Remove parentheses in equation (1).
(2) $\underline{-33x - 12y = -153}$	Remove parentheses in equation (2).
$-23x +\ \ 0\ =\ -69$	Add the equations.
$-23x =\ -69$	Simplify.
$x = \dfrac{-69}{-23}$	Divide by -23 to isolate x.
$x = 3$	Simplify.

Now we know that José earns $3 per hour mowing lawns. To find y, we can substitute 3 for x in equation (2) and check our solution in equation (1).

(2) $11x + 4y = 51$

$11(3) + 4y = 51$

$33 + 4y = 51$

$4y = 18$

$y = \dfrac{18}{4}$

$y = 4.5$

CHECK

(1) $5x + 6y = 42$

$5(3) + 6(4.5) \stackrel{?}{=} 42$

$15 + 27 \stackrel{?}{=} 42$

$42 = 42$

José earns $3 per hour mowing lawns and $4.50 per hour lifeguarding.

Now you try completing Example 3.

Example 3. In Mario's psychology class, the male students outnumber the female students by 10. If there are 46 students in the class, how many are male and how many are female?

Solution. Our problem gives us information about male students outnumbering female students *and* about the total number of students in the class.

We Think	**We Write**
We must find *two* unknown quantities, so we use two variables.	Let x = number of females
	y = number of males
We know that the male students *outnumber* the female students by 10.	(1) $y = x + 10$
We know the *total* number of students in the class is _____ .	(2) $x + y =$ _____

We can use substitution to solve this system.

$$(1) \quad y = x + 10$$
$$(2) \quad x + y = 46$$

We substitute _____ for y in equation (2) and solve for x.

$$(2) \quad x + (\text{_____}) = 46$$

Now we can substitute _____ for x in equation (1) and find y.

$$(1) \quad y = x + 10$$

There are _____ females and _____ males in Mario's psychology class.

Check your work on page 601. ▶

Example 4. When Marty jogs around the perimeter of a rectangular field, she covers a distance of 540 yards. When she jogs around *another* rectangular field that is twice as wide and 60 yards longer, she covers a distance of 860 yards. Find the dimensions of the smaller field.

Solution. We have information about the *perimeter* of each of the two fields. Recall that the perimeter of a rectangle is found by $P = 2$ lengths $+ 2$ widths.

We Think	**We Write**
We must find *two* dimensions for the smaller field, so we use two variables.	Let l = length of small field (yards)
	w = width of small field (yards)
We know the *perimeter* of the smaller field is 540 yards.	(1) $2l + 2w = 540$

	We Think		*We Write*

The larger field is *twice as wide* and 60 yards *longer* than the smaller field.

Let $2 \cdot w =$ width of large field (yards)

$l + 60 =$ length of large field (yards)

We know the *perimeter* of the larger field is 860 yards.

$$2(l + 60) + 2(2w) = 860$$
$$2l + 120 + 4w = 860$$
$$(2) \qquad 2l + 4w = 740$$

Now we have a system of two equations that we can solve by eliminating *l*.

$$(1) \quad 2l + 2w = \quad 540$$
$$(2) \quad 2l + 4w = \quad 740$$

$$(1) \quad -2l - 2w = -540 \qquad \text{Multiply equation (1) by } -1.$$
$$(2) \quad \underline{2l + 4w = \quad 740} \qquad \text{Equation (2).}$$
$$ 0 + 2w = \quad 200 \qquad \text{Add the equations.}$$
$$ w = \quad 100 \qquad \text{Solve for } w.$$

The width of the smaller field is 100 yards, and we can substitute 100 for *w* in equation (1) to find the length, *l*. Then you should check the solution in equation (2).

$$(1) \qquad 2l + 2w = 540$$
$$2l + 2(100) = 540$$
$$2l + 200 = 540$$
$$2l = 340$$
$$l = 170$$

CHECK:

$$(2) \qquad\qquad 2l + 4w = 740$$
$$2(\underline{}) + 4(\underline{}) \overset{?}{=} 740$$

You complete the check.

Check your work on page 601. ▶

The smaller field is 170 yards long and 100 yards wide.

To use a system of equations to solve a problem stated in words, we have followed these steps.

1. Decide what the two unknown quantities are and identify them, using a different variable to represent each quantity.
2. From the information in the problem, write *two* different equations containing the two variables.
3. Solve the resulting system by the method of your choice (substitution or elimination).
4. Check your solution in the original word problem.

▶ **Examples You Completed** _____

Example 3. In Mario's psychology class, the male students outnumber the female students by 10. If there are 46 students altogether in the class, how many are male and how many are female?

Solution. Our problem gives us information about male sutdents outnumbering female students *and* about the total number of students in the class.

We Think	*We Write*
We must find *two* unknown quantities, so we use two variables.	Let x = number of females y = number of males
We know that the male students *outnumber* the female students by 10.	(1) $y = x + 10$
We know the *total* number of students in the class is 46.	(2) $x + y = 46$

$$(1) \ y = x \cdot + \ 10$$
$$(2) \ x + y = 46$$

We substitute $x + 10$ for y in equation (2) and solve for x.

$$(2) \quad x + (x + 10) = 46$$
$$2x + 10 = 46$$
$$2x = 36$$
$$x = 18$$

Now we can substitute 18 for x in equation (1) and find y.

$$(1) \quad y = x + 10$$
$$y = 18 + 10$$
$$y = 28$$

There are 18 females and 28 males in Mario's psychology class.

Example 4 (CHECK)

$$(2) \qquad 2l + 4w = 740$$
$$2(170) + 4(100) \overset{?}{=} 740$$
$$340 + 400 \overset{?}{=} 740$$
$$740 = 740$$

EXERCISE SET 11.4

Change each word problem to a system of equations containing two variables. Solve each system by the better method (substitution or elimination).

_____ 1. One number is 9 more than twice the other. The sum of the two numbers is 6. Find the numbers.

_____ 2. The sum of two numbers is 10. One number is 14 less than 3 times the other. Find the numbers.

_____ 3. The length of a rectangle is 5 feet more than twice the width. If the perimeter of the rectangle is 58 feet, find the dimensions.

_____ 4. For her flower bed, Nancy needed 3 times as many boxes of petunias as she did marigolds. If she bought 12 boxes of flowers, how many boxes of each type did she buy?

_____ 5. The cashier at Mary's Market has a stack of 25 bills. Some are 5-dollar bills and some are 1-dollar bills. The total value of the stack of bills is $77. How many bills of each denomination does the cashier have?

_____ 6. The length of a carpet is 20 feet less than 3 times the width. The difference between the length and the width is 4 feet. Find the dimensions of the carpet.

_____ 7. Jay sold a box of books at a yard sale for $51. He sold some of the books for $2 each and some for $3 each. If there were 20 books in the box, how many did he sell at each price?

_____ 8. When Hector went to see his father he made part of the trip by bus and the other part by taxi. The time he traveled by taxi was twice the time he rode in the bus. He averaged 40 miles per hour on the bus and 30 miles per hour in the taxi. If the total trip was 25 miles, how much time did he spend traveling by bus and how much time by taxi?

_____ 9. A. J. has money invested in two different companies. The sum of twice the amount invested in Abell's Alarm Service and the amount invested in Betti's Boutique is $11,000. The amount invested in Betti's Boutique is $2000 more than the amount invested in Abell's Alarm Service. Find the amount of each investment.

_____ 10. The sum of 3 times one number and 4 times a second number is -17. The difference between twice the first number and the second number is 18. Find the two numbers.

_____ 11. At a sidewalk sale Kris spent $72 for three pairs of jeans and two tops. Kelly bought five pairs of jeans and three tops from the same racks for $116. What was the price of each pair of jeans and each top?

_____ 12. For a concert, the sum of 3 times the number of tickets sold in advance and the number of tickets sold at the door was 2825. The advance tickets sold for $8 and tickets at the door cost $10. If the total ticket receipts were $16,700, how many tickets of each type were sold?

_____ 13. The sum of the measures of the three angles of any triangle is 180 degrees. One angle measures 90 degrees and one of the remaining angles measures 6 degrees less than 3 times the other. Find the measure of each angle.

_____ 14. The length of a rectangular bulletin board is 7 inches more than twice the width. If the perimeter is 224 inches, find the dimensions of the bulletin board.

_____ 15. At the Record Mart going-out-of-business sale, all records were sold at the same price and all tapes were sold at the same price. Cynthia bought 6 records and 4 tapes for $58. Netta bought 5 records and 3 tapes for $46. What is the sale price of each record and each tape?

_____ 16. The treasurer of the Lion's Club has a stack of 50 bills. Some are 5-dollar bills and some are 10-dollar bills. The total value of the stack of bills is $390. How many bills of each type does he have?

_____ 17. A blouse and skirt cost $49. The skirt cost $7 more than the blouse. What is the cost of each?

_____ 18. The perimeter of a room is 54 feet. Twice the length increased by three times the width is 66 feet. What are the dimensions of the room?

_____ 19. A druggist wishes to put 500 grains of a certain blood pressure medicine into 3-grain and 2-grain capsules. If he fills 200 capsules, how many capsules of each kind does he fill?

_____ 20. In making a trip of 180 miles, Ronald traveled part of the way by train and the rest of the distance by bus. Two-thirds of the distance traveled by train equals $\frac{5}{6}$ of the distance by bus. How far did he travel by train and how far by bus?

☆ Stretching the Topics _____

Use a system of equations to solve each problem.

_____ 1. Coletta has invested a total of $16,000, part at 8 percent per year and the rest at 10 percent per year. How much does she have invested at each rate if her simple interest at the end of 1 year is $8\frac{3}{4}$ percent of the total amount?

_____ 2. The formula for the cost y of producing x telephones is $y = \$25x + \$2000$. The formula for the revenue y from selling x telephones is $y = \$40x - \$550$. The company will _break even_ when the cost and the revenue are the same. Find the number of telephones that must be sold for the company to break even.

Check your answers in the back of your book.

If you can solve the problems in **Checkup 11.4**, you are ready to do the **Review Exercises** for Chapter 11.

✔ CHECKUP 11.4

Use a system of equations in two variables to solve each problem.

_____ 1. Lenora works part-time as a waitress and part-time as a window washer. When she works as a waitress for 5 hours and washes windows for 9 hours, she earns $61. When she washes windows for 3 hours and waits tables for 12 hours, she earns $63. How much does she earn per hour at each job?

_____ 2. The Fashion Outlet sells all its shirts at one price and all its ties at another price. Pat bought three shirts and two ties for $49 and Jon bought four shirts and three ties for $67. What was the price of each shirt and each tie?

_____ 3. The width of a table is 20 inches less than the length. If the perimeter is 200 inches, find the dimensions of the table.

_____ 4. The sum of 5 times one number and twice the second is 1. The difference between the first and the second is 17. Find the numbers.

_____ 5. A total of 2825 concert tickets was sold and total ticket receipts were $25,898. If advance tickets sold for $8 and tickets at the door sold for $10, how many tickets of each type were sold?

Check your answers in the back of your book.

If You Missed Problems:	You Should Review Example:
1, 2	2
3, 4	3
5	1

Summary

When we find all ordered pairs that satisfy two equations containing two variables, we are **solving a system of equations**. In this chapter we discussed three methods for solving systems.

Method	Useful When
Graphing	Common point is easy to identify from the graphs of the equations.
Substitution	One of the variables is easily isolated in one of the equations.
Elimination (addition)	We can make the coefficient of one of the variables in one of the equations the *opposite* of the coefficient of the same variable in the other equation.

No matter which method we used, we discovered three possible situations that could occur in solving a system of two first-degree equations.

Type of System	Solution	Graphs	Signal
Independent and consistent	One ordered pair	Lines cross at only one point	Solving algebraically yields one x-value and one y-value
Inconsistent	None	Lines are parallel	Solving algebraically yields a false statement
Dependent	Any solution for one equation is a solution for the other	Lines are the same	Solving algebraically yields a statement that is *always* true

Having learned to solve systems of equations algebraically, we put our methods to use in solving problems stated in words.

☐ Speaking the Language of Algebra _____

Complete each statement with the appropriate word or phrase.

1. In solving a system of two first-degree equations, we are seeking ordered pairs that _____ both equations.

2. When one variable is easily isolated in one of the equations in a system, we can solve the system by the method of _____ .

3. If the method of elimination yields a false statement, we know the system has _____ solution(s).

4. If the method of graphing yields two straight lines crossing at only one point, we know the system has _____ solution(s).

Check your answers in the back of your book.

△ Writing About Mathematics

Write your response to each question in complete sentences.

1. Explain what you are looking for when you solve a system of equations.

2. Explain why the method of graphing is often not a very efficient method for solving a system of equations.

3. Write a word problem that would lead you to write this system of equations.

$$2x + 6y = 90$$
$$3x + 2y = 65$$

REVIEW EXERCISES for Chapter 11

Solve each system by graphing.

_____ 1. $2x - y = 1$
$\quad\quad x + 2y = 8$

_____ 2. $5x - 6y = 4$
$\quad\quad\quad\quad x = 2$

_____ 3. $x - 3y = 7$
$\quad\quad x - 2y = 0$

_____ 4. $3x + y = -1$
$\quad\quad 2x + y = 0$

Solve each system by the substitution method.

_____ 5. $6x + y = -3$
$\quad\quad\quad y = 3$

_____ 6. $9x - 5y = 12$
$\quad\quad\quad\quad y = -3x$

_____ 7. $4x - y = 0$
$\quad\quad 3x - 2y = 10$

_____ 8. $x - 3y = 8$
$\quad\quad -2x + 6y = 20$

_____ 9. $x = 2y + 7$
$\quad\quad 2x - y = 23$

_____ 10. $x - y = -5$
$\quad\quad\quad x - 9y = -13$

Solve each system by the elimination method.

_____ 11. $2x + y = 12$
$\quad\quad\quad 3x - y = 13$

_____ 12. $x + 3y = 9$
$\quad\quad\quad 4x + 5y = 15$

_____ 13. $2x - y = -6$
$\quad\quad\quad -4x + 2y = -5$

_____ 14. $3x + 11y = 0$
$\quad\quad\quad -2x + 3y = -5$

_____ 15. $-6x + 5y = 9$
$\quad\quad\quad 24x - 20y = -36$

_____ 16. $5x + 6y = 17$
$\quad\quad\quad 8x - 9y = -10$

Solve each system by any method you choose.

_____ 17. $3x - y = -8$
$\quad\quad\quad -5x + 4y = 32$

_____ 18. $x - 2y = 0$
$\quad\quad\quad 6x - 13y = 5$

_____ 19. $4x + 15y = -18$
$\quad\quad\quad 6x - 5y = 28$

_____ 20. $2x - 3y = 4$
$\quad\quad\quad 5x + 4y = 13$

Change each word problem to a system of equations containing two variables. Solve each system by the better method (substitution or elimination).

_____ 21. The sum of two numbers is 22. The difference between twice the first number and the second number is 8. Find both numbers.

_____ 22. The length of a rectangular field is 10 meters more than the width. The perimeter of the field is 116 meters. Find the length and width of the field.

_____ 23. Chico has savings accounts at two different banks. The amount at the Union Bank is $100 less than 3 times the amount at the National Bank. The sum of the two accounts is $1700. Find the amount in each savings account.

_____ 24. Marian bought some collectible bottles at the flea market for $76. She paid $5 each for some of the bottles and $3 each for the others. If she bought 20 bottles altogether, how many did she buy at each price?

Check your answers in the back of your book.

If You Missed Exercises:	You Should Review Examples:	
1, 3, 4	SECTION 11.1	1
2		3
5, 6, 9	SECTION 11.2	1
7, 10		2, 3
8		5
11, 12	SECTION 11.3	1, 2
13, 15		5, 6
14, 16		3, 4
17–20	SECTION 11.2	1–5
	SECTION 11.3	1–4
21–24	SECTION 11.4	1–3

If you have completed the **Review Exercises** and corrected your errors, you are ready to take the **Practice Test** for Chapter 11.

PRACTICE TEST for Chapter 11

Solve each system by graphing.

SECTION	EXAMPLE

_____ 1. $2x + y = 4$
$\quad\ \ x - 3y = -5$

11.1   1

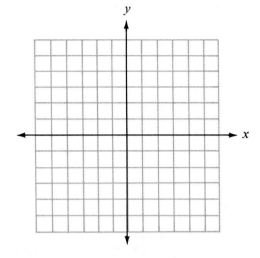

_____ 2. $x + 5 = 0$
$\quad\ \ 2x + 3y = -1$

11.1   3

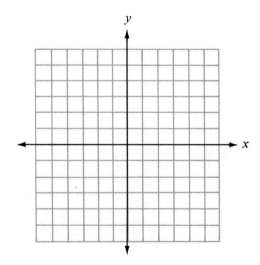

Solve each system by the substitution method.

_____ 3. $-4x + y = -3$
$\quad\ \ \ 7x - 2y = 4$

11.2   3

_____ 4. $x = 3y - 1$
$\quad\ \ 2x - 6y = 3$

11.2   4

Solve each system by the elimination method.

_____ 5. $\ \ 6x + 5y = -7$
$\quad\ \ -2x + 3y = -7$

11.3   2

_____ 6. $3x - 3y = 6$
$\quad\ \ 5y = 5x - 10$

11.3   6

	SECTION	EXAMPLES

Solve each system by any method you choose.

_____ 7. $x = 3y - 2$

  $3x - 5y = 10$

	SECTION	EXAMPLES
7.	11.2	2
	11.3	2

_____ 8. $4x - 3y = -1$

  $-5x + 2y = -4$

11.3   3

_____ 9. $6x - 7y = 8$

  $7x - 6y = 5$

11.3   3

_____ 10. $8x + 5y = 2$

  $3x - 7y = 1$

11.3   3

Use a system of equations in two variables to solve each problem.

_____ 11. When Ben mows the Smiths' yard for 3 hours and digs the Harmons' garden for 2 hours, he earns $18. When he mows for 5 hours and digs for 4 hours, he earns $33. What is Ben's hourly wage at each job?

11.4   1, 2

_____ 12. The perimeter of Martin's cornfield is 320 yards. If Martin doubles the width of the field and increases its length by 30 yards, the perimeter will be 500 yards. Find the original dimensions of the cornfield.

11.4   4

Check your answers in the back of your book.

SHARPENING YOUR SKILLS after Chapters 1–11

_____ **1.** Compare -2.39 and -2.184 using $<$ or $>$.   2.1

_____ **2.** Evaluate $\dfrac{(m-7)^2}{2n}$ when $m=3$ and $n=5$.   5.1

_____ **3.** Find the restrictions on the variable for $\dfrac{y-3}{2y^2-y-15}$.   8.1

_____ **4.** Write an inequality that says $-\frac{2}{3}$ is between -1 and $-\frac{1}{5}$.   9.1

Perform the indicated operations and simplify.

_____ **5.** $|-9-3|-|-15|$ 2.1

_____ **6.** $\left(-\dfrac{2}{3}\right)^2 \div \dfrac{45}{14}$ 2.4

_____ **7.** $-4[2a-(5a-4)]+3[(2a-3)+7]$ 3.3

_____ **8.** $(0.8x^4y^3z^2)(-5xy^2z^3)$ 6.2

_____ **9.** $2a(a-5)(a^2+2)$ 6.3

_____ **10.** $\dfrac{3x-6}{x-6} \cdot \dfrac{2}{x-2}$ 8.2

_____ **11.** $\dfrac{3x}{4x-12} - \dfrac{2x}{5x-15}$ 8.4

Rewrite each expression as a product of factors.

_____ **12.** $81x^4-16$ 7.2

_____ **13.** $7x^3y-175xy^3$ 7.2

_____ **14.** $2ax+2bx-3ay-3by$ 7.1

_____ **15.** $x^2y^2-8xyz-9z^2$ 7.3

Solve each equation or inequality.

_____ **16.** $5=3(x-2)-5(x+1)$ 4.2

_____ **17.** $3x^3-27x^2=0$ 7.5

_____ **18.** $4y(3y-4)=5y(y+1)$ 7.5

_____ **19.** $\dfrac{18}{7x}=\dfrac{3}{x+7}$ 8.6

_____ **20.** $5y - 6 < 11y - 6$ 9.2

_____ **21.** $\dfrac{x}{2} - 4 > -1$ and $6 - \dfrac{x}{3} \leq 5$ 9.3

22. Graph $x - 2y = 2$ by the intercepts method. Label the intercepts.   10.3

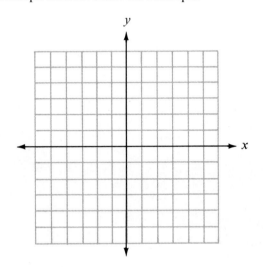

_____ **23.** Find the slope of the line through the points $(-3, 2)$ and $(2, 5)$.   10.4

_____ **24.** A plumber charges a flat rate of $25 per visit plus $15 per hour for labor. If materials cost $58 and Ms. Carrell's total bill was $128, for how many hours did the plumber work?   4.3

_____ **25.** The number of rows of seats in the Little Theater is 5 more than the number of seats per row. If there are 300 seats in the theater, how many rows are there?   7.6

Check your answers in the back of your book.

12

Working with Square Roots

A ladder that is 20 feet long is leaning against the wall of a building. If the foot of the ladder is 10 feet from the bottom of the wall, how far up the wall is the top of the ladder?

It appears that this problem might be solved using the Pythagorean Theorem, but the solution is not in the set of rational numbers that we have been using. After working in this chapter you will be able to find the solution.

In earlier chapters we learned to recognize numbers that were perfect squares of rational numbers. Now we turn our attention to numbers that are squares of numbers that are not rational. In this chapter we

1. Define and simplify principal square roots.

2. Add and subtract radical expressions.

3. Multiply and divide radical expressions.

4. Rationalize denominators.

5. Solve quadratic equations with irrational solutions.

6. Switch from words to quadratic equations.

12.1 Working with Radical Expressions

We have talked often about *squares* of numbers and variables in our work with polynomials and factoring.

Finding Square Roots

In determining whether a certain number was a square, we searched for a number that could be multiplied times itself to give the original number. Such a number is called a **square root** of the original number. We may say that

3 is a square root of 9	because	$3^2 =$	9
-3 is a square root of 9	because	$(-3)^2 =$	9
4 is a square root of 16	because	$4^2 =$	16
-4 is a square root of 16	because	$(-4)^2 =$	16
1 is a square root of 1	because	$1^2 =$	1
-1 is a square root of 1	because	$(-1)^2 =$	1

To avoid confusion, mathematicians have agreed that

> The **principal square root** of a positive number is always the *positive* square root.

From now on, we shall use the phrase *square root* to mean the *principal* (positive) square root. Mathematicians have invented a symbol to represent the principal square root of a number. That symbol is called a **radical** and it looks like this: $\sqrt{}$. The quantity under the radical is called the **radicand**.

The expression $\sqrt{9}$ stands for the principal square root of 9. We may write

$$\sqrt{9} = 3 \quad \text{because} \quad 3^2 = 9$$
$$\sqrt{25} = 5 \quad \text{because} \quad 5^2 = 25$$
$$\sqrt{81} = 9 \quad \text{because} \quad 9^2 = 81$$
$$\sqrt{1} = 1 \quad \text{because} \quad 1^2 = 1$$
$$\sqrt{0} = 0 \quad \text{because} \quad 0^2 = 0$$

In general, we agree that if b is not negative, then

$$\boxed{\sqrt{a} = b \quad \text{means} \quad b^2 = a}$$

In this radical statement, a is the *radicand* and b is the *square root*.

What happens if we try to find the square root of a *negative* number? What is the square root of -9? We must find a number that will multiply times itself to give -9. Will 3 work? No, because $3 \cdot 3 = 9$. Will -3 work? No, because $(-3)(-3) = 9$. Indeed, there is no possible real number square root for -9. As a matter of fact, *it is impossible to find the square root of any negative number among the real numbers.*

The square root of a negative number is *not* a real number.

Let's practice finding some square roots. You complete Example 1.

Example 1. Find $\sqrt{36}$.

Solution

$$\sqrt{36} = \underline{\qquad}$$

because $(\underline{\qquad})^2 = 36$.

Check your work on page 621. ▶

Example 2. Find $\sqrt{\dfrac{4}{9}}$.

Solution

$$\sqrt{\frac{4}{9}} = \frac{2}{3}$$

because $\left(\dfrac{2}{3}\right)^2 = \dfrac{2^2}{3^2}$

$$= \frac{4}{9}$$

Now complete Example 3.

Example 3. Find $\sqrt{1}$.

Solution

$$\sqrt{1} = \underline{\qquad}$$

because $(\underline{\qquad})^2 = 1$.

Check your work on page 621. ▶

Example 4. Find $\sqrt{-16}$.

Solution. We can find no real number whose square is -16.

It is worthwhile to note what happens when we square a square root.

Example 5. Find $(\sqrt{9})^2$.

Solution

$$(\sqrt{9})^2 = (\sqrt{9})(\sqrt{9})$$
$$= (3)(3)$$
$$(\sqrt{9})^2 = 9$$

You complete Example 6.

Example 6. Find $(\sqrt{36})^2$.

Solution

$$(\sqrt{36})^2 = (\sqrt{36})(\sqrt{36})$$
$$= (\underline{\qquad})(\underline{\qquad})$$
$$(\sqrt{36})^2 = \underline{\qquad}$$

Check your work on page 621. ▶

From these examples, we agree that

$$(\sqrt{a})^2 = (\sqrt{a})(\sqrt{a}) = a \qquad \text{provided that } a \text{ is not negative.}$$

This rule allows us to compute the square of a square root without finding the square root first.

Example 7. Find $(\sqrt{17})^2$.

Solution

$$(\sqrt{17})^2 = 17$$

You complete Example 8.

Example 8. Find $(\sqrt{53})^2$.

Solution

$$(\sqrt{53})^2 = \underline{\quad\quad}$$

Check your work on page 621. ▶

To work with the square root of a variable expression, it is absolutely necessary that the variable expression represent a positive quantity or zero. Why? Because we have no way to find the square root of a negative quantity.

> Throughout this chapter, we assume that variable expressions under square root radicals do *not* represent negative quantities.

Once we have agreed on this very important condition, we may simplify radical expressions containing variables.

Example 9. Find $\sqrt{x^2}$, $\sqrt{y^4}$, and $\sqrt{\dfrac{a^2}{4}}$.

Solution

$$\sqrt{x^2} = x \quad \text{because} \quad (x)^2 = x^2$$

$$\sqrt{y^4} = y^2 \quad \text{because} \quad (y^2)^2 = y^4$$

$$\sqrt{\frac{a^2}{4}} = \frac{a}{2} \quad \text{because} \quad \left(\frac{a}{2}\right)^2 = \frac{a^2}{4}$$

Now complete Example 10.

Example 10. Find $(\sqrt{x^2})^2$, $(\sqrt{y})^2$, and $(\sqrt{5x})^2$.

Solution

$$(\sqrt{x^2})^2 = x^2$$

$$(\sqrt{y})^2 = \underline{\quad\quad}$$

$$(\sqrt{5x})^2 = \underline{\quad\quad}$$

Check your work on page 621. ▶

⫸ Trial Run

Simplify each radical expression.

_____ 1. $\sqrt{64}$

_____ 2. $\sqrt{100}$

_____ 3. $(\sqrt{5})^2$

_____ 4. $\sqrt{y^2}$

_____ 5. $\sqrt{\dfrac{9}{x^2}}$

_____ 6. $(\sqrt{2y})^2$

Answers are on page 622.

So far we have practiced finding square roots that were **rational numbers**. Recall that rational numbers are numbers that can be expressed as *fractions of integers*.

$$\sqrt{9} = 3 \quad \text{is rational because} \quad 3 = \frac{3}{1}$$

$$\sqrt{\frac{4}{25}} = \frac{2}{5} \quad \text{is rational because} \quad \frac{2}{5} = \frac{2}{5}$$

We do not have to look very far in the set of natural numbers {1, 2, 3, 4, . . .} before we find a number whose square root is *not* rational. For example,

$$\sqrt{2} \text{ is not rational}$$

You could hunt forever and never find a rational number that can be multiplied times itself to give an answer of 2. This does *not* mean that $\sqrt{2}$ does not exist. It merely means that $\sqrt{2}$ is *not rational*; $\sqrt{2}$ cannot be expressed as a quotient of integers.

A real number that is *not* rational is called an **irrational number**. Its value can be approximated, but it cannot be expressed as a fraction of integers. When we approximate irrational square roots, we use the symbol $\doteq$ to mean "is approximately equal to." (A table of square roots appears inside the back cover of your book.)

Irrational Number	Approximate Value	Check
$\sqrt{2}$	$\sqrt{2} \doteq 1.414$	$(1.414)^2 \doteq 1.999396 \doteq 2$
$\sqrt{3}$	$\sqrt{3} \doteq 1.732$	$(1.732)^2 = 2.999824 \doteq 3$
$\sqrt{5}$	$\sqrt{5} \doteq 2.236$	$(2.236)^2 = 4.999696 \doteq 5$

In order to represent *exact* values of irrational square roots, we must leave them in radical form. Let's look at some square roots and decide whether they are rational or irrational numbers. We shall simplify those square roots that are rational.

Square Root	Simplest Form	Rational or Irrational?
$\sqrt{0}$	0	Rational
$\sqrt{1}$	1	Rational
$\sqrt{2}$	$\sqrt{2}$	Irrational
$\sqrt{3}$	$\sqrt{3}$	Irrational
$\sqrt{4}$	2	Rational
$\sqrt{7}$	$\sqrt{7}$	Irrational
$\sqrt{9}$	3	Rational
$\sqrt{19}$	$\sqrt{19}$	Irrational
$\sqrt{100}$	10	Rational

⫸ Trial Run

Tell whether each radical expression is a rational or irrational number.

_____ 1. $\sqrt{49}$ _____ 2. $\sqrt{13}$

_____ 3. $\sqrt{0}$ _____ 4. $\sqrt{\dfrac{9}{16}}$

Answers are on page 622.

Simplifying Radical Expressions

Let's consider a method for simplifying radical expressions such as $\sqrt{36}$. We know that

$$\sqrt{36} = 6 \quad \text{and} \quad \sqrt{4} \cdot \sqrt{9} = 2 \cdot 3$$
$$= 6$$

so here we conclude that

$$\sqrt{36} = \sqrt{4} \cdot \sqrt{9}$$

Do you suppose that the square root of a product is *always* equal to the product of the square roots of the factors? Look at another example.

$$\sqrt{144} = 12 \quad \text{and} \quad \sqrt{9} \cdot \sqrt{16} = 3 \cdot 4$$
$$= 12$$

so here we conclude that

$$\sqrt{144} = \sqrt{9} \cdot \sqrt{16}$$

Although we have verified this rule for only two examples, it is indeed true that

---

The square root of a product of factors is equal to the product of the square roots of the factors.

$$\sqrt{a \cdot b} = \sqrt{a} \cdot \sqrt{b} \quad \text{provided that } a \text{ and } b \text{ are not negative.}$$

We make use of this rule in simplifying radical expressions.

Example 11. Simplify $\sqrt{4x^2}$.

Solution

$$\sqrt{4x^2} = \sqrt{4 \cdot x^2}$$
$$= \sqrt{4} \cdot \sqrt{x^2}$$
$$= 2 \cdot x$$
$$= 2x$$

You complete Example 12.

Example 12. Simplify $\sqrt{9a^2b^4}$.

Solution

$$\sqrt{9a^2b^4}$$
$$= \sqrt{9 \cdot a^2 \cdot b^2 \cdot b^2}$$
$$= \sqrt{\underline{\quad}} \cdot \sqrt{\underline{\quad}} \cdot \sqrt{\underline{\quad}} \cdot \sqrt{\underline{\quad}}$$
$$= \underline{\quad\quad}$$

Check your work on page 621. ▶

Sometimes a radical will not be removed completely in the simplification process. The best that we can do is to rewrite the radicand as a product of factors, some of which are *perfect squares*. The factors whose roots are *not* rational must then be left under the radical.

Example 13. Simplify $\sqrt{12}$.

Solution. Remember that we wish to write 12 as a product of factors, where at least one factor is a perfect square.

$$\sqrt{12} = \sqrt{4 \cdot 3}$$
$$= \sqrt{4} \cdot \sqrt{3}$$
$$= 2\sqrt{3}$$

This is simplest radical form.

You try Example 14.

Example 14. Simplify $\sqrt{98}$.

Solution. We wish to write _____ as a product of factors, where at least one factor is a _____ _____ .

$$\sqrt{98} = \sqrt{49 \cdot 2}$$
$$= \sqrt{\underline{}} \cdot \sqrt{\underline{}}$$
$$= \underline{} \sqrt{\underline{}}$$

Check your work on page 622. ▶

Example 15. Simplify $\sqrt{50x^3}$.

Solution

$$\sqrt{50x^3}$$
$$= \sqrt{25 \cdot 2 \cdot x^2 \cdot x}$$
$$= \sqrt{25} \cdot \sqrt{2} \cdot \sqrt{x^2} \cdot \sqrt{x}$$
$$= 5 \cdot \sqrt{2} \cdot x \cdot \sqrt{x}$$
$$= 5 \cdot x \cdot \sqrt{2} \cdot \sqrt{x}$$
$$= 5x\sqrt{2 \cdot x}$$
$$= 5x\sqrt{2x}$$

You try completing Example 16.

Example 16. Simplify $\sqrt{108ab^4}$.

Solution

$$\sqrt{108ab^4}$$
$$= \sqrt{36 \cdot 3 \cdot a \cdot b^2 \cdot b^2}$$
$$= \sqrt{\underline{}} \cdot \sqrt{3} \cdot \sqrt{\underline{}} \cdot \sqrt{\underline{}} \cdot \sqrt{\underline{}}$$
$$= 6 \cdot \sqrt{3} \cdot \sqrt{a} \cdot b \cdot b$$
$$= 6 \cdot b \cdot b \cdot \sqrt{3} \cdot \sqrt{a}$$
$$= 6 \underline{} \sqrt{\underline{} \cdot \underline{}}$$
$$= \underline{} \sqrt{\underline{}}$$

Check your work on page 622. ▶

A square root radical is said to be in **simplest form** when the radicand contains no perfect square factors. If the exponent on a factor in the radicand is greater than 1, then some simplification of the square root is necessary.

Let's try to simplify a radical in which the radicand is a fraction.

$$\sqrt{\frac{4}{9}} = \frac{2}{3} \quad \text{and} \quad \frac{\sqrt{4}}{\sqrt{9}} = \frac{2}{3}$$

Therefore, it seems that

$$\sqrt{\frac{4}{9}} = \frac{\sqrt{4}}{\sqrt{9}}$$

Indeed, it is always possible to state that

The square root of a quotient is equal to the quotient of the square roots.

$$\sqrt{\frac{a}{b}} = \frac{\sqrt{a}}{\sqrt{b}} \quad \text{provided that } a \text{ is not negative and } b \text{ is positive.}$$

This rule gives us another tool for simplifying radicals.

Try to complete Example 17.

Example 17. Simplify $\sqrt{\dfrac{25x^2}{49y^2}}$.

Solution

$$\sqrt{\dfrac{25x^2}{49y^2}} = \dfrac{\sqrt{25x^2}}{\sqrt{49y^2}}$$

$$= \dfrac{\sqrt{25}\cdot\sqrt{x^2}}{\sqrt{49}\cdot\sqrt{y^2}}$$

$$= \underline{\qquad}$$

Check your work on page 622. ▶

Example 18. Simplify $\sqrt{\dfrac{32x^3}{y^4}}$.

Solution

$$\sqrt{\dfrac{32x^3}{y^4}} = \dfrac{\sqrt{32x^3}}{\sqrt{y^4}}$$

$$= \dfrac{\sqrt{16\cdot 2\cdot x^2\cdot x}}{\sqrt{y^2\cdot y^2}}$$

$$= \dfrac{4x\sqrt{2x}}{y^2}$$

Our rule for quotients can also be used in reverse. That is, we may sometimes wish to make use of the fact that

$$\dfrac{\sqrt{a}}{\sqrt{b}} = \sqrt{\dfrac{a}{b}}$$

Suppose, for example, that we wish to simplify $\dfrac{\sqrt{50x^3}}{\sqrt{2x}}$.

$$\dfrac{\sqrt{50x^3}}{\sqrt{2x}} = \sqrt{\dfrac{50x^3}{2x}} \qquad \text{Rule for quotients of radicals.}$$

$$= \sqrt{25x^2} \qquad \text{Reduce the fraction as usual.}$$

$$= 5x \qquad \text{Simplify the radical.}$$

In general, we note that a quotient of two radicals should be written as a single radical whenever the numerator and denominator contain common factors. Once the numerator and denominator have no more common factors, the fraction should again be considered as the quotient of two separate radicals.

Example 19. Simplify $\dfrac{\sqrt{20x^3y^4}}{\sqrt{125xy}}$.

Solution. Since the numerator and denominator contain common factors, we write

$$\dfrac{\sqrt{20x^3y^4}}{\sqrt{125xy}}$$

$$= \sqrt{\dfrac{20x^3y^4}{125xy}} \qquad \begin{array}{l}\text{Rewrite as}\\ \text{one radical.}\end{array}$$

$$= \sqrt{\dfrac{4x^2y^3}{25}} \qquad \begin{array}{l}\text{Reduce the}\\ \text{fraction.}\end{array}$$

$$= \dfrac{\sqrt{4x^2y^3}}{\sqrt{25}} \qquad \begin{array}{l}\text{Rewrite as}\\ \text{separate radicals.}\end{array}$$

$$= \dfrac{\sqrt{4\cdot x^2\cdot y^2\cdot y}}{\sqrt{25}} \qquad \begin{array}{l}\text{Look for perfect}\\ \text{square factors.}\end{array}$$

$$= \dfrac{2xy\sqrt{y}}{5} \qquad \begin{array}{l}\text{Remove square}\\ \text{roots of perfect}\\ \text{square factors.}\end{array}$$

Now you try Example 20.

Example 20. Simplify $\dfrac{\sqrt{18a^7c}}{\sqrt{6a^6c}}$.

Solution. Since the numerator and denominator contain common factors, we write

$$\dfrac{\sqrt{18a^7c}}{\sqrt{6a^6c}} = \sqrt{\dfrac{18a^7c}{6a^6c}}$$

$$= \sqrt{\underline{\qquad}}$$

Check your work on page 622. ▶

⫸ Trial Run

Write each radical expression in simplest form.

_____ **1.** $\sqrt{16x^2}$

_____ **2.** $\sqrt{18}$

_____ **3.** $\sqrt{27a^3}$

_____ **4.** $\sqrt{20x^3y^4}$

_____ **5.** $\sqrt{\dfrac{9x^2}{64y^2}}$

_____ **6.** $\sqrt{\dfrac{50x^3}{y^4}}$

_____ **7.** $\dfrac{\sqrt{2x^5y}}{\sqrt{98xy}}$

_____ **8.** $\dfrac{\sqrt{169a^4b^2}}{\sqrt{36c^2}}$

Answers are on page 622.

▶ Examples You Completed

Example 1. Find $\sqrt{36}$.

Solution

$$\sqrt{36} = 6$$

because $(6)^2 = 36$.

Example 3. Find $\sqrt{1}$.

Solution

$$\sqrt{1} = 1$$

because $(1)^2 = 1$.

Example 6. Find $(\sqrt{36})^2$.

Solution

$$(\sqrt{36})^2 = (\sqrt{36})(\sqrt{36})$$
$$= (6)(6)$$
$$(\sqrt{36})^2 = 36$$

Example 8. Find $(\sqrt{53})^2$.

Solution

$$(\sqrt{53})^2 = 53$$

Example 10. Find $(\sqrt{x^2})^2$, $(\sqrt{y})^2$, and $(\sqrt{5x})^2$.

Solution

$$(\sqrt{x^2})^2 = x^2$$
$$(\sqrt{y})^2 = y$$
$$(\sqrt{5x})^2 = 5x$$

Example 12. Simplify $\sqrt{9a^2b^4}$.

Solution

$$\sqrt{9a^2b^4} = \sqrt{9 \cdot a^2 \cdot b^2 \cdot b^2}$$
$$= \sqrt{9} \cdot \sqrt{a^2} \cdot \sqrt{b^2} \cdot \sqrt{b^2}$$
$$= 3 \cdot a \cdot b \cdot b$$
$$= 3ab^2$$

Example 14. Simplify $\sqrt{98}$.

Solution. We wish to write 98 as a product of factors, where at least one factor is a perfect square.

$$\sqrt{98} = \sqrt{49 \cdot 2}$$
$$= \sqrt{49} \cdot \sqrt{2}$$
$$= 7\sqrt{2}$$

Example 16. Simplify $\sqrt{108ab^4}$.

Solution

$$\sqrt{108ab^4} = \sqrt{36 \cdot 3 \cdot a \cdot b^2 \cdot b^2}$$
$$= \sqrt{36} \cdot \sqrt{3} \cdot \sqrt{a} \cdot \sqrt{b^2} \cdot \sqrt{b^2}$$
$$= 6 \cdot \sqrt{3} \cdot \sqrt{a} \cdot b \cdot b$$
$$= 6 \cdot b \cdot b \cdot \sqrt{3} \cdot \sqrt{a}$$
$$= 6b^2 \sqrt{3 \cdot a}$$
$$= 6b^2\sqrt{3a}$$

Example 17. Simplify $\sqrt{\dfrac{25x^2}{49y^2}}$.

Solution

$$\sqrt{\frac{25x^2}{49y^2}} = \frac{\sqrt{25x^2}}{\sqrt{49y^2}}$$
$$= \frac{\sqrt{25} \cdot \sqrt{x^2}}{\sqrt{49} \cdot \sqrt{y^2}}$$
$$= \frac{5x}{7y}$$

Example 20. Simplify $\dfrac{\sqrt{18a^7c}}{\sqrt{6a^6c}}$.

Solution. Since the numerator and denominator contain common factors, we write

$$\frac{\sqrt{18a^7c}}{\sqrt{6a^6c}} = \sqrt{\frac{18a^7c}{6a^6c}}$$
$$= \sqrt{3a}$$

Answers to Trial Runs

page 616   **1.** 8   **2.** 10   **3.** 5   **4.** y **5.** $\dfrac{3}{x}$ **6.** $2y$

page 617   **1.** Rational   **2.** Irrational   **3.** Rational   **4.** Rational

page 621   **1.** $4x$ **2.** $3\sqrt{2}$ **3.** $3a\sqrt{3a}$ **4.** $2xy^2\sqrt{5x}$ **5.** $\dfrac{3x}{8y}$ **6.** $\dfrac{5x\sqrt{2x}}{y^2}$ **7.** $\dfrac{x^2}{7}$ **8.** $\dfrac{13a^2b}{6c}$

EXERCISE SET 12.1

Decide whether each expression is a rational or irrational number.

———— **1.** $\sqrt{25}$ ———— **2.** $\sqrt{49}$

———— **3.** $\sqrt{5}$ ———— **4.** $\sqrt{3}$

———— **5.** $\sqrt{1}$ ———— **6.** $\sqrt{0}$

———— **7.** $\sqrt{8}$ ———— **8.** $\sqrt{12}$

———— **9.** $\sqrt{\dfrac{25}{36}}$ ———— **10.** $\sqrt{\dfrac{1}{4}}$

———— **11.** $\sqrt{65}$ ———— **12.** $\sqrt{35}$

———— **13.** $\sqrt{\dfrac{2}{3}}$ ———— **14.** $\sqrt{\dfrac{5}{7}}$

———— **15.** $\sqrt{225}$ ———— **16.** $\sqrt{196}$

Simplify each radical expression. Assume variable quantities are not negative.

———— **17.** $\sqrt{144}$ ———— **18.** $\sqrt{169}$

———— **19.** $\sqrt{a^4}$ ———— **20.** $\sqrt{a^6}$

———— **21.** $(\sqrt{8})^2$ ———— **22.** $(\sqrt{10})^2$

———— **23.** $(\sqrt{x^5})^2$ ———— **24.** $(\sqrt{a^3})^2$

———— **25.** $\sqrt{25b^2}$ ———— **26.** $\sqrt{64c^2}$

———— **27.** $\sqrt{\dfrac{x^2}{36}}$ ———— **28.** $\sqrt{\dfrac{y^2}{49}}$

———— **29.** $\sqrt{45}$ ———— **30.** $\sqrt{40}$

———— **31.** $\sqrt{600}$ ———— **32.** $\sqrt{128}$

———— **33.** $\sqrt{75}$ ———— **34.** $\sqrt{108}$

———— **35.** $\sqrt{44}$ ———— **36.** $\sqrt{80}$

———— **37.** $\sqrt{x^2}$ ———— **38.** $\sqrt{x^4}$

———— **39.** $\sqrt{y^5}$ ———— **40.** $\sqrt{y^3}$

———— **41.** $\sqrt{x^2y^3}$ ———— **42.** $\sqrt{x^3y^2}$

———— **43.** $\sqrt{32a^3}$ ———— **44.** $\sqrt{18b^3}$

_____ 45. $\sqrt{4x^2y}$

_____ 46. $\sqrt{9xy^2}$

_____ 47. $\sqrt{81a^5b^4}$

_____ 48. $\sqrt{100a^3b^6}$

_____ 49. $\sqrt{192x^2y^4z^3}$

_____ 50. $\sqrt{243x^4y^5z^2}$

_____ 51. $\sqrt{180a^7b^3c}$

_____ 52. $\sqrt{98a^5b^4c}$

_____ 53. $\sqrt{\dfrac{4x^2}{25}}$

_____ 54. $\sqrt{\dfrac{9x^2}{121}}$

_____ 55. $\sqrt{\dfrac{32a^3}{b^4}}$

_____ 56. $\sqrt{\dfrac{75a}{b^2}}$

_____ 57. $\sqrt{\dfrac{40x^3}{9y^2}}$

_____ 58. $\sqrt{\dfrac{27x^5}{4y^2}}$

_____ 59. $\sqrt{\dfrac{20a^3b^3}{500ab}}$

_____ 60. $\sqrt{\dfrac{35a^5b^5}{28a^3b^3}}$

_____ 61. $\dfrac{\sqrt{75x^3}}{\sqrt{3x}}$

_____ 62. $\dfrac{\sqrt{80x^5}}{\sqrt{5x}}$

_____ 63. $\dfrac{\sqrt{15x^7}}{\sqrt{5x^3}}$

_____ 64. $\dfrac{\sqrt{56x^5}}{\sqrt{7x^4}}$

_____ 65. $\dfrac{\sqrt{3xy^3}}{\sqrt{363xy^2}}$

_____ 66. $\dfrac{\sqrt{5xy^2}}{\sqrt{720y}}$

☆ Stretching the Topics ———————————————

Simplify each radical expression. Assume variable quantities are not negative.

_____ 1. $\sqrt{\dfrac{24x^{10}y^2}{y^6z^{28}}}$

_____ 2. $-7xy^4\sqrt{63x^2y^7}$

_____ 3. $\sqrt{0.16x^4y^2(a-b)^2}$

Check your answers in the back of your book.

If you can simplify the radical expressions in **Checkup 12.1**, you are ready to go on to Section 12.2.

✓ **CHECKUP 12.1**

Simplify each radical expression. Assume variable quantities are not negative.

_____ 1. $\sqrt{49}$

_____ 2. $\sqrt{9a^2}$

_____ 3. $(\sqrt{12})^2$

_____ 4. $\sqrt{80}$

_____ 5. $\sqrt{75y^4}$

_____ 6. $\sqrt{45a^3b^2}$

_____ 7. $\sqrt{\dfrac{25x^2}{36y^2}}$

_____ 8. $\sqrt{\dfrac{32x^5}{y^6}}$

_____ 9. $\dfrac{\sqrt{50a^5b^3}}{\sqrt{2ab}}$

_____ 10. $\dfrac{\sqrt{15a^5}}{\sqrt{5a}}$

Check your answers in the back of your book.

If You Missed Problems:	You Should Review Examples:
1	1–3
2	9
3	5–8
4–6	13–16
7, 8	17, 18
9, 10	19, 20

12.2 Operating with Radical Expressions

So far, we have spent time recognizing radical expressions and learning how to simplify them. As with any kind of algebraic expressions, we must now discuss the basic operations of addition, subtraction, and multiplication. We discuss division in Section 12.3.

Adding and Subtracting Radical Expressions

First we turn our attention to learning how to compute sums and differences of radical expressions such as $2\sqrt{3} + 5\sqrt{3}$. From our knowledge of combining like terms, it seems logical to conclude that the sum will be

$$2\sqrt{3} + 5\sqrt{3} = 7\sqrt{3}$$

The distributive property can be used to verify that this result is reasonable.

$$2\sqrt{3} + 5\sqrt{3} = (2 + 5)\sqrt{3}$$
$$= 7\sqrt{3}$$

The key fact to realize here is that the *radicals matched exactly*. As in our work with combining like terms, the rule for combining radical expressions demands that the radicals be *exactly alike*. They must both be square roots of the same radicand.

Combining Like Radicals. To combine **like radicals**, we combine the numerical coefficients and keep the same radical part.

Example 1. Find
$17\sqrt{x} - 12\sqrt{x} + \sqrt{x}$.

Solution

$$17\sqrt{x} - 12\sqrt{x} + \sqrt{x}$$
$$= 17\sqrt{x} - 12\sqrt{x} + 1\sqrt{x}$$
$$= 6\sqrt{x}$$

You try Example 2.

Example 2. Find
$3\sqrt{7} + 6\sqrt{7} - 10\sqrt{7}$.

Solution

$$3\sqrt{7} + 6\sqrt{7} - 10\sqrt{7}$$
$$=$$
$$= \underline{\quad} \sqrt{7}$$

Check your work on page 633. ▶

Let's try to simplify $\sqrt{18a} - \sqrt{8a} + \sqrt{2a}$. Notice that the radicals are all square roots, but the radicands are all *different*. Before we decide whether these radicals can be combined, we must try to *simplify* each radical completely.

$$\sqrt{18a} - \sqrt{8a} + \sqrt{2a} = \sqrt{9 \cdot 2a} - \sqrt{4 \cdot 2a} + \sqrt{2a}$$
$$= 3\sqrt{2a} - 2\sqrt{2a} + 1\sqrt{2a}$$
$$= 2\sqrt{2a}$$

Try combining the radicals in Example 3.

Example 3. Find $\sqrt{50x} - 6\sqrt{32x}$.

Solution

$$\sqrt{50x} - 6\sqrt{32x}$$
$$= \sqrt{25 \cdot 2x} - 6\sqrt{16 \cdot 2x}$$
$$= 5\sqrt{2x} - 6 \cdot 4\sqrt{2x}$$

Check your work on page 633. ▶

Example 4. Find $6 + \sqrt{27} + \sqrt{48}$.

Solution

$$6 + \sqrt{27} + \sqrt{48}$$
$$= 6 + \sqrt{9 \cdot 3} + \sqrt{16 \cdot 3}$$
$$= 6 + 3\sqrt{3} + 4\sqrt{3}$$
$$= 6 + 7\sqrt{3}$$

Notice that these last two terms cannot be combined. They are *not like terms*.

⫸ Trial Run

Find each sum or difference.

_____ 1. $2\sqrt{3} + 5\sqrt{3} - 9\sqrt{3}$

_____ 2. $8\sqrt{x} - 7\sqrt{x} + \sqrt{x}$

_____ 3. $\sqrt{32a} - 5\sqrt{18a}$

_____ 4. $11 + \sqrt{75} - \sqrt{12}$

_____ 5. $\dfrac{\sqrt{8}}{\sqrt{2}} + \sqrt{50} - \sqrt{72}$

Answers are on page 633.

Multiplying Radical Expressions

The methods needed for multiplying radical expressions are similar to those used for multiplying polynomials. As you remember, multiplying polynomials required use of the associative and distributive properties and the rules for combining like terms.

We discussed multiplying radicals in our work with simplifying radicals. At that time we agreed that

$$\sqrt{a} \cdot \sqrt{b} = \sqrt{a \cdot b} \qquad \text{provided that } a \text{ and } b \text{ are not negative}$$

See if you can complete Example 5.

Example 5. Find $\sqrt{2} \cdot \sqrt{5} \cdot \sqrt{10}$.

Solution

$$\sqrt{2} \cdot \sqrt{5} \cdot \sqrt{10}$$
$$= \sqrt{2 \cdot 5 \cdot 10}$$
$$= \sqrt{}$$
$$= \underline{}$$

Check your work on page 633. ▶

Example 6. Find $\sqrt{3a} \cdot \sqrt{3a}$.

Solution

$$\sqrt{3a} \cdot \sqrt{3a}$$
$$= (\sqrt{3a})^2$$
$$= 3a$$

Example 6 illustrates a fact that we have used before and that soon will be even more useful.

$$\sqrt{a} \cdot \sqrt{a} = (\sqrt{a})^2 = a \qquad \text{provided that } a \text{ is not negative}$$

Example 7. Find
$$\sqrt{2x - 1} \cdot \sqrt{2x - 1}$$

Solution

$$\sqrt{2x - 1} \cdot \sqrt{2x - 1}$$
$$= (\sqrt{2x - 1})^2$$
$$= 2x - 1$$

Now you complete Example 8.

Example 8. Find
$$\sqrt{y + 6} \cdot \sqrt{y + 6}$$

Solution

$$\sqrt{y + 6} \cdot \sqrt{y + 6}$$
$$=$$

Check your work on page 633. ▶

Example 9. Find $5\sqrt{2a} \cdot \sqrt{8a}$.

Solution

$$5\sqrt{2a} \cdot \sqrt{8a}$$
$$= 5\sqrt{2a \cdot 8a}$$
$$= 5\sqrt{16a^2}$$
$$= 5 \cdot 4a$$
$$= 20a$$

You complete Example 10.

Example 10. Find $(-3\sqrt{x})(-6\sqrt{x})$.

Solution

$$(-3\sqrt{x})(-6\sqrt{x})$$
$$= -3(-6)\sqrt{x} \cdot \sqrt{x}$$

Check your work on page 633. ▶

To multiply a single radical times a sum or difference, we use the *distributive property* and then simplify.

Example 11. Find $\sqrt{3}(5 + \sqrt{3})$.

Solution

$$\sqrt{3}(5 + \sqrt{3})$$
$$= \sqrt{3} \cdot 5 + \sqrt{3} \cdot \sqrt{3}$$
$$= 5\sqrt{3} + 3$$

Example 12. Find $2\sqrt{2}(\sqrt{3} + \sqrt{8})$.

Solution

$$2\sqrt{2}(\sqrt{3} + \sqrt{8})$$
$$= 2\sqrt{2} \cdot \sqrt{3} + 2\sqrt{2} \cdot \sqrt{8}$$
$$= 2\sqrt{2 \cdot 3} + 2\sqrt{2 \cdot 8}$$
$$= 2\sqrt{6} + 2\sqrt{16}$$
$$= 2\sqrt{6} + 2 \cdot 4$$
$$= 2\sqrt{6} + 8$$

⏵ Trial Run

Find each product.

_____ 1. $\sqrt{2} \cdot \sqrt{5}$

_____ 2. $\sqrt{5 + a} \cdot \sqrt{5 + a}$

_____ 3. $4\sqrt{3a} \cdot \sqrt{12a}$

_____ 4. $(-2\sqrt{x})(5\sqrt{3x})$

_____ 5. $\sqrt{2}(3 - \sqrt{2})$

_____ 6. $3\sqrt{2}(\sqrt{5} + \sqrt{18})$

Answers are on page 633.

Let's spend a moment reviewing the multiplication of binomials to prepare ourselves for multiplying sums and differences containing radicals. Recall how we found products of binomials using the FOIL method or formulas for special products.

FOIL Method	$(A + B)^2 = A^2 + 2AB + B^2$ $(A - B)^2 = A^2 - 2AB + B^2$	$(A + B)(A - B) = A^2 - B^2$
$(x + 3)(x + 7)$ $= x^2 + 7x + 3x + 21$ $= x^2 + 10x + 21$	$(x + 5)^2$ $= x^2 + 2(5x) + 5^2$ $= x^2 + 10x + 25$	$(x + 6)(x - 6)$ $= x^2 - 6^2$ $= x^2 - 36$

We can use the same methods for multiplying sums and/or differences containing radicals.

$$(2 + \sqrt{3})(4 + \sqrt{2}) = 2 \cdot 4 + 2 \cdot \sqrt{2} + \sqrt{3} \cdot 4 + \sqrt{3} \cdot \sqrt{2}$$

$$= 8 + 2\sqrt{2} + 4\sqrt{3} + \sqrt{6}$$

Here there were no like terms to combine.

Example 13. Find $(\sqrt{6} + 1)^2$.

Solution

$$(\sqrt{6} + 1)^2$$
$$= (\sqrt{6})^2 + 2 \cdot 1 \cdot \sqrt{6} + 1^2$$
$$= 6 + 2\sqrt{6} + 1$$
$$= 7 + 2\sqrt{6}$$

Now you try Example 14.

Example 14. Find $(8 - \sqrt{y})^2$.

Solution

$$(8 - \sqrt{y})^2$$
$$= 8^2 - 2 \cdot 8 \cdot \sqrt{y} + (\sqrt{y})^2$$
$$= \underline{\quad} - \underline{\quad} \sqrt{y} + \underline{\quad}$$

Check your work on page 633. ▶

Example 15. Find

$$(5 + \sqrt{x})(3 + 4\sqrt{x})$$

Solution

$$(5 + \sqrt{x})(3 + 4\sqrt{x})$$
$$= 5 \cdot 3 + 5 \cdot 4\sqrt{x} + 3\sqrt{x} + 4\sqrt{x} \cdot \sqrt{x}$$
$$= 15 + 20\sqrt{x} + 3\sqrt{x} + 4x$$
$$= 15 + 23\sqrt{x} + 4x$$

Example 16. Find

$$(6 + \sqrt{17})(6 - \sqrt{17})$$

Solution

$$(6 + \sqrt{17})(6 - \sqrt{17})$$
$$= 6^2 - (\sqrt{17})^2$$
$$= 36 - 17$$
$$= 19$$

Example 16 is especially interesting because we multiplied two irrational quantities and our product was a *rational* number. Notice that the original quantities were the *sum and difference of the same two terms*. Irrational numbers such as $6 + \sqrt{17}$ and $6 - \sqrt{17}$ are sometimes called **real conjugates** of each other.

The product of an irrational number and its real conjugate is always a *rational* number.

This fact will be useful to us in Section 12.3.

Example 17. Find the missing irrational numbers.

Irrational Number	Real Conjugate
$1 + \sqrt{2}$	_____
$-3 - \sqrt{5}$	_____
_____	$\sqrt{7} - 1$
_____	$2\sqrt{x} + \sqrt{13}$

Check your work on page 633. ▶

Example 18. Find

$$(\sqrt{5} + 2\sqrt{7})(\sqrt{5} - 2\sqrt{7})$$

Solution

$$(\sqrt{5} + 2\sqrt{7})(\sqrt{5} - 2\sqrt{7})$$
$$= (\sqrt{5})^2 - (2\sqrt{7})^2$$
$$= 5 - 4(7)$$
$$= 5 - 28$$
$$= -23$$

You try Example 19.

Example 19. Find

$$(\sqrt{y} + 3)(\sqrt{y} - 3)$$

Solution

Check your work on page 633. ▶

⫸ Trial Run

Find each product.

_____ **1.** $(1 + \sqrt{2})(2 - \sqrt{3})$

_____ **2.** $(5 - 2\sqrt{5})(2 + \sqrt{5})$

_____ **3.** $(2 - 3\sqrt{7})(2 + 3\sqrt{7})$

_____ **4.** $(\sqrt{x} + 8)(\sqrt{x} - 8)$

_____ **5.** $(\sqrt{x} + 2)(\sqrt{x} - 4)$

_____ **6.** $(\sqrt{x} - 5)^2$

Answers are on page 633.

▶ **Examples You Completed** _____

Example 2. Find $3\sqrt{7} + 6\sqrt{7} - 10\sqrt{7}$.

Solution

$$3\sqrt{7} + 6\sqrt{7} - 10\sqrt{7}$$
$$= (3 + 6 - 10)\sqrt{7}$$
$$= -1\sqrt{7}$$
$$= -\sqrt{7}$$

Example 3. Find $\sqrt{50x} - 6\sqrt{32x}$.

Solution

$$\sqrt{50x} - 6\sqrt{32x}$$
$$= \sqrt{25 \cdot 2x} - 6\sqrt{16 \cdot 2x}$$
$$= 5\sqrt{2x} - 6 \cdot 4\sqrt{2x}$$
$$= 5\sqrt{2x} - 24\sqrt{2x}$$
$$= -19\sqrt{2x}$$

Example 5. Find $\sqrt{2} \cdot \sqrt{5} \cdot \sqrt{10}$.

Solution

$$\sqrt{2} \cdot \sqrt{5} \cdot \sqrt{10}$$
$$= \sqrt{2 \cdot 5 \cdot 10}$$
$$= \sqrt{100}$$
$$= 10$$

Example 8. Find $\sqrt{y + 6} \cdot \sqrt{y + 6}$.

Solution

$$\sqrt{y + 6} \cdot \sqrt{y + 6}$$
$$= (\sqrt{y + 6})^2$$
$$= y + 6$$

Example 10. Find $(-3\sqrt{x})(-6\sqrt{x})$.

Solution

$$(-3\sqrt{x})(-6\sqrt{x})$$
$$= -3(-6)\sqrt{x} \cdot \sqrt{x}$$
$$= 18(\sqrt{x})^2$$
$$= 18x$$

Example 14. Find $(8 - \sqrt{y})^2$.

Solution

$$(8 - \sqrt{y})^2$$
$$= 8^2 - 2 \cdot 8\sqrt{y} + (\sqrt{y})^2$$
$$= 64 - 16\sqrt{y} + y$$

Example 17. Find the missing irrational numbers.

Irrational Number	Real Conjugate
$1 + \sqrt{2}$	$1 - \sqrt{2}$
$-3 - \sqrt{5}$	$-3 + \sqrt{5}$
$\sqrt{7} + 1$	$\sqrt{7} - 1$
$2\sqrt{x} - \sqrt{13}$	$2\sqrt{x} + \sqrt{13}$

Example 19. Find $(\sqrt{y} + 3)(\sqrt{y} - 3)$.

Solution

$$(\sqrt{y} + 3)(\sqrt{y} - 3)$$
$$= (\sqrt{y})^2 - 3^2$$
$$= y - 9$$

Answers to Trial Runs _____

page 628   **1.** $-2\sqrt{3}$ **2.** $2\sqrt{x}$ **3.** $-11\sqrt{2a}$ **4.** $11 + 3\sqrt{3}$ **5.** $2 - \sqrt{2}$

page 630   **1.** $\sqrt{10}$ **2.** $5 + a$ **3.** $24a$ **4.** $-10x\sqrt{3}$ **5.** $3\sqrt{2} - 2$ **6.** $3\sqrt{10} + 18$

page 632   **1.** $2 - \sqrt{3} + 2\sqrt{2} - \sqrt{6}$ **2.** $\sqrt{5}$ **3.** -59 **4.** $x - 64$ **5.** $x - 2\sqrt{x} - 8$
6. $x - 10\sqrt{x} + 25$

EXERCISE SET 12.2

Perform the indicated operations and simplify. Assume variable quantities are not negative.

_____ 1. $2\sqrt{3} + \sqrt{3}$

_____ 2. $5\sqrt{6} - \sqrt{6}$

_____ 3. $5\sqrt{x} - 2\sqrt{x} + 8\sqrt{x}$

_____ 4. $9\sqrt{x} - 5\sqrt{x} - 6\sqrt{x}$

_____ 5. $\sqrt{27} - \sqrt{300}$

_____ 6. $\sqrt{32} - \sqrt{98}$

_____ 7. $\sqrt{54a} + \sqrt{24a}$

_____ 8. $\sqrt{125a} + \sqrt{80a}$

_____ 9. $\sqrt{81} - \sqrt{12} - \sqrt{48}$

_____ 10. $\sqrt{49} - \sqrt{45} - \sqrt{180}$

_____ 11. $4\sqrt{12x} + 3\sqrt{75x}$

_____ 12. $5\sqrt{72x} + 3\sqrt{18x}$

_____ 13. $5\sqrt{28} - 4\sqrt{50}$

_____ 14. $4\sqrt{27} - 3\sqrt{20}$

_____ 15. $\dfrac{\sqrt{288}}{\sqrt{4}} + 3\sqrt{50} - \sqrt{200}$

_____ 16. $\dfrac{\sqrt{360}}{\sqrt{8}} - 2\sqrt{405} + \sqrt{80}$

_____ 17. $\sqrt{75x^2} - \sqrt{300x^2}$

_____ 18. $\sqrt{72x^2} - \sqrt{32x^2}$

_____ 19. $\sqrt{3} \cdot \sqrt{7}$

_____ 20. $\sqrt{5} \cdot \sqrt{11}$

_____ 21. $\sqrt{2} \cdot \sqrt{12} \cdot \sqrt{6}$

_____ 22. $\sqrt{4} \cdot \sqrt{8} \cdot \sqrt{2}$

_____ 23. $\sqrt{11a} \cdot \sqrt{11a}$

_____ 24. $\sqrt{17a} \cdot \sqrt{17a}$

_____ 25. $\sqrt{x-1} \cdot \sqrt{x-1}$

_____ 26. $\sqrt{x+5} \cdot \sqrt{x+5}$

_____ 27. $\sqrt{15a} \cdot \sqrt{3a}$

_____ 28. $\sqrt{14a} \cdot \sqrt{2a}$

_____ 29. $3\sqrt{50} \cdot \sqrt{2}$

_____ 30. $\sqrt{27} \cdot \sqrt{3}$

_____ 31. $(5\sqrt{18x})(-2\sqrt{8x})$

_____ 32. $(-2\sqrt{27x})(4\sqrt{3x})$

_____ 33. $(\sqrt{5x})(3\sqrt{10x})$

_____ 34. $(5\sqrt{3x})(\sqrt{21x})$

_____ 35. $\sqrt{5}(2 - \sqrt{5})$

_____ 36. $\sqrt{11}(2 - \sqrt{11})$

_____ 37. $2\sqrt{3}(\sqrt{2} + \sqrt{12})$

_____ 38. $3\sqrt{5}(\sqrt{2} + \sqrt{20})$

_____ 39. $-3(\sqrt{x} - 2\sqrt{5} + \sqrt{6})$

_____ 40. $-5(\sqrt{x} - 4\sqrt{2} + \sqrt{7})$

_____ 41. $\sqrt{x}(\sqrt{x} - 16)$

_____ 42. $\sqrt{y}(\sqrt{y} - 9)$

_____ 43. $\sqrt{6}(\sqrt{6} - 5) - 5(\sqrt{6} - 5)$

_____ 44. $\sqrt{2}(\sqrt{2} - 11) - 11(\sqrt{2} - 11)$

_____ 45. $\sqrt{x}(\sqrt{x} - 5) + 5(\sqrt{x} - 5)$

_____ 46. $\sqrt{a}(\sqrt{a} - 7) + 7(\sqrt{a} - 7)$

_____ 47. $4(3 - \sqrt{11}) - \sqrt{11}(3 - \sqrt{11})$

_____ 48. $2(7 - \sqrt{5}) + \sqrt{5}(7 - \sqrt{5})$

_____ 49. $2\sqrt{x}(\sqrt{x} - 1) - 3(\sqrt{x} + 1)$

_____ 50. $3\sqrt{x}(\sqrt{x} - 2) - 5(\sqrt{x} + 2)$

_____ 51. $(5 - \sqrt{2})(1 - \sqrt{3})$ _____ 52. $(4 - \sqrt{3})(1 + \sqrt{5})$

_____ 53. $(7 - 2\sqrt{5})(3 + \sqrt{5})$ _____ 54. $(4 - \sqrt{3})(9 + 2\sqrt{3})$

_____ 55. $(2 - \sqrt{3x})(4 + \sqrt{3x})$ _____ 56. $(5 - \sqrt{2x})(3 + \sqrt{2x})$

_____ 57. $(\sqrt{3} - \sqrt{7})(\sqrt{3} + \sqrt{7})$ _____ 58. $(\sqrt{10} - \sqrt{5})(\sqrt{10} + \sqrt{5})$

_____ 59. $(\sqrt{x} - 10)(\sqrt{x} + 10)$ _____ 60. $(\sqrt{x} - 9)(\sqrt{x} + 9)$

_____ 61. $(\sqrt{y} - \sqrt{7})^2$ _____ 62. $(\sqrt{y} - \sqrt{11})^2$

_____ 63. $(a\sqrt{2} + 5)^2$ _____ 64. $(a\sqrt{3} + 4)^2$

☆ Stretching the Topics

Perform the indicated operation and simplify.

_____ 1. $(5\sqrt{6} + 2\sqrt{3})^2$

_____ 2. $(\sqrt{x + y} - \sqrt{x - y})(\sqrt{x + y} + \sqrt{x - y})$

_____ 3. $(\sqrt{3} - \sqrt{5})^2 (\sqrt{3} + \sqrt{5})^2$

Check your answers in the back of your book.

If you can simplify the expressions in **Checkup 12.2**, you are ready to go on to Section 12.3.

CHECKUP 12.2

Perform the indicated operation and simplify.

_____ 1. $\sqrt{24x} + \sqrt{54x}$

_____ 2. $\sqrt{12} + \sqrt{75} - \sqrt{48}$

_____ 3. $\dfrac{\sqrt{54}}{\sqrt{2}} - 5\sqrt{12} + \sqrt{300}$

_____ 4. $\sqrt{15a} \cdot \sqrt{3a}$

_____ 5. $\sqrt{y+6} \cdot \sqrt{y+6}$

_____ 6. $\sqrt{5}(7 - \sqrt{10})$

_____ 7. $2\sqrt{2}(\sqrt{3} + \sqrt{6})$

_____ 8. $(\sqrt{x} - \sqrt{5})^2$

_____ 9. $(3 + 2\sqrt{5})(4 - 3\sqrt{5})$

_____ 10. $(\sqrt{x} - \sqrt{3})(\sqrt{x} + \sqrt{3})$

Check your answers in the back of your book.

If You Missed Problems:	You Should Review Examples:
1–3	1–4
4, 5	5–8
6, 7	11, 12
8	13, 14
9, 10	15–18

12.3 Rationalizing Denominators

Mathematicians generally agree that a fraction containing a radical in the denominator is not in simplest form. Fractions such as

$$\frac{1}{\sqrt{3}} \qquad \frac{2\sqrt{2}}{\sqrt{5}} \qquad \frac{5}{1 + \sqrt{3}} \qquad \frac{3 + \sqrt{7}}{5 - \sqrt{10}}$$

are *not* in simplest form because each contains a radical in the denominator.

We seek a way to change each *irrational* denominator to a *rational* number. Recall that we are allowed to multiply the numerator and denominator of a fraction by the same nonzero number. Let's see if we can choose such a number for the fraction $\frac{1}{\sqrt{3}}$ to make its denominator *rational*. We want to multiply the irrational denominator ($\sqrt{3}$) by some number that will leave no radical in the denominator. Remembering that $\sqrt{3} \cdot \sqrt{3} = 3$, let's multiply numerator and denominator by $\sqrt{3}$ and simplify.

$$\frac{1}{\sqrt{3}} = \frac{1}{\sqrt{3}} \cdot \frac{\sqrt{3}}{\sqrt{3}}$$
$$= \frac{1 \cdot \sqrt{3}}{\sqrt{3} \cdot \sqrt{3}}$$
$$= \frac{\sqrt{3}}{3}$$

Notice that our denominator is now a *rational* number. Indeed, the numerator now contains a radical, but that is acceptable.

This process is called **rationalizing the denominator** because it changes the denominator into a rational number.

Rationalizing Denominators of One Term. To rationalize a denominator that is a single term containing a simplified radical, $\sqrt{a}$, we multiply numerator and denominator by that radical, $\sqrt{a}$.

See if you can rationalize the denominator in Example 1.

Example 1. Rationalize the denominator of $\frac{4}{\sqrt{2}}$.

Solution

$$\frac{4}{\sqrt{2}} = \frac{4}{\sqrt{2}} \cdot \frac{\sqrt{2}}{\sqrt{2}}$$
$$= \frac{4 \cdot \sqrt{2}}{\sqrt{2} \cdot \sqrt{2}}$$

Did you remember to reduce your fraction? Check your work on page 641. ▶

Example 2. Rationalize the denominator of $\frac{2\sqrt{2}}{\sqrt{5}}$.

Solution

$$\frac{2\sqrt{2}}{\sqrt{5}} = \frac{2\sqrt{2}}{\sqrt{5}} \cdot \frac{\sqrt{5}}{\sqrt{5}}$$
$$= \frac{2\sqrt{2} \cdot \sqrt{5}}{\sqrt{5} \cdot \sqrt{5}}$$
$$= \frac{2\sqrt{10}}{5}$$

You complete Example 3.

Example 3. Rationalize the denominator of $\sqrt{\dfrac{5}{3}}$.

Solution

$$\sqrt{\dfrac{5}{3}} = \dfrac{\sqrt{5}}{\sqrt{3}}$$

$$= \dfrac{\sqrt{5}}{\sqrt{3}} \cdot \dfrac{\sqrt{\rule{1.5em}{0pt}}}{\sqrt{\rule{1.5em}{0pt}}}$$

Check your work on page 641. ▶

Example 4. Rationalize the denominator of $\dfrac{\sqrt{6} - 2}{\sqrt{6}}$.

Solution

$$\dfrac{\sqrt{6} - 2}{\sqrt{6}} = \dfrac{\sqrt{6} - 2}{\sqrt{6}} \cdot \dfrac{\sqrt{6}}{\sqrt{6}}$$

$$= \dfrac{(\sqrt{6} - 2)\sqrt{6}}{\sqrt{6} \cdot \sqrt{6}}$$

$$= \dfrac{\sqrt{6} \cdot \sqrt{6} - 2\sqrt{6}}{6}$$

$$= \dfrac{6 - 2\sqrt{6}}{6}$$

$$= \dfrac{\overset{1}{\cancel{2}}(3 - \sqrt{6})}{\underset{3}{\cancel{6}}}$$

$$= \dfrac{3 - \sqrt{6}}{3}$$

Notice we factored the numerator before reducing the fraction.

⫸ Trial Run

Rationalize each denominator and simplify.

_____ 1. $\dfrac{1}{\sqrt{5}}$ _____ 2. $\dfrac{5}{\sqrt{3}}$ _____ 3. $\dfrac{2\sqrt{11}}{\sqrt{6}}$

_____ 4. $\sqrt{\dfrac{7}{2}}$ _____ 5. $\dfrac{1 + \sqrt{2}}{\sqrt{3}}$ _____ 6. $\dfrac{\sqrt{10} - 4}{\sqrt{2}}$

Answers are on page 641.

Rationalizing denominators containing *two* terms requires a different kind of rationalizing factor. Suppose that we try to rationalize the denominator of $\dfrac{5}{7 + \sqrt{3}}$.

Recall from our multiplication examples that when we multiplied an irrational *sum* of two numbers (such as $7 + \sqrt{3}$) times an irrational *difference* of the same two numbers (such as $7 - \sqrt{3}$), the result was a *rational* number. Let's try rationalizing our denominator by multiplying numerator and denominator by $7 - \sqrt{3}$ (the *real conjugate* of $7 + \sqrt{3}$).

$$\dfrac{5}{7 + \sqrt{3}} = \dfrac{5}{7 + \sqrt{3}} \cdot \dfrac{7 - \sqrt{3}}{7 - \sqrt{3}}$$

Multiply numerator and denominator by $7 - \sqrt{3}$ (the conjugate of $7 + \sqrt{3}$).

$$= \frac{5(7 - \sqrt{3})}{(7 + \sqrt{3})(7 - \sqrt{3})}$$ Rewrite the product as one fraction.

$$= \frac{5(7 - \sqrt{3})}{7^2 - (\sqrt{3})^2}$$ Multiply factors in the denominator. Remember that $(A + B)(A - B) = A^2 - B^2$.

$$= \frac{5(7 - \sqrt{3})}{49 - 3}$$ Simplify terms in the denominator.

$$= \frac{5(7 - \sqrt{3})}{46}$$ Simplify the denominator.

$$= \frac{35 - 5\sqrt{3}}{46}$$ Remove parentheses in numerator.

We have successfully rationalized the denominator.

Rationalizing Denominators of Two Terms. To rationalize an irrational denominator containing a sum or difference of two terms, multiply the numerator and denominator by the *real conjugate* of the original denominator.

In rationalizing denominators, it is wise to wait until the denominator is a rational number before multiplying the factors in the numerator. Otherwise, you might overlook a common factor that would allow you to reduce the fraction.

Now complete Example 5.

Example 5. Rationalize the denominator of $\frac{\sqrt{3}}{5 - \sqrt{2}}$.

Solution. The real conjugate of $5 - \sqrt{2}$ is $5 + \sqrt{2}$.

$$\frac{\sqrt{3}}{5 - \sqrt{2}} = \frac{\sqrt{3}}{5 - \sqrt{2}} \cdot \frac{5 + \sqrt{2}}{5 + \sqrt{2}}$$

$$= \frac{\sqrt{3}(5 + \sqrt{2})}{(5 - \sqrt{2})(5 + \sqrt{2})}$$

$$= \frac{\sqrt{3}(5 + \sqrt{2})}{5^2 - (\sqrt{2})^2}$$

Check your work on page 641. ▶

Example 6. Rationalize the denominator of $\frac{\sqrt{3}}{9 + \sqrt{3}}$.

Solution. The real conjugate of $9 + \sqrt{3}$ is $9 - \sqrt{3}$.

$$\frac{\sqrt{3}}{9 + \sqrt{3}} = \frac{\sqrt{3}}{9 + \sqrt{3}} \cdot \frac{9 - \sqrt{3}}{9 - \sqrt{3}}$$

$$= \frac{\sqrt{3}(9 - \sqrt{3})}{(9 + \sqrt{3})(9 - \sqrt{3})}$$

$$= \frac{\sqrt{3}(9 - \sqrt{3})}{9^2 - (\sqrt{3})^2}$$

$$= \frac{\sqrt{3}(9 - \sqrt{3})}{81 - 3}$$

$$= \frac{\sqrt{3}(9 - \sqrt{3})}{78}$$

$$= \frac{9\sqrt{3} - 3}{78}$$

$$= \frac{\overset{1}{\cancel{3}}(3\sqrt{3} - 1)}{\underset{26}{\cancel{78}}}$$

$$= \frac{3\sqrt{3} - 1}{26}$$

Example 7. Rationalize the denominator of $\dfrac{8}{\sqrt{5} - \sqrt{7}}$.

Solution. The real conjugate of $\sqrt{5} - \sqrt{7}$ is $\sqrt{5} + \sqrt{7}$.

$$\frac{8}{\sqrt{5} - \sqrt{7}} = \frac{8}{\sqrt{5} - \sqrt{7}} \cdot \frac{\sqrt{5} + \sqrt{7}}{\sqrt{5} + \sqrt{7}}$$

$$= \frac{8(\sqrt{5} + \sqrt{7})}{(\sqrt{5} - \sqrt{7})(\sqrt{5} + \sqrt{7})}$$

$$= \frac{8(\sqrt{5} + \sqrt{7})}{(\sqrt{5})^2 - (\sqrt{7})^2}$$

$$= \frac{8(\sqrt{5} + \sqrt{7})}{5 - 7}$$

$$= \frac{\overset{-4}{\cancel{8}}(\sqrt{5} + \sqrt{7})}{\underset{1}{\cancel{-2}}}$$

$$= -4(\sqrt{5} + \sqrt{7})$$

$$= -4\sqrt{5} - 4\sqrt{7}$$

You complete Example 8.

Example 8. Rationalize the denominator of $\dfrac{\sqrt{7} - 2}{\sqrt{7} + 2}$.

Solution. The real conjugate of $\sqrt{7} + 2$ is $\sqrt{7} - 2$.

$$\frac{\sqrt{7} - 2}{\sqrt{7} + 2} = \frac{\sqrt{7} - 2}{\sqrt{7} + 2} \cdot \frac{\sqrt{7} - 2}{\sqrt{7} - 2}$$

$$= \frac{(\sqrt{7} - 2)^2}{(\sqrt{7} + 2)(\sqrt{7} - 2)}$$

$$= \frac{(\sqrt{7})^2 - 2 \cdot 2\sqrt{7} + 2^2}{(\sqrt{7})^2 - 2^2}$$

Check your work on page 641. ▶

⫸ Trial Run

Rationalize each denominator and simplify.

_____ **1.** $\dfrac{2}{7 + \sqrt{5}}$

_____ **2.** $\dfrac{-3}{\sqrt{2} + 1}$

_____ **3.** $\dfrac{\sqrt{5}}{4 - \sqrt{5}}$

_____ **4.** $\dfrac{3 - \sqrt{5}}{3 + \sqrt{5}}$

Answers are on page 641.

▶ Examples You Completed

Example 1. Rationalize the denominator of $\frac{4}{\sqrt{2}}$.

Solution

$$\frac{4}{\sqrt{2}} = \frac{4}{\sqrt{2}} \cdot \frac{\sqrt{2}}{\sqrt{2}}$$

$$= \frac{4 \cdot \sqrt{2}}{\sqrt{2} \cdot \sqrt{2}}$$

$$= \frac{\overset{2}{\cancel{4}}\sqrt{2}}{\underset{1}{\cancel{2}}}$$

$$= 2\sqrt{2}$$

Example 3. Rationalize the denominator of $\sqrt{\frac{5}{3}}$.

Solution

$$\sqrt{\frac{5}{3}} = \frac{\sqrt{5}}{\sqrt{3}}$$

$$= \frac{\sqrt{5}}{\sqrt{3}} \cdot \frac{\sqrt{3}}{\sqrt{3}}$$

$$= \frac{\sqrt{5} \cdot \sqrt{3}}{\sqrt{3} \cdot \sqrt{3}}$$

$$= \frac{\sqrt{15}}{3}$$

Example 5. Rationalize the denominator of $\frac{\sqrt{3}}{5 - \sqrt{2}}$.

Solution. The real conjugate of $5 - \sqrt{2}$ is $5 + \sqrt{2}$.

$$\frac{\sqrt{3}}{5 - \sqrt{2}} = \frac{\sqrt{3}}{5 - \sqrt{2}} \cdot \frac{5 + \sqrt{2}}{5 + \sqrt{2}}$$

$$= \frac{\sqrt{3}(5 + \sqrt{2})}{(5 - \sqrt{2})(5 + \sqrt{2})}$$

$$= \frac{\sqrt{3}(5 + \sqrt{2})}{5^2 - (\sqrt{2})^2}$$

$$= \frac{\sqrt{3}(5 + \sqrt{2})}{25 - 2}$$

$$= \frac{\sqrt{3}(5 + \sqrt{2})}{23}$$

$$= \frac{5\sqrt{3} + \sqrt{6}}{23}$$

Example 8. Rationalize the denominator of $\frac{\sqrt{7} - 2}{\sqrt{7} + 2}$.

Solution. The real conjugate of $\sqrt{7} + 2$ is $\sqrt{7} - 2$.

$$\frac{\sqrt{7} - 2}{\sqrt{7} + 2} = \frac{\sqrt{7} - 2}{\sqrt{7} + 2} \cdot \frac{\sqrt{7} - 2}{\sqrt{7} - 2}$$

$$= \frac{(\sqrt{7} - 2)^2}{(\sqrt{7} + 2)(\sqrt{7} - 2)}$$

$$= \frac{(\sqrt{7})^2 - 2 \cdot 2\sqrt{7} + 2^2}{(\sqrt{7})^2 - 2^2}$$

$$= \frac{7 - 4\sqrt{7} + 4}{7 - 4}$$

$$= \frac{11 - 4\sqrt{7}}{3}$$

Answers to Trial Runs

page 638   1. $\frac{\sqrt{5}}{5}$ 2. $\frac{5\sqrt{3}}{3}$ 3. $\frac{\sqrt{66}}{3}$ 4. $\frac{\sqrt{14}}{2}$ 5. $\frac{\sqrt{3} + \sqrt{6}}{3}$ 6. $\sqrt{5} - 2\sqrt{2}$

page 640   1. $\frac{7 - \sqrt{5}}{22}$ 2. $-3\sqrt{2} + 3$ 3. $\frac{4\sqrt{5} + 5}{11}$ 4. $\frac{7 - 3\sqrt{5}}{2}$

EXERCISE SET 12.3

Rationalize the denominators. Write your answers in simplest form.

_____ 1. $\dfrac{3}{\sqrt{3}}$ _____ 2. $\dfrac{5}{\sqrt{5}}$

_____ 3. $\dfrac{2}{\sqrt{7}}$ _____ 4. $\dfrac{5}{\sqrt{6}}$

_____ 5. $\dfrac{4}{\sqrt{2}}$ _____ 6. $\dfrac{9}{\sqrt{3}}$

_____ 7. $\dfrac{12}{\sqrt{10}}$ _____ 8. $\dfrac{21}{\sqrt{6}}$

_____ 9. $\dfrac{9}{\sqrt{32}}$ _____ 10. $\dfrac{8}{\sqrt{27}}$

_____ 11. $\dfrac{\sqrt{5}}{\sqrt{8}}$ _____ 12. $\dfrac{\sqrt{7}}{\sqrt{12}}$

_____ 13. $\dfrac{\sqrt{10}}{\sqrt{5}}$ _____ 14. $\dfrac{\sqrt{6}}{\sqrt{2}}$

_____ 15. $\sqrt{\dfrac{1}{2}}$ _____ 16. $\sqrt{\dfrac{1}{3}}$

_____ 17. $\sqrt{\dfrac{4}{5}}$ _____ 18. $\sqrt{\dfrac{9}{7}}$

_____ 19. $\sqrt{\dfrac{3}{x}}$ _____ 20. $\sqrt{\dfrac{2}{x}}$

_____ 21. $\dfrac{1+\sqrt{2}}{\sqrt{5}}$ _____ 22. $\dfrac{2+\sqrt{3}}{\sqrt{2}}$

_____ 23. $\dfrac{\sqrt{8}-6}{\sqrt{2}}$ _____ 24. $\dfrac{\sqrt{10}-5}{\sqrt{5}}$

_____ 25. $\dfrac{\sqrt{2}+\sqrt{5}}{\sqrt{10}}$ _____ 26. $\dfrac{\sqrt{3}+\sqrt{2}}{\sqrt{6}}$

_____ 27. $\dfrac{1}{2+\sqrt{3}}$ _____ 28. $\dfrac{1}{5-\sqrt{2}}$

_____ 29. $\dfrac{-12}{3-\sqrt{6}}$ _____ 30. $\dfrac{-6}{2+\sqrt{7}}$

_____ 31. $\dfrac{\sqrt{2}}{1 - \sqrt{2}}$

_____ 32. $\dfrac{\sqrt{3}}{2 - \sqrt{3}}$

_____ 33. $\dfrac{\sqrt{8}}{\sqrt{3} + 1}$

_____ 34. $\dfrac{\sqrt{12}}{\sqrt{5} - 1}$

_____ 35. $\dfrac{2\sqrt{3}}{\sqrt{5} + \sqrt{2}}$

_____ 36. $\dfrac{2\sqrt{5}}{\sqrt{11} - \sqrt{3}}$

_____ 37. $\dfrac{2 + \sqrt{2}}{\sqrt{2} + 1}$

_____ 38. $\dfrac{3 + \sqrt{5}}{\sqrt{5} - 2}$

_____ 39. $\dfrac{\sqrt{6} + 2}{\sqrt{6} - 2}$

_____ 40. $\dfrac{\sqrt{7} + 1}{\sqrt{7} - 1}$

☆ Stretching the Topics

Rationalize the denominators.

_____ 1. $\sqrt{\dfrac{a^2}{4(a + b)}}$

_____ 2. $\dfrac{3\sqrt{x}}{5\sqrt{x} - 2\sqrt{y}}$

_____ 3. $\dfrac{2\sqrt{x - y}}{\sqrt{x - y} - 2}$

Check your answers in the back of your book.

If you can rationalize the denominators in **Checkup 12.3**, you are ready to go on to Section 12.4.

CHECKUP 12.3

Rationalize the denominators. Write your answers in simplest form.

_____ **1.** $\dfrac{6}{\sqrt{6}}$ _____ **2.** $\dfrac{20}{\sqrt{10}}$ _____ **3.** $\dfrac{7}{\sqrt{12}}$

_____ **4.** $\dfrac{\sqrt{6}}{\sqrt{15}}$ _____ **5.** $\sqrt{\dfrac{8}{3}}$ _____ **6.** $\dfrac{\sqrt{12}-10}{\sqrt{2}}$

_____ **7.** $\dfrac{\sqrt{5}+\sqrt{15}}{\sqrt{20}}$ _____ **8.** $\dfrac{\sqrt{3}}{6-\sqrt{3}}$

_____ **9.** $\dfrac{5+\sqrt{11}}{\sqrt{11}-3}$ _____ **10.** $\dfrac{\sqrt{7}-2}{\sqrt{7}+2}$

Check your answers in the back of your book.

If You Missed Problems:	You Should Review Examples:
1–4	1, 2
5	3
6, 7	4
8	5–7
9, 10	8

12.4 Solving More Quadratic Equations

Recall from Chapter 7 that a **quadratic equation** is an equation of the form

$$ax^2 + bx + c = 0 \qquad \text{where } a \neq 0$$

In that chapter we learned to solve such equations by factoring and using the Zero Product Rule.

$$x^2 + 3x - 4 = 0$$
$$(x + 4)(x - 1) = 0$$
$$x + 4 = 0 \quad \text{or} \quad x - 1 = 0$$
$$x = -4 \quad \text{or} \quad x = 1$$

$$x^2 - 9 = 0$$
$$(x + 3)(x - 3) = 0$$
$$x + 3 = 0 \quad \text{or} \quad x - 3 = 0$$
$$x = -3 \quad \text{or} \quad x = 3$$

Every quadratic equation in Chapter 7 could be solved by means of factoring to yield *rational* solutions. Unfortunately, it is not true that all quadratic equations have solutions that are rational numbers. We turn our attention now to methods that can be used for solving *any* quadratic equation.

Solving Quadratic Equations of the Form ax² = c

Some quadratic equations do not contain a first-degree term. For example,

$$x^2 - 2 = 0$$

is a quadratic (second-degree) equation containing *no first-degree term*. We would like to factor the left-hand side of this equation, but no rational factors will "work." Let's rewrite our equation by adding 2 to both sides.

$$x^2 - 2 = 0$$
$$x^2 = 2$$

Now we are looking for a number x whose square is the number 2.

Certainly, $x = \sqrt{2}$ is a solution because it satisfies the original equation.

$$x^2 = 2$$
$$(\sqrt{2})^2 \overset{?}{=} 2$$
$$(\sqrt{2})(\sqrt{2}) \overset{?}{=} 2$$
$$2 = 2$$

But another solution is $x = -\sqrt{2}$ because it also satisfies the original equation.

$$x^2 = 2$$
$$(-\sqrt{2})^2 \overset{?}{=} 2$$
$$(-\sqrt{2})(-\sqrt{2}) \overset{?}{=} 2$$
$$2 = 2$$

Thus, the solutions for $x^2 = 2$ are $x = \sqrt{2}$ or $x = -\sqrt{2}$.

As before, we discover that a second-degree equation usually has *two* solutions.

You try Example 1.

Example 1. Solve $x^2 = 7$.

Solution

$$x^2 = 7$$
$$x = \sqrt{\underline{\quad}} \quad \text{or}$$
$$x = -\sqrt{\underline{\quad}}$$

Check your work on page 650. ▶

Complete Example 2.

Example 2. Solve $2x^2 - 22 = 0$.

Solution

$$2x^2 - 22 = 0$$
$$2x^2 = \underline{\quad}$$
$$x^2 = \underline{\quad}$$
$$x = \underline{\quad} \quad \text{or} \quad x = -\underline{\quad}$$

Check your work on page 650. ▶

Solving Equations of the Form $x^2 = C$. If $x^2 = C$ and C is *not* negative, the solutions are

$$x = \sqrt{C} \quad \text{or} \quad x = -\sqrt{C}$$

If $x^2 = C$ and C is negative, there is no real number solution.

To save time and space when solving quadratic equations of the form $x^2 = C$ (where C is not negative), we may start by writing both solutions at the same time, using the notation

$$x = \pm\sqrt{C}$$

This shortcut allows us to simplify both solutions at once, but we shall usually conclude by stating the solutions separately.

Example 3. Solve $3x^2 - 5 = 0$.

Solution

$$3x^2 - 5 = 0$$

$$3x^2 = 5$$

$$x^2 = \frac{5}{3}$$

$$x = \pm\sqrt{\frac{5}{3}} \qquad \text{Solve for } x.$$

$$= \frac{\pm\sqrt{5}}{\sqrt{3}} \qquad \begin{array}{l}\text{Separate the}\\\text{radicals.}\end{array}$$

$$= \frac{\pm\sqrt{5}}{\sqrt{3}} \cdot \frac{\sqrt{3}}{\sqrt{3}} \qquad \begin{array}{l}\text{Rationalize}\\\text{the denominator.}\end{array}$$

$$x = \frac{\pm\sqrt{15}}{3} \qquad \text{Multiply.}$$

$$x = \frac{\sqrt{15}}{3} \quad \text{or} \quad x = -\frac{\sqrt{15}}{3}$$

Write both solutions.

You try completing Example 4.

Example 4. Solve $2x^2 - 11 = 2$.

Solution

$$2x^2 - 11 = 2$$

$$2x^2 = 13$$

$$x^2 = \frac{13}{2}$$

$$x = \underline{\hspace{1cm}} \quad \text{or} \quad x = \underline{\hspace{1cm}}$$

Check your work on page 651. ▶

Example 5. Solve $x^2 + 1 = 0$.

Solution

$$x^2 + 1 = 0$$

$$x^2 = -1$$

There is no real number solution.

Complete Example 6.

Example 6. Solve $x^2 + 7 = 5$.

Solution

Check your work on page 651. ▶

◗ **Trial Run** ━━━━━━━━━━━━━━━━━━━━━━━━━━━━━━━━━━

Solve.

_____ 1. $x^2 - 21 = 0$ _____ 2. $3x^2 = 15$

_____ 3. $2x^2 - 16 = 0$ _____ 4. $5 = 2x^2 - 6$

Answers are on page 651.

━━━

Using the Quadratic Formula

We know how to solve a quadratic equation that is in standard form

$$ax^2 + bx + c = 0$$

as long as the left-hand side can be factored, but suppose that it cannot be factored by our usual methods. Fortunately, mathematicians have derived a formula that allows us to solve any quadratic equation.

┌───┐
│ **Quadratic Formula.** If $ax^2 + bx + c = 0$, the solutions can be found by │
│   │
│ $$x = \dfrac{-b \pm \sqrt{b^2 - 4ac}}{2a}$$ │
└───┘

Let's practice using this formula with an equation whose solutions can also be found by factoring; then we can be more convinced that the formula "works." Solve the equation

$$x^2 + 2x - 3 = 0$$

Approach 1. Use factoring.

$$x^2 + 2x - 3 = 0$$
$$(x + 3)(x - 1) = 0$$
$$x + 3 = 0 \quad \text{or} \quad x - 1 = 0$$
$$x = -3 \quad \text{or} \quad x = 1$$

Approach 2. Use the quadratic formula.

$$x^2 + 2x - 3 = 0$$

We carefully note that here

$$a = 1 \quad b = 2 \quad \text{and} \quad c = -3$$

and we substitute these values into the quadratic formula.

$$x = \frac{-b \pm \sqrt{b^2 - 4ac}}{2a}$$

$$x = \frac{-2 \pm \sqrt{2^2 - 4(1)(-3)}}{2(1)} \qquad \begin{array}{l}\text{Substitute 2 for } b, \\ 1 \text{ for } a, \text{ and } -3 \text{ for } c.\end{array}$$

$$= \frac{-2 \pm \sqrt{4 + 12}}{2} \qquad \begin{array}{l}\text{Simplify terms in} \\ \text{the radicand.}\end{array}$$

$$= \frac{2 \pm \sqrt{16}}{2} \qquad \begin{array}{l}\text{Find the sum in the} \\ \text{radicand.}\end{array}$$

$$x = \frac{-2 \pm 4}{2} \qquad \text{Simplify the radical.}$$

$$x = \frac{-2 + 4}{2} \quad \text{or} \quad x = \frac{-2 - 4}{2} \qquad \begin{array}{l}\text{Separate the two} \\ \text{solutions.}\end{array}$$

$$= \frac{2}{2} \qquad\qquad = \frac{-6}{2} \qquad \text{Simplify each numerator.}$$

$$x = 1 \qquad \text{or} \quad x = -3 \qquad \text{Reduce each fraction.}$$

Notice that both approaches yielded the same two solutions. Of course, factoring was a much quicker method to use in this example. We shall *always solve by factoring when possible*; otherwise, we shall use the quadratic formula.

Example 7. Solve $x^2 + 5x + 1 = 0$.

Solution. We cannot factor $x^2 + 5x + 1$, so we must use the quadratic formula. Here

$$a = 1 \quad b = 5 \quad \text{and} \quad c = 1$$

$$x = \frac{-b \pm \sqrt{b^2 - 4ac}}{2a}$$

$$x = \frac{-5 \pm \sqrt{5^2 - 4(1)(1)}}{2(1)} \qquad \begin{array}{l}\text{Substitute 5 for } b, \\ 1 \text{ for } a, \text{ and } 1 \text{ for } c.\end{array}$$

$$= \frac{-5 \pm \sqrt{25 - 4}}{2} \qquad \begin{array}{l}\text{Simplify terms in} \\ \text{the radicand.}\end{array}$$

$$x = \frac{-5 \pm \sqrt{21}}{2} \qquad \begin{array}{l}\text{Find the difference in} \\ \text{the radicand.}\end{array}$$

$$x = \frac{-5 + \sqrt{21}}{2} \quad \text{or} \quad x = \frac{-5 - \sqrt{21}}{2} \qquad \begin{array}{l}\text{Separate the two} \\ \text{solutions.}\end{array}$$

These solutions are irrational.

Try using the quadratic formula in Example 8.

Example 8. Solve $2x^2 - 7x + 2 = 0$.

Solution

$$2x^2 - 7x + 2 = 0$$

We note that

$$a = 2 \quad b = -7 \quad c = 2$$

and we substitute into the quadratic formula.

$$x = \frac{-b \pm \sqrt{b^2 - 4ac}}{2a}$$

$$x = \frac{-(-7) \pm \sqrt{(-7)^2 - 4(2)(2)}}{2(2)}$$

$x = \underline{\hspace{1cm}}$ or $x = \underline{\hspace{1cm}}$

Check your work on page 651. ▶

Example 9. Solve $x^2 + 2x = 1$.

Solution. The equation must be rewritten in standard form.

$$x^2 + 2x - 1 = 0$$

We note that

$$a = 1 \quad b = 2 \quad c = -1$$

and we substitute into the quadratic formula.

$$x = \frac{-b \pm \sqrt{b^2 - 4ac}}{2a}$$

$$x = \frac{-2 \pm \sqrt{2^2 - 4(1)(-1)}}{2(1)}$$

$$= \frac{-2 \pm \sqrt{4 + 4}}{2}$$

$$= \frac{-2 \pm \sqrt{8}}{2}$$

$$= \frac{-2 \pm \sqrt{4 \cdot 2}}{2}$$

$$= \frac{-2 \pm 2\sqrt{2}}{2}$$

$$= \frac{\overset{1}{\cancel{2}}(-1 \pm \sqrt{2})}{\underset{1}{\cancel{2}}}$$

$$x = -1 \pm \sqrt{2}$$

$$x = -1 + \sqrt{2}$$

$$\text{or} \quad x = -1 - \sqrt{2}$$

In Example 9, notice that we simplified the radical and then *factored* the numerator in order to reduce the fraction.

⫸ Trial Run

Write each equation in standard form $ax^2 + bx + c = 0$. Then give the values of a, b, and c.

_____ **1.** $2x^2 = 3x - 2$ _____ **2.** $x^2 - 5x + 2 = 7x + 5$

Use the quadratic formula to solve each equation.

_____ **3.** $x^2 + x - 3 = 0$ _____ **4.** $2x^2 - 3x - 1 = 0$

_____ **5.** $2y^2 + 4y = -1$

Answers are on page 651.

▶ Examples You Completed

Example 1. Solve $x^2 = 7$.

Solution

$$x^2 = 7$$
$$x = \sqrt{7} \quad \text{or} \quad x = -\sqrt{7}$$

Example 2. Solve $2x^2 - 22 = 0$.

Solution

$$2x^2 - 22 = 0$$
$$2x^2 = 22$$
$$x^2 = 11$$
$$x = \sqrt{11} \quad \text{or} \quad x = -\sqrt{11}$$

Example 4. Solve $2x^2 - 11 = 2$.

Solution

$$2x^2 - 11 = 2$$

$$2x^2 = 13$$

$$x^2 = \frac{13}{2}$$

$$x = \pm \sqrt{\frac{13}{2}}$$

$$= \pm \frac{\sqrt{13}}{\sqrt{2}}$$

$$= \pm \frac{\sqrt{13}}{\sqrt{2}} \cdot \frac{\sqrt{2}}{\sqrt{2}}$$

$$x = \pm \frac{\sqrt{26}}{2}$$

$$x = \frac{\sqrt{26}}{2} \quad \text{or} \quad x = -\frac{\sqrt{26}}{2}$$

Example 6. Solve $x^2 + 7 = 5$.

Solution

$$x^2 + 7 = 5$$

$$x^2 = -2$$

There is no real number solution.

Example 8. Solve $2x^2 - 7x + 2 = 0$.

Solution

$$2x^2 - 7x + 2 = 0$$

We note that

$$a = 2$$

$$b = -7$$

$$c = 2$$

and substitute into the quadratic formula.

$$x = \frac{-b \pm \sqrt{b^2 - 4ac}}{2a}$$

$$x = \frac{-(-7) \pm \sqrt{(-7)^2 - 4(2)(2)}}{2(2)}$$

$$= \frac{7 \pm \sqrt{49 - 16}}{4}$$

$$x = \frac{7 \pm \sqrt{33}}{4}$$

$$x = \frac{7 + \sqrt{33}}{4} \quad \text{or} \quad x = \frac{7 - \sqrt{33}}{4}$$

Answers to Trial Runs

page 647   **1.** $x = \sqrt{21}$ or $x = -\sqrt{21}$ **2.** $x = \sqrt{5}$ or $x = -\sqrt{5}$ **3.** $x = 2\sqrt{2}$ or $x = -2\sqrt{2}$
4. $x = \frac{\sqrt{22}}{2}$ or $x = -\frac{\sqrt{22}}{2}$

page 650   **1.** $2x^2 - 3x + 2 = 0$; $a = 2$, $b = -3$, $c = 2$ **2.** $x^2 - 12x - 3 = 0$; $a = 1$, $b = -12$,
$c = -3$ **3.** $x = \frac{-1 \pm \sqrt{13}}{2}$ **4.** $x = \frac{3 \pm \sqrt{17}}{4}$ **5.** $y = \frac{-2 \pm \sqrt{2}}{2}$

Name _____ **Date** _____

EXERCISE SET 12.4

Solve each quadratic equation. Write answers in simplest radical form and reduce if possible.

_____ **1.** $x^2 = 12$ _____ **2.** $x^2 = 32$

_____ **3.** $9a^2 = 5$ _____ **4.** $4a^2 = 7$

_____ **5.** $5x^2 - 1 = 0$ _____ **6.** $6x^2 - 5 = 0$

_____ **7.** $6y^2 - 15 = 0$ _____ **8.** $8y^2 - 28 = 0$

_____ **9.** $3y^2 - 8 = 7$ _____ **10.** $5y^2 - 17 = 8$

_____ **11.** $9 = 11 - a^2$ _____ **12.** $8 = 13 - a^2$

_____ **13.** $4x^2 + 3 = 3$ _____ **14.** $6x^2 + 8 = 8$

_____ **15.** $a^2 - 80 = 0$ _____ **16.** $a^2 - 75 = 0$

_____ **17.** $x^2 + x - 3 = 0$ _____ **18.** $x^2 + 6x + 1 = 0$

_____ **19.** $a^2 = 5a + 5$ _____ **20.** $a^2 = 2a + 2$

_____ **21.** $x^2 - 3 = 4x$ _____ **22.** $x^2 - 2 = 3x$

_____ **23.** $y^2 = 2y + 1$ _____ **24.** $y^2 = 4y + 2$

_____ **25.** $a^2 + 7a = 1$ _____ **26.** $a^2 + 5a = 4$

_____ **27.** $3x^2 - 2x - 2 = 0$ _____ **28.** $5x^2 - 3x - 1 = 0$

_____ **29.** $4 - 6y = -y^2$ _____ **30.** $4z - 1 = -z^2$

_____ **31.** $2z = 15 - 2z^2$ _____ **32.** $7z = 6z^2 - 1$

_____ **33.** $x^2 + 9 = 2$ _____ **34.** $y^2 + 10 = 7$

☆ Stretching the Topics ──────────────

Solve each equation.

_____ **1.** $(x - 3)(3x + 5) = (2x + 7)(x + 1)$

_____ **2.** $x^2 - 3\sqrt{2}\,x - 4 = 0$

_____ **3.** $\sqrt{3}\,x^2 - 6x + 2\sqrt{3} = 0$

Check your answers in the back of your book.

If you can solve the quadratic equations in **Checkup 12.4**, you are ready to go on to Section 12.5.

✓ CHECKUP 12.4

Solve each quadratic equation. Write answers in simplest radical form and reduce if possible.

_____ **1.** $x^2 = 2$

_____ **2.** $4a^2 = 20$

_____ **3.** $y^2 - 32 = 0$

_____ **4.** $9x^2 - 80 = 100$

_____ **5.** $9 - 3x^2 = 9$

_____ **6.** $x^2 + 4x - 4 = 0$

_____ **7.** $x^2 = 3x - 1$

_____ **8.** $3x^2 - 6x + 2 = 0$

_____ **9.** $2x^2 - 5x = 1$

_____ **10.** $9x^2 - 7 = 6x$

Check your answers in the back of your book.

If You Missed Problems:	You Should Review Examples:
1–3	1, 2
4, 5	3–6
6	7
7–10	8, 9

12.5 Switching from Word Statements to Quadratic Equations

Now that we have the tools for solving any quadratic equation, let's practice switching from words to quadratic equations. Do not be surprised if our equations have irrational solutions. To find approximations for such irrational numbers, we may use the table of square roots inside the back cover of this book.

Example 1. A carpenter wishes to construct a square tabletop with an area of 8 square feet. How long should each side of the tabletop be?

Solution

We Think	*We Write*
We don't know the length of a side.	Let x = length of each side.
The area is found by *squaring* the length of a side; the area is 8 square inches.	x^2 = area of top (square feet) $x^2 = 8$

We can solve this second-degree equation by the method we have learned.

$$x^2 = 8$$
$$x = \pm \sqrt{8}$$
$$= \pm \sqrt{4 \cdot 2}$$
$$= \pm 2\sqrt{2}$$
$$x = 2\sqrt{2} \quad \text{or} \quad x = -2\sqrt{2}$$

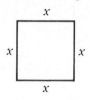

We reject the negative value for length and the solution is $x = 2\sqrt{2}$ feet. If we wish an approximation for the length, we may use $\sqrt{2} \doteq 1.414$. Then

$$x \doteq 2(1.414)$$
$$\doteq 2.828$$

Each side of the tabletop should be approximately 2.828 feet long.

You try completing Example 2.

Example 2. This spring Martina enlarged her square garden by doubling the width and tripling the length. If the area of the enlarged garden is 72 square yards, what were the approximate dimensions of her square garden?

Solution

We Think	*We Write*
We do not know the length of a side of the square garden.	Let x = length of side of square garden (yards)
Martina *doubled* the width and *tripled* the length.	$2x$ = width of new garden $3x$ = length of new garden

The new garden's area is found by *multiply-ing* length times width.

$$\text{Area} = (2x)(3x)$$
$$= 6x^2$$

The new garden's area is 72 square yards. $6x^2 = 72$

Now we solve this equation by the methods we have learned.

$6x^2 = 72$	New area is 72 square yards.
$x^2 = 12$	Isolate x^2.
$x = \pm \sqrt{12}$	Solve for x.
$x = \pm \sqrt{\underline{\quad} \cdot \underline{\quad}}$	Factor the radicand.
$x = \pm \underline{\quad} \sqrt{3}$	Remove square root.
$x = \underline{\quad} \sqrt{3}$ or $x = - \underline{\quad} \sqrt{3}$	

We reject the negative value for the length of a side and approximate the positive value using $\sqrt{3} \doteq 1.732$.

$$x = \underline{\quad} \sqrt{3}$$
$$\doteq \underline{\quad} (1.732)$$
$$\doteq \underline{\quad}$$

The original garden measured approximately _____ yards on each side.

Check your work on page 658. ▶

Let's solve the problem stated at the beginning of this chapter using the Pythagorean Theorem.

Example 3. A ladder that is 20 feet long is leaning against the wall of a building. If the foot of the ladder is 10 feet from the bottom of the wall, how far up the wall is the top of the ladder?

Solution. We can let

$$x = \text{distance up wall}\quad \text{(feet)}$$

From the Pythagorean Theorem we know that

$$x^2 + 10^2 = 20^2$$
$$x^2 + 100 = 400$$
$$x^2 = 300$$
$$x = \pm \sqrt{300}$$
$$= \pm \sqrt{100 \cdot 3}$$
$$x = \pm 10\sqrt{3}$$
$$x = 10\sqrt{3}\quad \text{or}\quad x = -10\sqrt{3}$$

Once again we reject the negative value and conclude that the top of the ladder is $10\sqrt{3}$ feet up the wall. If we wish an approximation for this distance, we may use $\sqrt{3} \doteq 1.732$. Then

$$x = 10\sqrt{3}$$

$$\doteq 10(1.732)$$

$$x \doteq 17.32$$

The top of the ladder is approximately 17.32 feet up the wall.

Example 4. One irrational number is 6 less than another irrational number. The product of the two numbers is -7. Find the irrational numbers.

Solution. We can let

$$x = \text{first irrational number}$$

$$x - 6 = \text{second irrational number}$$

$$x(x - 6) = \text{product of numbers}$$

Our equation must be

$$x(x - 6) = -7$$

$$x^2 - 6x = -7$$

$$x^2 - 6x + 7 = 0$$

We solve by the quadratic formula. Here $a = 1$, $b = -6$, and $c = 7$.

$$x = \frac{-b \pm \sqrt{b^2 - 4ac}}{2a}$$

$$x = \frac{-(-6) \pm \sqrt{(-6)^2 - 4(1)(7)}}{2(1)}$$

$$= \frac{6 \pm \sqrt{36 - 28}}{2}$$

$$= \frac{6 \pm \sqrt{8}}{2}$$

$$= \frac{6 \pm \sqrt{4 \cdot 2}}{2}$$

$$= \frac{6 \pm 2\sqrt{2}}{2}$$

$$= \frac{\overset{1}{\cancel{2}}(3 \pm \sqrt{2})}{\underset{1}{\cancel{2}}}$$

$$x = 3 \pm \sqrt{2}$$

There are two solutions to our equation and each solution gives a different pair of irrational numbers.

First pair	*Second pair*
$x = 3 + \sqrt{2}$	$x = 3 - \sqrt{2}$
$x - 6 = 3 + \sqrt{2} - 6$	$x - 6 = 3 - \sqrt{3} - 6$
$= -3 + \sqrt{2}$	$= -3 - \sqrt{2}$

So the pairs of irrational numbers that satisfy the conditions of our problem are

$$3 + \sqrt{2} \quad \text{and} \quad -3 + \sqrt{2}$$

or

$$3 - \sqrt{2} \quad \text{and} \quad -3 - \sqrt{2}$$

▶ **Example You Completed** ————————————————————

Example 2 (*Solution*)

$$6x^2 = 72$$
$$x^2 = 12$$
$$x = \pm \sqrt{12}$$
$$x = \pm \sqrt{4 \cdot 3}$$
$$x = \pm 2\sqrt{3}$$
$$x = 2\sqrt{3} \quad \text{or} \quad x = -2\sqrt{3}$$
$$x \doteq 2(1.732)$$
$$\doteq 3.464$$

The original garden measured approximately 3.464 yards on each side.

EXERCISE SET 12.5

Solve and write the answers in simplest radical form, unless directed otherwise.

_____ 1. The area of a square is 50 square centimeters. Find the length of a side of the square.

_____ 2. The area of a square is 98 square inches. Find the length of a side of the square.

_____ 3. The length of the Lathams' patio is the same as the width. If it takes 432 square feet of outdoor carpet to cover the patio, find the length of one side of the patio. (Approximate answer to the nearest tenth.)

_____ 4. Rosa's family room is a square. If it takes 675 square feet of carpeting to cover the floor, find the length of one side of the room. (Approximate answer to the nearest tenth.)

_____ 5. In a right triangle, the hypotenuse is 8 feet and one side is 4 feet. Find the length of the other side.

_____ 6. If a flower bed in the shape of a right triangle has one side that measures 6 feet and the hypotenuse is 12 feet, find the length of the other side.

_____ 7. One leg of a right triangle is twice as long as the other leg and the hypotenuse is 15 meters long. Find the length of each leg of the triangle. (Approximate answer to the nearest hundredth.)

_____ 8. One leg of a right triangle is 3 times as long as the other leg and the hypotenuse is 50 meters long. Find the length of each leg of the triangle. (Approximate answer to the nearest hundredth.)

_____ 9. The width of a rectangle is 6 meters less than the length. If the area of the rectangle is 10 square meters, find the length of the rectangle.

_____ 10. The length of a rectangle is 2 meters more than the width. If the area of the rectangle is 15 square meters, find the width of the rectangle.

_____ 11. One irrational number is 2 less than another irrational number. The product of the two numbers is 9. Find the irrational numbers.

_____ 12. One irrational number is 2 more than another irrational number. The sum of the squares of the two numbers is 5. Find the irrational numbers.

_____ 13. One irrational number is 4 more than another irrational number. The sum of the squares of the two numbers is 9. Find the irraitonal numbers.

_____ 14. One irrational number is 4 less than another irrational number. The product of the two numbers is 9. Find the irrational numbers.

_____ 15. A rectangle is 3 feet by 5 feet. If both dimensions are increased by the same amount, the area will be 30 square feet. Find the number of feet by which each dimension is increased.

_____ 16. A rectangle is 4 feet by 6 feet. If both dimensions are increased by the same amount, the area will be 50 square feet. Find the number of feet by which each dimension is increased.

_____ 17. The dimensions of a picture, including its frame, are 8 inches and 10 inches. Find the width of the frame if the area of the picture inside the frame is 50 square inches. (Approximate answer to the nearest tenth.)

_____ 18. The dimensions of a rectangular linen tablecloth, including its lace border, are 5 feet by 7 feet. Find the width of the lace if the area of the linen before adding the border was 27 square feet. (Approximate answer to the nearest tenth.)

☆ Stretching the Topics _____

_____ 1. One leg of a right triangle is twice as long as the other and the hypotenuse is 5 feet more than the longer leg. Find the length of each side of the triangle.

_____ 2. Each side of a square is 2 inches shorter than a diagonal of the square. Find the length of the side and the area of the square.

Check your answers in the back of your book.

If you can do **Checkup 12.5**, you are ready to do the **Review Exercises** for Chapter 12.

✓ CHECKUP 12.5

Solve and write the answers in simplest radical form, unless directed otherwise.

_____ **1.** The area of a square is 32 square inches. Find the length of a side of the square.

_____ **2.** If one side of a square is decreased by 2 meters and the other side is increased by 3 meters, the resulting rectangle has an area of 32 square meters. Find the length of a side of the original square. (Approximate answer to the nearest hundredth.)

_____ **3.** One irrational number is 4 more than another irrational number. The product of the two numbers is 1. Find the irrational numbers.

_____ **4.** In a right triangle one leg is 12 feet and the hypotenuse is twice the other leg. Find the length of the other leg. (Approximate answer to the nearest tenth.)

Check your answers in the back of your book.

If You Missed Problem:	You Should Review Example:
1	1
2	2
3	4
4	3

Summary

In this chapter we discussed square roots of numbers and agreed to call the positive square root of a positive number a the **principal square root of a**, denoted by $\sqrt{a}$. We noted that the square root of zero is zero and the square root of a negative number is *not* a real number. The square root of a positive number is always real, but it may be rational ($\sqrt{81} = 9$) or irrational ($\sqrt{13}$).

To multiply, divide, and/or simplify radical expressions, we used two important properties of radicals.

Property	Symbols	Examples
Multiplication property for radicals	$\sqrt{a} \cdot \sqrt{b} = \sqrt{ab}$ $(a \geq 0, b \geq 0)$	(1) $\sqrt{12a} \cdot \sqrt{3a} = \sqrt{36a^2} = 6a$ (2) $\sqrt{98a^3b^2} = \sqrt{49 \cdot 2 \cdot a^2 \cdot a \cdot b^2}$ $= 7ab\sqrt{2a}$
Division property for radicals	$\dfrac{\sqrt{a}}{\sqrt{b}} = \sqrt{\dfrac{a}{b}}$ $(a \geq 0, b > 0)$	(1) $\dfrac{\sqrt{300a^3}}{\sqrt{3a}} = \sqrt{\dfrac{300a^3}{3a}}$ $= \sqrt{100a^2} = 10a$ (2) $\sqrt{\dfrac{9x^2}{4}} = \dfrac{\sqrt{9x^2}}{\sqrt{4}} = \dfrac{3x}{2}$

We found that square roots can be combined in a sum or difference only if their radicands match exactly. Then we combine them in the same way that we combine like terms.

We learned that fractions containing square roots in the denominator are not considered to be in simplest form. Ridding the denominator of an irrational square root is called **rationalizing the denominator**.

If the Denominator Is	We Can Rationalize the Denominator by	Examples
A single simplified radical term, $\sqrt{a}$	Multiplying numerator and denominator by the same radical, $\sqrt{a}$.	$\dfrac{3}{\sqrt{5}} = \dfrac{3}{\sqrt{5}} \cdot \dfrac{\sqrt{5}}{\sqrt{5}}$ $= \dfrac{3\sqrt{5}}{5}$
An irrational sum or difference of two terms	Multiplying numerator and denominator by the real conjugate of the original denominator.	(1) $\dfrac{\sqrt{7}}{\sqrt{7}+2} = \dfrac{\sqrt{7}}{\sqrt{7}+2} \cdot \dfrac{\sqrt{7}-2}{\sqrt{7}-2}$ $= \dfrac{\sqrt{7}(\sqrt{7}-2)}{7-4}$ $= \dfrac{7 - 2\sqrt{7}}{3}$ (2) $\dfrac{3}{6-\sqrt{5}} = \dfrac{3}{6-\sqrt{5}} \cdot \dfrac{6+\sqrt{5}}{6+\sqrt{5}}$ $= \dfrac{3(6+\sqrt{5})}{36-5}$ $= \dfrac{18 + 3\sqrt{5}}{31}$

Finally, we developed methods for solving second-degree (or quadratic) equations that cannot be solved by factoring.

Equation	Solution	Examples
$x^2 = C$ (C is not negative)	$x = \sqrt{C}$ or $x = -\sqrt{C}$	$3x^2 = 15$ $x^2 = 5$ $x = \sqrt{5}$ or $x = -\sqrt{5}$
$ax^2 + bx + c = 0$	$x = \dfrac{-b \pm \sqrt{b^2 - 4ac}}{2a}$	$x^2 + 3x + 1 = 0$ $(a = 1, b = 3, c = 1)$ $x = \dfrac{-3 \pm \sqrt{3^2 - 4(1)(1)}}{2(1)}$ $x = \dfrac{-3 \pm \sqrt{5}}{2}$ $x = \dfrac{-3 + \sqrt{5}}{2}$ or $x = \dfrac{-3 - \sqrt{5}}{2}$

We found we could use these methods to solve word problems that might not have rational solutions.

☐ Speaking the Language of Algebra

Complete each statement with the appropriate word or phrase.

1. The positive square root of a positive number is called the _____ _____ _____ of that number.

2. The square root of a negative number is not a _____ number.

3. In the expression $\sqrt{a}$, the symbol $\sqrt{}$ is called the _____ , and a is called the _____ .

4. The number $\sqrt{7}$ is an example of a number that is real but _____ .

5. The process by which we rid a denominator of radicals is called _____ _____ _____ .

6. For the irrational number $\sqrt{3} + 2$, the irrational number $\sqrt{3} - 2$ is called the _____ _____ .

7. If $x^2 = C$ (and C is not negative), the equation's two solutions are $x =$ _____ or $x =$ _____ .

8. If $ax^2 + bx + c = 0$, the solutions can be found by $x =$ _____ . This is called the _____ _____ .

Check your answers in the back of your book.

△ Writing About Mathematics _____

Write your response to each question in complete sentences.

1. Explain why it is impossible to find among the real numbers the square root of any negative number.

2. Explain the difference between the methods you would use to rationalize a denominator containing one radical term and a denominator containing an irrational sum or difference of two terms.

3. Show and explain the steps you would use to solve the equation $3x^2 + 4x - 2 = 0$ by the quadratic formula.

REVIEW EXERCISES for Chapter 12

Perform the indicated operation and/or simplify. Assume variable quantities are positive.

_____ 1. $\sqrt{144}$

_____ 2. $\sqrt{64a^2}$

_____ 3. $(\sqrt{13})^2$

_____ 4. $\sqrt{45y^4}$

_____ 5. $\sqrt{48a^3b^2}$

_____ 6. $\sqrt{\dfrac{27x^5}{y^4}}$

_____ 7. $\dfrac{\sqrt{98a^7b^3}}{\sqrt{2ab}}$

_____ 8. $\dfrac{\sqrt{63a^7}}{\sqrt{7a}}$

_____ 9. $\sqrt{20x} + \sqrt{80x} - \sqrt{45x}$

_____ 10. $\sqrt{21a} \cdot \sqrt{7a}$

_____ 11. $(3\sqrt{10})(-4\sqrt{15})$

_____ 12. $\sqrt{2}(3 - \sqrt{14})$

_____ 13. $5\sqrt{3}(\sqrt{2} + \sqrt{8})$

_____ 14. $\sqrt{5}(\sqrt{3} - 2) - 4(\sqrt{5} + 3)$

_____ 15. $(4 - 3\sqrt{6})(1 + 2\sqrt{6})$

_____ 16. $(\sqrt{a} - \sqrt{5})(\sqrt{a} + \sqrt{5})$

Rationalize each denominator and write the answer in simplest form.

_____ 17. $\dfrac{15}{\sqrt{3}}$

_____ 18. $\dfrac{\sqrt{15}}{\sqrt{12}}$

_____ 19. $\sqrt{\dfrac{7}{5}}$

_____ 20. $\dfrac{\sqrt{12} - 21}{\sqrt{3}}$

_____ 21. $\dfrac{\sqrt{7} + \sqrt{21}}{\sqrt{14}}$

_____ 22. $\dfrac{\sqrt{11}}{5 - \sqrt{11}}$

_____ 23. $\dfrac{4 + \sqrt{3}}{\sqrt{3} - 2}$

_____ 24. $\dfrac{\sqrt{13} - 2}{\sqrt{13} + 2}$

Solve.

_____ 25. $y^2 - 48 = 0$

_____ 26. $9a^2 = 18$

_____ 27. $7 - 4x^2 = 9$

_____ 28. $x^2 + 6x - 3 = 0$

_____ 29. $x^2 = 4x - 1$

_____ 30. $5x^2 - 10x + 2 = 0$

_____ 31. $2x^2 - 3x = 1$

_____ 32. $3x^2 = 4x + 2$

_____ 33. In a right triangle, the lengths of the two legs are the same and the hypotenuse is 6 centimeters. Find the length of each leg.

_____ 34. The length of a rectangle is 4 feet more than the width. If the area of the rectangle is 2 square feet, find its length.

_____ **35.** One irrational number is 4 more than another irrational number. The product of the numbers is 1. Find the irrational numbers.

Check your answers in the back of your book.

If You Missed Exercises:	You Should Review Examples:	
1, 2	SECTION 12.1	1, 2, 9
3		5, 6
4, 5		15, 16
6		18
7, 8		19, 20
9	SECTION 12.2	3
10, 11		9, 10
12–14		11, 12
15, 16		13–16
17, 18	SECTION 12.3	1, 2
19		3
20, 21		4
22		5, 6
23, 24		8
25–27	SECTION 12.4	1–4
28, 29		7, 9
30–32		8, 9
33	SECTION 12.5	3
34, 35		4

If you have completed the **Review Exercises** and corrected your errors, you are ready to take the **Practice Test** for Chapter 12.

PRACTICE TEST for Chapter 12

Perform the indicated operation and/or simplify. Assume variable quantities are positive. SECTION   EXAMPLES

_____ 1. $\sqrt{\dfrac{16}{25}}$ 12.1   2

_____ 2. $(\sqrt{3a})^2$ 12.1   10

_____ 3. $\sqrt{81a^2}$ 12.1   11

_____ 4. $\sqrt{32}$ 12.1   13, 14

_____ 5. $\sqrt{48a^3b^4}$ 12.1   16

_____ 6. $\sqrt{\dfrac{32x^3}{y^4}}$ 12.1   18

_____ 7. $\dfrac{\sqrt{500x^7y^3}}{\sqrt{5x^5y}}$ 12.1   19, 20

_____ 8. $2\sqrt{160z} + \sqrt{40z} - 6\sqrt{10z}$ 12.2   3

_____ 9. $(-5\sqrt{10a})(\sqrt{30a})$ 12.2   9, 10

_____ 10. $4\sqrt{6}(\sqrt{12} - \sqrt{3})$ 12.2   11, 12

_____ 11. $(\sqrt{y} - \sqrt{10})^2$ 12.2   14

_____ 12. $(4 + \sqrt{x})(5 - 3\sqrt{x})$ 12.2   15

_____ 13. $(7 - 2\sqrt{6})(7 + 2\sqrt{6})$ 12.2   16

_____ 14. $(\sqrt{a} - 5)(\sqrt{a} + 5)$ 12.2   19

Rationalize each denominator and simplify.

_____ 15. $\dfrac{20}{\sqrt{10}}$ 12.3   1

_____ 16. $\sqrt{\dfrac{5}{3}}$ 12.3   3

_____ 17. $\dfrac{3 + \sqrt{5}}{\sqrt{5}}$ 12.3   4

_____ 18. $\dfrac{\sqrt{7}}{4 + \sqrt{7}}$ 12.3   5, 6

		SECTION	EXAMPLES

———— 19. $\dfrac{6}{\sqrt{2} - \sqrt{3}}$ 12.3   7

———— 20. $\dfrac{1 - \sqrt{10}}{1 + \sqrt{10}}$ 12.3   8

Solve.

———— 21. $x^2 = 72$ 12.4   1

———— 22. $7x^2 - 21 = 0$ 12.4   2

———— 23. $x^2 + 5x - 1 = 0$ 12.4   7

———— 24. $x^2 + 8x = -10$ 12.4   9

———— 25. The width of a rectangular tablecloth is 2 feet less than the length. The area of the tablecloth is 11 square feet. Find the length of the tablecloth.   12.5   2, 4

Check your answers in the back of your book.

SHARPENING YOUR SKILLS after Chapters 1–12

SECTION

_____ 1. Find 115% of $630.

1.3

_____ 2. Compare -3.98 and -4.21 using $<$ or $>$.

2.1

_____ 3. Evaluate $\dfrac{5b(a^2 - 3a + 1)}{(b - c)^2}$ when $a = 1$, $b = -1$, and $c = 2$.

4.1

_____ 4. Is -2 a solution for $3(2x - 1) + 5 = 7 - 5x$?

4.1

_____ 5. Find the restrictions on the variable for the expression $\dfrac{x^2 - 3x - 28}{2x^2 - 15x + 7}$.

8.1

_____ 6. Does the ordered pair $(\frac{1}{2}, -\frac{5}{4})$ satisfy the equation $3x - 2y = 1$?

10.1

Perform the indicated operations. Write fractional answers in reduced form.

_____ 7. $\dfrac{-8(7 - 2) - 4(-2)}{-4(-12 + 10)}$

2.3

_____ 8. $-[5x - 5(x - 4)] - 5[2(x - 2) - 2x]$

3.3

_____ 9. $7a^3b(-ab^2)^4$

5.1

_____ 10. $\left(\dfrac{-x^{-2}y^3}{2xy^{-4}}\right)^{-3}$

5.3

_____ 11. $-x^2y^3(2x + 4xy^4) + 5x^3y(y^2 - 2y^6)$

6.3

_____ 12. $(3x - 2y)^2$

6.3

_____ 13. $(x - 5)(x^2 + 5x + 25)$

6.3

_____ 14. $(x^4 - 16) \div (x - 2)$

6.4

_____ 15. $\dfrac{18x^2}{34y^3} \div \dfrac{9y^2}{17x}$

8.2

_____ 16. $\dfrac{15x^2 - 47x + 36}{5x^2 - 24x + 27} \cdot \dfrac{x^2 + 5x - 24}{3x^2 + 2x - 8}$

8.2

_____ 17. $\dfrac{5x - 15}{5x^2 - 16x + 3} - \dfrac{2}{5x^2 - x}$

8.4

Rewrite each expression as a product of factors.

_____ 18. $a^3b^2 - 14a^2b^3 + 4ab^4$ 7.3

_____ 19. $63m^2 + 38mn - 16n^2$ 7.3

Solve.

_____ 20. $3[x - (2x - 1)] = -2(x - 3) - 6$ 4.2

_____ 21. $4x(x - 1) - 40 = 3(x^2 + 5) + 2x$ 7.5

_____ 22. $\dfrac{x}{x - 9} + \dfrac{1}{9} = \dfrac{9}{x - 9}$ 8.6

_____ 23. $\dfrac{-x}{2} + 1 \geq \dfrac{2x}{5} - 2$ 9.2

_____ 24. Solve the system by substitution.   11.2

$$3x - 9y = -13$$
$$2x + y = 3$$

_____ 25. Solve the system by elimination.   11.3

$$9x - 5y = 8$$
$$4x + 3y = 14$$

26. Graph $y = 3x + 4$ using the slope-intercept method.   10.4

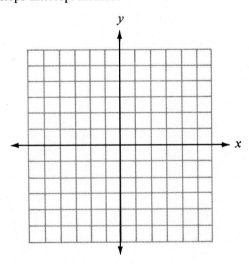

27. Graph $x + 2 = 0$ by inspection.

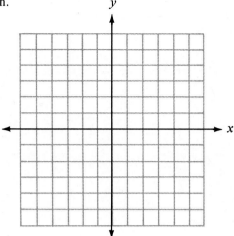

Solve.

_____ 28. The support beams of a barn form a right triangle. The longest beam is 17 feet and the shortest beam is 7 feet less than the other. Find the length of the piece of lumber that would be needed to cut all three beams.

7.6

_____ 29. A car rental agency charges customers $35.50 per day plus $0.25 per mile driven. If Lena's car rental bill was $134.25 for 1 day, find how many miles she drove the car.

4.3

_____ 30. Ray and Glenn are planning to drive to Canada, sharing the time at the wheel. Ray drives at one constant rate and Glenn drives at another constant rate. If Ray drives for 7 hours and Glenn for 4 hours, they can travel 680 miles. If Ray drives 5 hours and Glenn for 6 hours, they can travel 690 miles. How fast does each one drive?

11.4

Check your answers in the back of your book.

Answers

This section includes the answers to Odd-Numbered Exercises, Stretching the Topics, Check-ups, Speaking the Language of Algebra, Review Exercises, Practice Tests, and Sharpening Your Skills.

CHAPTER 1

Exercise Set 1.1 (page 18)
1.

0 1 2 3 4 5 6 7 8 9 10 11 12 13 14 15

3. $A(2)$; $B(6)$; $C(9)$; $D(12)$ 5. $12 > 7$ 7. $0 < 9$
9. $22 < 55$ 11. $32 = 15 + 17$ 13. $35 = 7 \cdot 5$
15. 6   17. 12   19. 48   21. 160   23. 2
25. 47   27. 5   29. 64   31. 5   33. 30
35. 64   37. 37   39. 2   41. 28   43. Undefined
45. 28   47. 37   49. 96   51. 0   53. 31
55. $2 \cdot 5^2$ 57. $3 \cdot 5 \cdot 7$ 59. $2 \cdot 3^2 \cdot 7^2$

Stretching the Topics
1. 1   2. 0   3. 13

Checkup 1.1 (page 21)
1. 80   2. 100   3. 40   4. 1   5. 6   6. 23
7. 92   8. 85   9. 0   10. Undefined

Exercise Set 1.2 (page 35)
1. $\frac{1}{5}$ 3. $\frac{7}{5}$ 5. $\frac{9}{16}$ 7. $\frac{5}{7}$ 9. $\frac{3}{2}$ 11. $\frac{1}{7}$
13. $\frac{2}{21}$ 15. 3   17. $\frac{2}{5}$ 19. $\frac{1}{15}$ 21. 9   23. $\frac{14}{15}$
25. $\frac{7}{3}$ 27. $\frac{1}{15}$ 29. $\frac{55}{4}$ 31. $\frac{9}{10}$ 33. $\frac{42}{49}$
35. $\frac{96}{36}$ 37. $\frac{36}{99}$ 39. $\frac{51}{45}$ 41. $\frac{63}{72}$ 43. $\frac{45}{5}$
45. $\frac{93}{252}$ 47. $\frac{4}{5}$ 49. $\frac{1}{3}$ 51. $\frac{5}{13}$ 53. $\frac{1}{11}$ 55. 1
57. 0   59. $\frac{8}{15}$ 61. $\frac{2}{9}$ 63. $\frac{11}{5}$ 65. 1   67. $\frac{47}{90}$
69. $\frac{19}{15}$

Stretching the Topics
1. $\frac{43}{60}$ 2. $\frac{43}{150}$ 3. $\frac{1}{10}$

Checkup 1.2 (page 38)
1. $\frac{8}{9}$ 2. $\frac{40}{56}$ 3. $\frac{2}{21}$ 4. $\frac{3}{5}$ 5. 15   6. $\frac{117}{80}$
7. $\frac{10}{21}$ 8. $\frac{13}{3}$ 9. $\frac{29}{56}$ 10. $\frac{41}{35}$

Exercise Set 1.3 (page 47)
1. $\frac{4}{5}$ 3. $\frac{11}{20}$ 5. $\frac{7}{40}$ 7. $\frac{3}{250}$ 9. 2.300

11. $15.00   13. 13.200   15. 0.75   17. 0.63
19. 0.83   21. 0.53   23. 0.36   25. 1.53
27. $25.80   29. 5.285   31. 2.614   33. 0.02106
35. 0.02616   37. 3.9   39. 0.203   41. 0.19
43. 0.371   45. 1.50   47. 0.005   49. 9%
51. 710%   53. 3.2%   55. 175%   57. 61.1
59. 6.48   61. 4050   63. 498   65. 222
67. 937.8   69. 312.5

Stretching the Topics
1. 3873.75   2. $625   3. $1.50

Checkup 1.3 (page 49)
1. $\frac{3}{125}$ 2. 43.500   3. 0.875   4. 0.0325
5. 12.5%   6. 5.862   7. $7.02   8. 0.224
9. 0.008   10. 27.625

Exercise Set 1.4 (page 58)
1. $49.17   3. 285 miles   5. $1495   7. $7000
9. 430   11. 21   13. $\frac{3}{10}$ 15. $7\frac{8}{15}$ 17. 7 m
19. 251.25   21. $21.25   23. 5343.61 sq m
25. $5.39   27. 35.75   29. 1800

Stretching the Topics
1. $105.66   2. $13.39

Checkup 1.4 (page 61)
1. $\frac{13}{20}$ 2. $673.69   3. 29.8   4. $148.47
5. $101.75

Speaking the Language of Algebra (page 65)
1. Whole numbers   2. Sum; difference
3. Factors; product   4. Dividend; divisor; quotient
5. Proper; improper   6. Common factors
7. Invert; multiply   8. Same denominator
9. Whole number; fractional   10. Per hundred; two; left

Review Exercises (page 67)
1. $3 < 19$ 2. $43 = 32 + 11$ 3. $91 = 13 \cdot 7$
4. 57   5. 72   6. 76   7. 72   8. 35   9. 8

10. Undefined   **11.** 0   **12.** $\frac{3}{5}$ **13.** $\frac{1}{8}$ **14.** $\frac{1}{27}$
15. $\frac{3}{8}$ **16.** $\frac{32}{75}$ **17.** $\frac{4}{9}$ **18.** 0.3125   **19.** 13.1
20. 17.1665   **21.** $177.49   **22.** 0.088   **23.** 5.2
24. 9%   **25.** 0.127   **26.** $660   **27.** $50
28. $7275   **29.** $\frac{13}{20}$ **30.** 6.5 lb

Practice Test for Chapter 1 (page 69)
1. 24 = 13 + 11   **2.** 55 = 11 · 5   **3.** 192
4. 102   **5.** 37   **6.** Undefined   **7.** 0   **8.** $\frac{10}{21}$
9. $\frac{1}{18}$ **10.** 15   **11.** $\frac{31}{36}$ **12.** $\frac{11}{12}$ **13.** 0.375
14. 7.14   **15.** 19.485   **16.** 0.8225   **17.** 5.02
18. $22.95   **19.** $5.47   **20.** 4.05   **21.** $4.49
22. $337\frac{1}{2}$ sq ft

CHAPTER 2

Exercise Set 2.1 (page 82)
1.

3. $A(-5)$; $B(-2)$; $C(0)$; $D(3)$
5. (a) $5 < 12$ **(b)** $-4 > -10$ **(c)** $8 > -11$
   (d) $-3 < 7$
7. (a) $2 + (-7)$ **(b)** $-9 + 6$
   (c) $-6 + (-20)$ **(d)** $10 + 8$
9. (a) $6 + [3 + (-2)]$ **(b)** $(-5 + 4) + (-3)$
   (c) $-4 + [9 + (-2)]$ **(d)** $[7 + (-2)] + 11$
11. 3   **13.** 5   **15.** -9 **17.** -8 **19.** 0
21. 22   **23.** -12 **25.** 12   **27.** -6 **29.** -7
31. 13   **33.** 20   **35.** 8   **37.** -17 **39.** -14
41. 18   **43.** -21 **45.** 0   **47.** -11 **49.** 10
51. -12.7 **53.** -12 **55.** 24   **57.** 3   **59.** 3
61. Undefined   **63.** -7 **65.** 0   **67.** -8 **69.** 0

Stretching the Topics
1. 65   **2.** 36   **3.** 120

Checkup 2.1 (page 85)
1. $3 > -15$ **2.** $-5 + 17$ **3.** $[-9 + (-3)] + 7$
4. -12 **5.** -5 **6.** 8   **7.** -1 **8.** -15
9. 0   **10.** 1

Exercise Set 2.2 (page 90)
1. $12 + (-9) = 3$ **3.** $11 + (-27) = -16$
5. $-15 + (-7) = -22$ **7.** $12 + 3 = 15$
9. $-9 + 7 = -2$ **11.** 10   **13.** 25   **15.** -11
17. -9 **19.** 3   **21.** -12 **23.** -10 **25.** -6
27. 0   **29.** 1   **31.** 15   **33.** -2 **35.** -6
37. 40   **39.** 3   **41.** -16 **43.** -7 **45.** 13
47. 2   **49.** 2   **51.** Undefined   **53.** -21
55. 0   **57.** -26

Stretching the Topics
1. -28 **2.** 78   **3.** 18.9

Checkup 2.2 (page 92)
1. 7   **2.** 21   **3.** -22 **4.** -10 **5.** -18
6. 6   **7.** 1   **8.** -8 **9.** -1 **10.** 23

Exercise Set 2.3 (page 104)
1. (a) $-8(-9)$ **(b)** $-12(5)$ **(c)** $6 \cdot 0$
   (d) $7(-3)$
3. (a) $3(4 \cdot 2)$ **(b)** $(6 \cdot 5)(-7)$ **(c)** $[-7(-3)](10)$
   (d) $9[-2(-8)]$
5. (a) $-3(7) + (-3)(2)$ **(b)** $5(-8) - 5(3)$
   (c) $4 \cdot 7 - 4(-2)$ **(d)** $-10(-9) + (-10)(8)$
7. -30 **9.** 42   **11.** -24 **13.** 150   **15.** 0
17. 45   **19.** -80 **21.** 37   **23.** 64   **25.** -25
27. -27 **29.** -1 **31.** -288 **33.** -4
35. -2 **37.** 5   **39.** -2 **41.** -7 **43.** -9
45. -12 **47.** Undefined   **49.** -1 **51.** 1
53. -4 **55.** -7 **57.** $-\dfrac{1}{6}$ **59.** -13 **61.** -3
63. 51   **65.** 200   **67.** 1   **69.** -71

Stretching the Topics
1. 63   **2.** -256 **3.** -34

Checkup 2.3 (page 107)
1. $5(-3)$ **2.** $[(-3)(2)](-7)$
3. $-5(-2) + (-5)(8)$ **4.** -30 **5.** -5
6. -2 **7.** -60 **8.** -4 **9.** -2 **10.** $-\dfrac{2}{3}$

Exercise Set 2.4 (page 117)
1. 8.3   **3.** 5.7   **5.** $-\dfrac{2}{3}$ **7.** $-\dfrac{3}{4}$ **9.** $\dfrac{8}{21}$
11. $-\dfrac{14}{15}$ **13.** -6 **15.** 4.31   **17.** $-\dfrac{3}{7}$
19. -8.8786 **21.** 30,104.3   **23.** $-\dfrac{27}{64}$ **25.** 1.44
27. $\dfrac{1}{625}$ **29.** -0.7 **31.** 0.22   **33.** $\dfrac{1}{7}$ **35.** $\dfrac{5}{2}$
37. 1.47   **39.** $\dfrac{2}{5}$

Stretching the Topics
1. $\dfrac{16}{25}$ **2.** -0.0002854 **3.** $-\dfrac{7}{48}$

Checkup 2.4 (page 119)
1. $\dfrac{4}{5}$ **2.** $-\dfrac{1}{3}$ **3.** 1.71   **4.** $-\dfrac{17}{30}$ **5.** $\dfrac{1}{40}$
6. 4.3   **7.** $-\dfrac{27}{125}$ **8.** 0.1   **9.** -2.8
10. -1.58

Exercise Set 2.5 (page 124)
1. $87 - 14 = 73$ **3.** $1605 - 312 = 1293$
5. $\dfrac{\$5.86}{2} = \$2.93$ **7.** $53(4.5) = 238.5$ miles
9. $\dfrac{2}{3}(25) = 16\dfrac{2}{3}$ **11.** $3(75¢) - 7(25¢) = 50¢$ won
13. $\$1.75 + \$1.10(7) = \$9.45$
15. $2(-\$5) + \$7 - \$3 = -\$6$; she owes him $6.
17. $2\left(150\dfrac{3}{4} + 250\dfrac{2}{3}\right) - 3\dfrac{1}{3} - 6\dfrac{1}{2} = 793$

19. $2(\$3.85) + 4(\$4.90) = \$27.30$

21. $0.055(\$1.69 + \$1.25 + \$0.45) = \$0.19$;
($\$1.69 + \$1.25 + \$0.45) + \$0.19 = \$3.58$;
$\$10.00 - \$3.58 = \$6.42$

23. $27(2) - 2(13) - 1.5(10) = 13$

Stretching the Topics
1. $\$61.59$ loss   **2.** $\$12,668$

Checkup 2.5 (page 126)
1. $\$785.28 - (\$119.61 + \$28.75) = \$636.92$

2. $3(20) - 2(10) = 40$ ft

3. $\dfrac{\$625,823 - \$125,729}{3} = \$166,698$

4. $48 + \dfrac{2}{9} + \dfrac{1}{8} + 0 - \dfrac{1}{6} - \dfrac{2}{3} = 47\dfrac{37}{72}$ ft

5. $\dfrac{-\$5(3) + \$7(2) - \$2.50(4)}{9} = -\$1.22$; loss of $\$1.22$

Speaking the Language of Algebra (page 129)
1. Integers   **2.** Integer; nonzero integer   **3.** Irrational
4. Absolute value of a; distance; 0   **5.** Negative
6. Positive; negative   **7.** Zero
8. Addition, multiplication; subtraction, division
9. Opposite   **10.** Positive; negative
11. Positive; positive; negative   **12.** A; n

Review Exercises (page 131)
1.

2. $-6 + (-5)$ **3.** $-3 + (-2 + 5)$ **4.** $5(-4)$
5. $[-7(-8)](9)$ **6.** $-2(-6) + (-2)(9)$
7. (a) $4 < 14$ **(b)** $-5 > -11$ **(c)** $9 > -12$
(d) $-9 < 3$
8. 13   **9.** -18 **10.** 3   **11.** -5 **12.** 0
13. 38   **14.** -16 **15.** 9   **16.** 120   **17.** 1
18. 105   **19.** -9 **20.** -7 **21.** 14   **22.** 3
23. 0   **24.** 8   **25.** Undefined   **26.** 1.38

27. $-\dfrac{13}{12}$ **28.** -0.21 **29.** -42.2 **30.** $-\dfrac{5}{6}$

31. Won $\$3$ **32.** $\$232.42$ **33.** $\$102\dfrac{1}{4}$ **34.** $\dfrac{25}{72}$

35. $0.\overline{3}$ in. above normal

Practice Test for Chapter 2 (page 133)
1.

2. $5(-3)$ **3.** $(-4 + 9) + (-2)$
4. $-4(11) + (-4)(-2)$
5. (a) $-15 < -1$ **(b)** $0 > -3$ **(c)** $6 > 5.9$
(d) $\dfrac{1}{3} < \dfrac{1}{2}$
6. -8 **7.** -13 **8.** 7   **9.** 0   **10.** 42
11. -30 **12.** -16 **13.** -2 **14.** $-\dfrac{23}{3}$ **15.** 0

16. Undefined   **17.** -5.45 **18.** $-\dfrac{35}{24}$ **19.** -0.96

20. -140 **21.** $\dfrac{189}{640}$ **22.** 1.75 lb   **23.** $\dfrac{1}{3}$ c

24. $\$230.90$

CHAPTER 3

Exercise Set 3.1 (page 143)
1. $y + 10$ **3.** $-3m$ **5.** $a - 6.5$ **7.** $\dfrac{t}{-15}$
9. $x + (-9)$ or $x - 9$ **11.** $3.2q$ **13.** $n - 6.3$
15. $\dfrac{2x}{7}$ **17.** $\dfrac{1}{6}k + 9$ **19.** $\dfrac{p}{7} - 5$ **21.** $9 - 4x$
23. $-4(x + 1.5)$ **25.** $7; -6$ **27.** $-3; 0$
29. $42; -49$ **31.** $-32; 40$ **33.** $-4; 0$
35. $1; -3$ **37.** $\$2.50y; \$25; \$62.50$
39. $75 - x; 66.3; 61.8$ **41.** $200 - 3n; 180.5; 110$
43. $\$50 + \$12n; \$122; \$194$
45. $(587.39 - 6x)$ dollars; $\$437.39; \$287.39$
47. $\dfrac{\$x}{32}; \$3; \$5.50$ **49.** $2N - 3; 27; 47$

Stretching the Topics
1. $\dfrac{4y - 3x}{15}$ **2.** -7 **3.** $2108 - \dfrac{1}{3}x; 2084$ sq ft

Checkup 3.1 (page 146)
1. $x + (-3)$ or $x - 3$ **2.** $11 - a$ **3.** $-6m$
4. $2b + 5$ **5.** $-3; 0$ **6.** -4; undefined
7. $-21; 0$ **8.** $1; 3$ **9.** $\dfrac{\$x}{3}; \$50,000; \$120,000$
10. $0.06(\$2500) + \$500x; \$1650; \$7650$

Exercise Set 3.2 (page 151)
1. $3x; -5$ **3.** $\frac{1}{2}a; -7b; 9$ **5.** $1; -5m; 6n$
7. $13x; -5y; 2z; -7$ **9.** -19 (constant); $4x$ (variable)
11. $2x, 3y$ (variable)
13. 5.2 (constant); $-4a, -7b$ (variable)
15. $1, -4$ (constant); $7m, -5n, 14p$ (variable)
17. 3 is coefficient of x
19. $\frac{1}{4}$ is coefficient of x; -12 is coefficient of y
21. 3 is coefficient of a; 4 is coefficient of b
23. 0.3 is coefficient of m; -1 is coefficient of n; -2 is coefficient of p
25. $8a$ and $-3a$ **27.** $3h$ and $5h$; $-2k$ and $5k$
29. $\frac{2}{5}a$ and $3a$; -6 and -9
31. $3x$ and $-1.2x$; $-0.24y$ and $4y$; 1 and -7 **33.** $-2a$
35. $3x$ **37.** $9 - 0.5y$ **39.** $15m + 3$
41. $7h + k - 4$ **43.** $11x + 5y + 7z$
45. $-8a + 10b + 7$ **47.** $m - 11n + 4p$
49. $-v - 6w$

Stretching the Topics
1. $-0.8a + 1.79b + 0.22c$ **2.** $-\frac{7}{10}x - \frac{17}{3}y + \frac{7}{4}z$
3. $20m + 20n - 5p$

Checkup 3.2 (page 153)
1. $3; -7x; 2y$

2. 11, −13 (constant); −4a, 7a (variable)
3. 3 is coefficient of x; −1 is coefficient of y; −4 is
    coefficient of z
4. 5x and −7x; −2y and 2y; 4 and −1   **5.** −4a
6. 4x **7.** 5 + 8y **8.** m + 2   **9.** −x + 7y − 6
10. −5b + 2c

Exercise Set 3.3 (page 161)
1. 8x **3.** −1.5a **5.** −9n **7.** 6x **9.** −3y
11. 30k **13.** −5x **15.** 5x **17.** a **19.** m
21. −5y **23.** −n **25.** 11x **27.** 7x
29. $\frac{1}{4}a - \frac{1}{4}b$ **31.** −3x − 6y **33.** 9x − 3y
35. −20a + 28b **37.** −18m − 27n
39. 3x − 12y **41.** −$\frac{1}{2}x$ + 5y **43.** 3a + 7b − 5.4
45. x + 1   **47.** 7a + 5b **49.** −6m
51. 9x − 4y **53.** 11m − 20   **55.** −7x + 10y
57. 4a + 14   **59.** 10x − 15y **61.** −38a + 6b
63. −20x

Stretching the Topics
1. 3.2x − 0.7y + 0.3   **2.** −9a + 11b
3. 90x + 135

Checkup 3.3 (page 163)
1. −32x **2.** 40y **3.** −4x **4.** −a **5.** −5x
6. −10x + 15y **7.** 2x − 8   **8.** 12a − 15b
9. −6a + 6   **10.** 49n

Speaking the Language of Algebra (page 165)
1. Numerical coefficient; numerical coefficient; constant
2. Numerical coefficients   **3.** Like terms
4. Distributive   **5.** Innermost

Review Exercises (page 167)
1. 6x **2.** y + 9   **3.** 3m + 10   **4.** 23.7 − 4k
5. 9(x + 7)   **6.** −7; −21   **7.** −8; undefined
8. −6; 0   **9.** 17   **10.** 0; −4   **11.** 4h; 14 miles
12. $\frac{\$x}{3}$; $70,000   **13.** $200 + 19.50d$; $258.50
14. 3x (variable); −12 (constant)
15. −2x, 7y (variable)   **16.** 4a, −3b (variable); 2 (constant)
17. 3m, 4n, −p (variable); −2, −9 (constant)
18. 3 is coefficient of x
19. 7 is coefficient of x; −2 is coefficient of y
20. −3 is coefficient of a; 1 is coefficient of b
21. 1 is coefficient of m; −1 is coefficient of n;
    2 is coefficient of p
22. −2x **23.** 5m + 2   **24.** −7a − 13b + 3
25. 4x + 5z **26.** −20x **27.** 24y **28.** −6x
29. a **30.** −3x **31.** 28a − 35b **32.** −7x + 2y
33. −4x + 7   **34.** −16m **35.** −9a − 6
36. 20x − 50y

Practice Test for Chapter 3 (page 169)
1. m − 5   **2.** 3x + 9   **3.** 10(b + 4.6)
4. 2; −16   **5.** −14; −16   **6.** $\frac{19}{3}$
7. 33.79n$; $168.95   **8.** $200 + 4.10x$; $323
9. $\frac{\$180}{m}$; $4   **10.** −4$n$ **11.** −3x − 7   **12.** −16y
13. −4y + 2z **14.** 6x **15.** −12n **16.** 6x + 2

17. −18x + 12y **18.** 4y − 7x **19.** −12a + 12
20. −42x − 14y

Sharpening Your Skills after Chapters 1–3 (page 171)
1. $\frac{7}{8}$ **2.** 0.83   **3.** $\frac{8}{27}$ **4.** 3.15
5. 2 · 3 − 7 > −8   **6.** Undefined   **7.** $\frac{61}{40}$
8. 0.5664   **9.** 9   **10.** 27   **11.** 12   **12.** 4
13. 16   **14.** 6   **15.** $\dfrac{-13 + 12}{7} = -\dfrac{1}{7}$

CHAPTER 4

Exercise Set 4.1 (page 185)
1. Yes   **3.** Yes   **5.** Yes   **7.** No   **9.** Yes
11. No   **13.** x = 7   **15.** x = 12   **17.** y = 2.6
19. y = −8   **21.** m = 5   **23.** a = 10
25. k = 6   **27.** x = 0   **29.** m = −$\frac{5}{2}$
31. y = −8   **33.** a = −7   **35.** x = 0
37. x = 7   **39.** x = −36   **41.** x = −$\frac{35}{2}$
43. x = 0.6   **45.** x = 0   **47.** x = −6.5
49. x = 12   **51.** x = −4   **53.** x = −56
55. x = 0   **57.** x = −40   **59.** x = 63
61. x = −6   **63.** x = −$\frac{27}{2}$

Stretching the Topics
1. $x = \dfrac{7}{12}$ **2.** $x = -1.5$ **3.** $y = d - c$
4. $y = \dfrac{d}{a}, a \neq 0$

Checkup 4.1 (page 187)
1. Yes   **2.** x = −1   **3.** x = 7   **4.** x = 1.2
5. x = 0   **6.** x = 9   **7.** x = −12   **8.** x = 0
9. x = 24   **10.** x = −60

Exercise Set 4.2 (page 199)
1. x = 7   **3.** x = 10   **5.** y = −5   **7.** x = 1
9. a = −9   **11.** m = −$\frac{13}{2}$ **13.** x = −13
15. x = −3   **17.** x = 13   **19.** y = 0   **21.** a = 8
23. x = 10   **25.** x = 6   **27.** x = 0
29. x = −10   **31.** x = 12   **33.** a = −4
35. y = 16   **37.** x = −4   **39.** x = 5
41. x = 16   **43.** x = −3   **45.** x = $\frac{35}{2}$
47. x = 1   **49.** x = 5   **51.** x = 2   **53.** x = 0.7
55. x = −2   **57.** x = $\frac{15}{7}$ **59.** x = −4
61. x = −2   **63.** x = −4   **65.** x = 0
67. x = −11   **69.** x = 3   **71.** x = −2
73. x = 1   **75.** x = −3   **77.** $d = \dfrac{C}{\pi}$
79. $y = \dfrac{3 - x}{2}$ **81.** $t = \dfrac{I}{pr}$ **83.** $a = P - b - c$
85. $p = \dfrac{k - c}{2c}$ **87.** $x = b$ **89.** $n = \dfrac{l - a + d}{d}$
91. $r = \dfrac{A - P}{Pt}$ **93.** $h = \dfrac{3V}{B}$ **95.** $x = \dfrac{2y + 1}{3}$
97. $d = \dfrac{l - a}{n - 1}$ **99.** $x = \dfrac{5y - 6}{5}$

Stretching the Topics

1. $x = -0.5$ **2.** $x = -2$ **3.** $M = \dfrac{6V - hB - hb}{4h}$

Checkup 4.2 (page 202)

1. $x = 5$ **2.** $x = -1$ **3.** $y = -6$ **4.** $x = -3$

5. $x = 4$ **6.** $x = -8$ **7.** $x = 6$ **8.** $y = \dfrac{13}{6}$

9. $y = 8$ **10.** $r = \dfrac{A - P}{Pt}$

Exercise Set 4.3 (page 209)

1. Lydia, 5 lb; Lyle, 7 lb   **3.** 9 hr
5. Julia, 15 tickets; Sara, 60 tickets   **7.** 13 m
9. $0.55   **11.** $36   **13.** 12 lb   **15.** 31 by 36 in.
17. $1200   **19.** 66   **21.** $1980   **23.** $420,000
25. $415

Stretching the Topics

1. 14, 13, 12, 11 in.   **2.** $43,000

Checkup 4.3 (page 211)

1. 14   **2.** 9 by 27 in.   **3.** 24   **4.** 80   **5.** 180 bu

Speaking the Language of Algebra (page 213)

1. Sentence   **2.** Solution; satisfies   **3.** Isolate
4. Substitute; original

Review Exercises (page 215)

1. Yes   **2.** No   **3.** No   **4.** Yes   **5.** $x = \bar{1}0$
6. $y = -14$ **7.** $m = 4$ **8.** $a = 2$ **9.** $x = 8$
10. $x = -18$ **11.** $x = -7$ **12.** $x = 0$
13. $x = -3$ **14.** $y = -14$ **15.** $x = \frac{19}{2}$
16. $x = 1$ **17.** $x = -3$ **18.** $x = 4$
19. $x = -6$ **20.** $x = 4$ **21.** $x = 4$
22. $x = -10$ **23.** $y = 5$ **24.** $x = \frac{17}{4}$
25. $a = 2$ **26.** $x = -19$ **27.** $x = -3$
28. $x = 0$ **29.** $x = 11$ **30.** $a = \dfrac{P - c}{2}$

31. 45   **32.** 4 by 12 ft   **33.** 4

Practice Test for Chapter 4 (page 217)

1. Yes   **2.** No   **3.** $y = 13$ **4.** $x = 13$
5. $x = -10$ **6.** $x = 4$ **7.** $x = 8$ **8.** $x = -7$
9. $x = \frac{9}{2}$ **10.** $x = 1$ **11.** $x = -15$
12. $y = 15 - 10x$
13. Harvey, 37 tickets; Martha, 148 tickets
14. 2 by 5 in.   **15.** 83 mi

Sharpening Your Skills after Chapters 1–4 (page 219)

1. -86 **2.** $\frac{8}{15}$ **3.** 104   **4.** $\frac{15}{8}$ **5.** -11.37
6. 16   **7.** -7 **8.** $-2x - 3y$ **9.** $-6x$
10. $12x + 28y$ **11.** 49   **12.** -3 **13.** $\frac{2}{3} < \frac{3}{4}$
14. -10 **15.** $500n$, $6000

CHAPTER 5

Exercise Set 5.1 (page 231)

1. $3x^2y^2$ **3.** $-a^4$ **5.** $(-2x)^3$ **7.** $3x^3 - 4y^3$

9. $8 \cdot a \cdot a \cdot a$ **11.** $(3a)(3a)(3a)$
13. $-1 \cdot x \cdot x \cdot x \cdot x$ **15.** $\frac{2}{3} \cdot x \cdot x \cdot x \cdot y \cdot y \cdot y$
17. -27 **19.** -16 **21.** 4   **23.** 64   **25.** 1
27. x^5 **29.** a^7 **31.** y^9 **33.** $-2a^{12}$ **35.** x^7y^5
37. $2m^8$ **39.** $\frac{1}{2}x^9y^4$ **41.** $0.5x^4y^4z^5$ **43.** $3x^9yz^2$
45. a^6 **47.** x^5y^5 **49.** $49x^2$ **51.** $\frac{1}{9}y^4$
53. $-64x^3$ **55.** x^{18} **57.** $81x^4y^6$ **59.** $16x^4y^{12}$
61. $\frac{1}{4}x^5y^6$

Stretching the Topics

1. $-\frac{1}{8}x^{14}y^{16}z^{18}$ **2.** $42x^4y^3$ **3.** $x^{12}y^5z^9$
4. $72a^{19}b^{14}c^4$

Checkup 5.1 (page 233)

1. 16   **2.** $-\frac{1}{3}x^4$ **3.** $0.5 \cdot x \cdot x \cdot x \cdot y \cdot y$
4. -216 **5.** x^8 **6.** $-y^6$ **7.** $-5x^3y^7$ **8.** a^{18}
9. $\frac{4}{9}x^2$ **10.** $-8x^3y^3$

Exercise Set 5.2 (page 242)

1. x^7 **3.** a^4 **5.** 1   **7.** $\dfrac{1}{y^2}$ **9.** $4x^3$ **11.** $-3y^5$

13. $\dfrac{1}{3x^2}$ **15.** $-\dfrac{3}{a^2}$ **17.** $-\dfrac{x^2}{5y^3}$ **19.** $\dfrac{7}{ab^2}$

21. $-\dfrac{x}{y^6}$ **23.** $-\dfrac{2z^2}{y^2}$ **25.** $\dfrac{7a^3b^2}{6c^2}$ **27.** $\dfrac{y^4}{x}$

29. $-\dfrac{y^7}{3x}$ **31.** $\dfrac{x^3}{8}$ **33.** $-\dfrac{27}{y^3}$ **35.** $\dfrac{x^8}{y^4}$

37. $-\dfrac{x^{10}}{y^{15}}$ **39.** $\dfrac{16x^8}{y^4}$ **41.** $\dfrac{25a^6}{4b^4}$ **43.** $-\dfrac{32x^{10}y^5}{243z^5}$

45. $\dfrac{9b^6}{a^4}$ **47.** $-\dfrac{z^9}{27x^6}$ **49.** $4x^2$

Stretching the Topics

1. $-\dfrac{xy^2}{2z^{13}}$ **2.** $-ab$ **3.** $\dfrac{11}{12}x^2$

Checkup 5.2 (page 244)

1. x^5 **2.** $-\dfrac{1}{a^3}$ **3.** $\dfrac{3}{x^2}$ **4.** $-\dfrac{a^2}{2b^4}$ **5.** $\dfrac{x}{z^3}$

6. xy^4 **7.** $\dfrac{x^5}{32}$ **8.** $\dfrac{16}{a^4}$ **9.** $-\dfrac{x^{12}}{8y^3}$ **10.** $\dfrac{64a^9}{b^6}$

Exercise Set 5.3 (page 256)

1. $8x^6$ **3.** a^3 **5.** $\dfrac{5}{y^3}$ **7.** $-\dfrac{27a^6}{b^3}$ **9.** $\dfrac{1}{9x^2}$

11. $-\dfrac{1}{125}$ **13.** $\dfrac{25}{16}$ **15.** $\dfrac{1}{36x^2}$ **17.** $\dfrac{6}{x^2}$ **19.** 4

21. $\dfrac{1}{a^3}$ **23.** $2x^4$ **25.** $-\dfrac{5}{x^3}$ **27.** $\dfrac{1}{a^{12}}$ **29.** y

31. $\dfrac{9}{x^3}$ **33.** $\dfrac{1}{64x^3}$ **35.** $\dfrac{27}{a^{12}}$ **37.** $\dfrac{y^{10}}{64}$ **39.** $\dfrac{25a^6}{b^{10}}$

41. $\dfrac{x^{10}y^6}{81}$ **43.** $\dfrac{1}{7}$ **45.** x^{15} **47.** $\dfrac{3}{5a^{10}}$ **49.** $\dfrac{1}{3a^7b^{12}}$

51. $-\dfrac{2x^2}{3}$ **53.** $\dfrac{27}{8x^3}$ **55.** a^2x^4 **57.** $\dfrac{x^{12}}{y^6}$ **59.** $\dfrac{9}{x^8y^2}$

Stretching the Topics

1. $\dfrac{3y^8}{2x^8}$ **2.** $\dfrac{4b^{20}}{27a^{16}}$ **3.** y^{10}

Checkup 5.3 (page 258)

1. $\dfrac{1}{32x^5}$ **2.** a^4 **3.** $\dfrac{3}{y^2}$ **4.** 2   **5.** $-3x^7$ **6.** $\dfrac{a^2}{2}$

7. $-\dfrac{x^4}{3}$ **8.** $\dfrac{3}{x^8}$ **9.** $9x^4$ **10.** $\dfrac{1}{27a^9b^6}$

Speaking the Language of Algebra (page 261)

1. Base; exponent; a; n **2.** Parentheses   **3.** Add
4. Multiply   **5.** Factor
6. Subtract; denominator; numerator
7. Numerator; denominator   **8.** Reciprocal
9. Numerator   **10.** 1

Review Exercises (page 263)

1. $-2x^2y^3$ **2.** $\left(\dfrac{1}{3}a\right)^3 = \dfrac{a^3}{27}$ **3.** $-3 \cdot m \cdot m \cdot m \cdot n \cdot n$

4. $(0.5a)(0.5a)(0.5a)$ **5.** -27 **6.** $\dfrac{1}{16}$ **7.** -2

8. 1   **9.** x^{10} **10.** $-2a^9$ **11.** $9x^5y^8$ **12.** m^6

13. $-27y^3$ **14.** $x^6y^4z^2$ **15.** $3a^{13}b^6$ **16.** $\dfrac{1}{27}x^3$

17. $\dfrac{1}{x^3}$ **18.** $-5x^9$ **19.** $-\dfrac{3x^3}{y^6}$ **20.** $-\dfrac{y^2}{x^3}$

21. $\dfrac{8x^7}{y}$ **22.** $-\dfrac{27a^6}{8b^3}$ **23.** $\dfrac{16a^{12}}{b^{20}}$ **24.** $\dfrac{1}{25x^2}$

25. $\dfrac{4}{y^3}$ **26.** $\dfrac{1}{x^4}$ **27.** $4a^2$ **28.** $\dfrac{8}{x^{11}}$ **29.** $\dfrac{1}{9x^2}$

30. $-7x^8$ **31.** $\dfrac{a^3}{4}$ **32.** $\dfrac{16}{81x^4}$ **33.** $4y^2$ **34.** $\dfrac{x^4}{4}$

Practice Test for Chapter 5 (page 265)

1. -64 **2.** -3 **3.** 4   **4.** x^9 **5.** $-2y^8$
6. $4x^7y^3z^4$ **7.** a^{12} **8.** $x^{12}y^6$ **9.** $-27x^3$

10. $16a^{16}b^{12}$ **11.** $5a^7b^3$ **12.** $\dfrac{1}{x^8}$ **13.** $-2x^5$

14. $-\dfrac{3x^5}{y^4}$ **15.** $-\dfrac{b^2}{a^3}$ **16.** $-32x^2y^{11}$ **17.** $-\dfrac{27a^{15}}{64b^6}$

18. $\dfrac{1296x^{12}}{y^{16}}$ **19.** $\dfrac{7}{x^4}$ **20.** $-\dfrac{5}{w^5}$ **21.** $\dfrac{1}{81a^4}$

22. $\dfrac{x^8}{25y^{12}}$ **23.** $-\dfrac{3}{x^4}$ **24.** $\dfrac{4x^5}{y^5}$ **25.** $-\dfrac{3a^5b^5}{4}$

Sharpening Your Skills after Chapters 1–5 (page 267)

1. 2   **2.** $-20b$ **3.** $2x - y + z$ **4.** 0.85

5. -0.96 **6.** $\dfrac{2}{21}$ **7.** No   **8.** $8(x + 13)$

9. $x = 22$ **10.** $x = 40$ **11.** $x = 8$ **12.** $x = 3$

13. $w = \dfrac{P - 2l}{2}$ **14.** Ethel, 80 cans; Sam, 210 cans

15. 4 by 16 ft

CHAPTER 6

Exercise Set 6.1 (page 275)

1. Monomial   **3.** Binomial   **5.** Trinomial
7. Binomial   **9.** Binomial   **11.** Trinomial
13. Polynomial of four terms   **15.** $4x^2 - 2x + 4$

17. $5a - 3c$ **19.** $-4x^2 - 2x - 9$ **21.** $-12x^2$
23. $-8xy$ **25.** $6a$ **27.** $-0.8ab^3$ **29.** $-6x^5y^4$
31. $5x^2 - 15x + 10$ **33.** $-6a + 4b - c$
35. $-8a^2 + 8a - 4$ **37.** $3x^2 - 2x + 1$
39. $-x + 2y$ **41.** $4x - 17y$
43. $0.4x^2 - 2.7x + 1$ **45.** $-5y^2 - 12y + 21$
47. 1   **49.** $-y^2 - 3$

Stretching the Topics

1. $-7h + 30k$ **2.** $-3xy - 3xz$ **3.** $4a + 2b$

Checkup 6.1 (page 277)

1. Trinomial   **2.** Monomial   **3.** $3x^2 - 9$ **4.** $12xy$
5. $12a^3b^2$ **6.** $2x^2 - 4x + 1$ **7.** $-3a + 4b + 7c$
8. $-5x^2 + 11x + 20$ **9.** $-2x^2 - 5x + 6$
10. $2y^2 - 4y + 10$

Exercise Set 6.2 (page 283)

1. $10x^5$ **3.** $2x^8$ **5.** $10a^{12}$ **7.** $-0.9y^4$
9. $-6m^8$ **11.** $-12x^7y^5$ **13.** $2x^7y^4$ **15.** $42x^4y^3$
17. $35x^3y^4z^5$ **19.** $-24x^5yz^2$ **21.** $20x^7$
23. $-200y^{23}$ **25.** $-24a^4b^7$ **27.** $10x + 15$
29. $-14x^2 + 28$ **31.** $6x^2 - 12x$ **33.** $2y^3 - 5y^2$
35. $-a^4 + 2a^2$ **37.** $36x^4 - 16x^2$
39. $-21a^5 + 9a^4$ **41.** $10x^2y + 4xy^2$
43. $-11x^3y^2 + 4x^2y^3$ **45.** $14a^5b^2 - 2a^3b^4$
47. $2m^3 - 10m^2 + 12m$ **49.** $15x^4y^4 - 5x^3y^3 - 20x^2y^2$
51. $-3a^2b + 2a^3b^2 - 4a^4b^3$
53. $32m^4n - 8m^3n^2 + 4m^2n^3 - 12mn^4$ **55.** $4x^3y$
57. $x^4 + 2x^2 + 24$ **59.** $9m^2 - n^2$ **61.** $4ax - 5axy$
63. $7x^2y - 10xy^2$

Stretching the Topics

1. $-3m^5n^2 - 7m^4n^3 + 2m^3n^4$
2. $3x^3y^2 - 7x^2y^3 + 7x^2y^4$
3. $4a^4 - 4a^2b^2 + 22b^3$

Checkup 6.2 (page 285)

1. $-21x^{11}$ **2.** $8a^{12}$ **3.** $-2x^4y^4z^3$
4. $-2a^4 + 14a^3 - 10a^2$ **5.** $15x^2y - 6xy^2 + 12xy$
6. $12x^4y^4 - 18x^3y^3 + 24x^2y^2$
7. $-8m^4n + 2m^3n^2 - 4m^2n^3 + 2mn^4$ **8.** $-x^3y$
9. $15x^2 + xy + y^2$ **10.** $7ax$

Exercise Set 6.3 (page 293)

1. $x^2 + x - 6$ **3.** $x^2 + 13x + 36$
5. $x^2 - 0.8x + 0.15$ **7.** $y^2 - 18y + 81$
9. $a^2 + 0.4a + 0.04$ **11.** $x^2 - 16$
13. $y^2 - \dfrac{1}{9}$ **15.** $12y^2 - 23y + 5$
17. $-6a^2 - 11a + 35$ **19.** $36x^2 - 25$
21. $18 - 13y + 2y^2$ **23.** $2a^2 - 11ab + 5b^2$
25. $2x^2 - 5xy - 3y^2$ **27.** $-2x^2 + 5xy - 3y^2$
29. $a^2 + 6ab + 9b^2$ **31.** $0.04 - 0.25x^2$
33. $\dfrac{1}{4}x^2 - xy + y^2$ **35.** $ac + ad - 2bc - 2bd$
37. $a^3 - 3a^2 - 4a + 12$ **39.** $6a^4 - 31a^2 + 5$
41. $25a^4 - 20a^2 + 4$ **43.** $a^6 - 2a^3 - 3$
45. $x^2y^2 - 5xy - 50$ **47.** $a^2b^2 - 4c^2$
49. $9y^2 - 12yz + 4z^2$ **51.** $36a^4 - b^2$
53. $100 + 10x^2 + \dfrac{1}{4}x^4$ **55.** $x^3 - 5x^2 + 7x - 3$

57. $b^3 - 125$ **59.** $x^4 + 3x^3 - 2x^2 - 12x - 8$
61. $2 + x + x^2 + 6x^3$ **63.** $0.3x^3 - 2.1x^2 + 3x$
65. $-49y^2 + y^4$

Stretching the Topics

1. $6x^3 - 19x^2y + 16xy^2 - 4y^3$ **2.** $x^3 - 3x^2 + 3x - 1$
3. $2x^2 - xy - 3y^2$ sq in.

Checkup 6.3 (page 295)

1. $x^2 + 3x - 10$ **2.** $2y^2 - 15y + 18$
3. $25 - 20y + 4y^2$ **4.** $4a^2 + 20ab + 25b^2$
5. $25 - 4a^2$ **6.** $y^2 - \frac{9}{25}$
7. $a^3 - 6a^2 + 7a - 42$ **8.** $3x^4 - 7x^2 - 20$
9. $x^3 + 8$ **10.** $-3x + 0.4x^2 + 0.2x^3$

Exercise Set 6.4 (page 302)

1. $x + 5$ **3.** $8x - 7$ **5.** $4 - 3x$ **7.** $5x^2y - 2$
9. $2d - 3 - \frac{1}{4d}$ **11.** $-y^2 + ay + b$
13. $-m^2 - 3mn + 5n^2$ **15.** $3b + 4c - \frac{d}{2}$
17. $x + 7$ **19.** $a - 4$ **21.** $z^2 - 3z + 2$
23. $x^2 - 3x + 9 - \frac{54}{x + 3}$ **25.** $4x - 7$
27. $3x^3 + x^2 + 4x + 1$ **29.** $3x^2 + 3x - 1 + \frac{1}{x - 1}$

Stretching the Topics

1. $x^4 + 2x^3y + 4x^2y^2 + 8xy^3 + 16y^4$
2. $3a^3 + 4a^2b + 5ab^2 + 2b^3$ **3.** $4x; 8x + 6xy^3$

Checkup 6.4 (page 304)

1. $2x^2 - \frac{1}{3}$ **2.** $-a + 4b + \frac{3b^2}{a}$
3. $2y + 3 - \frac{3}{y + 4}$ **4.** $x^2 + 2x - 1$
5. $3x^3 + x^2 + 3x + 1$ **6.** $a - 3$

Exercise Set 6.5 (page 310)

1. $4l - 14; l^2 - 7l$ **3.** $12x; 9x^2$
5. $3x + 15; \frac{x^2 + 7x}{2}$ **7.** $5x^2$ **9.** $x^2 + x$
11. $x^2 + 12x$ **13.** $x^2 + 5x - 45$ (or $x^2 + 23x + 81$)
15. $x^2 + 4x$ **17.** $(w^2 + 15w)$ sq rods   **19.** $4w + 30$
21. $(x^2 + 8x - 20)$ sq ft   **23.** $2w^2 + 6w$
25. $\frac{x^2 + 9x}{2}$ sq ft   or $\frac{x^2 - 9x}{2}$ sq ft
27. $5x^2 + 20x + 25$ **29.** $5x^2 + 12x$

Stretching the Topics

1. $\frac{1}{4}l^2 - 3l + 9$ **2.** $2t - \frac{1}{t}$ mph   **3.** $n^2 + 4n + 3$

Checkup 6.5 (page 312)

1. $12x; 9x^2$ **2.** $16w - 6; 7w^2 - 3w$
3. $4x + 10; \frac{x^2 + 7x}{2}$ **4.** $2x^2 + 14x + 49$
5. $4x^2 + 2x + 1$

Speaking the Language of Algebra (page 314)

1. Monomial, one; binomial, two; trinomial, three
2. Combine like terms   **3.** Distributive   **4.** Binomials
5. $A^2 + 2AB + B^2$ **6.** $A^2 - B^2$ **7.** Monomial
8. Long division

Review Exercises (page 315)

1. Binomial   **2.** Binomial   **3.** Monomial
4. Trinomial   **5.** $3x^2 + 3x - 3$ **6.** $-2a^2 - 3a$
7. $-14xy$ **8.** $6x^2$ **9.** $-3.2a^2b^3$
10. $-5x^2 - 15x + 10$ **11.** $-x^2 + 2x - 1$
12. $1.6x^2 - 0.6x + 0.4$ **13.** $2x - 6y$
14. $-7a - 2b$ **15.** $-3x^2 - 4x - 14$ **16.** x^{10}
17. $-6a^8$ **18.** $-18x^5y^8$ **19.** m^6 **20.** $81y^4$
21. $x^6y^4z^2$ **22.** $3a^{13}b^6$ **23.** $\frac{1}{8}x^3$ **24.** $4x - 2.4$
25. $21y^2 + 14y$ **26.** $3m^4 - 12m^3n - 15m^2n^2$
27. $35x^2y - 55xy^2$ **28.** $-6a^2b + 3a^3b^2 - 4a^4b^3$
29. $2x^2y + xy^2$ **30.** $x^2 + x - 72$
31. $12y^2 - 5y - 2$ **32.** $4a^2 - 9b^2$
33. $x^2 - 10xy + 25y^2$ **34.** $4x^2y^2 - 4xy + 1$
35. $m^4 + 2m^2 - 15$ **36.** $6x^4 + 2x^2y^2 - 20y^4$
37. $27x^3 - 36x^2 + 12x$ **38.** $2x^3 - 18x$
39. $-36a^2 + 16a^4$ **40.** $y^3 - 4y^2 - y + 4$
41. $x^2 - 2x - \frac{2}{3}$ **42.** $-x^2 - 15xy + 5y$
43. $x + 7$ **44.** $x - 6 + \frac{29}{x + 4}$
45. $x^2 - 4x + 5 - \frac{1}{x + 1}$ **46.** $4x^3 + 2x^2 + \frac{1}{2x - 1}$
47. $y - 1 + \frac{2}{y^2 + y + 1}$ **48.** $(l^2 - 10l)$ sq ft
49. $5x^2 + 8x + 4$ **50.** $-2x^2 + x$

Practice Test for Chapter 6 (page 317)

1. Monomial   **2.** Trinomial   **3.** $6x + 6y - z$
4. $-2.1a^2$ **5.** $9x^2 - 2x + 16$ **6.** $-15y^{10}$
7. $-20x^5y^3$ **8.** $-5a^2c + 15ac^2 + 5c^3$
9. $12x^2y - 24xy^2 + 24x^2y^2$ **10.** $x^2 + x - 110$
11. $12x^2 - 23x + 10$ **12.** $a^2 - 14a + 49$
13. $x^2 - \frac{4}{9}$ **14.** $20a^2 + 7ac - 3c^2$
15. $y^4 + 0.2y^2 - 0.35$ **16.** $16x^2 - 8xz + z^2$
17. $x^3 + 8$ **18.** $-2y^3 + 162y$
19. $-48x^2 + 24xy - 3y^2$ **20.** $3x^2 - 5x - 4$
21. $-\frac{1}{y} + \frac{2}{x} + \frac{5}{6x^2}$ **22.** $2x - 7$
23. $x^2 + 3x + 4 + \frac{3}{x - 3}$ **24.** $10x^2 + 12x + 4$
25. $x^2 + 6x + 9$ **26.** $x^2 + 6x + 5$

Sharpening Your Skills after Chapters 1–6 (page 319)

1. 0.625 **2.** $-3 + [4 + (-10)]$ **3.** 3 **4.** Yes
5. -6 **6.** $\frac{31}{36}$ **7.** $\frac{8}{7}$ **8.** $6a + 4b + c$
9. $-36x + 42y$ **10.** $9x^6y^5$ **11.** $-\frac{8x^2}{9y^3z^3}$ **12.** -7

13. $x = 5$ **14.** $x = 8$ **15.** 6 by 9 ft

CHAPTER 7

Exercise Set 7.1 (page 328)
1. $7(x + 2)$ **3.** $7(3x - 4)$ **5.** $6(7 - y)$
7. $6(a^2 - 6)$ **9.** $-6(2x - 3y)$ **11.** $15(2x^2 + y^2)$
13. $7(x^2 - 3x - 1)$ **15.** $5(2y^2 + 3y - 7)$
17. $7(4 - a + 5a^2)$ **19.** $10(5x^2 - xy + 2y^2)$
21. $2(x^3 - 3x^2 + 2x + 4)$ **23.** $x(x + 1)$
25. $-y(y^2 - y + 2)$ **27.** $2b(b^2 - 3b - 2)$
29. $b^3(7 + 2b^2 - 3b^4)$ **31.** $3x(x^2 - 2xy + 4y^2)$
33. $-a^2b(a^2 - 8ab + 5b^2)$ **35.** $2xy(2x^2y^2 - 10xy + 1)$
37. $x^3y^2z(1 - 2xy - x^2y^2)$ **39.** $9x(7x + 8)$
41. $20(5a^3 - 3b^3)$ **43.** $16b^3(1 + 4b)$
45. $6a(a^2 - 7a - 13)$ **47.** $5xy(x^2 + 16y^2)$
49. $-7b^2(1 + 6b - 5b^2)$ **51.** $xy(9x^2 - 12xy + 20y^2)$
53. $8a^2(9a^2 - 6a - 1)$ **55.** $4xy(3x^2 - 2xy + y^2)$
57. $9xyz(x^3 + 2x^2y - 3xy^2 - 5y^3)$ **59.** $(5 + c)(a + b)$
61. $(x - 5)(x^2 + 6)$ **63.** $(7a - 4b)(10a + 3)$
65. $(y - 2)(3x - 1)$ **67.** $(y^2 + 4)(2x + 3)$
69. $(x - 2)(3x^2 + 1)$

Stretching the Topics
1. $(m - n)(2m - 2n - 1)$ **2.** $3x^{2m}(x^5 + 3)$
3. $\frac{1}{6}h(B + b + 4M)$

Checkup 7.1 (page 330)
1. $3(x + 5)$ **2.** $7(5 - 2y)$ **3.** $-x(x - 1)$
4. $5(3x^2 - 2xy + 5y^2)$ **5.** $8a(a^2 - 2a - 5)$
6. $3ab(3a^2b^2 - 9ab + 1)$ **7.** $-8m(m^2 + 3)$
8. $x^2yz(x^2 - xy + y^2)$ **9.** $(m + n)(5 + p)$
10. $(x + 2y)(3x - 1)$

Exercise Set 7.2 (page 335)
1. $(x + 5)(x - 5)$ **3.** $(3y + 1)(3y - 1)$
5. $(7 + a)(7 - a)$ **7.** $(2a + 11)(2a - 11)$
9. $(5b + \frac{2}{3})(5b - \frac{2}{3})$ **11.** $(x + y)(x - y)$
13. $(a + 0.6b)(a - 0.6b)$ **15.** $(2x + 15y)(2x - 15y)$
17. $(mn + 12)(mn - 12)$ **19.** $(3ab + c)(3ab - c)$
21. $(0.8xy + 0.5z)(0.8xy - 0.5z)$
23. $(x^2 + 9)(x + 3)(x - 3)$
25. $(5x^2 + 11)(5x^2 - 11)$
27. $(15 + a^2b^2)(15 - a^2b^2)$ **29.** $3(x + 5)(x - 5)$
31. $11(x + 2y)(x - 2y)$ **33.** $y(4 + 3x)(4 - 3x)$
35. $6m(m + 5)(m - 5)$ **37.** $b(7a + \frac{1}{2}b)(7a - \frac{1}{2}b)$
39. $2y^2(x + 6y)(x - 6y)$
41. $3(m^2 + 9)(m + 3)(m - 3)$
43. $2xy(x + 2y)(x - 2y)$ **45.** $3xy(3xy + 2)(3xy - 2)$
47. $a^2(a^2 + 5)(a^2 - 5)$ **49.** $z^2(17xy + 1)(17xy - 1)$

Stretching the Topics
1. $(a + b + 3c)(a + b - 3c)$
2. $(2x + 3z)(2x - 2y - 3z)$
3. $(x^a + 5y^b)(x^a - 5y^b)$

Checkup 7.2 (page 337)
1. $(x + 8)(x - 8)$ **2.** $(5 + y)(5 - y)$
3. $(3a + 10)(3a - 10)$ **4.** $(x + \frac{5}{7}y)(x - \frac{5}{7}y)$
5. $(x^2 + 4)(x + 2)(x - 2)$ **6.** $3(x + 0.5y)(x - 0.5y)$

7. $7y^2(x + 2y)(x - 2y)$
8. $5(m^2 + 9n^2)(m + 3n)(m - 3n)$
9. $4xy(x + 10y)(x - 10y)$ **10.** $3a(1 + 9a)(1 - 9a)$

Exercise Set 7.3 (page 350)
1. $(x - 4)(x - 3)$ **3.** $(x + 0.5)(x + 0.4)$
5. $(x - 12)^2$ **7.** $(a - 7)(a + 2)$ **9.** $(x + 4)^2$
11. $(x - 0.7)(x + 0.5)$ **13.** $(x + 9)(x - 8)$
15. $(x - 2y)^2$ **17.** $(x - \frac{1}{5}y)^2$
19. $(m + 3n)(m - 2n)$ **21.** $(13 - m)(3 - m)$
23. $(xy + 0.5)(xy - 0.3)$ **25.** $(x^2 - 3)(x^2 - 2)$
27. $3(x - 3)^2$ **29.** $-2y(y - 5)(y + 2)$
31. $4(m^2 + 3n^2)(m^2 - 2n^2)$ **33.** $xy(x - 10y)(x - 8y)$
35. $2a^2b(a - 16b)(a + 2b)$ **37.** $(2x + 1)(x + 3)$
39. $(3a - 5)^2$ **41.** $(5y + 4)(2y - 3)$
43. $(4m - 3)(2m + 5)$ **45.** $(3x + 8y)(2x - 7y)$
47. $(9x - 2y)^2$ **49.** $(11 + 6a)^2$
51. $5(3c - d)(2c - 9d)$ **53.** $-2(2y - 7)^2$
55. Prime   **57.** $a^2b^2(4b - 1)(b - 6)$
59. $2xy(3x + 7)(4x - 5)$ **61.** $(0.3k - 5h)^2$
63. $(x^2 - 5)(x + 2)(x - 2)$

Stretching the Topics
1. $(a + b - 7)(a + b + 2)$
2. $xy(x + 2y)(x - 2y)(x + 7y)(x - 7y)$
3. $(5a^x + 7b^y)(a^x - 3b^y)$

Checkup 7.3 (page 352)
1. $(x - 8y)(x + 5y)$ **2.** $(7 - x)(3 - x)$
3. $(x + 1.1)(x - 0.2)$ **4.** $5(x + y)(x - 3y)$
5. $(x - 6)^2$ **6.** $x(x - \frac{1}{3})^2$ **7.** $2xy(x + 8y)(x - y)$
8. $(5x - 2)(x - 3)$ **9.** $-y(4 - 3y)(2 + y)$
10. $2ab(2a + 5b)(3a - b)$

Exercise Set 7.4 (page 355)
1. $(x - 6)(x - 5)$ **3.** $6(c - d)(7c - 4d)$
5. $(5x + y)(4a - 3)$ **7.** $(7x - 3y)^2$
9. $(4b + 11)(4b - 11)$ **11.** $4a(3a^2b - 4a + 5b^3)$
13. $2a(a + 10)(a - 10)$ **15.** $(ab - 12c)(ab + c)$
17. $x^2(3x^2 + 5y^2)^2$ **19.** $(13 + 2a)^2$
21. $-8(3x^2 - x + 7)$ **23.** $(a^2 + 9)(a - 3)$
25. $(x^2 + 7)(x + 3)(x - 3)$ **27.** $14a(a - 1)^2$
29. $xyz(x - 5y)(x - 3y)$ **31.** $-x^2(x - 17)(x - 2)$
33. $3(7x^2 - 3x + 1)$ **35.** $(x - 9y)^2$
37. $(5x - 9y)(2x + 3y)$ **39.** $-5(6y^2 + 45y - 8)$
41. $a^2b^2(2b - 7)^2$ **43.** $(x - 4)(x + 3)(x - 3)$
45. $4x(x - 4y)(x - y)$ **47.** $a^2(a^2 + 4)(a + 2)(a - 2)$
49. $-y^2(11 + 4x)(11 - 4x)$

Stretching the Topics
1. $-ab^2(4a^2 + 3)(3a^2 - 7)$ **2.** $(a + b - 5c)^2$
3. $(5x - y)^2$

Checkup 7.4 (page 357)
1. $(3x - 7)^2$ **2.** $9(2x^2 - 7x - 1)$
3. $4(x + 3)(x - 3)$ **4.** $(9x^2 + y^2)(3x + y)(3x - y)$
5. $(x - 1)(x + 3)(x - 3)$ **6.** $(y - 9)(y + 6)$
7. $(3m + 2)(3m - 8)$ **8.** $b(8a + 5b)(8a - 5b)$
9. $3xy(10x^2 - 6x - 33)$
10. $10y^2(x^2 + 12y^2)(x + 2y)(x - 2y)$

Exercise Set 7.5 (page 368)

1. $x = 3, x = 2$ **3.** $x = -0.2, x = 0.6$
5. $a = -13, a = -5$ **7.** $x = 0, x = \frac{7}{8}$
9. $x = 7, x = -9$ **11.** $y = 3, y = -1$
13. $x = 4, x = 2$ **15.** $x = \frac{5}{4}, x = -\frac{5}{4}$
17. $y = -6, y = -3$ **19.** $m = 0, m = -1.5$
21. $z = -9, z = 6$ **23.** $x = 0, x = 5$ **25.** $z = 15$
27. $y = 10, y = -10$ **29.** $n = 0, n = 2$
31. $y = -11, y = 3$ **33.** $x = \frac{2}{5}, x = -3$
35. $x = 0.3, x = -0.6$ **37.** $y = 9, y = -9$
39. $x = 0, x = \frac{5}{2}$ **41.** $x = 9, x = -8$
43. $a = -5, a = 2$ **45.** $x = 8$ **47.** $x = \frac{1}{5}, x = 2$
49. $a = 12, a = -12$ **51.** $y = 0, y = -17$
53. $y = 20, y = -5$ **55.** $x = 0.9, x = 0.5$
57. $x = -2, x = -3$ **59.** $y = \frac{2}{3}, y = -\frac{2}{3}$
61. $x = 3$ **63.** $y = 0, y = 1$ **65.** $x = 9, x = -3$
67. $y = 6, y = -6$ **69.** $x = 5, x = -5$
71. $a = \frac{2}{7}, a = \frac{1}{3}$ **73.** $x = 8, x = -6$ **75.** $x = 5$
77. $x = -1$ **79.** $y = 0, y = \frac{4}{5}$

Stretching the Topics

1. $m = -\frac{3}{5}, m = -2$ **2.** $x = -\frac{1}{3}, x = -2$
3. $x = 3, x = -3$

Checkup 7.5 (page 370)

1. $x = 0, x = 13$ **2.** $a = 9, a = -7$
3. $y = \frac{4}{5}, y = -\frac{4}{5}$ **4.** $x = -7, x = -3$
5. $x = 0, x = 0.4$ **6.** $x = 8, x = -5$
7. $x = 3, x = 2$ **8.** $a = 3$ **9.** $x = 8, x = -8$
10. $y = -5, y = -2$

Exercise Set 7.6 (page 376)

1. 4, 9 or −7, −2   **3.** 8 by 9 ft   **5.** 5 or −7
7. 9 cm   **9.** 8 cm, 15 cm   **11.** 8 by 16 ft
13. 150 by 200 ft   **15.** 5 ft, 12 ft, 13 ft
17. 7, 8 or −8, −7   **19.** −6, −5 or 5, 6   **21.** 4, 6

Stretching the Topics

1. 16 in.   **2.** $S = 2\pi r(r + h)$; 828.96 sq in.   **3.** 2, 4

Checkup 7.6 (page 378)

1. 5 by 7 cm   **2.** 15 ft   **3.** 60 ft
4. −6, −2 or 4, 8   **5.** −3, −2 or 2, 3

Speaking the Language of Algebra (page 380)

1. Factoring   **2.** Common factor
3. Difference of two squares   **4.** Perfect square trinomial
5. Prime   **6.** Quadratic; second-degree
7. Zero product

Review Exercises (page 381)

1. $5(2x - 3)$ **2.** $y^2(2y - \frac{3}{5})$ **3.** $2xy(5x + 2y)$
4. $-a^2b(3 - 2ab + 4a^2b^2)$ **5.** $(x + y)(4 + z)$
6. $(5 + y)(5 - y)$ **7.** $3(m + 4n)(m - 4n)$
8. $2x^2(3x + 2y)(3x - 2y)$ **9.** $(a - 8b)^2$
10. $(x - 5)(x - 3)$ **11.** $(xy - 5)(xy + 2)$
12. $2(2a^2 - 10ab + b^2)$ **13.** $(3x - 7)(2x + 1)$
14. $-2(x - 3y)(x + y)$ **15.** $5(2m + 11)(m - 4)$
16. $(8x - 3)(x - 1)$ **17.** $3(y^2 + 25)(y + 2)(y - 2)$
18. $(x - 4)(x + 1)(x - 1)$

19. $5x(x^2 + 1)(x - 1)(x + 1)$ **20.** $x = 7, x = -8$
21. $x = 13, x = -13$ **22.** $x = -7$
23. $m = 0, m = -\frac{2}{3}$ **24.** $a = 9, a = -9$
25. $z = -13, z = -4$ **26.** $y = -3, y = -8$
27. $y = 0, y = -2.3$ **28.** $x = 21, x = 3$
29. $x = 9, x = -9$ **30.** $x = 10, x = -3$
31. 6 by 10 ft   **32.** 15 cm   **33.** 9, 13 or −13, −9
34. 12 ft   **35.** 5, 7 or −3, −1

Practice Test for Chapter 7 (page 383)

1. $-5(2x - 5)$ **2.** $11x(3x - 7)$
3. $3y^2(2y^3 + y^2 - 5)$ **4.** $-2ab(4a^2 - 7ab + 4b^2)$
5. $(a - 3)(y + 5)$ **6.** $(4x + 5a)(4x - 5a)$
7. $(0.6 + ab)(0.6 - ab)$ **8.** $7(x + 1)(x - 1)$
9. $(a + 4b)(a + 5b)$ **10.** $(x - 2)(x - 11)$
11. $(10 - x)(2 + x)$ **12.** $(2x - 3)(10x + 1)$
13. $-3b(a - 4b)^2$ **14.** $(x + 3)(x - 3)(x + 1)(x - 1)$
15. $x = 0, x = -11$ **16.** $x = 0, x = -1.4$
17. $n = 5, n = -5$ **18.** $a = -7, a = -10$
19. $y = \frac{5}{11}, y = -\frac{5}{11}$ **20.** $x = 4$
21. $y = 4, y = -2$ **22.** $x = 10, x = -2$
23. 4 by 10 m   **24.** 13 yd
25. 3, 10 or −10, −3

Sharpening Your Skills after Chapters 1–7 (page 385)

1. $-\frac{64}{125}$ **2.** 15.336   **3.** $\frac{8}{9}$ **4.** −6   **5.** $-7.2k$
6. $9x + 8$ **7.** $-8x^8y^4$ **8.** $3y^2 - 8yz + 4z^2$
9. $-45x^7y^7$ **10.** $-5a^2b + 2a^3b^2 - 7a^4b^3$
11. $8a^2 - 2ab - 15b^2$ **12.** $-y^4 - 2y^3 + 15y^2$
13. $-5x^2 + 4xy - \frac{4}{3}y^2$ **14.** $4x + 3 + \dfrac{1}{x - 5}$
15. $x = 36$ **16.** $x = -2$ **17.** $x = 40$ **18.** 432
19. $3l^2 - 10l$; 525 sq m   **20.** $t = \dfrac{A - p}{pr}$

CHAPTER 8

Exercise Set 8.1 (page 397)

1. $x \neq 0$ **3.** $x \neq -1$ **5.** $x \neq 0$ **7.** $y \neq 8$
9. $x \neq \frac{1}{2}$ **11.** $2x^3$ **13.** $-\dfrac{a}{4}$ **15.** $\dfrac{3x^2}{y^2}$
17. $-\dfrac{3b^2}{4a^2}$ **19.** $\dfrac{5}{4x^2z}$ **21.** $\dfrac{x + y}{3}$ **23.** $a + 5b$
25. $9y - 7$ **27.** $\dfrac{2}{16 + x}$ **29.** $\dfrac{6(a^2 + b^2)}{5}$
31. $5(x^2 + y^2)$ **33.** $\dfrac{7x^2}{x + 1}$ **35.** $\dfrac{-4a - 5c}{6}$
37. $\dfrac{2x^2 + 3x + 1}{10}$ **39.** $\dfrac{a + b}{x + y}$ **41.** $\dfrac{x - y}{2(x + y)}$
43. $\dfrac{1}{2}$ **45.** $-\dfrac{1}{3}$ **47.** $\dfrac{x - y}{3}$ **49.** $\dfrac{x - 7}{4}$
51. $\dfrac{3}{x + y}$ **53.** $\dfrac{-7}{3(a - b)}$ **55.** $\dfrac{x - 2y}{3}$
57. $\dfrac{x + 5}{x - 4}$ **59.** $\dfrac{x + 4}{x + 5}$ **61.** $\dfrac{5(x + 3)}{x - 7}$
63. $\dfrac{x + 5y}{x + y}$ **65.** $\dfrac{2(a + b)}{a - 3b}$ **67.** $\dfrac{(x + 5)(x - 5)}{x^2 + 25}$

69. $\dfrac{1}{2b-1}$ **71.** $\dfrac{(x^2+4y^2)(x+2y)}{x+4y}$ **73.** $-\dfrac{1}{9}$

75. $-(x+5)$

Stretching the Topics

1. $m \neq 0,\ m \neq -\dfrac{3}{2},\ m \neq -4$ **2.** $\dfrac{-2b(a-c)}{3(a+c)}$

3. $2x(b+c)^{n-1}$

Checkup 8.1 (page 400)

1. $x \neq 0$ **2.** $x \neq \dfrac{1}{2}$ **3.** $\dfrac{x^3}{y^2}$ **4.** $-\dfrac{1}{4x}$

5. $4y-5$ **6.** $\dfrac{1}{5x-4y}$ **7.** $\dfrac{a+b}{4(a-b)}$

8. $\dfrac{-1}{3(x+y)}$ **9.** $\dfrac{m+2n}{2}$ **10.** $\dfrac{x+4}{x+5}$

Exercise Set 8.2 (page 406)

1. $\dfrac{x}{y^2}$ **3.** x **5.** $\dfrac{105y}{x^2}$ **7.** b **9.** $-\dfrac{x}{9}$ **11.** $4xz$

13. $\dfrac{5b^5}{a^2}$ **15.** $\dfrac{18y^2}{x^2}$ **17.** $\dfrac{4b^4}{9a}$ **19.** $\dfrac{5x^6}{6y^{11}}$

21. $\dfrac{2}{x+3}$ **23.** $\dfrac{3}{8x}$ **25.** $\dfrac{2b}{a}$ **27.** $\dfrac{7}{3}$ **29.** $x-3y$

31. $\dfrac{-2(x-2)}{3(x+2)}$ **33.** $\dfrac{5y(x+3)}{8(x-3)}$ **35.** $\dfrac{x+4}{x+6}$

37. $\dfrac{x+5y}{5}$ **39.** $y(x-7)$ **41.** $\dfrac{x+5}{49}$

43. $\dfrac{1}{2(x-9)}$ **45.** $\dfrac{-3}{x-2}$ **47.** $\dfrac{1}{x+3}$

49. $\dfrac{5(a-b)}{a^2b^2}$ **51.** 1

Stretching the Topics

1. $\dfrac{243x^7}{40}$ **2.** $\dfrac{(a+b)^2(a-b)^2}{a^3}$

3. $\dfrac{20x(x+y)}{3y(x+2y)(2x+3y)}$

Checkup 8.2 (page 408)

1. $-\dfrac{14x^4}{z}$ **2.** $\dfrac{10a^3c^2}{7b}$ **3.** $-\dfrac{27z^2}{28x^2y}$ **4.** $\dfrac{2}{y}$

5. $\dfrac{2(x-3)}{3(x+3)}$ **6.** -1 **7.** $\dfrac{x-8}{y}$ **8.** $\dfrac{1}{3(x+6)}$

9. $\dfrac{(x-11)(x-8)}{(x+3)^2}$ **10.** $\dfrac{2n-1}{n-6}$

Exercise Set 8.3 (page 413)

1. $\dfrac{2x}{3x}$ **3.** $\dfrac{-3x}{15x^2}$ **5.** $\dfrac{8xy}{2x^2y^2}$ **7.** $\dfrac{9xy^2}{21y^3}$ **9.** $\dfrac{-30a^2}{15abc}$

11. $\dfrac{3xy}{3y}$ **13.** $\dfrac{-3x^2y}{3y^3}$ **15.** $\dfrac{-8a^2b^4}{4a^2b^2}$ **17.** $\dfrac{8x-8y}{12(x-y)}$

19. $\dfrac{25x^2}{10x^3}$ **21.** $\dfrac{-4a-4b}{a(a+b)}$ **23.** $\dfrac{3x^2}{x(x+y)}$

25. $\dfrac{2ab-2a}{(b+1)(b-1)}$ **27.** $\dfrac{6x^3+12x^2}{(x-5)(x+2)}$

29. $\dfrac{x^3+5x^2}{x^2(x+3)}$ **31.** $\dfrac{a^2-5a+4}{(a-3)(a-4)}$ **33.** $\dfrac{-3xy+3y}{x(x-1)^2}$

35. $\dfrac{x^2-4xy+3y^2}{2(x-3y)(x+y)}$

Stretching the Topics

1. $\dfrac{3x^3-x^2-10x}{x^2(2x-3)(3x+5)}$ **2.** $\dfrac{5x-5y}{(a+b)(x-y)}$

3. $\dfrac{-3x^3-6x^2-3x-6}{(x^2+1)(x+2)(x-2)}$

Checkup 8.3 (page 415)

1. $\dfrac{54a^2b}{9a^3b^2}$ **2.** $\dfrac{-25xyz^2}{10y^2z^3}$ **3.** $\dfrac{-3xy^2}{18y^4}$ **4.** $\dfrac{16a^4b^2}{16a^2b^2}$

5. $\dfrac{x^2-3x-10}{(x+2)(x+2)}$ **6.** $\dfrac{2a^4-a^3b}{a^3(a-3b)}$ **7.** $\dfrac{2a^2b}{ab(a+b)}$

8. $\dfrac{5x^2-30x}{(x+6)(x-6)}$ **9.** $\dfrac{x^2+2x-8}{(x+4)(x-1)}$

10. $\dfrac{a^2-b^2}{5(a+b)^2}$

Exercise Set 8.4 (page 424)

1. $3x$ **3.** $\dfrac{5a-6}{7}$ **5.** $\dfrac{1}{x}$ **7.** $\dfrac{2x-1}{y}$ **9.** $-\dfrac{2}{b}$

11. $\dfrac{9y}{x+3y}$ **13.** 3 **15.** $\dfrac{2x}{2x+5}$ **17.** $\dfrac{x}{x-2}$

19. $3(x-5)$ **21.** $\dfrac{2a-3}{a}$ **23.** $\dfrac{1}{x-2y}$

25. $\dfrac{5x-3y}{30}$ **27.** $\dfrac{15x+23}{5x^2}$ **29.** $\dfrac{7y-1}{y}$

31. $\dfrac{2(2-x^2)}{x}$ **33.** $\dfrac{5y+2x}{xy}$ **35.** $\dfrac{2(18y-25x)}{15xy}$

37. $\dfrac{4y-5x}{x^2y^2}$ **39.** $\dfrac{-3x^2-8y^2}{12x^2y^2}$ **41.** $\dfrac{7x+9}{(x+2)^2}$

43. $\dfrac{-2x-32}{(x+4)(x-4)}$ **45.** $\dfrac{x^2+x-6}{3(x+1)}$

47. $-\dfrac{11}{3(x+1)}$ **49.** $\dfrac{2x+24}{(x+2)(x-2)}$

51. $\dfrac{x^2-x-5}{x(x+5)(x-5)}$ **53.** $\dfrac{1}{x^2}$ **55.** $\dfrac{3x+35}{(x+7)^2(x-7)}$

57. $\dfrac{4x}{(x+4)(x-2)}$ **59.** $\dfrac{2x-5}{x(2x-1)}$

61. $\dfrac{2x-36}{(x-9)^2(x+9)}$ **63.** $\dfrac{x^2-2}{(2x-3)(x+2)(x+1)}$

Stretching the Topics

1. $\dfrac{-5n+24}{(n-5)(n-3)}$ **2.** $\dfrac{3x^3+21x^2-36}{x(x+7)(x+3)(x-3)}$

3. $\dfrac{-2}{x-3y}$

Checkup 8.4 (page 427)

1. $-\dfrac{x}{2}$ **2.** $\dfrac{2a-1}{b}$ **3.** $x-8$ **4.** $\dfrac{5}{2x}$

5. $\dfrac{3y+2x}{xy^2}$ **6.** $\dfrac{11x-21}{x(x-3)}$ **7.** $\dfrac{15}{2(x+1)}$

8. $\dfrac{3x+25}{(x+5)^2(x-5)}$ **9.** $\dfrac{1}{x^2}$ **10.** $\dfrac{17}{(x-9)(x+7)}$

Exercise Set 8.5 (page 434)

1. $\dfrac{4}{7}$ 3. $\dfrac{x^5}{4y}$ 5. $\dfrac{3x}{x+3y}$ 7. $\dfrac{5}{4(x-4)}$

9. $\dfrac{x^2}{x-7}$ 11. $\dfrac{3}{x^2}$ 13. $\dfrac{2(5x-2)}{x^2}$ 15. 6

17. $\dfrac{x-7}{x}$ 19. $\dfrac{1}{x-1}$ 21. $\dfrac{x}{5x+1}$ 23. $\dfrac{3}{4}$

25. $\dfrac{3+x}{3x}$ 27. $\dfrac{5(2x-5)}{x}$ 29. $\dfrac{9(x+5)}{x(x+9)}$

Stretching the Topics

1. $\dfrac{5x(x-y)}{(x-4y)(x+2y)}$ 2. $\dfrac{8(y-2)^2}{(y+3)(y-3)}$

3. $\dfrac{-(x+1)}{2(x-12)}$

Checkup 8.5 (page 436)

1. $\dfrac{3}{x^3y}$ 2. $\dfrac{9x^3y^4}{2}$ 3. $\dfrac{4x^2}{3x+5}$ 4. $\dfrac{(x+3)(x+1)}{3}$

5. $\dfrac{3x}{x-8}$ 6. $\dfrac{x+4}{x}$ 7. $\dfrac{5a+4}{a}$ 8. $\dfrac{x+9}{9x}$

9. $\dfrac{4(x-1)}{x}$ 10. $\dfrac{4(x-6)}{x(x-4)}$

Exercise Set 8.6 (page 442)

1. $x=\dfrac{7}{2}$ 3. $x=4$ 5. $x=-\dfrac{14}{5}$ 7. $y=-15$
9. $x=\dfrac{90}{11}$ 11. $x=\dfrac{33}{2}$ 13. $y=-1$ 15. $x=3$
17. $x=-5$ 19. $z=-3$ 21. $x=3$
23. $x=\dfrac{13}{3}$ 25. $x=-8$ 27. $x=\dfrac{44}{9}$
29. No solution   31. $x=-\dfrac{23}{4}$ 33. $x=-\dfrac{1}{10}$
35. $a=\dfrac{38}{7}$ 37. $b=-\dfrac{1}{5}$ 39. $x=-\dfrac{2}{5}$
41. $x=-41$ 43. $x=-\dfrac{17}{3}$ 45. $a=\dfrac{3}{5}$
47. No solution   49. $x=0$ 51. $x=5,\ x=8$
53. $x=5,\ x=-\dfrac{3}{2}$ 55. $y=4,\ y=-3$
57. $x=2,\ x=\dfrac{1}{3}$ 59. $x=3$

Stretching the Topics

1. $x=-11$ 2. $x=\dfrac{abck}{bc-2ac+3ab}$

3. $n=\dfrac{Ir}{E-IR}$

Checkup 8.6 (page 445)

1. $y=18$ 2. $x=-2$ 3. $x=-11$
4. $y=-23$ 5. $a=\dfrac{23}{2}$ 6. $x=\dfrac{55}{17}$
7. No solution   8. $x=3$ 9. $x=-1$
10. $x=1,\ x=-2$

Exercise Set 8.7 (page 454)

1. $2800   3. Louise, 56 mph; husband, 50 mph
5. José, 20; Rick, 16   7. 165 mi   9. 2500
11. 12 by 18 ft   13. 3   15. $3\frac{1}{3}$ hr   17. $3\frac{3}{7}$ hr
19. $900

Stretching the Topics

1. 1   2. 30 days   3. 2 mph

Checkup 8.7 (page 456)

1. $240,000   2. 100

---

3. Faye, 6 mph; daughter, 4 mph   4. 2   5. $2\frac{2}{5}$ hr

Speaking the Language of Algebra (page 459)

1. Rational algebraic expression   2. 0; undefined
3. Numerators; denominators   4. Denominator
5. Single fraction   6. Least common denominator
7. Extraneous   8. Proportion

Review Exercises (page 461)

1. $\dfrac{1}{2}$ 2. $-\dfrac{1}{5}$ 3. $x\neq-4$ 4. $x\neq\dfrac{1}{2}$ 5. $2x^5$

6. $\dfrac{5x^2y^2}{4}$ 7. $a+3b$ 8. $\dfrac{8x^4}{x^2+1}$ 9. $-\dfrac{1}{5}$

10. $\dfrac{x-4}{5x}$ 11. $\dfrac{1}{5a-6b}$ 12. $\dfrac{-3}{x+y}$ 13. $\dfrac{x-5}{2x-3}$

14. $x+y$ 15. $\dfrac{7y^2}{6}$ 16. $9abc$ 17. $10b^4$

18. $\dfrac{y^2(x+4)}{8z(x-4)}$ 19. $\dfrac{5x+2}{5x-2}$ 20. $\dfrac{1}{2}$ 21. $\dfrac{15xy}{35y^2}$

22. $\dfrac{3y^2+6y}{(y+2)(y-2)}$ 23. $\dfrac{2x+3}{y}$ 24. $\dfrac{x-5}{3x-1}$

25. $\dfrac{32x+1}{4x^2}$ 26. $\dfrac{-7x+24}{6x}$

27. $\dfrac{-4x^2-10xy+3y^2}{12x^2y^2}$ 28. $\dfrac{-2x}{3(x-3)}$

29. $\dfrac{2a+3}{(a+2)(a+1)}$ 30. $\dfrac{-3x+17}{(x+6)(x-3)}$

31. $\dfrac{5}{2ab^5c^2}$ 32. $\dfrac{x+7}{7x}$ 33. $x=16$

34. $x=-15$ 35. No solution   36. $a=1$
37. $a=-6$ 38. $x=1,\ x=3$ 39. 12
40. 25   41. $1\frac{1}{5}$ hr   42. 65 mph

Practice Test for Chapter 8 (page 465)

1. $\dfrac{7}{8}$ 2. $y\neq-\dfrac{1}{3}$ 3. $-\dfrac{y}{4x^3}$ 4. $\dfrac{6x^2}{x+2}$

5. $\dfrac{2x}{x+5}$ 6. $\dfrac{-(x-y)}{3}$ 7. $\dfrac{2b^3}{5a^2}$ 8. $\dfrac{18}{25x^2y^4}$

9. $\dfrac{2x}{5}$ 10. 1   11. $\dfrac{(x+1)(2x+1)}{3x^4}$ 12. $\dfrac{x-1}{a}$

13. $\dfrac{x-1}{x+2}$ 14. $\dfrac{5a-9ab+60b}{15a^2b^2}$

15. $\dfrac{-60}{(y+5)(y-5)}$ 16. $\dfrac{a-8}{a(a-3)(a+4)}$

17. $x=\dfrac{21}{2}$ 18. $x=5$ 19. $x=-1$ 20. 12

21. 90   22. $2\frac{2}{9}$ hr

Sharpening Your Skills after Chapters 1–8 (page 467)

1. Undefined   2. 2   3. $5x+85y$ 4. $-8x^8y^9$

5. $\dfrac{729b^{10}}{784a^{10}}$ 6. $-36a^5b^5$ 7. $4a^2-12ab+9b^2$

8. $x^2-0.25y^2$ 9. $-2x^3+6x^2y+20xy^2$
10. $5xy(3x-y)(x-4y)$ 11. $2(x+7)(x-7)$
12. $(x-5y)^2$ 13. $-5(3x-2)(x+6)$ 14. $x=0$
15. $y=\dfrac{13}{7}$ 16. $y=0,\ y=3$ 17. 5 in. and 12 in.
18. 38

Chapter 9

Exercise Set 9.1 (page 474)

1. < **3.** > **5.** = **7.** < **9.** < **11.** >
13. > **15.** $-11 < -5 < 0$ **17.** $-2 < 3 < 7$
19. $-\frac{4}{5} < -\frac{2}{5} < 0$

21.

23.

25.

27.

29.

31.

33.

35.

37.

39.

Stretching the Topics

1.

2.

3.

Checkup 9.1 (page 476)

1. < **2.** = **3.** > **4.** > **5.** $-5 < -2 < 0$
6. $\frac{3}{2} < 3 < \frac{7}{2}$
7.

8.

9.

10.

Exercise Set 9.2 (page 484)

1. $1 < 9$ **3.** $6 > -5$ **5.** $8 > -12$
7. $-6 < -2$ **9.** $-7 > -14$ **11.** $-2 < 1$
13. $13 > 1$ **15.** $4 > -8$
17. $x \leq 4$

19. $x > -5$

21. $a < -\frac{7}{2}$

23. $a \geq 5$

25. $x < -3$

27. $z \geq -1$

29. $y < 2$

31. $x \geq 2$

33. $x \geq \frac{9}{2}$ **35.** $y < 0$ **37.** $x > 12$ **39.** $a < -1$
41. $x \leq \frac{1}{2}$ **43.** $x < -\frac{5}{4}$ **45.** $z > -3$
47. $x \leq 12$ **49.** $x \leq -\frac{1}{5}$ **51.** $-2 < x < 3$
53. $-4 \leq x \leq \frac{1}{2}$ **55.** $-3 \leq x \leq 4$
57. $-4 \leq x < 16$ **59.** $-1 \leq x \leq \frac{2}{3}$

Stretching the Topics
1. $x \geq \frac{8}{5}$ **2.** $y < \frac{51}{7}$ **3.** $-\frac{31}{2} \leq x < -\frac{23}{2}$

Checkup 9.2 (page 486)
1. $3 > -4$ **2.** $-2 < 3$
3. $x \geq -2$

4. $a \geq -3$

5. $x > 3$

6. $-1 \leq y \leq 2$

7. $a \leq \frac{5}{3}$ **8.** $x > -\frac{5}{4}$ **9.** $y > -\frac{1}{2}$
10. $2 \leq x \leq 6$

Exercise Set 9.3 (page 493)
1. $-2 \leq x \leq 3$ **3.** $x > 0$ **5.** $x < 5$
7. $x > 1$ or $x < -2$ **9.** No solution **11.** $\frac{1}{2} < x < 4$
13. $x > 0$ **15.** $-2 < x < 2$ **17.** $2 \leq x < 3$
19. $x \geq -5$ **21.** $x < -1$ or $x > 0$ **23.** $x > 4$
25. $x < 3$ **27.** $x \leq 3$ **29.** $x > 3$
31. $5 \leq x < 11$ **33.** $x < -4$ or $x > 13$
35. $x \geq 3$ **37.** $x \geq -8$

Stretching the Topics
1. No solution **2.** All real numbers **3.** $x > 4$

Checkup 9.3 (page 495)
1. $x > 3$ **2.** $x > -2$ **3.** $a \leq -2$ or $a \geq 4$
4. No solution **5.** All real numbers **6.** $0 < x < 4$
7. $-4 < x < 4$ **8.** $-1 < x < \frac{3}{2}$ **9.** $x > 5$
10. $x < \frac{3}{2}$ or $x > 4$

Exercise Set 9.4 (page 500)
1. At least 8 ft **3.** Fewer than 176
5. No more than $14\frac{1}{2}$ ft **7.** $2 < n < 5$
9. No more than 300 **11.** $w \leq 24$
13. $2400 \leq x \leq 5000$ **15.** No more than 19 by 26 ft
17. $w \leq 5\frac{2}{3} m$ **19.** $4 \leq j \leq 10$

Stretching the Topics
1. $\$10,000 < S \leq \$25,000$ **2.** $w \leq 54$ ft

Checkup 9.4 (page 502)
1. No more than 4 **2.** At least 4
3. No more than $54\frac{1}{2}$ by 109 ft **4.** $950 \leq x \leq 1200$
5. At least $\$50.75$

Speaking the Language of Algebra (page 504)
1. Inequalities **2.** $2 < x < 5$ **3.** Left
4. Multiply; divide; negative **5.** $3n \geq 18.7$

Review Exercises (page 505)
1. $<$ **2.** $>$ **3.** $-10 < -7 < 3$ **4.** $0 < \frac{5}{3} < 4$
5.

6.

7.

8.

9. $4 < 13$ **10.** $-21 > -30$ **11.** $-5 < 2$
12. $-9 > -15$
13. $x \leq 4$

14. $y < 5$

15. $a \geq -8$

16. $x > 6$

17. $x > -5$ **18.** $x \geq 2$ **19.** $x > -\frac{9}{5}$
20. $x \leq \frac{18}{5}$ **21.** $y \leq -6$ **22.** $-\frac{2}{3} \leq y \leq 2$
23. $2 < x \leq 6$ **24.** $-3 \leq x \leq 3$ **25.** $x > 3$
26. $x \geq 6$ **27.** $x < -1$ or $x > 3$ **28.** At least 15
29. No more than 12 ft by 24 ft   **30.** $300 \leq x \leq 500$

Practice Test for Chapter 9 (page 507)
1. $-1 < -\frac{1}{2} < \frac{1}{4}$

2.

3.

4. $x \leq 2$

5. $x > 5$

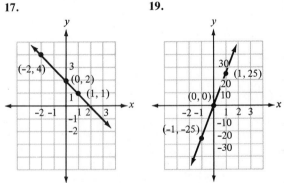

6. $x > -9$ **7.** $x > 0$ **8.** $x < \frac{1}{2}$ **9.** $a \leq -17$
10. $y \leq -\frac{72}{29}$ **11.** $-2 \leq x \leq -1$ **12.** $1 < x < 4$
13. $-8 \leq x \leq 0$ **14.** $x < -5$
15. $x < -11$ or $x > 5$ **16.** Fewer than 80
17. Between 15 and $19\frac{1}{2}$ feet

Sharpening Your Skills after Chapters 1–9 (page 509)
1. 62.5%   **2.** $\frac{4}{5} > \frac{5}{7}$ **3.** $442.50   **4.** 33
5. $x \neq 4, x \neq -4$ **6.** $2x - 2y$ **7.** $18x^4y^6$
8. $x^3 - 1$ **9.** $-24x + 4x^2 + 4x^3$ **10.** $\frac{m - 3n}{2}$
11. $\frac{3(x - 7)}{4(x + 7)}$ **12.** $\frac{13}{4(x + 1)}$ **13.** $\frac{1}{xy^3(x + 2)}$
14. $8ab(2ab - 1)(ab - 1)$
15. $6(m^2 + 4)(m + 2)(m - 2)$
16. $-xy(x + 3y)(x + 5y)$ **17.** $x = 9$ **18.** $y = 4$
19. $x = 3, x = -13$ **20.** $x = 0, x = 9$
21. $x = -2$ **22.** $a = -\frac{16}{3}$ **23.** 7 by 21 ft
24. 9, 10 or $-10, -9$
25. Louise, 56 mph; husband, 50 mph

CHAPTER 10

Exercise Set 10.1 (page 519)
1. Yes   **3.** No   **5.** Yes   **7.** No   **9.** No
11. Yes   **13.** Answers will vary.
15. Answers will vary.   **17.** Answers will vary.
19. Answers will vary.   **21.** Answers will vary.

23. Answers will vary.   **25.** Answers will vary.
27. Answers will vary.   **29.** $y = 4x + 8$
31. $x + y = 28$ **33.** $l = 2w - 5$
35. $x + 5y = 100$ **37.** $2l + 2w = 48$
39. $x + y = 12,500$ **41.** $3.20x + 4.80y = 147.20$
43. $c = 5d - 2$ **45.** $3x + 2y = 20$

Stretching the Topics
1. Yes   **2.** $x = 4, y = 13; x = 6, y = 19; y = 3x + 1$
3. $0.085x + 0.12y = 0.10(x + y)$

Checkup 10.1 (page 521)
1. No   **2.** No   **3.** Yes   **4.** No
5. Answers will vary.   **6.** Answers will vary.
7. Answers will vary.   **8.** $y = 2x - 500$
9. $2x + 3y = 17$ **10.** $2l + 2w = 196$

Exercise Set 10.2 (page 532)
1–8.   **9.** $A(2, 5)$

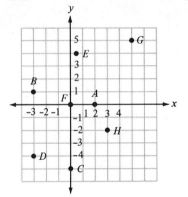

11. $C(2, -1)$ **13.** $E(0, 0)$ **15.** $G(-3, -3)$

17.   **19.**

21.

23.

25.

27.

Stretching the Topics

1.

2.

3.

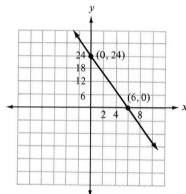

Checkup 10.2 (page 533)

1–3.   **4.** $A(-2, 0)$

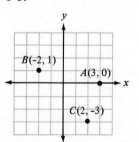

5. $B(4, 2)$ **6.** $C(3, -2)$

7.   **8.**

9.

10.

13.

15.

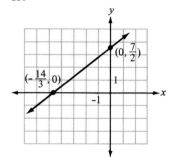

Exercise Set 10.3 (page 545)

1.

3.

17.

19.

5.

7.

21.

23.

9.

11.

25.

27.

A16   ANSWERS

29.

31.

3.

33.

35.

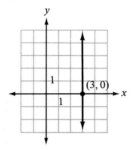

Checkup 10.3 (page 546)

1.

2.

Stretching the Topics

1.

3.

4.

5.

6.

2.

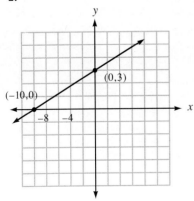

Exercise Set 10.4 (page 559)

1. $\dfrac{5}{3}$ **3.** Undefined   **5.** -1 **7.** 0   **9.** 3

11.

13.

15.

17.

19.

21.

23.

25.

27.

29.

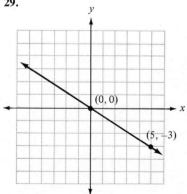

Stretching the Topics

1. $m = -\dfrac{2}{3}$ **2.** $m = \dfrac{7c}{4k}$ **3.** $x = 5$

Checkup 10.4 (page 560)

1. -1 **2.** 0

3.

4.

5.

6.

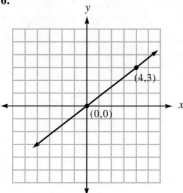

Speaking the Language of Algebra (page 561)
1. 3; −2   **2.** *x*-intercept; *y*-intercept   **3.** Origin

4. Straight line   **5.** $\dfrac{y_2 - y_1}{x_2 - x_1}$ **6.** 0; $y = k$, a constant

7. Undefined; $x = k$, a constant   **8.** Slope; *y*-intercept

Review Exercises (page 563)
1. Yes   **2.** Yes   **3.** Answers will vary.

4. Answers will vary.   **5.** Answers will vary.

6. Answers will vary.

7. $l = 3w - 5$; answers will vary.

8. $5x + 3y = 45$; answers will vary.

10. $A(-4, 1)$; $B(2, 0)$; $C(3, -4)$; $D(4, 3)$

9.

11.

12.

13.

14.

15.

16.

17.

18.

19.

20.

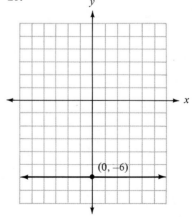

21. 1   **22.** 0   **23.** $-\dfrac{7}{4}$ **24.** Undefined

25.

26.

27.

28.

29.

30.

31.

32.

33.

34.

6.

7.

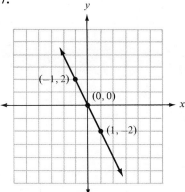

Practice Test for Chapter 10 (page 565)
1. No   **2.** Yes   **3.** $y = \frac{1}{2}x + 15$
4. $0.25x + 0.15y = 15.35$
5.

8.

9.

10.

11.

12.

13. $\dfrac{1}{8}$ **14.**

15.

16.

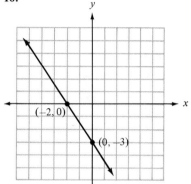

Sharpening Your Skills after Chapters 1–10 (page 569)

1. 0.193   **2.** 0.09   **3.** $-\dfrac{1}{18}$ **4.** 0.15

5. $16x - 12y$ **6.** $-\dfrac{3}{2}y$ **7.** $\dfrac{16b^{14}}{a^7}$ **8.** $-\dfrac{8y^3}{x^5}$

9. $40x^{11}y^3$ **10.** $-5axy$ **11.** $x^4 - 9$

12. $x - 5 - \dfrac{4}{x}$ **13.** $x^2 - 2x + 3 - \dfrac{1}{x + 2}$

14. $\dfrac{1}{2x + 3y}$ **15.** $\dfrac{3(3y - 1)}{2xy(3y + 1)}$ **16.** $-\dfrac{12y^2z^3}{5x^2}$

17. $\dfrac{13x - 8}{5x(x + 4)}$ **18.** $y = 4$ **19.** $x = 7, x = -3$

20. $y = -4, y = -5$ **21.** $a = 1$ **22.** $b = -\dfrac{1}{11}$

23. $x > -\dfrac{4}{5}$ **24.** $1 \le y \le 4$ **25.** $x < 3$

26. $P = \dfrac{A}{1 + rt}$

27. $x \ge -2$

28. At least 7

29. 5 by 9 cm   **30.** $x = 2$

CHAPTER 11

Exercise Set 11.1 (page 577)
1. $(1, 3)$ **3.** $(-2, 6)$ **5.** $(-2, -12)$
7. No solution (parallel lines)   **9.** $(2, 0)$
11. All points on the line $2x - y = 2$ **13.** $(2, 3)$
15. $(7, 4)$ **17.** $(4, 1)$ **19.** No solution (parallel lines)

Stretching the Topics
1. $(5, -3)$ **2.** All points on the line $2x - 3y = 0$

Checkup 11.1 (page 578)
1. $(1, -1)$ **2.** No solution (parallel lines)
3. $(-3, 4)$ **4.** All points on the line $-x + 2y = 4$

Exercise Set 11.2 (page 584)
1. $(2, 3)$ **3.** $(-1, -2)$ **5.** $(6, 19)$ **7.** $(17, 5)$
9. All ordered pairs satisfying $y = 2x - 4$
11. $(-5, -9)$ **13.** $(8, -\frac{9}{2})$ **15.** $(\frac{24}{5}, \frac{6}{5})$
17. $(3, \frac{7}{3})$ **19.** No solution   **21.** $(1, 3)$
23. $(-23, -10)$ **25.** $(\frac{1}{10}, \frac{17}{10})$

Stretching the Topics
1. $(0.05, -0.1)$ **2.** $(-6, \frac{13}{3})$ **3.** $(2a, a)$

Checkup 11.2 (page 585)
1. $(3, 6)$ **2.** No solution   **3.** $(6, \frac{5}{2})$ **4.** $(2, 2)$
5. $(\frac{5}{3}, \frac{4}{3})$

Exercise Set 11.3 (page 594)
1. $(1, -1)$ **3.** $(-1, -1)$ **5.** $(4, \frac{7}{4})$ **7.** $(2, -2)$
9. $(-\frac{1}{2}, -1)$ **11.** $(\frac{1}{2}, \frac{3}{4})$ **13.** $(2, 1)$
15. $(-\frac{11}{3}, 4)$ **17.** No solution   **19.** $(0, -\frac{5}{2})$
21. $(-\frac{2}{3}, \frac{11}{4})$ **23.** $(1, 1)$
25. All ordered pairs satisfying $14x + 7y = 7$
27. $(1, -\frac{1}{3})$ **29.** $(1, 1)$

Stretching the Topics
1. $(a, b) = (4, 9)$ **2.** $(-\frac{5}{7}, -\frac{36}{7})$ **3.** $k = 6$

Checkup 11.3 (page 595)
1. $(2, 0)$ **2.** $(a, b) = (3, -2)$ **3.** No solution
4. $(\frac{4}{3}, -\frac{1}{3})$ **5.** $(2, 7)$

Exercise Set 11.4 (page 602)
1. -1 and 7 **3.** 8 by 21 ft
5. 13 five-dollar bills; 12 one-dollar bills
7. 9 books at \$2; 11 books at \$3
9. \$3000 in Abell's; \$5000 in Betti's
11. \$16 for jeans; \$12 for tops
13. 24°; 66°   **15.** \$5 for records; \$7 for tapes
17. Blouse, \$21; skirt, \$28   **19.** 100 of each

Stretching the Topics
1. \$10,000 at 8%; \$6000 at 10%   **2.** 170

Checkup 11.4 (page 604)
1. \$5 as waitress; \$4 as window washer
2. Shirts, \$13; ties, \$5   **3.** 40 by 60 in.   **4.** 5 and -12
5. 1176 advance tickets; 1649 at door

Speaking the Language of Algebra (page 605)
1. Satisfy   **2.** Substitution   **3.** No   **4.** One

Review Exercises (page 607)
1. $(2, 3)$ **2.** $(2, 1)$ **3.** $(-14, -7)$
4. $(-1, 2)$ **5.** $(-1, 3)$ **6.** $(\frac{1}{2}, -\frac{3}{2})$
7. $(-2, -8)$ **8.** No solution   **9.** $(13, 3)$
10. $(-4, 1)$ **11.** $(5, 2)$ **12.** $(0, 3)$
13. No solution   **14.** $(\frac{55}{31}, -\frac{15}{31})$
15. All ordered pairs satisfying $-6x + 5y = 9$
16. $(1, 2)$ **17.** $(0, 8)$ **18.** $(-10, -5)$
19. $(3, -2)$ **20.** $(\frac{55}{23}, \frac{6}{23})$ **21.** 10 and 12
22. 24 by 34 m
23. \$1250 at Union; \$450 at National
24. 8 bottles at \$5; 12 bottles at \$3

Practice Test for Chapter 11 (page 609)
1. $(1, 2)$ **2.** $(-5, 3)$ **3.** $(2, 5)$ **4.** No solution
5. $(\frac{1}{2}, -2)$ **6.** All ordered pairs satisfying $3x - 3y = 6$
7. $(10, 4)$ **8.** $(2, 3)$ **9.** $(-1, -2)$ **10.** $(\frac{19}{71}, -\frac{2}{71})$
11. \$3 for mowing; \$4.50 for digging   **12.** 60 by 100 yd

Sharpening Your Skills after Chapters 1–11 (page 611)
1. $-2.39 < -2.184$ **2.** $\frac{8}{5}$ **3.** $y \neq 3, y \neq -\frac{5}{2}$
4. $-1 < -\frac{2}{3} < -\frac{1}{5}$ **5.** -3 **6.** $\dfrac{56}{405}$
7. $18a - 4$ **8.** $-4x^5y^5z^5$
9. $2a^4 - 10a^3 + 4a^2 - 20a$ **10.** $\dfrac{6}{x - 6}$
11. $\dfrac{7x}{20(x - 3)}$ **12.** $(9x^2 + 4)(3x + 2)(3x - 2)$
13. $7xy(x + 5y)(x - 5y)$ **14.** $(a + b)(2x - 3y)$
15. $(xy - 9z)(xy + z)$ **16.** $x = -8$
17. $x = 0, x = 9$ **18.** $y = 0, y = 3$ **19.** $x = 42$
20. $y > 0$ **21.** $x > 6$

22.   **23.** $m = \dfrac{3}{5}$

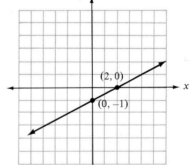

24. 3   **25.** 20

CHAPTER 12

Exercise Set 12.1 (page 623)
1. Rational   **3.** Irrational   **5.** Rational
7. Irrational   **9.** Rational   **11.** Irrational
13. Irrational   **15.** Rational   **17.** 12   **19.** a^2

21. 8   **23.** x^5 **25.** $5b$ **27.** $\dfrac{x}{6}$ **29.** $3\sqrt{5}$
31. $10\sqrt{6}$ **33.** $5\sqrt{3}$ **35.** $2\sqrt{11}$ **37.** x
39. $y^2\sqrt{y}$ **41.** $xy\sqrt{y}$ **43.** $4a\sqrt{2a}$ **45.** $2x\sqrt{y}$
47. $9a^2b^2\sqrt{a}$ **49.** $8xy^2z\sqrt{3z}$ **51.** $6a^3b\sqrt{5abc}$
53. $\dfrac{2x}{5}$ **55.** $\dfrac{4a\sqrt{2a}}{b^2}$ **57.** $\dfrac{2x\sqrt{10x}}{3y}$
59. $\dfrac{ab}{5}$ **61.** $5x$ **63.** $x^2\sqrt{3}$ **65.** $\dfrac{\sqrt{y}}{11}$

Stretching the Topics
1. $\dfrac{2x^5\sqrt{6}}{y^2z^{14}}$ **2.** $-21x^2y^7\sqrt{7y}$ **3.** $0.4x^2y(a-b)$

Checkup 12.1 (page 625)
1. 7   **2.** $3a$ **3.** 12   **4.** $4\sqrt{5}$ **5.** $5y^2\sqrt{3}$
6. $3ab\sqrt{5a}$ **7.** $\dfrac{5x}{6y}$ **8.** $\dfrac{4x^2\sqrt{2x}}{y^3}$ **9.** $5a^2b$
10. $a^2\sqrt{3}$

Exercise Set 12.2 (page 634)
1. $3\sqrt{3}$ **3.** $11\sqrt{x}$ **5.** $-7\sqrt{3}$ **7.** $5\sqrt{6a}$
9. $9-6\sqrt{3}$ **11.** $23\sqrt{3x}$ **13.** $10\sqrt{7}-20\sqrt{2}$
15. $11\sqrt{2}$ **17.** $-5x\sqrt{3}$ **19.** $\sqrt{21}$ **21.** 12
23. $11a$ **25.** $x-1$ **27.** $3a\sqrt{5}$ **29.** 30
31. $-120x$ **33.** $15x\sqrt{2}$ **35.** $2\sqrt{5}-5$
37. $2\sqrt{6}+12$ **39.** $-3\sqrt{x}+6\sqrt{5}-3\sqrt{6}$
41. $x-16\sqrt{x}$ **43.** $31-10\sqrt{6}$ **45.** $x-25$
47. $23-7\sqrt{11}$ **49.** $2x-5\sqrt{x}-3$
51. $5-5\sqrt{3}-\sqrt{2}+\sqrt{6}$ **53.** $11+\sqrt{5}$
55. $8-2\sqrt{3x}-3x$ **57.** -4 **59.** $x-100$
61. $y-2\sqrt{7y}+7$ **63.** $2a^2-10a\sqrt{2}+25$

Stretching the Topics
1. $162+60\sqrt{2}$ **2.** $2y$ **3.** 4

Checkup 12.2 (page 636)
1. $5\sqrt{6x}$ **2.** $3\sqrt{3}$ **3.** $3\sqrt{3}$ **4.** $3a\sqrt{5}$
5. $y+6$ **6.** $7\sqrt{5}-5\sqrt{2}$ **7.** $2\sqrt{6}+4\sqrt{3}$
8. $x-2\sqrt{5x}+5$ **9.** $-18-\sqrt{5}$ **10.** $x-3$

Exercise Set 12.3 (page 642)
1. $\sqrt{3}$ **3.** $\dfrac{2\sqrt{7}}{7}$ **5.** $2\sqrt{2}$ **7.** $\dfrac{6\sqrt{10}}{5}$ **9.** $\dfrac{9\sqrt{2}}{8}$
11. $\dfrac{\sqrt{10}}{4}$ **13.** $\sqrt{2}$ **15.** $\dfrac{\sqrt{2}}{2}$ **17.** $\dfrac{2\sqrt{5}}{5}$
19. $\dfrac{\sqrt{3x}}{x}$ **21.** $\dfrac{\sqrt{5}+\sqrt{10}}{5}$ **23.** $2-3\sqrt{2}$
25. $\dfrac{2\sqrt{5}+5\sqrt{2}}{10}$ **27.** $2-\sqrt{3}$ **29.** $-4(3+\sqrt{6})$
31. $-\sqrt{2}-2$ **33.** $\sqrt{6}-\sqrt{2}$
35. $\dfrac{2\sqrt{15}-2\sqrt{6}}{3}$ **37.** $\sqrt{2}$ **39.** $5+2\sqrt{6}$

Stretching the Topics
1. $\dfrac{a\sqrt{a+b}}{2(a+b)}$ **2.** $\dfrac{15x+6\sqrt{xy}}{25x-4y}$

3. $\dfrac{2x-2y+4\sqrt{x-y}}{(x-y)-4}$

Checkup 12.3 (page 644)
1. $\sqrt{6}$ **2.** $2\sqrt{10}$ **3.** $\dfrac{7\sqrt{3}}{6}$ **4.** $\dfrac{\sqrt{10}}{5}$ **5.** $\dfrac{2\sqrt{6}}{3}$
6. $\sqrt{6}-5\sqrt{2}$ **7.** $\dfrac{1+\sqrt{3}}{2}$ **8.** $\dfrac{2\sqrt{3}+1}{11}$
9. $13+4\sqrt{11}$ **10.** $\dfrac{11-4\sqrt{7}}{3}$

Exercise Set 12.4 (page 652)
1. $x=\pm 2\sqrt{3}$ **3.** $a=\pm\dfrac{\sqrt{5}}{3}$ **5.** $x=\pm\dfrac{\sqrt{5}}{5}$
7. $y=\pm\dfrac{\sqrt{10}}{2}$ **9.** $y=\pm\sqrt{5}$ **11.** $a=\pm\sqrt{2}$
13. $x=0$ **15.** $a=\pm 4\sqrt{5}$ **17.** $x=\dfrac{-1\pm\sqrt{13}}{2}$
19. $x=\dfrac{5\pm 3\sqrt{5}}{2}$ **21.** $x=2\pm\sqrt{7}$
23. $y=1\pm\sqrt{2}$ **25.** $a=\dfrac{-7\pm\sqrt{53}}{2}$
27. $x=\dfrac{1\pm\sqrt{7}}{3}$ **29.** $y=3\pm\sqrt{5}$
31. $z=\dfrac{-1\pm\sqrt{31}}{2}$ **33.** No real solutions

Stretching the Topics
1. $x=\dfrac{13\pm\sqrt{257}}{2}$ **2.** $x=\dfrac{3\sqrt{2}\pm\sqrt{34}}{2}$
3. $x=\sqrt{3}\pm 1$

Checkup 12.4 (page 653)
1. $x=\pm\sqrt{2}$ **2.** $a=\pm\sqrt{5}$ **3.** $y=\pm 4\sqrt{2}$
4. $x=\pm 2\sqrt{5}$ **5.** $x=0$ **6.** $x=-2\pm 2\sqrt{2}$
7. $x=\dfrac{3\pm\sqrt{5}}{2}$ **8.** $x=\dfrac{3\pm\sqrt{3}}{3}$
9. $x=\dfrac{5\pm\sqrt{33}}{4}$ **10.** $x=\dfrac{1\pm 2\sqrt{2}}{3}$

Exercise Set 12.5 (page 659)
1. $5\sqrt{2}$ cm   **3.** 20.8 ft   **5.** $4\sqrt{3}$ ft
7. 6.71 m, 13.42 m   **9.** $(3+\sqrt{19})$ m
11. $1+\sqrt{10}$ and $-1+\sqrt{10}$ or $1-\sqrt{10}$ and $-1-\sqrt{10}$
13. $\dfrac{-4+\sqrt{2}}{2}$ and $\dfrac{4+\sqrt{2}}{2}$ or $\dfrac{-4-\sqrt{2}}{2}$ and $\dfrac{4-\sqrt{2}}{2}$
15. $(-4+\sqrt{31})$ ft   **17.** 0.9 in.

Stretching the Topics
1. $(10+5\sqrt{5})$ ft; $(20+10\sqrt{5})$ ft; $(25+10\sqrt{5})$ ft
2. $(4+2\sqrt{2})$ in.; $(24+16\sqrt{2})$ sq in.

Checkup 12.5 (page 661)
1. $4\sqrt{2}$ in.   **2.** 5.68 m
3. $-2+\sqrt{5}$ and $2+\sqrt{5}$ or $-2-\sqrt{5}$ and $2-\sqrt{5}$
4. 6.9 ft

Speaking the Language of Algebra (page 664)
1. Principal square root   **2.** Real
3. Radical; radicand   **4.** Irrational
5. Rationalizing the denominator   **6.** Real conjugate
7. $x = \sqrt{C}$ or $x = -\sqrt{C}$
8. $\dfrac{-b \pm \sqrt{b^2 - 4ac}}{2a}$; quadratic formula

Review Exercises (page 667)
1. 12   **2.** 8a   **3.** 13   **4.** $3y^2\sqrt{5}$ **5.** $4ab\sqrt{3a}$
6. $\dfrac{3x^2\sqrt{3x}}{y^2}$ **7.** $7a^3b$ **8.** $3a^3$ **9.** $3\sqrt{5x}$
10. $7a\sqrt{3}$ **11.** $-60\sqrt{6}$ **12.** $3\sqrt{2} - 2\sqrt{7}$
13. $15\sqrt{6}$ **14.** $\sqrt{15} - 6\sqrt{5} - 12$
15. $-32 + 5\sqrt{6}$ **16.** $a - 5$ **17.** $5\sqrt{3}$
18. $\dfrac{\sqrt{5}}{2}$ **19.** $\dfrac{\sqrt{35}}{5}$ **20.** $2 - 7\sqrt{3}$
21. $\dfrac{\sqrt{2} + \sqrt{6}}{2}$ **22.** $\dfrac{5\sqrt{11} + 11}{14}$
23. $-(11 + 6\sqrt{3})$ **24.** $\dfrac{17 - 4\sqrt{13}}{9}$
25. $y = \pm 4\sqrt{3}$ **26.** $a = \pm \sqrt{2}$
27. No real solutions   **28.** $x = -3 \pm 2\sqrt{3}$
29. $x = 2 \pm \sqrt{3}$ **30.** $x = \dfrac{5 \pm \sqrt{15}}{5}$
31. $x = \dfrac{3 \pm \sqrt{17}}{4}$ **32.** $x = \dfrac{2 \pm \sqrt{10}}{3}$
33. $3\sqrt{2}$ cm   **34.** $(2 + \sqrt{6})$ ft
35. $-2 + \sqrt{5}$ and $2 + \sqrt{5}$ *or* $-2 - \sqrt{5}$ and $2 - \sqrt{5}$

Practice Test for Chapter 12 (page 669)
1. $\dfrac{4}{5}$ **2.** $3a$ **3.** $9a$ **4.** $4\sqrt{2}$ **5.** $4ab^2\sqrt{3a}$
6. $\dfrac{4x\sqrt{2x}}{y^2}$ **7.** $10xy$ **8.** $4\sqrt{10z}$ **9.** $-50a\sqrt{3}$
10. $12\sqrt{2}$ **11.** $y - 2\sqrt{10y} + 10$
12. $20 - 7\sqrt{x} - 3x$ **13.** 25 **14.** $a - 25$
15. $2\sqrt{10}$ **16.** $\dfrac{\sqrt{15}}{3}$ **17.** $\dfrac{3\sqrt{5} + 5}{5}$
18. $\dfrac{4\sqrt{7} - 7}{9}$ **19.** $-6(\sqrt{2} + \sqrt{3})$
20. $\dfrac{2\sqrt{10} - 11}{9}$ **21.** $x = \pm 6\sqrt{2}$ **22.** $x = \pm \sqrt{3}$
23. $x = \dfrac{-5 \pm \sqrt{29}}{2}$ **24.** $x = -4 \pm \sqrt{6}$
25. $(1 + 2\sqrt{3})$ ft

Sharpening Your Skills after Chapters 1–12 (page 671)
1. \$724.50   **2.** $-3.98 > -4.21$ **3.** $\frac{5}{9}$ **4.** No
5. $x \neq 7, x \neq \frac{1}{2}$ **6.** No   **7.** -4 **8.** 0   **9.** $7a^7b^9$
10. $-\dfrac{8x^9}{y^{21}}$ **11.** $3x^3y^3 - 14x^3y^7$
12. $9x^2 - 12xy + 4y^2$ **13.** $x^3 - 125$
14. $x^3 + 2x^2 + 4x + 8$ **15.** $\dfrac{x^3}{y^5}$ **16.** $\dfrac{x + 8}{x + 2}$

17. $\dfrac{5x - 2}{x(5x - 1)}$ **18.** $ab^2(a^2 - 14ab + 4b^2)$
19. $(9m + 8n)(7m - 2n)$ **20.** $x = 3$
21. $x = 11, x = -5$ **22.** No solution   **23.** $x \leq \frac{10}{3}$
24. $(\frac{2}{3}, \frac{5}{3})$ **25.** $(2, 2)$
26.

27.

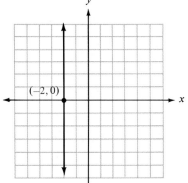

28. 40 ft   **29.** 395   **30.** Ray, 60 mph; Glenn, 65 mph

Index